THIRD EDITION

INTRODUCTION TO FLUID MECHANICS

JAMES E. A. JOHN
Dean, College of Engineering
University of Massachusetts

WILLIAM L. HABERMAN
Engineering Consultant
Rockville, Maryland

PRENTICE HALL, Englewood Cliffs, NJ 07632

Library of Congress Cataloging-in-Publication Data

JOHN, JAMES E. A.
   Introduction to fluid mechanics / James E.A. John,
William L. Haberman.—3rd ed.

     p. cm.
   Includes index.
   ISBN 0-13-483967-6
   1. Fluid mechanics.   I. Haberman, William L.
   II. Title.
  TA357.J63   1988   87-20996
  620.1'06—dc19   CIP

Editorial/production supervision: Colleen Brosnan
Cover design: Lundgren Graphics, Ltd.
Cover photo: U.S. Windpower, San Francisco, CA; Ed Linton, photographer
Manufacturing buyer: Gordon Osbourne

© 1988, 1980, 1971 by Prentice Hall
A Division of Simon & Schuster
Englewood Cliffs, New Jersey 07632

Printed in the United States of America
10 9 8 7 6 5 4 3 2 1

ISBN 0-13-483967-6   025

Prentice-Hall International (UK) Limited, *London*
Prentice-Hall of Australia Pty. Limited, *Sydney*
Prentice-Hall Canada Inc., *Toronto*
Prentice-Hall Hispanoamericana, S.A., *Mexico*
Prentice-Hall of India Private Limited, *New Delhi*
Prentice-Hall of Japan, Inc., *Tokyo*
Simon & Schuster Asia Pte. Ltd., *Singapore*
Editora Prentice-Hall do Brasil, Ltda., *Rio de Janeiro*

INTRODUCTION TO FLUID MECHANICS

To Connie
and
Elizabeth, Jimmy, Thomas, and Constance

To the memory of my wife,
Florence H. Haberman

CONTENTS

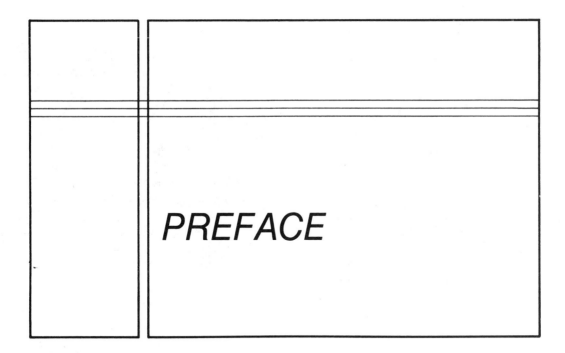

PREFACE

This textbook is intended to provide the undergraduate engineering student with an understanding of the basic principles of fluid mechanics. Our approach is to keep the derivation of the fundamental equations and principles at an uncomplicated mathematical level; the results of these derivations will always be tied to a real physical situation. Emphasis will be placed on applications of current interest to the engineer. Numerous worked-out examples are provided throughout to illustrate the utility of the fundamental equations in describing various physical situations.

The first five chapters serve as a core, to be required of all engineering students of fluid mechanics. The remaining chapters are devoted to applications of the core material, with the particular chapters to be studied depending on the department. For example, the civil engineer would be primarily interested in Chapters 6, 10, 14 and 15; the mechanical engineer, Chapters 7, 9, 11, 12, 13, 15 and 16; the aerospace engineer, Chapters 7, 8, 9, 11, 13, 15 and 16. The text is arranged so that each of the foregoing groups of chapters can be taken independently; intervening chapters can be omitted with no loss of continuity.

Whereas the Second Edition used SI units exclusively, we made the decision to use both English and SI units in this edition of the text. Outside the United States, use of SI units is widespread, whereas in the United States, English units are predominant. It will probably be some time before conversion is a fact. However, for those companies involved with international markets, conversion to SI has already been accomplished or will have to be accomplished in the very near future. Certainly, every engineering graduate coming out of college today should be familiar with, have a feel for, and be able to work with SI units. In contrast, it is a rarity in the United States to see a pressure gauge with readings in kPa; most road signs are expressed in miles and mph; quarts and gallons are still common units. With English units used extensively in the United States, but with the change to SI taking place or already having occurred in many industries, it is necessary to present engineering subject matter in both English and SI units.

In the Third Edition, we have greatly increased the number of problems at the ends of the chapters, with problems presented in both English and SI units. Likewise, many worked-out examples are presented throughout the text in both SI and English units.

We would like to thank Oretta Taylor for her help in preparing portions of the manuscript for the Third Edition. We also thank our many colleagues and students for their many helpful suggestions over the years which have helped us in the writing of this textbook.

James E. A. John
University of Massachusetts
Amherst, Massachusetts

William L. Haberman
Rockville, Maryland

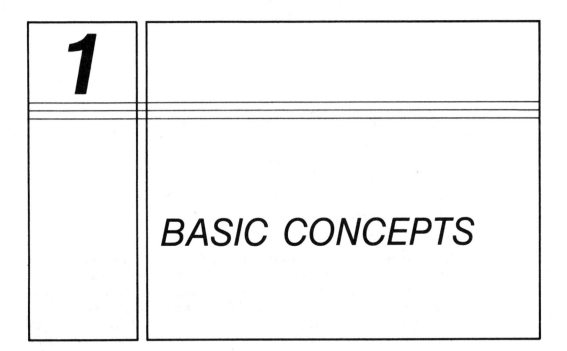

BASIC CONCEPTS

1.1 INTRODUCTION

The subject of fluid mechanics is of great significance to man; the passage of blood through our veins, the falling of the rain through the atmosphere, the currents in the oceans and the seas are all examples of fluid flow. One is interested in studying fluid mechanics in order to utilize and control the effects of fluid flow for the benefit of society. It is the purpose of this textbook to acquaint the student with laws describing the behavior of fluids in motion and to indicate the application of these laws. The student has become familiar with the field of particle and solid body dynamics; in Section 1.2 we shall indicate in what ways fluid mechanics is different from solid mechanics.

1.2 SOLIDS VERSUS FLUIDS

In order to distinguish between a solid medium and a fluid medium, we shall examine the response of each substance to an applied shear force. Consider a solid bar with one end rigidly attached and the other free. Apply a torque to the free end, as shown in Figure 1.1. As long as the resultant stress does not exceed the yield stress of the material, the bar will twist until an equilibrium position is reached, with the final position dependent on the magnitude of the applied torque and the elastic properties of the bar.

Let us place a fluid between two concentric cylinders, with the inner cylinder held fixed, and apply a torque to the outer cylinder (Figure 1.2).

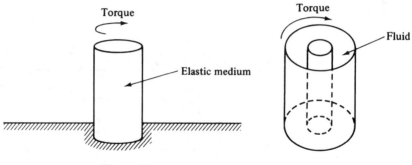

Figure 1.1   Figure 1.2

When the shear force is applied to the outer cylinder, the fluid is incapable of reaching an equilibrium position; instead the outer cylinder will continue to rotate as long as the shear force is maintained. The magnitude of the angular velocity will depend on the magnitude of the applied torque and on the properties of the fluid. This response to an applied shear force forms the basis of the definition of a fluid, namely, a substance that is unable to resist the application of a shear force without undergoing a continuing deformation.

1.3 VELOCITY FIELD OF FLUID AND RIGID BODIES

When a rigid body is in motion, there is a definite relationship between the velocities of the various particles that make up the body. For example, if the body is translating, all particles must have the same velocity (Figure 1.3). If the body is rotating about an axis, the velocities of the particles are linearly dependent on the distance from the axis of rotation (Figure 1.4).

When we consider the motion of the particles in a fluid body, however, we find that no such simple relationship exists between the velocities of the various particles. Unlike those in a rigid body, the particles in a fluid body are not rigidly attached to each other. Therefore, the relative motion of adjacent fluid particles becomes more complex. In order to describe completely

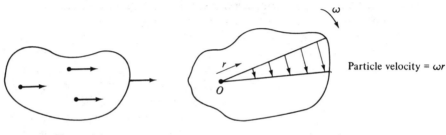

Figure 1.3   Figure 1.4

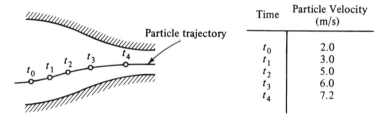

Time	Particle Velocity (m/s)
t_0	2.0
t_1	3.0
t_2	5.0
t_3	6.0
t_4	7.2

Figure 1.5

the motion of a fluid, information about the velocity of each particle of the fluid is necessary. Each fluid particle in the flow has a definite value of velocity at every instant. As the particle moves in the flow field, its velocity changes with location and time. In rigid body dynamics, it is customary to follow a particle as it moves about in space. This approach could also be taken in describing the motion of a fluid particle as it moves about in the flow field. However, in describing the motion of a fluid, it is more convenient to use an alternate approach, that is, to define the flow field by specifying the velocity-time history at each point in the field.

The difference between the two approaches can be seen in the following example, in which water flows continuously through a converging nozzle as shown in Figure 1.5. First, let us follow a fluid particle as it moves through the nozzle. The position and velocity of the particle at successive equal time intervals are shown. It can be seen that the particle accelerates as it moves through the nozzle; in other words, the particle velocity increases with time.

Another way of describing the same flow field is to specify the particle velocity at each point as a function of time. When the velocity at each point in the flow field does not change with respect to time, the flow is said to be steady. Figure 1.6 gives the velocity-time history at point A and at point B in the flow field for the case of steady flow. Note that even in steady flow, a fluid particle can accelerate as it moves through the nozzle. Again, it is emphasized that steady flow implies that the fluid velocity at a point remains constant with time. Here, for steady flow, the velocity at point A is always 3 m/s (meters per second), and the velocity at point B is always 7.2 m/s.

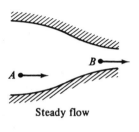

Steady flow

Time	Velocity of Fluid Particle at A (m/s)	Velocity of Fluid Particle at B (m/s)
t_0	3	7.2
t_1	3	7.2
t_2	3	7.2
t_3	3	7.2
t_4	3	7.2

Figure 1.6

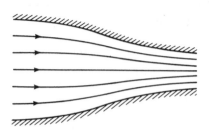

Figure 1.7

The second approach is preferred in fluid mechanics, for the description of flow at given locations is usually desired. For example, in a study of river flow about a bridge pier, the flow velocities at points in the vicinity of the pier and the resulting forces on the pier are of interest. To try to determine the flow at a given location from the particle location-time history of the first approach would be a far more complex task.

In order to obtain an indication of the flow pattern in a velocity field, we make use of *streamlines*. A streamline is a line drawn in the fluid at a given instant of time such that there is no flow across that line. Thus, at the given instant, the velocity of every fluid particle on the streamline is in the direction tangent to the line. By considering a sufficient number of such streamlines in the flow field, at a given instant, we obtain the flow pattern in the flow field, that is, a description of the direction of the velocities of the fluid particles. For steady flow, the streamline pattern does not change with time. The streamline pattern for flow through the nozzle of Figures 1.5 and 1.6, for steady flow, appears in Figure 1.7.

1.4 UNITS AND FLOW PROPERTIES

In any engineering subject, considerable care must be taken to use a consistent set of units. In this text, we will use both SI (Système International) and English units. The dimensions used in fluid mechanics are those of mass, force, length, time, and temperature. In SI, the unit of mass is the kilogram (kg), the unit of length is the meter (m), the unit of time is the second (s), and the unit of absolute temperature is the degree Kelvin, or, simply, the kelvin (K). Force and mass are related by Newton's law of motion, which states that force is proportional to mass times acceleration:

$$F = kma$$

where k is a proportionality constant.

According to the SI, the newton (N) is the force required to give a mass of 1 kg an acceleration of 1 m/s². In this case,

$$k = \frac{F}{ma} = 1 \text{ N} \cdot \text{s}^2/\text{kg} \cdot \text{m}$$

In the English system, the unit of mass is the slug, the unit of length is the foot (ft), the unit of time is the second (s), and the unit of absolute temperature is the degree Rankine (°R). The unit of force is the pound force (lbf), which is the force required to give a mass of 1 slug an acceleration of 1 ft/s². In this case,

$$k = \frac{F}{ma} = 1 \text{ lbf} \cdot \text{s}^2/\text{slug} \cdot \text{ft}$$

Another unit of mass in the English system is the pound mass: 1 lbm weighs 1 lbf on the surface of the earth, where the acceleration due to gravity is 32.17 ft/s². Since weight is a force, we can determine k as follows:

$$k = \frac{F}{ma} = \frac{1 \text{ lbf}}{1 \text{ lbm} \cdot 32.17 \text{ ft/s}^2}$$

Since 1 lbf will give 1 slug an acceleration of 1 ft/s² and 1 lbf will give 1 lbm an acceleration of 32.17 ft/s², it follows that 1 slug = 32.17 lbm.

For consistency in this text, we will use k equal to unity for both the SI and the English systems. Hence, Newton's law will have the form $F = ma$ for both systems. This means we will be using the slug as the primary unit of mass in the English system. For problems specifying lbm as the mass unit, we must be careful to convert to slugs before substitution into the appropriate momentum or energy equation. Since thermodynamic data in the literature for the English system are expressed in units of lbm, we will adhere to that convention. Therefore, again, care must be used to convert to slugs.

EXAMPLE 1.1
Determine the resultant acceleration of the mass if

(a) A force of 10 N is applied to a mass of 10 kg.
(b) A force of 10 lbf is applied to a mass of 10 slugs.
(c) A force of 10 lbf is applied to a mass of 10 lbm.

Solution

(a)
$$F = ma$$
$$a = \frac{10 \text{ N}}{10 \text{ kg}} = \underline{1.0 \text{ m/s}^2}$$

(b)
$$F = ma$$
$$a = \frac{10 \text{ lbf}}{10 \text{ slugs}} = \underline{1.0 \text{ ft/s}^2}$$

(c)
$$F = ma$$
$$m = \frac{10 \text{ lbm}}{32.17 \text{ lbm/slug}} = 0.3108 \text{ slugs}$$
$$a = \frac{10 \text{ lbf}}{0.3108 \text{ slugs}} = \underline{32.17 \text{ ft/s}^2}$$
∎

Again, note that in the English system, for $F = ma$, mass must be expressed in slugs with force in lbf and acceleration in ft/s².

It must be recognized that weight is a force, namely, the force with which a mass is attracted to the earth or some other body. According to the law of conservation of mass, the mass of a body remains constant, independent of its distance from the earth's surface. However, the weight of a body will decrease as it is moved away from the earth's surface.

EXAMPLE 1.2

(a) The payload of a spacecraft has a mass of 100 kg on the earth's surface. Determine the weight and mass of the payload at a point in space at which $g = 0.8$ m/s^2.

(b) An object weighs 100 pounds on the earth's surface. Determine the mass and weight of the object on the moon's surface, where the acceleration due to gravity is one-sixth of that on the earth's surface.

Solution

(a) According to the law of conservation of mass, the mass of the payload in space is the same as its mass on the earth's surface, and so $m = \underline{100 \text{ kg}}$ in space. Weight $= mg$, so that the weight of the payload in space is equal to (100 kg) (0.8 m/s^2) = $\underline{80 \text{ N}}$.

(b) In the English system, $F = ma$ with m in slugs. Therefore, $m = 100$ lbf/(32.17 ft/s^2) = 3.108 slugs. (Note that 3.108 slugs = 100 lbm.) On the moon's surface, the object's mass is still 3.108 slugs or $\underline{100 \text{ lbm}}$. The weight of the object on the moon's surface will be:

$$F = ma = (3.108 \text{ slugs}) \left(\frac{32.17}{6} \text{ ft/s}^2 \right) = \underline{16.66 \text{ lbf}} \quad \blacksquare$$

The following multiplying prefixes will be used throughout this text in conjunction with the various units in the SI:

Factor	Prefix	Symbol
10^6	mega	M
10^3	kilo	k
10^{-2}	centi	c
10^{-3}	milli	m
10^{-6}	micro	μ

For example, 1 kN = 1000 newtons, 1 mm = 0.001 meter, 1 μg = 0.000001 gram.

We discussed velocity and the velocity field in Section 1.3; other flow properties are of interest as well. Two such properties are pressure and density. The *mean pressure* over a plane area in a fluid is defined as the ratio of the normal force acting on the area to the area. The pressure at a point in the fluid is the limit that the mean pressure approaches as the area is reduced to a very small size around the point. Since pressure is defined as force per unit area, pressure in SI units is given in newtons per square meter, or N/m^2. One N/m^2 is called 1 pascal (Pa). For comparison, 1 standard atmosphere is equal to 101,325 Pa or 101.325 kPa or 0.101325 MPa.

In the English system, pressure is expressed as lbf per square foot (psf) or as lbf per square inch (psi). Note that one standard atmosphere is equal to 14.70 psi or 2116 psf.

When pressure is given relative to zero pressure, it is called absolute pressure. For example, suppose a pressure gage connected to a compressed air tank registers 50 kPa. If the local atmospheric pressure is 100 kPa, the absolute pressure inside the tank is 150 kPa. Unless otherwise specified, pressures given in pascals throughout this text will refer to absolute pressures. In the English system, gage pressure is denoted as psfg or psig, absolute pressure as psfa or psia. For example, suppose a pressure gage connected to a compressed air tank registers a pressure of 50 psig. If the local atmospheric pressure is 14.5 psia, the absolute pressure of the air inside the tank is 50 + 14.5 = 64.5 psia.

The mean density of a fluid is defined as the ratio of a given mass of the fluid to the volume which that mass occupies. The density at a point is the limit that the mean density approaches as the volume is reduced to a very small size around the point. It follows that the units of density, or mass per unit volume, are kilograms per cubic meter or kg/m^3. In the English system the units of density are slugs per cubic foot ($slug/ft^3$) or, alternatively, pounds mass per cubic foot (lbm/ft^3).

The ratio of the density of a substance to that of water at 4°C (39°F) and atmospheric pressure is called the *specific gravity* of the substance. As defined, specific gravity is a pure number, with no physical dimensions. For example, the specific gravity of gasoline at 15°C is 0.75. Since the density of water at 4°C is 1000 kg/m^3, the density of gasoline at 15°C is 750 kg/m^3.

EXAMPLE 1.3

The density of water is measured to be 1000 kg/m^3 on the earth's surface, where the acceleration due to gravity is 9.81 m/s^2. Determine the density of water at a point in space where $g = 5.0$ m/s^2. Also calculate the weight of 1 cubic meter of water at this point in space.

Solution   According to the law of conservation of mass, the density remains constant at 1000 kg/m^3 anywhere in space. However, the weight of 1 cubic meter is found from:

$$
\begin{aligned}
\text{Weight} &= ma \\
&= (1000 \text{ kg})(5.0 \text{ m/s}^2) \\
&= 5000 \text{ kg} \cdot \text{m/s}^2 \\
&= 5000 \text{ N} \\
&= \underline{5.0 \text{ kN}}
\end{aligned}
$$

■

EXAMPLE 1.4

The density of water is determined to be 62.4 lbm/ft³ on the earth's surface, where $g = 32.2$ ft/s². Determine the density of water at a point in space where g is equal to 5.0 ft/s². Also calculate the weight of one cubic foot of water at this point in space.

Solution According to the law of conservation of mass, the density remains unchanged at 62.4 lbm/ft³ anywhere in space. However, the weight of one cubic foot is found from:

$$W = mg \qquad \text{with } m \text{ in slugs}$$

$$W = \frac{(62.4 \text{ lbm})(5.0 \text{ ft/s}^2)}{32.17 \text{ lbm/slug}}$$

$$= 9.70 \text{ lbf} \qquad \blacksquare$$

Energy is expressed in units of force times distance, i.e., in N · m in the SI, where 1 joule (J) = 1 N · m; and in ft-lbf in the English system, where 1 Btu = 778 ft-lbf.

EXAMPLE 1.5

An object of mass 10 lbm has a velocity of 100 ft/s. Determine the kinetic energy of the object.

Solution

$$\text{Kinetic energy} = \tfrac{1}{2}mV^2$$

$$= \frac{1}{2}\left(\frac{10 \text{ lbm}}{32.17 \text{ lbm/slug}}\right)(100^2 \text{ ft}^2/\text{s}^2)$$

But

$$1 \text{ lbf} = 1 \text{ slug} \cdot \text{ft/s}^2$$

and so the kinetic energy = 1554 ft-lbf. $\blacksquare$

Power, the rate of doing work, is given in the SI in units of joules per second, or watts (W), where 1 W = 1 J/s; in the English system, power is expressed in ft-lbf/s, in Btu/s, or in horsepower, where 1 horsepower = 550 ft-lbf/s.

Finally, let us consider units of temperature. On the SI Celsius temperature scale (°C), the freezing point of water is 0°C and the boiling point of water is 100°C, with both at standard atmospheric pressure. On the English Fahrenheit scale (°F), the freezing point of water is 32°F, while the boiling point is 212°F, again with both at standard atmospheric pressure. A comparison of the two scales is shown in Figure 1.8. Note that the magnitude of the degree Celsius is greater than the magnitude of the degree Fahrenheit by a factor of (212 − 32)/100 = 1.8. The conversion equations are

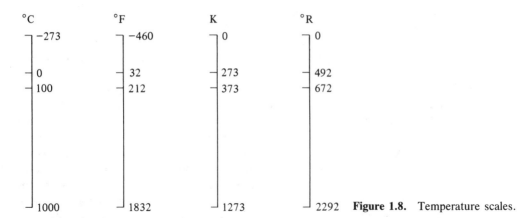

Figure 1.8. Temperature scales.

$$T(°F) = \frac{9}{5}T(°C) + 32$$

$$\text{or} \quad T(°C) = \frac{5}{9}[T(°F) - 32]$$

(1.1)

For example, a temperature of $-40°F$ is equal to the Celsius temperature of $\frac{5}{9}(-40 - 32) = -40°C$. A temperature of $1000°C$ is equal to the Fahrenheit temperature of $\frac{9}{5}(1000) + 32 = 1832°F$. The Celsius and Fahrenheit temperature scales have been defined in terms of the freezing and boiling points of water.

Associated with the Celsius scale is the *Kelvin* absolute temperature scale (K), with $K = °C + 273.15$ (usually rounded off to 273); note that the magnitude of the degree Kelvin (or the kelvin) is equal to the magnitude of the degree Celsius. Associated with the Fahrenheit scale is the *Rankine* absolute temperature scale (°R), with $°R = °F + 459.67$ (usually rounded off to 460); again, the magnitude of the degree Rankine is equal to the magnitude of the degree Fahrenheit.

Absolute zero is zero on both the Kelvin and the Rankine scale; thus conversion is given by

$$T(K) = \frac{5}{9}T(°R) \quad \text{and} \quad T(°R) = \frac{9}{5}(K)$$

For example, $90°R = 50K$ and $100K = 180°R$.

1.5 CONTINUUM

We have discussed the motion of the particles in a fluid body under the tacit assumption that the particles are distributed continuously throughout the fluid body. In reality, each fluid particle is composed of many finite-sized molecules with finite distance between the molecules. These molecules are in constant random motion and collision. For example, in air at standard conditions there are 2×10^{19} molecules in 1 cubic centimeter, with the mean distance between molecular collision of 6.35×10^{-6} cm. In order to describe the motion of each molecule, it would be necessary to resort to a statistical mechanics approach. In general, however, we are not interested in the motion of individual molecules but in the overall motion of the fluid. Therefore, we do not have to resort to the statistical approach but can treat the fluid as a continuous medium, also called a *continuum*.

In the continuum, or macroscopic, approach, we deal with average effects of the many molecules that make up the fluid particle. These average effects are readily measurable. Consider *pressure* defined at a point in a continuous medium as the time-averaged normal force exerted by the molecules on a unit of bounding surface. To be consistent with the continuum approach, this area is small, but its dimensions must be large enough compared with the distance between molecules to provide for a sufficient number of collisions to yield a representative time-averaged force. If the area were too small, it is conceivable that no molecules would impact the area and no force would be obtained. It is obvious that the pressure would have no meaning in this situation. Mathematically, we define pressure at a point in a continuum as

$$p = \lim_{A \to A^*} \frac{F}{A}$$

where F is the force normal to the surface A and A^* is the smallest area surrounding the point consistent with the continuum approach. It follows that a large number of molecular collisions still occur over the area A^*.

Density at a point in a continuum is defined in a similar fashion:

$$\rho = \lim_{V \to V^*} \frac{M}{V}$$

where M is the mass contained in volume V. Again, V^* is the smallest volume surrounding the point consistent with the continuum approach, and therefore it contains a large number of molecules. If the volume V^* were taken with linear dimensions comparable to molecular size and the volume were to lie between molecules, there would be no molecules in the volume and the density would be zero. Obviously, density would have no meaning in this situation.

With the exception of very-low-density gas flows where the mean distance between molecular collisions is comparable to the dimensions of surface of the body, it has been found that fluids behave as continuous media. For this reason the continuum approach will be used throughout this text.

1.6 LIQUIDS AND GASES

We have discussed some of the characteristics of fluids in general. Fluids can be further classified into *liquids* and *gases*. The difference between the two is that liquids occupy a definite volume, independent of the volume in

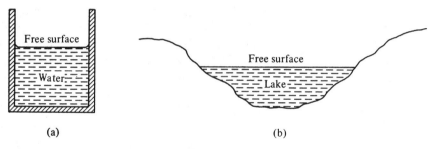

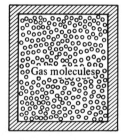

Figure 1.9

which they are contained, whereas gases expand to fill the entire volume of the container in which they are placed. For example, if water is poured into a vessel and the volume of water is not sufficient to fill the vessel, a free liquid surface will be formed, as shown in Figure 1.9. When a gas is allowed to enter an empty vessel, however, the gas molecules always fill the entire vessel (Figure 1.10).

The difference between liquids and gases can also be seen by examining their response to an applied pressure. For example, if the pressure over the water of Figure 1.9(a) is increased, only a very small change in the volume of the water results. The physical property that describes the change of volume with applied pressure at a given temperature is called the *coefficient of compressibility*, defined as

$$\beta = -\frac{1}{V}\left(\frac{\partial V}{\partial p}\right)_T$$

or its reciprocal, the *isothermal bulk modulus k*, defined as

$$k = -V\left(\frac{\partial p}{\partial V}\right)_T$$

For water at room temperature, the isothermal bulk modulus is approximately 2×10^4 atmospheres. In other words, the application of 20 atmospheres pressure to the water surface of Figure 1.9(a) would only decrease the water volume by 0.1%. We say that water, and liquids in general, are relatively incompressible.

Figure 1.10

For a gas, however, the application of a pressure can have a great effect on the gas volume. An *equation of state* is a relation between the pressure, temperature, and volume or density of a substance at equilibrium. Whereas an equation of state tends to be a complex relationship for a liquid, there is a simple equation of state, called the *perfect gas law*, for gases; this equation approximates the behavior of most real gases as long as the gases are not too near the condensation point and are not at very high pressures. For a perfect gas,

$$p\mathcal{V} = mRT$$

or

$$p = \rho RT \qquad (1.2)$$

where m is the mass of gas and R is a constant with magnitude dependent on the individual gas (values of R are given in Appendix A). In this equation, with p in Pa, T in K, and ρ in kg/m^3, R will be expressed in the units $J/kg \cdot K$. With p in lbf/ft^2, T in °R, and ρ in $slugs/ft^3$, R will be expressed in the units ft-lbf/slug°R. If mass were to be expressed in lbm, with p in lbf/ft^2 and T in °R, then R would be in units of ft-lbf/lbm°R.

For any perfect gas,

$$R = \frac{\overline{R}}{\overline{M}} \qquad (1.3)$$

where

$$\overline{R} = \text{universal gas constant}$$
$$= 8314.3 \text{ J/kg-mol} \cdot K$$
$$= 1545.3 \text{ ft-lbf/lbm-mol°R}$$
$$\text{and } \overline{M} = \text{molecular mass of gas}$$

EXAMPLE 1.6

A gage connected to a tank of oxygen registers 50 kPa, with the atmospheric pressure equal to 101 kPa. If the internal volume of the tank is 10 m^3 and the oxygen temperature is 30°C, determine the mass of oxygen in the tank.

Solution From Appendix A, $R = 0.2598$ kJ/kg $\cdot$ K. Absolute pressure of oxygen = 50 kPa + 101 kPa = 151 kPa. Therefore, with $p\mathcal{V} = mRT$,

$$m = \frac{p\mathcal{V}}{RT}$$

$$= \frac{(151 \times 10^3 \text{ N/m}^2)(10 \text{ m}^3)}{(259.8 \text{ N} \cdot \text{m/kg} \cdot \text{K})[(30 + 273.15)\text{K}]}$$

$$= \underline{19.17 \text{ kg}} \qquad \blacksquare$$

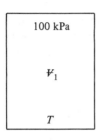

Figure 1.11

EXAMPLE 1.7

A perfect gas is known to have a molecular mass of 16.7. A tank is to be designed to contain 30 lbm of this gas at 50°F. Determine the internal tank volume if the gas pressure is to be 35 psia.

Solution   From Equation (1.3),

$$R = \frac{\overline{R}}{\overline{M}} = \frac{1545.3 \text{ ft-lbf}/\text{lbm-mol}°R}{16.7 \text{ lbm}/\text{lbm-mol}}$$

$$= 92.53 \text{ ft-lbf}/\text{lbm}°R$$

From Equation (1.2),

$$p\Psi = mRT$$

so that

$$\Psi = \frac{(30 \text{ lbm})(92.53 \text{ ft-lbf}/\text{lbm}°R)[(50 + 459.67)°R]}{(35 \text{ lbf}/\text{in}^2)(144 \text{ in}^2/\text{ft}^2)}$$

$$= \underline{280.7 \text{ ft}^3} \qquad \blacksquare$$

It can be seen that if the absolute pressure exerted on the gas of Figure 1.10 were increased from 100 to 400 kPa at constant temperature, the ratio of initial to final volume of the gas would be 4:1 (Figure 1.11), the gas undergoing a percentage volume change of $[(\Psi_2 - \Psi_1)/\Psi_1]100 = 300\%$! The difference between the compressibility of gases and liquids can be seen clearly from this example.

1.7 VISCOSITY

As discussed in Section 1.2, when a torque or shear force is applied to a solid elastic bar, the bar twists until an equilibrium position is established. It is found that the angular deformation due to this torsion is directly proportional to the applied shear stress; that is,

$$\tau = G\gamma$$

where τ is the shear stress (i.e., the shear force per unit area), G is the shear modulus of elasticity (a property of elastic material), and γ is the angular deformation.

When a shear force is similarly applied to a fluid medium (see Figure 1.2), no equilibrium position is reached, but rather the fluid continues to deform. In this case, the shear stress is found to be proportional to the rate of angular deformation:

$$\tau = \mu \frac{d\gamma}{dt}$$

with the coefficient μ called the *absolute* or *dynamic viscosity*, or sometimes just *viscosity*. It is apparent from this relationship that the greater the value of viscosity of a fluid, the larger the applied shear stress must be to obtain a given rate of angular deformation. The student is familiar with the "stickiness" of certain liquids. For example, heavy lubricating oils with high viscosity flow less easily than water, which has a much lower viscosity.

It will be useful to express the rate of angular deformation of a fluid in terms of flow velocity. Let us examine the simple example of the rotating cylinder shown in Figure 1.12. Viscous fluid particles adjacent to the cylinders adhere to the walls and move at the local wall velocity. At a given instant t_1, draw a line connecting all fluid particles lying on the radial line shown. At some subsequent time t_2, the fluid particle located adjacent to the outer cylinder has moved to the right with peripheral velocity u. On the other hand, the particle located at the inner cylinder remains at the same location, for the inner cylinder is held fixed. For small Δt and hence small $\Delta \gamma$, we obtain

$$\tan \Delta\gamma \sim \Delta\gamma = \frac{\Delta u\, \Delta t}{\Delta y}$$

where $\Delta\gamma$ is the angular deformation. Thus the angular rate of deformation $\Delta\gamma/\Delta t$ equals $\Delta u/\Delta y$. Written in differential form, the expression relating shear stress and velocity gradient du/dy is

$$\tau = \mu \frac{d\gamma}{dt} = \mu \frac{du}{dy} \tag{1.4}$$

For a large class of fluids, the coefficient of viscosity (μ) is independent of the velocity gradient. Such fluids are called *Newtonian fluids*. Most fluids familiar to us, such as water, air, and oil, behave as Newtonian fluids.

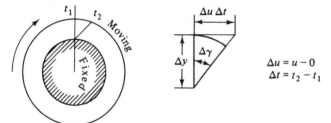

$$\Delta u = u - 0$$
$$\Delta t = t_2 - t_1$$

Figure 1.12

However, certain other fluids, such as blood, tar, and slurries, are non-Newtonian.

The viscosity is largely a function of temperature for most Newtonian fluids. The magnitude of the coefficient depends on the cohesive force between molecules and the momentum interchange between colliding molecules. The former is dominant for a liquid, so that as the temperature of a liquid is raised and the cohesive force between molecules decreases, the viscosity also decreases. However, the momentum interchange is dominant for a gas, and as the temperature of the gas is raised, providing for greater momentum interchange, the viscosity of the gas increases. The units of viscosity can be found from Equation (1.4). Rearranging, we see that

$$\mu = \frac{\tau}{du/dy}$$
$$= \frac{N/m^2}{m/(s \cdot m)}$$
$$= N \cdot s/m^2$$
$$= Pa \cdot s$$

or, in English units,

$$\mu = \frac{lbf/ft^2}{(ft/s)/ft}$$
$$= \frac{lbf \cdot s}{ft^2}$$

Typical curves of viscosity of various liquids and gases are shown in Figures 1.13(a) and 1.13(b) (p. 16) in SI units, and in Figures 1.13(c) and 1.13(d) (p. 17) in English units.

In dealing with problems in fluid flow, we shall frequently come across the ratio between dynamic or absolute viscosity and density; this ratio is called the *kinematic viscosity v*:

$$v = \frac{\mu}{\rho}$$

It can be seen that the units of kinematic viscosity, in the SI, are $[(N/m^2) \cdot s]/(kg/m^3)$, or, since $1 N = 1 kg \cdot m/s^2$, v has units of m^2/s. In the English system, v has units of $(lbf \cdot s/ft^2)/(slugs/ft^3)$, or, since $1 lbf = 1 slug \cdot ft/s^2$, v has units of ft^2/s. Since μ is dependent on temperature, it follows that v will also be dependent on temperature. Furthermore, for a gas, with density a function of pressure and temperature (e.g., $p = \rho RT$), v will vary significantly with both pressure and temperature. Curves of v versus T for liquids and gases are given in Figures 1.14a and 1.14b (p. 18) for SI units; curves of v versus T for liquids and gases are given in Figures 1.14c and 1.14d (p. 19) for English units.

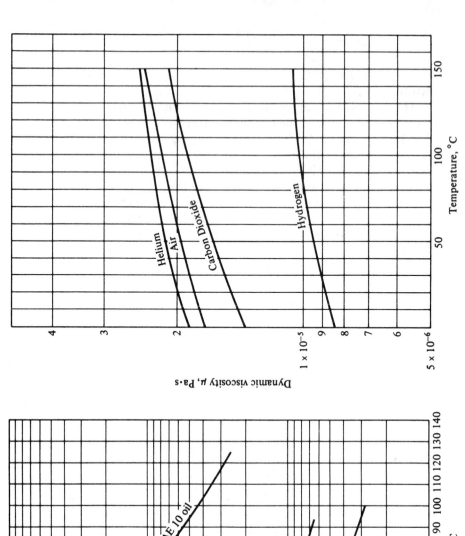

Figure 1.13(b). Dynamic (absolute) viscosities of several gases, SI units.

Figure 1.13(a). Dynamic (absolute) viscosities of several liquids, SI units.

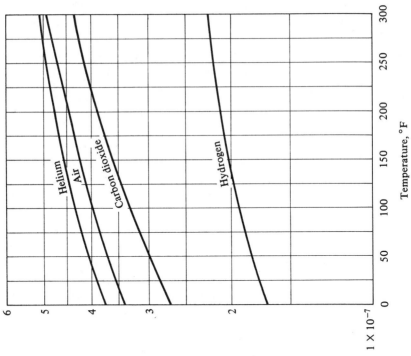

Figure 1.13(d). Dynamic (absolute) viscosities of several gases, English units.

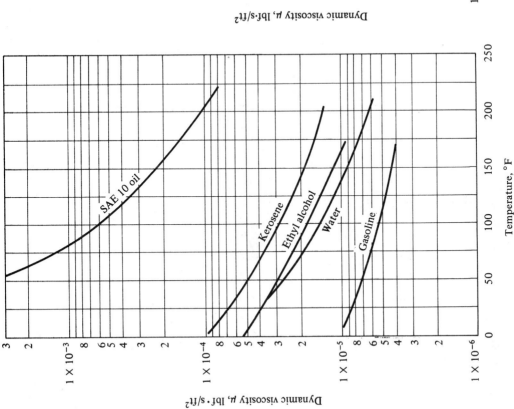

Figure 1.13(c). Dynamic (absolute) viscosities of several liquids, English units.

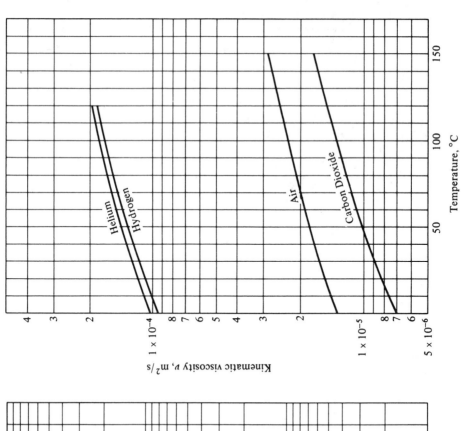

Figure 1.14(b). Kinematic viscosities of several gases, SI units (atmospheric pressure).

Figure 1.14(a). Kinematic viscosities of several liquids, SI units.

18

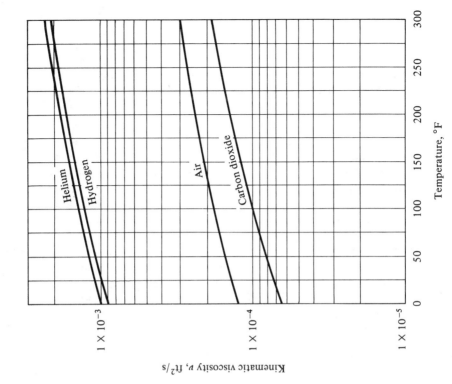

Figure 1.14(d). Kinematic viscosities of several gases, English units (atmospheric pressure).

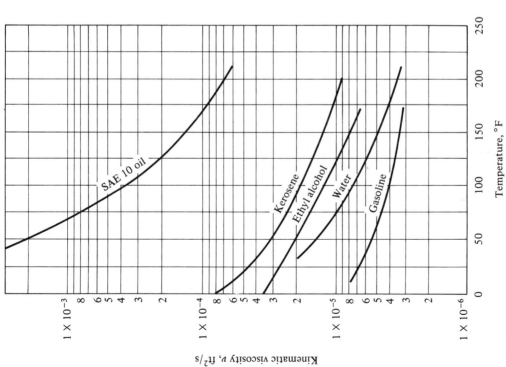

Figure 1.14(c). Kinematic viscosities of several liquids, English units.

19

1.8 OTHER PROPERTIES

Surface tension of a liquid is due to the forces of attraction between like molecules, called *cohesion*, and those between unlike molecules, called *adhesion*. In the interior of a liquid, the cohesive forces acting on a molecule due to its neighboring molecules are balanced out, since the molecule is surrounded by like molecules. Near a free surface, however, since the cohesive force between liquid molecules is much greater than that between an air molecule and a liquid molecule, there is a resultant force on a liquid molecule acting toward the interior of the liquid. This force, called *surface tension*, is proportional to the product of a surface tension coefficient σ and the length of the free surface. It is this force that holds a water droplet or a mercury globule together.

To illustrate, consider the equilibrium of the water droplet shown in cross section in Figure 1.15. The surface tension force acting on the surface of the drop is given by $\sigma 2\pi R$. This is balanced by the force due to the difference in pressure between the space inside and outside the droplet, $(p_i - p_o)\pi R^2$. Combining, we obtain

$$p_i - p_o = \frac{2\sigma}{R}$$

The surface tension coefficient for water at room temperature exposed to air is 0.07 N/m, so that for a droplet with a 0.25-cm radius, the pressure inside the drop will be higher than that outside by

$$\frac{2\sigma}{R} = \frac{2(0.07 \text{ N/m})}{0.0025 \text{ m}} = 56 \text{ Pa}$$

In English units, the surface tension coefficient of water is 5×10^{-3} lbf/ft, so that for a droplet with a 0.1-in radius, the pressure inside the drop is higher than the pressure outside by

$$\frac{2\sigma}{R} = \frac{2(5 \times 10^{-3} \text{ lbf/ft})}{(0.1/12) \text{ ft}} = 1.2 \text{ lbf/ft}^2$$

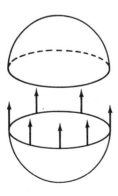

Figure 1.15. Cross section of water droplet showing surface tension force.

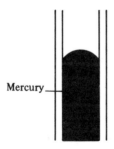

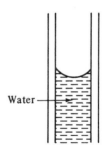

Figure 1.16. Mercury meniscus in glass tube.

Figure 1.17. Water meniscus in glass tube.

When a liquid is in contact with a solid surface, we must consider adhesion forces between solid and liquid as well as cohesive forces within the liquid. For example, the shape of a mercury meniscus in a glass tube is shown in Figure 1.16. In this case, the cohesive forces between mercury molecules are greater than the adhesive forces between mercury and glass. However, if water is placed in the glass tube, the meniscus takes on a different shape, as shown in Figure 1.17. In the latter case, the adhesive forces between water and glass are greater than the cohesive forces between the water molecules.

Except in certain specialized problems, such as those dealing with bubble formation, growth, and capillarity, surface tension forces are negligible in comparison with gravitational, viscous, and pressure forces.

The final fluid property that we shall mention in this chapter is *specific heat*. The specific heat c of a substance is the amount of heat that must be transferred into the substance to raise one unit mass by one unit temperature difference. In the SI, c has units of J/kg · K; in the English system, c has units of Btu/lbm°R. For example, the specific heat of water at room temperature is 4187 J/kg · K, or 1.0 Btu/lbm°R. For a gas, the amount of heat required depends on the nature of the process—for example, constant pressure or constant volume. The heat required to raise one unit mass of a gas by one unit temperature difference for a constant volume process is denoted by c_v; for a constant pressure process, it is c_p. For air at room temperature, in the SI, $c_v = 715$ J/kg · K and $c_p = 1000$ J/kg · K. In the English system, $c_v = 0.171$ Btu/lbm°R and $c_p = 0.240$ Btu/lbm°R.

EXAMPLE 1.8

(a) 5.0 kg of air is contained in a piston-cylinder arrangement as shown in Figure 1.18. Stops are placed above the piston so that it cannot move. 10 kJ of heat is then transferred into the air. Determine the resultant temperature rise for this constant-volume process. Assume that the air behaves as a perfect gas.

(b) 5.0 kg of air is contained in the same piston-cylinder arrangement

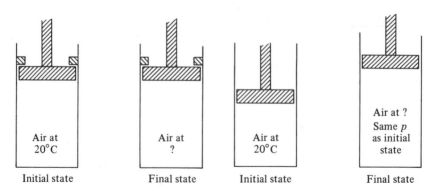

Figure 1.18   **Figure 1.19**

as in part (a), but the stops are removed. (See Figure 1.19.) 10 kJ of heat is then transferred into the air, and the piston is allowed to move upward in a constant-pressure process. Determine the resultant temperature rise, assuming that the air behaves as a perfect gas.

Solution

(a) In this case, the heat goes into raising the internal energy of the air, where $\Delta u = c_v \, \Delta T$. Therefore,

$$10 \text{ kJ} = mc\Delta T$$

or

$$\Delta T = \frac{10 \text{ kJ}}{(5 \text{ kg})(0.715 \text{ kJ/kg} \cdot \text{K})}$$

$$= \underline{2.80 \text{ K}} \qquad \text{(or the final air temperature is 22.8 K)}$$

(b) For a constant-pressure process,

$$10 \text{ kJ} = mc_p \, \Delta T$$

or

$$\Delta T = \frac{10 \text{ kJ}}{(5.0 \text{ kg})(1.0 \text{ kJ/kg} \cdot \text{K})}$$

$$= \underline{2.0 \text{ K}} \qquad \text{(or the final temperature is 22.0 K)} \qquad ■$$

PROBLEMS

1.1. The payload of a spacecraft has a mass of 33 kg on the earth's surface. Determine the weight and mass of the payload at a point in space at which $g = 3.0 \text{ m/s}^2$.

1.2. A person weighs 180 pounds on the earth's surface. Determine the weight and mass of the same person on the moon's surface, where the acceleration due to gravity is one-sixth of that on the earth's surface.

1.3. Calculate the change in volume of 1 liter of water if the pressure on the water is increased from atmospheric pressure to 10 atmospheres. Take the isothermal bulk modulus of water to be 2×10^4 atmospheres.

1.4. Calculate the change in volume of 1 kg of air if the pressure on the air is increased from atmospheric pressure to 10 atmospheres while the air temperature is maintained constant at 15°C. The initial volume of the air is 1 liter.

1.5. Find the kinetic energy of 10 ft³ of water moving at 3 ft/s. Take the density of water to be 62.4 lbm/ft³.

1.6. Hydrogen at 20°C is contained in a rigid container of internal volume 30 cm³, at an absolute pressure of 500 kPa. Determine the mass of hydrogen in the container.

1.7. Nitrogen is held in a piston-cylinder arrangement with an initial volume of 25 in³, initial temperature 100°F, initial pressure 25 psia. The piston is pushed down into the cylinder until the volume is reduced to 3 in³. The piston is sealed in the cylinder so that no nitrogen can enter or escape. Determine the final pressure of the nitrogen if the temperature is maintained at 100°F. Calculate the mass of nitrogen in the cylinder.

1.8. Show that the term $\frac{1}{2}\rho V^2$ has dimensions of pressure.

1.9. Show that the fraction $\rho VL/\mu$ is dimensionless. Here L is length and V is velocity.

1.10. In the apparatus of Figure 1.12, it is found that a force of 0.62 N applied to the outer cylinder is sufficient to give the outer cylinder a rotational speed of 2.2 rad/s, with the inner cylinder fixed. Determine the viscosity of the fluid between the cylinders. The radius of the outer cylinder is 20 cm; the radius of the inner cylinder is 10 cm; the axial length of the cylinders is 30 cm.

1.11. Estimate the force required to move the outer cylinder of Problem 1.10 at 2.2 rad/s for the following fluids at 25°C: SAE 10 oil, ethyl alcohol, and water.

1.12. Let the apparatus of Figure 1.12 have an outer cylinder radius of 10 in and an inner cylinder radius of 9 in, with water contained between the cylinders at 40°F. Determine the force required to give the outer cylinder a speed of 20 rpm (revolutions per minute). Assume the inner cylinder to be fixed.

1.13. Derive an expression for the isothermal bulk modulus of a perfect gas. Find the isothermal bulk modulus of air at 25°C and 1 atmosphere. Repeat for hydrogen.

1.14. Derive an expression for the coefficient of isothermal compressibility of a perfect gas. Determine the coefficient of isothermal compressibility of air and of hydrogen at 0°F.

1.15. An oil has a specific gravity of 0.86 and a dynamic viscosity 50% greater than that of kerosene at 4°C. Find the kinematic viscosity of the oil at 4°C.

1.16. Air is contained in a rigid container of 1.0 ft³ internal volume. The air pressure is initially 10 psia, the air temperature 68°F. Heat is added to bring the temperature to 100°F. Determine the final pressure in the tank, the mass of air in the tank, and the heat required (in Btu). Use the specific heat data given in Section 1.8.

1.17. Find the kinematic viscosity of air at 5 atmospheres and 25°C. Assume that air behaves as a perfect gas.

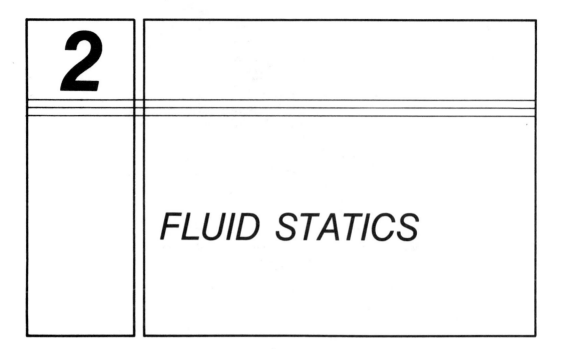

2

FLUID STATICS

2.1 INTRODUCTION

Fluid statics is the branch of fluid mechanics which deals with situations in which there is no relative motion between fluid elements. The fluid can be either at rest or in uniform motion. In this chapter, we will be particularly interested in problems in which the fluid is at rest.

In general, there are two types of forces that act on a fluid: surface forces and body forces. Surface forces are either normal or tangential to the surface and are exerted on the boundary of a fluid by direct contact. Surface forces include pressure and viscous shear. Body forces are external forces on a fluid developed without contact, such as gravity, and are dependent on the mass or volume of fluid present.

In a static fluid, there is no motion of one layer of fluid relative to an adjacent layer, so there are no viscous shear forces. Thus, the only forces we shall consider in a study of fluid statics are pressure forces and gravity.

A knowledge of fluid statics is necessary for the solution of many familiar problems, such as the determination of total water force on a dam, the calculation of pressure variation throughout the atmosphere, the analysis of a hydraulic braking system, or the determination of the stability of a partly submerged or floating body. In this chapter we shall first derive an expression for pressure variation throughout a static body of fluid. Then we shall apply this result to the calculation of the magnitude and location of fluid forces on submerged plane or curved surfaces. Finally, we shall determine the fluid forces on partially or totally submerged bodies and relate these forces to the stability of a floating body.

2.2 PRESSURE VARIATION THROUGHOUT A STATIC FLUID

The equilibrium of a static fluid requires that

$$\Sigma F = 0$$

First, we shall consider the balance of forces on an infinitesimal element in a static body of fluid in a gravitational field. Select the z axis in the direction of gravity, as shown in Figure 2.1. With no relative motion between fluid particles, there are no shear forces acting on the element, only normal forces (due to pressure) and the gravity force.

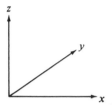

Figure 2.1

In the x direction, we have $\Sigma F_x = 0$. Gravity contributes no force in the x direction, so if the pressure on the left-hand face is p, as shown in Figure 2.2, the pressure on the right-hand face will be

$$p + \frac{\partial p}{\partial x}dx$$

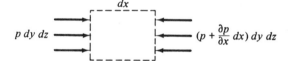

Figure 2.2. Forces in the x direction.

Summing forces, we obtain

$$p\,dy\,dz - \left(p + \frac{\partial p}{\partial x}dx\right)dy\,dz = 0$$

or

$$\frac{\partial p}{\partial x} = 0$$

Similarly, since gravity contributes no force in the y direction,

$$\frac{\partial p}{\partial y} = 0$$

However, in the z direction, gravity must be taken into account, as shown in Figure 2.3, so that

$$\Sigma F_z = 0$$

or

$$p \, dx \, dy - \left(p + \frac{\partial p}{\partial z} dz \right) dx \, dy - \rho g \, dx \, dy \, dz = 0$$

Simplifying, we get

$$\frac{dp}{dz} = -\rho g \qquad (2.1)$$

The preceding analysis shows that the pressure in a static body of fluid varies only in the direction of gravity.

For cases in which density is constant, Equation (2.1) can be integrated to yield

$$\int_1^2 dp = -\rho g \int_1^2 dz$$

or

$$p_2 - p_1 = \rho g (z_1 - z_2) \qquad (2.2)$$

If point 1 is taken at the free surface of a liquid, as shown in Figure 2.4, then

$$p_2 - p_1 = \rho g d$$

where d is the depth below the free surface.

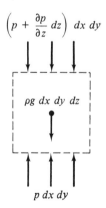

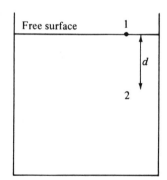

Figure 2.3. Forces in the z direction.

Figure 2.4. Pressure variation in a static liquid.

An application of Equation (2.2) is in the field of manometry, the measurement of pressure. A simple U-tube manometer is shown in Figure 2.5. The unknown pressure p_T in the tank containing a gas of density ρ_g is to be measured with a glass U tube containing a liquid of density ρ_L. The liquid in the tube will reach an equilibrium position where its weight will be balanced by the difference between the tank pressure and the local atmospheric pressure exerted on the liquid at D. Assuming all fluids to be of constant density, we have, from Equation (2.2),

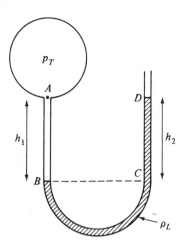

Figure 2.5. U-tube manometer.

$$p_A - p_B = -\rho_g g h_1$$

Since B and C are at the same elevations in a static fluid, the pressures at these points are equal. Finally,

$$p_C - p_D = \rho_L g h_2$$

Therefore,

$$p_A - p_D = p_A - p_{\text{atm}} = -\rho_g g h_1 + \rho_L g h_2$$

A common fluid used in a U-tube manometer is mercury, which has a relatively high density (the specific gravity of mercury at standard conditions is 13.6). Because of that high density, a mercury manometer can be used to measure relatively large pressure differences. With the density of mercury so much greater than the density of a gas, we can generally drop the term $\rho_g g h_1$ for a mercury manometer, with the resultant expression reducing to

$$p_T = p_{\text{atm}} + \rho_L g h_2 \qquad (2.3)$$

The use of a U-tube manometer, with one leg open to the atmosphere, thus allows a determination of the difference between an unknown absolute pressure and local atmospheric pressure; this difference is called *gage pressure* (see Section 1.4). For example, if the distance between the mercury levels in the manometer of Figure 2.5 were 10 cm, the tank pressure would be

$$p_T = p_{\text{atm}} + 13.6(1000) \text{ kg/m}^3 \times 9.81 \text{ m/s}^2 \times 0.1 \text{ m}$$

where 1000 kg/m^3 is the density of water; or

$$p_T - p_{\text{atm}} = 13.34 \text{ kPa}$$

We say that the tank pressure is thus 13.34 kPa gage pressure. If local atmospheric pressure is 101 kPa, then we say that the tank pressure is 101 + 13 = 114 kPa. When the pressure difference to be measured by the manometer is small, liquids having low specific gravities, such as water or alcohol, are used in the manometer tube.

If the difference in mercury levels in the manometer of Figure 2.5 were 12 in, the tank pressure would be

$$p_T = p_{atm} + 13.6(1.94) \text{ slugs/ft}^3 \times 32.17 \text{ ft/s}^2 \times \tfrac{12}{12} \text{ ft}$$

where 1.94 slugs/ft^3 is the density of water, derived as follows:

$$62.4 \text{ lbm/ft}^3 = \frac{62.4 \text{ lbm/ft}^3}{32.17 \text{ lbm/slug}} = 1.94 \text{ slugs/ft}^3$$

Thus

$$p_T - p_{atm} = 848.8 \text{ lbf/ft}^2 \text{ gage} = \frac{848.8 \text{ lbf/ft}^2}{144 \text{ in}^2/\text{ft}^2} = 5.89 \text{ psig}$$

A manometer can also be used to measure absolute pressure directly. If, instead of exposing one leg of the manometer to atmospheric pressure, we evacuate the space above that leg (Figure. 2.6), we have a device for measuring absolute pressure. Neglecting the density of the gas in the tank compared with the density of the manometer fluid, and with $p_B = 0$, we have

$$p_T = \rho_L g h \qquad\qquad (2.4)$$

with p_T now expressed directly as an absolute pressure. From Equation (2.4), we see that pressure can be measured in terms of an equivalent column of mercury. If the tank on leg A of Figure 2.6 were to be removed, with standard atmospheric pressure of 101.3 kPa (14.7 psia) exerted on this leg, the height of the mercury column would be 760 mm (29.92 in), as shown in Figure 2.7.

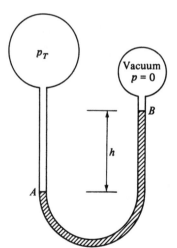

Figure 2.6. Absolute-pressure manometer.

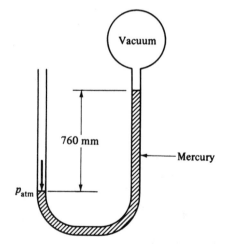

Figure 2.7. Atmospheric pressure in millimeters of mercury.

EXAMPLE 2.1

Determine the pressure p_g of the gas above the oil in Figure 2.8.

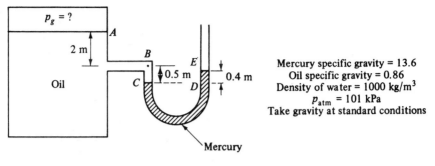

Figure 2.8

Solution   Applying Equation (2.2), $\Delta p = \rho g \, \Delta z$, we obtain

$$p_A - p_B = -(0.86 \times 1000 \text{ kg/m}^3)(9.81 \text{ m/s}^2)(2 \text{ m}) = -16.87 \text{ kPa}$$

$$p_B - p_C = -(0.86 \times 1000)(9.81)0.5 = -4.22 \text{ kPa}$$

$p_C = p_D$ (since C and D are at the same elevation in the same liquid)

$$p_D - p_E = (13.6 \times 1000)(9.81)0.4 = +53.37 \text{ kPa}$$

Adding, we get

$$p_A - p_E = 32.28 \text{ kPa}$$

Since p_E is equal to local atmospheric pressure, $p_A = 101 \text{ kPa} + 32 \text{ kPa} =$ 133 kPa absolute pressure.   ■

If the density of the fluid is not constant but varies with z, the integration of Equation (2.1) becomes more complex. An important example of a static fluid with varying density is the atmosphere; it is the intent of this section to indicate how to determine the pressure variation throughout the atmosphere.

In the troposphere, extending from sea level to about 11 km in altitude, the temperature variation with altitude can be approximated by

$$T = T_{\text{sea level}} - \lambda z$$

with the constant λ, termed the lapse rate, equal to about 6.5°C/km. In this region, air can be treated as a perfect gas, with $p = \rho RT$. In order to determine the pressure variation in the troposphere, substitute into Equation (2.1)

$$dp = -\rho g \, dz$$

$$= -\frac{p}{RT} g \, dz$$

$$= \frac{-pg \, dz}{R(T_{\text{sea level}} - \lambda z)}$$

or

$$R\frac{dp}{p} = -g\frac{dz}{T_{\text{sea level}} - \lambda z}$$

Integrating the above from $z = 0$ to $z = z$, we obtain

$$\int_{p\text{sea level}}^{p} R\frac{dp}{p} = -\int_{z=0}^{z=z} \frac{g\,dz}{T_{\text{sea level}} - \lambda z}$$

Neglecting any variation of g with altitude over the range of z being considered,

$$R\ln\frac{p}{p_{\text{sea level}}} = \frac{1}{\lambda}g\ln\frac{T_{\text{sea level}} - \lambda z}{T_{\text{sea level}}}$$

or

$$\frac{p}{p_{\text{sea level}}} = \left(1 - \frac{\lambda z}{T_{\text{sea level}}}\right)^{g/\lambda R}$$

For a sea level temperature of 288 K and $\lambda = 6.5°C/km$, we have for the pressure variation in the troposphere

$$\frac{p}{p_{\text{sea level}}} = \left(1 - \frac{0.0065z}{288}\right)^{9.81/0.0065R}$$

with z expressed in meters. For air, $R = 0.2870$ kJ/kg $\cdot$ K (Appendix A), so that

$$\frac{p}{p_{\text{sea level}}} = (1 - 0.0000226z)^{5.26} \qquad (2.5a)$$

This result is plotted in Figure 2.9a.

From about 11 km to 20 km, in the stratosphere, there exists an isothermal layer, with temperature equal to $-56.5°C$. Again, we can assume that air behaves as a perfect gas, with negligible change in g with altitude. From Equation (2.1),

$$dp = -\rho g\,dz$$

For a perfect gas, $p = \rho RT$; substituting for ρ, we find

$$\frac{dp}{p} = -g\frac{dz}{RT}$$

For $T = $ constant, we can integrate the above to obtain the pressure distribution in the isothermal layer:

$$\int_{p=11,000\text{m}}^{p} \frac{dp}{p} = -\frac{g}{RT}\int_{11,000\text{m}}^{z} dz$$

The lower limit of the integration can be obtained from Equation (2.5)—namely, $p = 22.5$ kPa at $z = 11,000$ m. We therefore have after integration

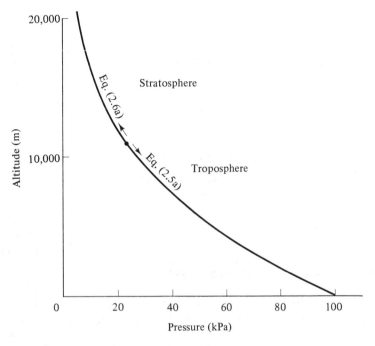

Figure 2.9(a). Pressure variation in the atmosphere, SI units.

$$\ln\frac{p}{22.5} = -\frac{9.81 \text{ m/s}^2 \times (z - 11{,}000) \text{ m}}{287 \text{ J/kg} \cdot \text{K} \times 216.65 \text{ K}}$$

where p is in kPa, or

$$\ln\frac{p}{22.5} = -(z - 11{,}000)(0.0001578) \tag{2.6a}$$

This result is also shown in Figure 2.9a.

In English units, the troposphere extends from sea level to about 36,000 ft in altitude, while the stratosphere extends from 36,000 ft to 100,000 ft in altitude. The lapse rate in the troposphere is 3.6°F/1000 ft. In English units, Equations (2.5a) and (2.6a) become, respectively,

$$\frac{p}{p_{\text{sea level}}} = (1 - 0.00000692z)^{5.21} \tag{2.5b}$$

and

$$\ln\frac{p}{473} = -(z - 36{,}000)(0.0000480) \tag{2.6b}$$

The variation of pressure with altitude in English units is shown in Figure 2.9b.

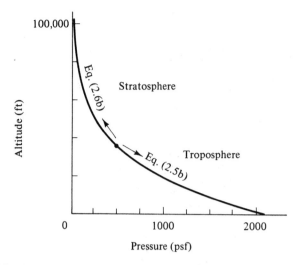

Figure 2.9(b). Pressure variation in the atmosphere, English units.

2.3 FORCES ON PLANE SUBMERGED AREAS

Consider the plane inclined area (Figure 2.10) submerged in an incompressible static liquid of uniform density ρ. It is desired to find the magnitude and location of the resultant hydrostatic force acting on the top side of the area.

For a fluid of constant density, the variation of pressure is given by Equation (2.2):

$$p - p_{\text{surface}} = \rho g d$$

with d the vertical distance below the free surface. Select a coordinate system with origin at the free surface and y axis directed along the plane area, with the area in the x-y plane. The force dF on an elemental area dA is given by

$$dF = p \, dA \qquad (2.7)$$
$$= (p_{\text{atm}} + \rho g y \sin \theta) \, dA$$

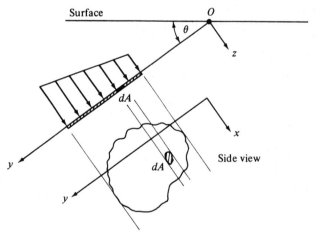

Figure 2.10. Plane submerged area.

The total force on the submerged area is then obtained by integration:

$$F = p_{atm}A + \rho g \sin\theta \iint y \, dA \qquad (2.8)$$

The integral on the right-hand side should be familiar from a study of statics. The y coordinate of the centroid of a plane area, y_c, is given by

$$y_c = \frac{\iint y \, dA}{A}$$

Therefore, Equation (2.8) reduces to

$$F = p_{atm}A + (\rho g \sin\theta)y_c A \qquad (2.9)$$

The right-hand side can be recognized as the pressure p_c at the centroid of the submerged area times the area, or

$$F = p_c A \qquad (2.10)$$

It can be seen that of the total force of Equation (2.9), the portion $p_{atm}A$ is due to the uniform atmospheric pressure acting on the liquid surface and the portion $(\rho g \sin\theta)y_c A$ is due to the liquid or hydrostatic force acting on the inclined area.

EXAMPLE 2.2

Calculate the resultant hydrostatic force of the liquid acting on the triangular area shown in Figure 2.11. The triangle's base b is 1 ft; its height h is 1.5 ft. The tank is filled with water of density 1.93 slugs/ft³.

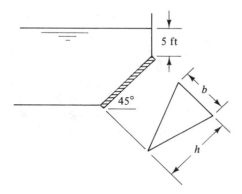

Figure 2.11

Solution   From Equation (2.9),

$$F_{water} = p_{atm}A + (\rho g \sin\theta)y_c A$$

Here, $y_c = 5/\sin 45° + 1.5/3 = 7.57$ ft, where y is measured from the free surface along the slanted area (see Figure 2.10) and the centroid of the triangle is one-third of the height measured from the base b. The area A is $bh/2 = 1(1.5)(2) = 0.75$ ft². Thus

$$F_{\text{water}} = (2116 \text{ lbf}/\text{ft}^2)(0.75 \text{ ft}^2)$$

$$+ (1.93 \text{ slugs}/\text{ft}^3)(32.17 \text{ lbm}/\text{slug}) \sin 45°(7.57 \text{ ft})(0.75 \text{ ft}^2)$$

$$= 1624.5 \text{ lbf} + 249.3 \text{ lbf}$$

$$= \underline{1873.8 \text{ lbf}}$$ ∎

We now wish to determine the point of application of this hydrostatic force; in other words, where would the force F_R of Figure 2.12 have to be exerted in order to hold the plane area in equilibrium? Taking moments about the x axis located at O, we find

$$y_{F_R} F_R = \iint (p - p_{\text{atm}})y \, dA$$

$$= \rho g \sin \theta \iint y^2 \, dA$$

or

$$y_{F_R} F_R = \rho g (\sin \theta) I_{xx}$$

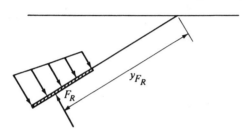

Figure 2.12.   Resultant of hydrostatic forces.

where I_{xx}, called the *second moment of the area about the x axis at O*, is defined as

$$I_{xx} = \iint y^2 \, dA$$

Since the term I_{xx} depends on the depth of the plane below the free surface, it is more convenient to write the above in terms of a second moment $I_{x'x'}$ taken about a parallel axis x' passing through the centroid of the inclined area (Figure 2.13). Using the parallel axis theorem from statics,

$$I_{xx} = I_{x'x'} + Ay_c^2$$

where $I_{x'x'}$ is taken about an origin at O'. Substituting, we obtain

$$y_{F_R} F_R = \rho g \sin \theta (I_{x'x'} + Ay_c^2)$$

or

$$y_{F_R} = \frac{\rho g (\sin \theta)(I_{x'x'} + Ay_c^2)}{\rho g (\sin \theta) y_c A}$$

$$= \frac{I_{x'x'}}{y_c A} + y_c$$ (2.11)

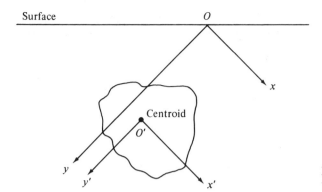

Figure 2.13. Axes passing through centroid.

The point of application of the resultant force on the inclined area is called the *center of pressure.* Since the term $I_{x'x'}/y_cA$ must be positive, it follows that the center of pressure always lies below the centroid of the inclined area.

We have found the y component of the location of the resultant force; we also wish to find the x component. Taking moments about the y axis, we obtain

$$F_R x_{FR} = \iint x \, dF$$

$$= \rho g \sin \theta \iint xy \, dA$$

The integral $\iint xy \, dA$ is called the product of inertia I_{xy} of the inclined area taken about the origin at O. Again we shall employ the parallel axis theorem so as to refer the moment of inertia to an origin O' located at the centroid of the inclined area:

$$I_{xy} = I_{x'y'} + Ax_c y_c$$

Substituting into the above,

$$x_{FR} = \frac{\rho g \sin \theta \iint xy \, dA}{\rho g (\sin \theta) y_c A}$$

$$= \frac{I_{x'y'} + Ax_c y_c}{y_c A}$$

$$= \frac{I_{x'y'}}{y_c A} + x_c \tag{2.12}$$

It can be seen that, by its definition, $I_{x'y'}$ can be either positive or negative. However, if either axis (x' or y') is an axis of symmetry, the value of $I_{x'y'}$ is zero. Fortunately, a great number of areas encountered are symmetrical about one or both of the axes; for these areas, with $I_{x'y'} = 0$, $x_{FR} = x_c$, and we must only determine y_{FR}.

The following example will illustrate these results. The moments of

TABLE 2.1 Moments of Inertia of Plane Areas About Their Centroids

Rectangle
$$I_{x'x'} = \frac{bh^3}{12}$$

Circle
$$I_{x'x'} = \frac{\pi d^4}{64}$$

Triangle
$$I_{x'x'} = \frac{bh^3}{36}$$

Trapezoid
$$I_{x'x'} = \frac{(b^2 + 4bc + c^2)h^3}{36(b + c)}$$

$$\frac{h}{3}\frac{b + 2c}{b + c}$$

Semicircle $\dfrac{2d}{3\pi}$
$$I_{x'x'} = \frac{d^4(9\pi^2 - 64)}{1152\pi}$$

Quadrant of circle $\dfrac{2d}{3\pi}$
$$I_{x'x'} = \frac{d^4(9\pi^2 - 64)}{2304\pi}$$

inertia $I_{x'x'}$ of several common plane areas about their centroids are provided in Table 2.1.

EXAMPLE 2.3
Calculate the force F required to hold the hinged door closed in Figure 2.14. The door is square, with side dimension 0.3 m. The tank is filled with water of density 998 kg/m³.

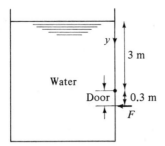

Figure 2.14

Solution For the case shown, atmospheric pressure acts on the water surface and also on the outside surface of the door; therefore the effect of atmospheric pressure cancels out, and we need only consider water forces. From Equation (2.9),

$$F_{water} = (\rho g \sin \theta) y_c A$$

where, for this case, $y_c = 3.15$ m, $\sin \theta = 1$, and $A = 0.09$ m². Substituting, we obtain

$$F_{water} = (998 \text{ kg/m}^3)(9.81 \text{ m/s}^2)(1)(3.15 \text{ m})(0.09 \text{ m}^2) = 2.776 \text{ kN}$$

To find the location of this force, use Equation (2.11):

$$y_{FR} = \frac{I_{x'x'}}{y_c A} + y_c$$

For a square, from Table 2.1,

$$I_{x'x'} = \frac{s^4}{12} = 0.000675 \text{ m}^4$$

As shown in Figure 2.14, the coordinate y is measured from the free surface, so, again, $y_c = 3.15$ m:

$$y_{FR} = \frac{0.000675 \text{ m}^4}{(3.15 \text{ m})(0.09 \text{ m}^2)} + 3.15 \text{ m}$$

$$= 0.00238 + 3.15$$

$$= 3.15238 \text{ m}$$

or slightly below the centroid of the door. Taking moments about the hinge, we find

$$F \times 0.3 \text{ m} = F_{\text{water}} \times 0.15238 \text{ m}$$

$$F = \frac{2.776 \text{ kN} \times 0.15238 \text{ m}}{0.3 \text{ m}}$$

$$= 1.410 \text{ kN} \qquad \blacksquare$$

EXAMPLE 2.4

Repeat the previous problem with a pressure over the water of 50 kPa (see Figure 2.15).

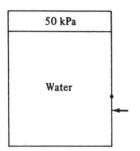

Figure 2.15

Solution   The total force on the door will now be increased over the previous case, since the pressure of 50 kPa denotes a value above local atmospheric pressure.

$$\text{Total } F = F_{\text{water}} + 50 \text{ kN/m}^2 \times 0.09 \text{ m}^2$$

$$= 2.776 \text{ kN} + 4.50 \text{ kN}$$

$$= 7.276 \text{ kN}$$

The water force acts at the same point as in the previous problem. However, the overpressure on the water surface gives a uniform pressure on the door (see Figure 2.16); this force can be considered to act, therefore, at the centroid of the door, 0.15 m below the hinge. Taking moments about the hinge,

$$F \times 0.3 \text{ m} = F_{\text{water}} \times 0.15238 \text{ m} + 4.50 \text{ kN}(0.15 \text{ m})$$

$$0.3F = 1.098 \text{ kN} \cdot \text{m}$$

$$F = \underline{3.66 \text{ kN}} \qquad \blacksquare$$

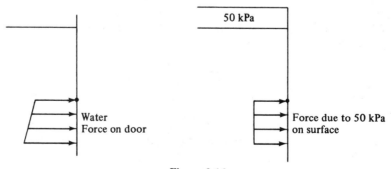

Figure 2.16

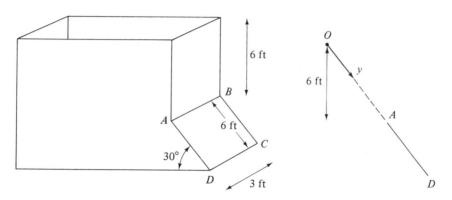

Figure 2.17   **Figure 2.18**

EXAMPLE 2.5

Calculate the water force and the location of this force on the slanted side wall *ABCD* of the tank shown in Figure 2.17. The tank is completely filled with water. Assume the density of water to be 62.4 lbm/ft².

Solution The force on the inclined area is equal to the pressure at the centroid times the area. From Equation (2.10)

$$F = p_c A$$

where $p_c = \rho g y_c \sin \theta$.

$$F = \left(\frac{62.4}{32.17} \text{ slugs/ft}^2\right)(32.17 \text{ ft/s}^2)[(6 + 3 \sin 30°)\text{ft}](18 \text{ ft}^2)(1 \text{ lbf} \cdot \text{s}^2/\text{slug} \cdot \text{ft})$$

$$= \underline{8424 \text{ lbf}}$$

The location of the force can be found from Equation (2.11):

$$y_{FR} = \frac{I_{x'x'}}{y_c A} + y_c$$

with the *y* coordinate measured from the free surface along the slanted area (see Figure 2.18).

$$y_c = \frac{6}{\sin 30°} + 3 = 15 \text{ ft}$$

$$I_{x'x'} = \frac{bh^3}{12} = \frac{(3 \text{ ft})(6^3 \text{ ft}^3)}{12} = 54 \text{ ft}^4$$

$$\frac{54}{15(18)} + 15 = \underline{15.2 \text{ ft}}$$

From symmetry, with $I_{x'y'} = 0$, $x_{FR} = x_c$, or the *x* location of the water force is halfway between sides *AD* and *BC*. ∎

2.4 HYDROSTATIC FORCES ON CURVED SUBMERGED AREAS

In order to determine the hydrostatic force on a curved submerged surface, consider the situation depicted in Figure 2.19. The total force on infinitesimal area dA is $p\, dA$, with the horizontal component of this force dF_H equal to $p\, dA \cos \theta$. But $dA \cos \theta$ is simply the projection of area dA on a vertical plane. It follows that the horizontal component of the hydrostatic force on a curved submerged surface is equal in magnitude and direction to the hydrostatic force on the vertical projection of that area.

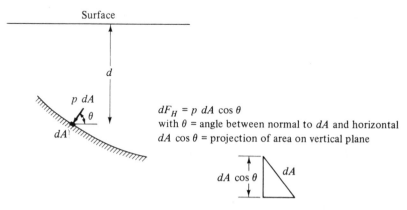

Figure 2.19. Horizontal component of force on submerged surface.

The vertical component of the force on dA is given by
$$dF_V = p\, dA \sin \theta$$
where $dA \sin \theta$ is simply the horizontal projection of dA. Since $p = \rho g d$, we have
$$dF_V = \rho g d\, dA \sin \theta$$
$$= \rho g\, d\mathcal{V}$$
with $d\mathcal{V}$ the volume of fluid lying directly above dA (Figure 2.20). But $\rho g\, d\mathcal{V}$ is simply the weight of fluid above dA. Integrating over the entire submerged surface, we obtain
$$F_V = \rho g \mathcal{V} \qquad (2.13)$$
with $\mathcal{V}$ the total volume of fluid above the submerged area. Therefore, the resultant vertical component of the hydrostatic force on a curved submerged area is equal to the weight of water directly above that area. In order to determine the effective line of action of the resultant vertical force component, select an axis 0, as shown in Figure 2.20. Taking moments about 0,
$$(dF_V)x = \rho g\, d\mathcal{V}\, x$$
Integrating over the entire surface, we obtain
$$F_V x_{F_V} = \rho g \iiint x\, d\mathcal{V}$$

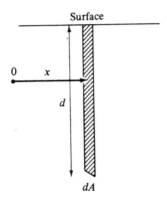

Figure 2.20. Vertical component of force on submerged surface.

with x_{F_V} the distance from 0 to the effective line of action of F_V. Since

$$F_V = \rho g \forall$$

we have

$$x_{F_V} = \frac{\iiint x \, d\forall}{\forall} \tag{2.14}$$

But Equation (2.14) is the definition of the centroid of a volume. The vertical component of the force acts through the centroid of the volume of fluid directly above the submerged area.

EXAMPLE 2.6

Determine the horizontal and vertical components of the hydrostatic force on the curved cylindrical area *ABCD* shown in Figure 2.21 as well as their location.

Solution The horizontal component is equal to the hydrostatic force on the vertical projection of the curved area. (See Figure 2.22.) The projection of *ABCD* is a rectangle 3 m × 1 m. From Figure 2.22,

$$F_H = p_c A_{\text{projected}} = [(1000 \text{ kg/m}^3)(9.81 \text{ m/s}^2)(2.5 \text{ m})] \, 3 \text{ m}^2$$

$$= (24.525 \text{ kPa})(3 \text{ m}^2)$$

$$= \underline{73.575 \text{ kN}}$$

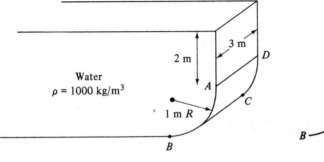

Figure 2.21

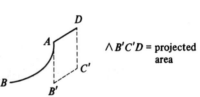

Figure 2.22

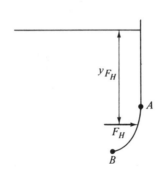

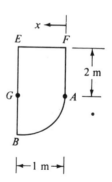

Figure 2.23   Figure 2.24

The line of action of F_H is given by

$$y_{F_H} = \frac{I_{x'x'}}{y_c A} + y_c$$

where

$$I_{x'x'} = \frac{bh^3}{12} = \frac{3(1)^3}{12} = 0.25 \text{ m}^4$$

$$y_{F_H} = \frac{0.25}{2.5(3)} + 2.5 = \underline{2.533 \text{ m}} \qquad \text{(see Figure 2.23)}$$

The vertical force on the submerged area is equal to the weight of water above the area. The total volume above $ABCD$ is equal to the area of $ABEF$ times the width of 3 m (see Figure 2.24):

$$\cancel{V} = \left[(2 \times 1) + \frac{\pi}{4}\right]3 = 8.356 \text{ m}^3$$

$$F_V = \rho g \cancel{V}$$

$$= (1000 \text{ kg/m}^3)(9.81 \text{ m/s}^2)(8.356 \text{ m}^3)$$

$$= \underline{81.97 \text{ kN}}$$

The vertical component acts through the centroid of the volume of water above the surface. To find the centroid of area $ABEF$, take moments about a vertical line through AF. The area $AGEF$ is 2 m², with its centroid at $x = 0.5$ m. From Table 2.1, the centroid of ABG is at

$$x = 1 - \frac{2d}{3\pi} = 1 - \frac{4}{3\pi} = 0.5756 \text{ m}$$

with

$$\text{Area } ABG = \frac{\pi}{4} = 0.7854 \text{ m}^2$$

Combining, the centroid of $ABGEF$ is at

$$x = \frac{2 \times 0.5 + 0.7854 \times 0.5756}{2.7854}$$

$$= 0.5213 \text{ m}$$

In other words, the vertical component acts at $x = 0.5213$ m.   ■

EXAMPLE 2.7
Calculate the force F required to hold the curved gate in equilibrium in Figure 2.25.

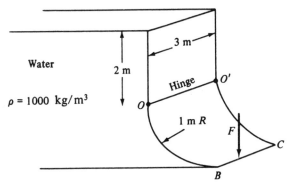

B   Figure 2.25

Solution   The horizontal component of the water force on the curved gate is equal to the force on the vertical projected area of $OBCO'$, or $F_H = p_c A$.

$$F_H = (1000 \text{ kg/m}^3 \times 9.81 \text{ m/s}^2 \times 2.5 \text{ m})(3 \times 1) \text{ m}^2 = 73.575 \text{ kN}$$

This force acts at a vertical distance y from the free surface of

$$y = y_c + \frac{I_{x'x'}}{y_c A}$$

$$= 2.5 + \frac{0.25}{2.5(3)} = 2.533 \text{ m} \qquad \text{(see results of Example 2.6)}$$

In order to calculate the vertical component of water force, first recognize that the pressure at each point on the gate $OO'CB$ is the same as if water were directly above $OO'CB$ (see Figure 2.26). Of course, in Case 1, the vertical component acts upward; in Case 2, it acts downward. The magnitude of this force, then, is equal to the weight of water over $OO'CB$ in Case 2,

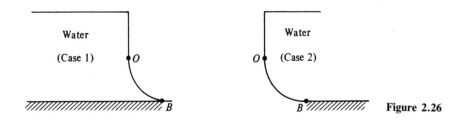

Figure 2.26

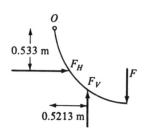

Figure 2.27   Figure 2.28

or

$$F_V = 1000 \times 9.81 \times 8.356 = 81.97 \text{ kN} \qquad \text{(see Example 2.6)}$$

This force acts upward, through the centroid of the crosshatched volume of Case 2 (see Figure 2.27), or at

$$x = 0.5213 \text{ m}$$

Finally, taking moments about the hinge, we obtain

$$0.5213F_V + 0.533F_H = 1F \qquad \text{(see Figure 2.28)}$$

or

$$F = \frac{(0.5213 \text{ m})(81.97 \text{ kN}) + (0.533 \text{ m})(73.575 \text{ kN})}{1 \text{ m}}$$

$$= \underline{81.95 \text{ kN}} \qquad \blacksquare$$

EXAMPLE 2.8

Determine the magnitude and location of the horizontal and vertical components of the hydrostatic force on the curved sidewall of the tank shown in Figure 2.29.

Solution   The horizontal component F_H is equal to the force on the vertical projection of the curved sidewall:

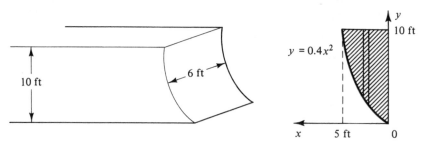

Figure 2.29

$$F_H = p_c A_{\text{projected}} = (1.94 \text{ slugs/ft}^3)(32.17 \text{ ft/s}^2)(5 \text{ ft})(60 \text{ ft}^2)$$
$$= \underline{18{,}723 \text{ lbf}}$$

The line of action of the horizontal component is given by Equation (2.11):

$$y_{F_H} = \frac{I_{x'x'}}{y_c A} + y_c$$

where

$$I_{x'x'} = \frac{bh^3}{12} = \frac{6(10)^3}{12} = 500 \text{ ft}^4$$

Thus

$$y_{F_H} = \frac{500}{5(60)} + 5 = \underline{6.67 \text{ ft}}$$

The vertical component F_V on the submerged area is equal to the weight of water above the area:

$$\cancel{V} = \left[50 - \int_0^5 0.4x^2 \, dx \right](6) = \left(50 - \frac{0.4}{3} x^3 \, \Big|_0^5 \right)(6)$$
$$= 33.33(6) = 200 \text{ ft}^3$$
$$F_V = \rho g \cancel{V} = (1.94 \text{ slugs/ft}^3)(32.17 \text{ ft/s}^2)(200 \text{ ft}^3) = \underline{12{,}482 \text{ lbf}}$$

The vertical component acts through the centroid of the water volume above the surface. To evaluate the centroid, take moments about a vertical line through the origin (see Figure 2.29). Hence

$$x_c = \frac{\iint x \, dA}{A} = \frac{\int_0^5 x(10 - 0.4x^2) \, dx}{A}$$

$$= \frac{\int_0^5 (10x - 0.4x^3) \, dx}{33.33} = \frac{[5x^2 - 0.1x^4]_0^5}{33.33}$$

$$= \frac{62.5}{33.33} = \underline{1.875 \text{ ft}} \qquad \text{(see Figure 2.30)} \qquad ■$$

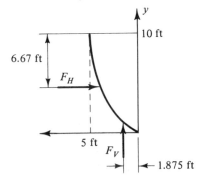

Figure 2.30

2.5 HYDROSTATIC FORCES ON SUBMERGED BODIES

We are now in a position to calculate the hydrostatic force on a completely submerged body, shown in Figure 2.31. Consider first a horizontal prism of cross-sectional area dA (Figure 2.31a). Since the pressure throughout a static body of fluid is only a function of depth below the free surface and not of horizontal position, the pressure at 1 is equal to the pressure at 2 and there is no horizontal force on the prism. Integrating over the entire surface of the body, it can be seen that the horizontal component of hydrostatic force on the submerged body is zero. Next, consider a vertical prism, again of cross-sectional area dA, shown in Figure 2.31b. The pressure at 1 is $\rho g d_1$, whereas the pressure at 2 is $\rho g d_2$. The vertical force on the prism is thus $\rho g(d_2 - d_1)\, dA$ or $\rho g\, d\mathcal{V}$, where $d\mathcal{V}$ is the differential volume of the prism. Integrating over the entire body, we obtain an expression for the total vertical force on the submerged body:

$$F_V = \iiint \rho g\, d\mathcal{V} \tag{2.15}$$

which is another way of saying that the vertical component of force on a body completely submerged in a static reservoir of fluid is equal to the weight of fluid displaced by the body.

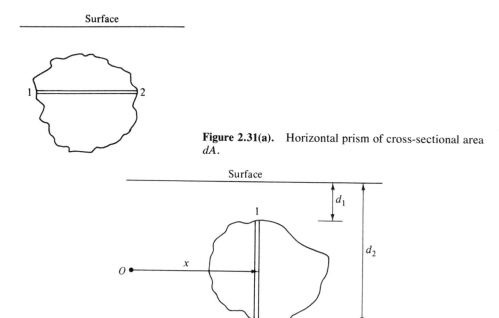

Figure 2.31(a). Horizontal prism of cross-sectional area dA.

Figure 2.31(b). Vertical prism of cross-sectional area dA.

In order to determine the line of action of the vertical force, F_V, select an axis O, as shown in Figure 2.31b. Taking moments, we find

$$F_V x_{F_V} = \iiint x\rho g \, d\mathcal{V}$$

or

$$x_{F_V} = \frac{\iiint x \, d\mathcal{V}}{\mathcal{V}}$$

In other words, the buoyant force acts through the centroid of the displaced volume of fluid, called the *center of buoyancy*.

If a body is floating at the surface of a liquid, with part of the body submerged below the liquid surface, as shown in Figure 2.32, we find that the vertical force on the differential prism is equal to

$$(\rho_L g d_1 + \rho_a g d_2) dA$$

Integrating over the entire body, we find that the total buoyant force is equal to the sum of the weights of air and liquid displaced by the body.

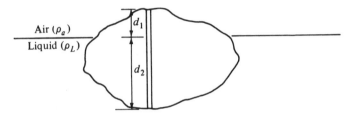

Figure 2.32.   Forces on a floating body.

Since the densities of air and liquid are different, however, the resultant buoyant force does not act through the centroid of the entire volume; the centroids of the volume of displaced air and displaced liquid must be calculated. With the density of a liquid so much greater than that of air, it is usual to neglect the contribution of the displaced air in comparison to that of the displaced liquid. In this case, the buoyant force acts through the centroid of the displaced volume of liquid.

EXAMPLE 2.9

What percentage of the total volume of an iceberg floats above the water surface? Assume the density of ice to be 57.2 lbm/ft^3, the density of water 62.4 lbm/ft^3.

Solution   At equilibrium, the weight of the iceberg is balanced by the buoyant force of the displaced water. In other words,

$$\rho_{ice} \mathcal{V}_{iceberg} = \rho_{water} \mathcal{V}_{submerged}$$

Solving,

$$\frac{\mathcal{V}_{submerged}}{\mathcal{V}_{iceberg}} = \frac{57.2}{62.4} = 0.917$$

Therefore only <u>8.3%</u> of the iceberg is above the surface.   ■

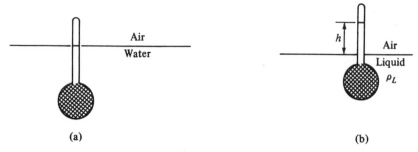

Figure 2.33. Hydrometer.

EXAMPLE 2.10

The hydrometer is a device used to measure the specific gravity of a liquid. As shown in Figure 2.33, the hydrometer consists of a weighted bulb and a stem of constant cross-sectional area. When floating in pure water, the hydrometer reaches the equilibrium position shown in Figure 2.33a, with V_0 the total volume submerged. When floating in a liquid of different density than water, the hydrometer will reach a new equilibrium position (Figure 2.33b). Obtain a correlation between h and specific gravity s. The cross-sectional area of the stem is A_s.

Solution In water, the hydrometer reaches an equilibrium position at which its weight W is balanced by the buoyant force $V_0 \rho_{\text{water}} g$. In other words,

$$W = \rho_{\text{water}} g V_0$$

In the second liquid, the weight must again be balanced by the buoyant force:

$$W = \rho_L g (V_0 - A_s h)$$

Equating the two expressions for W, we obtain

$$\rho_{\text{water}} g V_0 = \rho_L g (V_0 - A_s h)$$

But, by definition,

$$s = \frac{\rho_L}{\rho_{\text{water}}}$$

Therefore,

$$s = \frac{V_0}{V_0 - A_s h}$$

$$= \frac{1}{1 - A_s h / V_0} \quad \blacksquare$$

2.6 THE STABILITY OF SUBMERGED AND FLOATING BODIES

A body is in stable equilibrium if, when it is subjected to a small disturbance, forces are set up that tend to restore the body to equilibrium. Consider the submerged body shown in Figure 2.34. The buoyant force acts through the centroid of the submerged volume, called the center of buoyancy B. We shall assume that, for this body, the weight is distributed in such a way that

48

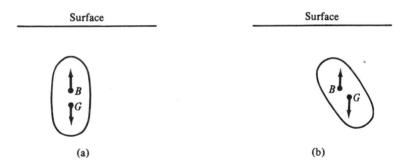

Figure 2.34. Stable submerged body.

the center of gravity G is below the center of buoyancy. In Figure 2.34b, the body has been given a disturbance, causing it to rotate counterclockwise. It can be seen that a clockwise moment is set up that tends to restore the body to equilibrium.

If the center of buoyancy of the submerged body, however, is below the center of gravity as shown in Figure 2.35, the situation becomes unstable. In this case, when the body is rotated counterclockwise, a counterclockwise moment is set up, tending to increase the angle of rotation. It follows that a submerged body is rotationally stable only if its center of gravity is below its center of buoyancy.

Figure 2.35. Unstable submerged body.

The preceding criterion for stability of a submerged body is not necessary for the stability of a floating body. Consider the ship depicted in Figure 2.36, with static equilibrium shown in Figure 2.36a. A counterclockwise rotation will cause the center of buoyancy to shift to the left, as shown in Figure 2.36b, setting up a moment that will restore equilibrium. Therefore, a floating

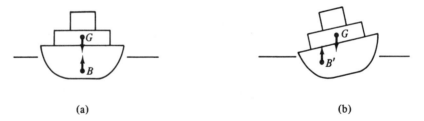

Figure 2.36. Stable floating body.

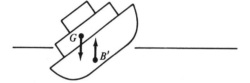

Figure 2.37. Unstable floating body.

body can be in equilibrium even though its center of gravity is above its center of buoyancy. If the body is tipped too far, however, as shown in Figure 2.37, the situation becomes unstable.

It is important to establish a criterion for the stability of a floating body. Returning to the situation just described, extend a line vertically upward through the displaced center of buoyancy B', as shown in Figure 2.38. We shall call the intersection of this line with the centerline of the ship the *metacenter M*. It follows that when M is above G, the situation is stable.

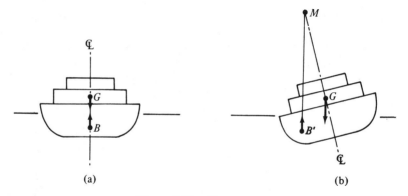

(a)   (b)

Figure 2.38. Metacenter.

The distance MG is called the *metacentric height*; this distance is positive for stability. It can be seen that for the unstable situation depicted in Figure 2.39, the metacenter is below the center of gravity; in other words, the metacentric height is negative.

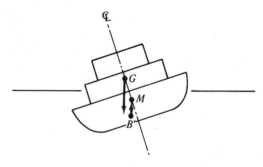

Figure 2.39. Metacenter for unstable floating body.

PROBLEMS

Unless otherwise specified, assume the density of water in the following problems to be 1000 kg/m^3 or 62.4 lbm/ft^3 (1.94 slugs/ft^3).

2.1. In the hydraulic press shown in Figure P2.1, a force of 100 N is exerted on the small piston. Determine the upward force on the large piston. The area of the small piston is 50 cm^2, the area of the large piston is 500 cm^2. The liquid in the press is oil, with a density of 900 kg/m^3.

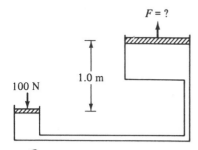

Figure P2.1

2.2. The big piston of the hydraulic press shown in Figure P2.2 has a diameter of 2 ft, while the diameter of the small piston is 1 in. The big piston has a mass of 5000 lbm and carries an external load of 10,000 lbf, while the small piston has a mass of 7.5 lbm. Determine the external force on the small piston necessary for equilibrium. Calculate the travel required by the small piston to raise the big piston by 0.1 in. Will the system be in equilibrium? If not, what additional action on the small piston will be required? The specific gravity of the hydraulic liquid is 0.9.

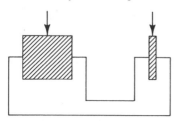

Figure P2.2

2.3. If $p_A - p_B = 1.0$ kPa, find the specific gravity of fluid X shown in Figure P2.3.

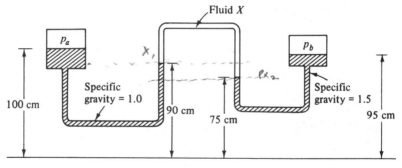

Figure P2.3

2.4. Calculate the absolute pressure above the water surface in the tank shown in Figure P2.4.

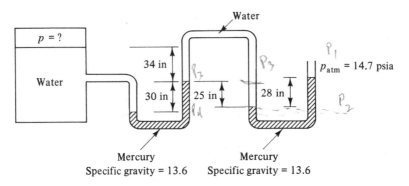

Figure P2.4

2.5. A triangular gate (1 m on each side) is located in a horizontal plane and is opened by a force applied perpendicular to the gate at its apex. The tank above this gate holds water having a depth of 2 m. The lower side of the gate is open to the atmosphere. Find the magnitude of the force required to open the gate.

2.6. A pneumatic lift in an automotive service station has a piston diameter of 9 in. Determine its maximum lifting capacity if the air supply pressure is 100 psig.

2.7. Calculate the force required to hold the lid on the circular opening of the closed tank shown in Figure P2.7. The diameter of the opening is 0.6 m, and the lid has a mass of 50 kg.

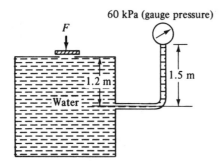

Figure P2.7

2.8. A mercury manometer reads 29.5 in on two different days when the ambient temperature registers at 80° and 90°F, respectively. Calculate the actual atmospheric pressure in psi.

2.9. Calculate the force F required to hold the rectangular gate closed (see Figure P2.9).

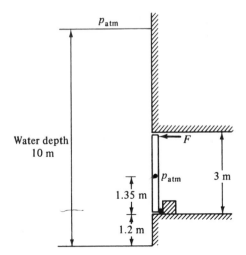

Figure P2.9

2.10. In problem 2.9, over what range of depths will the gate stay closed with no applied force?

2.11. A masonry dam (Figure P2.11) holds 10 ft of water. Masonry has an average density of 150 lbm/ft^3. How thick must the dam be if the coefficient of friction is 0.4? Will the dam wall be safe from overturning?

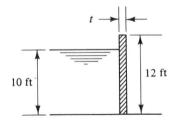

e P2.11

2.12. A tank, divided by a vertical partition, contains water on one side, nitric acid (specific gravity = 1.52) on the other, as shown in Figure P2.12. An opening in the bottom of the partition is closed by a rectangular door *AB*, hinged at *A*. If door and tank are each 0.6 m wide, find the force necessary to keep the door closed.

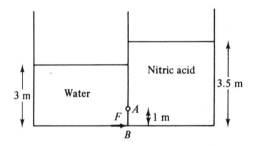

Figure P2.12

2.13. The mercury manometer shown in Figure P2.13 is connected to a pipe which carries hot water at 180°F. The elevation of point B is 10 ft above point A. Calculate the static pressure in the pipe in psi if the mercury reading is 50 in. Also calculate the static pressure in ft of water at 180°F.

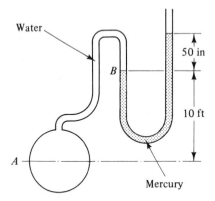

Figure P2.13

2.14. Calculate the depth of water d at which the conical plug in Figure P2.14 will start to leak. The mass of the plug is 1 kg.

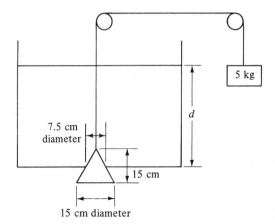

Figure P2.14

2.15. A manometer uses a liquid having a specific gravity of 0.8. The manometer reservoir diameter is $\frac{1}{4}$ in and the vertical measuring tube has a diameter of $\frac{5}{32}$ in. What must the distance between two marks on the measuring tube be to indicate a pressure difference of 1 in of water?

2.16. Calculate the force of the liquids on the end of the cylindrical tank shown in Figure P2.16.

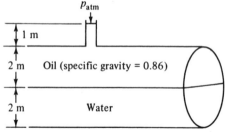

Figure P2.16

2.17. Determine the magnitude of the hydrostatic force acting on the surface shown in Figure P2.17. Find the location of the center of pressure.

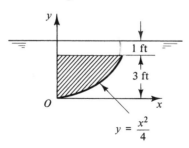

$$y = \frac{x^2}{4}$$

Figure P2.17

2.18. A tank is completely filled with a liquid of specific gravity 1.2 (Figure P2.18). What is the pressure at A if the difference in mercury level is 10 cm?

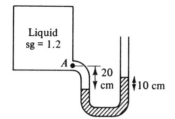

Figure P2.18

2.19. Two containers open to the atmosphere are filled with the same liquid (ρ = 700 kg/m³) to the same level (Figure P2.19). The two containers are connected by a pipe in which a frictionless piston of cross section A slides. How much work is done on the piston moving it a distance L = 0.1 m? Assume A = 0.05 m².

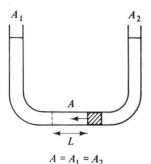

$A = A_1 = A_2$ **Figure P2.19**

2.20. The atmospheric pressure at the foot of Mount Rainier in the Cascade Range is measured to be 12 psia at a temperature of 77°F. Calculate the pressure at the mountain's peak 11,000 ft above.

2.21. Determine the boiling temperature of water atop Mount Rainier (see Problem 2.20).

2.22. In Figure P2.22 a hemispherical dome is to be located at the bottom of a body of water to allow for oceanographic observations. Calculate the force of the seawater on the dome.

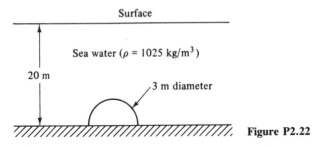

Figure P2.22

2.23. A tennis court "bubble" structure is in the shape of a spherical cap (base diameter = 100 ft, maximum height = 40 ft). Air blowers are used to maintain the pressure inside the bubble at $\frac{1}{2}$ in of water above atmospheric pressure. Determine the maximum utilizable material weight per square foot.

2.24. A rectangular barge 6 m × 3 m × 1.5 m deep sank 0.3 m in saltwater ($\rho = 1025 \text{ kg/m}^3$) when taking on cargo. Calculate the weight of the cargo.

2.25. The floodgate of a dam can be raised to discharge some of the water stored by the dam. The gate has a mass of 10,000 lbm. Calculate the force of the water on the closed floodgate and the initial lifting force required to raise the gate if the coefficient of static friction between the gate and its supports is 0.35.

2.26. Calculate the fluid force on the circular gate shown in Figure P2.26.

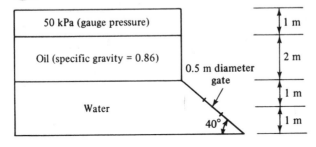

Figure P2.26

2.27. A rectangular open tank with vertical sides contains a liquid the density of which varies with depth of submergence y, according to the relation $\rho = 75 + 3y$. Calculate the force exerted on a tank wall 1 ft wide and 3 ft high.

2.28. The specific gravity of the cylinder shown in Figure P2.28 is 1.6. Find the specific gravity of the unknown fluid.

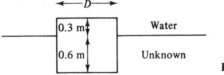

Figure P2.28

2.29. A cubical block, 3 m on a side, is in equilibrium in Figure P2.29. Find the density of the cube.

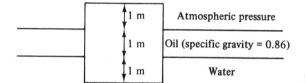

Figure P2.29

2.30. A water tank has a plane door 3 ft wide and 5 ft high located in a vertical wall. The door is hinged along its upper edge, located 3 ft below the surface of the water. The air pressure above the water is 5 psig. Atmospheric pressure acts on the outer surface of the door. Find the total resultant force acting on the door.

2.31. A gate 1 m wide and 2 m high is hinged at its upper edge (Figure P2.31). Find the force F required to open the gate.

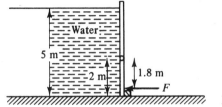

Figure P2.31

2.32. The circular access door in the vertical side of a water tank has a diameter of 1 ft. If the center of the door is located 25 ft below the water surface, calculate the total force on the door.

2.33. Find the water force exerted on a rectangular cover plate located on a vertical wall separating two bodies of water having different levels (Figure P2.33). The water levels are $H_1 = 3$ m and $H_2 = 2$ m, while the height of the plate is 0.5 m and the width of the plate is 0.6 m. The wall opening is rectangular, 0.4 m × 0.5 m.

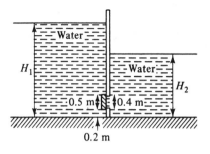

Figure P2.33

2.34. The sidewall of an open tank has the shape indicated in Figure P2.34. The tank is completely filled with water. Determine the force exerted on the sidewall and its point of application.

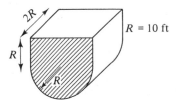

Figure P2.34

2.35. A valve is located in a square tube (Figure P2.35). The valve is closed and the water level on one side of the valve is half full, while on the other side the tube is completely filled. Find the turning moment required to hold the valve in this position if $a = 0.5$ m.

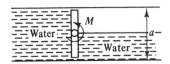

Figure P2.35

2.36. Repeat Problem 2.35, except make the tube circular, with a diameter of 0.5 m.

2.37. A cylindrical water tower 30 ft in diameter is filled to a depth of 60 ft. Calculate the water force on the vertical sides and tank bottom.

2.38. Calculate the horizontal and vertical force components and the line of action of these force components on the conical plug shown in Figure P2.38.

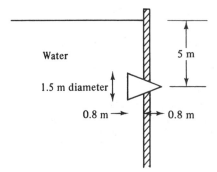

Figure P2.38

2.39. In Figure P2.39 determine the force F required to close the gate if the gate weighs 5000 lbf. The width of the gate is 10 ft.

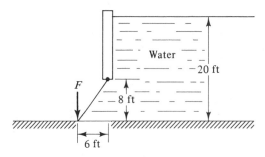

Figure P2.39

2.40. A concrete block rests on the bottom of a lake. The block is a cube, 1 m on a side, and the lake is 10 m deep. Calculate the force required to just lift the block off the lake bottom. Also, calculate the force required to hold the block at equilibrium at a depth of 5 m (see Figure P2.40). The density of concrete is 2400 kg/m^3.

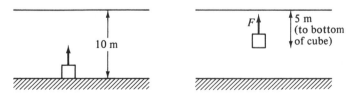

Figure P2.40

2.41. In Figure P2.41 a rectangular barge, 10 ft by 10 ft by 5 ft deep, reaches an equilibrium position, when empty, with 2 ft submerged. Determine the equilibrium position when towing a cubical block of concrete, 2 ft on a side. Assume the barge is floating in sea water (ρ = 64 lbm/ft^3). The density of concrete is 150 lbm/ft^3.

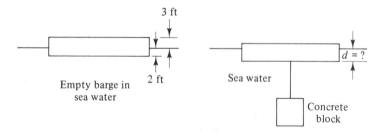

Figure P2.41

2.42. An observation balloon is filled with helium to atmospheric pressure. The balloon is spherical, with a radius of 1.0 m. If the payload carried by the balloon has a mass of 2.0 kg, determine the equilibrium altitude of the balloon. Assume that air and helium behave as perfect gases with R = 0.2870 kJ/kg · K for air, and R = 2.077 kJ/kg · K for helium.

2.43. A fluid has the property that its density increases with depth according to $\rho = \rho_0 + Kd$, with ρ_0 the density at the free surface, d the depth below the surface, and K a constant. The fluid fills a rectangular tank 6 ft deep, with cross section 10 ft by 10 ft as shown in Figure P2.43. Determine the total force on the bottom of the tank and the fluid forces on the sidewalls.

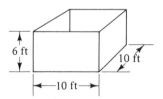

Figure P2.43

2.44. A wooden pole, 20 cm in diameter and 4 m long, floats vertically in water, with a 60-kg concrete block, 30 cm on a side, attached to the bottom of the pole as shown in Figure P2.44. Determine the equilibrium position of the pole. Determine the center of gravity and the center of buoyancy of the wood and concrete float system. Is the system stable? Take the specific gravity of wood to be 0.6.

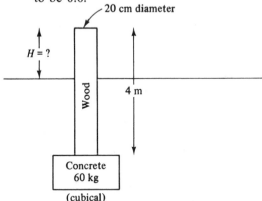

Figure P2.44

2.45. A vertical plane surface (3 ft × 3 ft) is submerged below the water surface (Figure P2.45). How far must it be submerged so the center of pressure is 1 in below the centroid of the area?

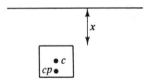

Figure P2.45

2.46. A homogeneous wooden cube (specific gravity = 0.65), 50 cm on a side, is placed into a water tank (150 cm square, water depth 300 cm). Calculate the change in the forces exerted on the tank walls due to the addition of the wooden cube.

2.47. A circular cylinder floats vertically in water, as shown in Figure P2.47. What will be its horizontal floating position?

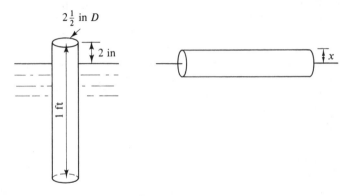

Figure P2.47

2.48. Determine the center of gravity and the center of buoyancy of the vertical cylinder of Figure P2.47. Is the system stable?

2.49. An open-ended tin can 5 cm in diameter is pushed down below the surface of a pool of water (Figure P2.49). Determine the force required to hold the can 1 meter below the surface. The can is 10 cm in length. Assume the air in the can is compressed isothermally as the can is pushed down, air temperature 20°C.

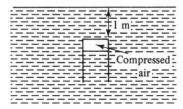

Figure P2.49

2.50. A chunk of ore weighs 10 lbm when weighed on a scale in air and weighs 7.5 lbm when submerged in water. Determine the density of the ore.

2.51. A hydrometer is to be built for which the calibration marks for specific gravities of 1.0 and 2.0 are 100 mm apart on a 5-mm diameter stem. How far from the 1.0 mark will the specific gravity marks of 1.25, 1.50, and 1.75 fall?

2.52. A hydrometer measures the specific gravity of liquids. The value of the specific gravity is determined by the level at which the stem of the hydrometer floats in the liquid. A mark for a specific gravity of 1 is obtained by floating the unit in distilled water. The immersed volume of the unit in distilled water is 1 in³. Determine the distance from the reference mark when the unit is submerged in a liquid of specific gravity 1.3. The diameter of the hydrometer's stem is $\frac{1}{4}$ in.

2.53. A wooden cylinder to which a lead weight is attached floats in water as shown in Figure P.2.53. Determine the equilibrium position of the cylinder, center of gravity, and center of buoyancy. Is the cylinder float stable?

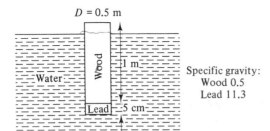

Specific gravity:
  Wood 0.5
  Lead 11.3

Figure P2.53

2.54. Archimedes used his classic immersion experiment to determine the gold purity of a crown. Take the volume of the crown at 10 in³. It weighs 3 lbm when suspended in water. Find the density of the crown material. Was it made of pure gold? (*Note*: The density of pure gold is 1206 lbm/ft³.)

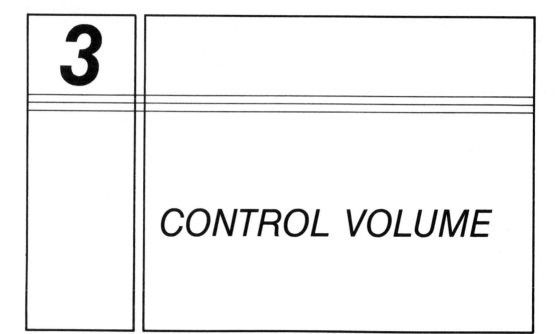

CONTROL VOLUME

3.1 FLUID MECHANICS VERSUS RIGID BODY DYNAMICS

In rigid body mechanics we are accustomed to describing the motion of a body in terms of its position versus time. As seen in Figure 3.1, as the body moves along its trajectory, we can write Newton's laws of motion to determine s versus t. For example, in the well-known case of free fall of a rigid body starting from rest, Newton's law yields

$$m\frac{d^2s}{dt^2} = mg$$

$$s = s_0 \qquad \text{at} \qquad t = 0$$

$$V = \frac{ds}{dt} = 0 \qquad \text{at} \qquad t = 0$$

where m = mass of body
g = acceleration due to gravity
V = velocity of body

Integrating twice, we obtain the relationship

$$s - s_0 = \tfrac{1}{2}gt^2$$

This approach, in which we write the equations of motion for a moving particle, is called the *Lagrangian approach*.

In fluid mechanics, however, it is desirable to adopt a different approach, that is, to observe the motion of the fluid particles as they pass a given

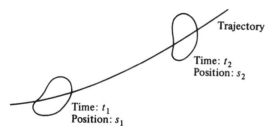

Figure 3.1

location in the flow field. Unlike a rigid body, as a fluid body moves from one position to the next, it usually deforms continuously. Therefore, in order to describe completely the motion of a fluid body, it is necessary to account for its deformation as well as its translation and rotation. Consequently, the rigid body dynamics approach is inherently cumbersome to pursue from a mathematical point of view. Furthermore, it is often necessary to determine the velocity and pressure distribution about a body with given size and shape. Information about the flow is required at specified locations in the flow field. Thus the analysis of the motion of the fluid particles as they pass given locations, called the *Eulerian* or *control volume approach*, is the more useful in fluid mechanics.

It is the purpose of this chapter to develop the relationship between the two approaches so that we can adapt the laws governing the motion of bodies or particles to the motion of fluid particles. These laws are related to the conservation of mass, momentum, and energy.

3.2 RELATIONSHIP BETWEEN CONTROL VOLUME APPROACH AND RIGID BODY DYNAMICS

The law of conservation of mass states that matter cannot be created or destroyed; or, expressing it as a time rate of change of total mass M of a system of particles,

$$\frac{dM}{dt} = 0 \tag{3.1}$$

Similarly, the law of conservation of momentum (Newton's second law) states that the time rate of change of momentum of a system of particles is equal to the sum of the externally applied forces; that is,

$$\Sigma \mathbf{F} = \frac{d(M\mathbf{V})}{dt} \tag{3.2}$$

where $M\mathbf{V}$ is the total linear momentum of the system. Finally, the law of conservation of energy states that the time rate of change of total energy possessed by a system of particles is equal to the rate of addition of heat energy to the system less the rate of work done by the system:

$$\frac{dE}{dt} = \frac{d\tilde{Q}}{dt} - \frac{dW}{dt} \qquad (3.3)$$

It is emphasized that all these laws involve time rates of change of mass, momentum, and energy.

We wish to apply these laws, which express the time rate of change of fluid quantities for a system of particles, to a fixed volume in a fluid, called a control volume. A *control volume* is defined as a volume fixed with respect to a coordinate system in a fluid flow field.

Figure 3.2 shows a closed surface in a flow field bounding a volume V_{t1} containing the system of fluid particles at time t_1. At some later time t_2 ($= t_1 + \Delta t$), this system of particles will have moved to a new position and will occupy a new and different volume V_{t2}. Let us signify the volumes A, B, C as shown in Figure 3.3. Here volume B is common to volumes V_{t1} and V_{t2}. Let us define the control volume to be volume V_{t1}. It should be noted that although fluid particles move into and out of the control volume, the control volume itself remains fixed in space. During the time interval Δt, some of the fluid particles that were contained in the control volume at time t_1 have left the control volume and occupy volume C, while others have entered the control volume to fill volume A.

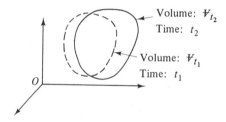

Figure 3.2

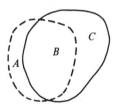

Figure 3.3

We wish to relate the rate of change of mass, momentum, and energy of the system particles as they move to the changes that occur in the fluid quantities inside the control volume. Let X equal the total flow quantity (mass, momentum, or energy) contained within a fluid volume, with x equal to the flow quantity per unit mass. Thus

$$X = \int x\rho \, dV$$

The total amount of X contained within volume V_{t1} (at time t_1) is composed of the quantity X_{At1} plus X_{Bt1}, where, for example, X_{At1} is the total X contained within volume A at time t_1. Similarly, the total amount of X contained within volume V_{t2} (at time t_2) is made up of the quantity X_{Bt2} plus X_{Ct2}. During the interval Δt, as the fluid mass moves from V_{t1} to V_{t2}, the change in X of the system of particles is given by

$$\Delta X = X_{Bt2} - X_{Bt1} + X_{Ct2} - X_{At1}$$

Rewriting the equation by adding and subtracting the quantity X_{At2}, we have

$$\Delta X = (X_{Bt2} + X_{At2}) - (X_{Bt1} + X_{At1}) + X_{Ct2} - X_{At2}$$

The rate of change of X for the system is then given by

$$\left.\frac{\Delta X}{\Delta t}\right|_{\text{system}} = \frac{X_{(A+B)t2} - X_{(A+B)t1}}{\Delta t} + \frac{X_{Ct2} - X_{At2}}{\Delta t}$$

To obtain the instantaneous rate of change of the quantity X, let Δt approach zero:

$$\lim_{\Delta t \to 0} \frac{\Delta X}{\Delta t} = \left.\frac{dX}{dt}\right|_{\text{system}} = \lim_{\Delta t \to 0} \frac{X_{(A+B)t2} - X_{(A+B)t1}}{\Delta t} + \lim_{\Delta t \to 0} \frac{X_{Ct2} - X_{At2}}{\Delta t} \quad (3.4)$$

The first term on the right-hand side is the time rate of change of X within the control volume $(A + B)$ (i.e., the rate at which X is stored within the control volume):

$$\lim_{\Delta t \to 0} \frac{X_{(A+B)t2} - X_{(A+B)t1}}{\Delta t} = \left.\frac{\partial X}{\partial t}\right|_{\substack{\text{control} \\ \text{volume}}} = \frac{\partial}{\partial t} \int x\rho \, dV \quad (3.5)$$

Next, we wish to obtain the limit expression for the second term on the right-hand side. The term X_{Ct2} is the quantity of X that leaves the control volume across surface S_2 in the time interval Δt, and X_{At2} is the amount that enters the control volume across surface S_1 in the same time interval (Figure 3.4). To obtain an expression for these quantities, consider a differential element of area dA on the control surface bounding the control volume as shown in Figure 3.5. The velocity of fluid at area element dA is V, with V_n

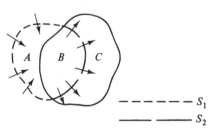

S_1 and S_2 together comprise the control surface
bounding the control volume

Figure 3.4

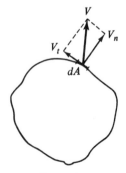

Figure 3.5

the velocity component normal to dA and V_t the component tangential to dA. We are interested in finding the differential amount δX that leaves across the area element dA in time Δt. Since the velocity component in the tangential direction carries no fluid across the area element, only the normal velocity

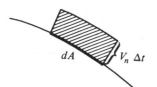

Figure 3.6

component is involved in evaluating the amount δX that crosses the element of surface area. The shaded volume shown in Figure 3.6 represents the volume of fluid that has crossed the fixed area element dA in time Δt. This volume is $dA(V_n \, \Delta t)$, which carries with it a differential mass $p \, dA \, V_n \, \Delta t$ and thereby a differential amount δX equal to

$$\delta X = x\rho \, dA \, V_n \, \Delta t$$

Therefore, the rate of efflux of X across the area element dA equals

$$\frac{\delta X}{\Delta t} = x\rho V_n \, dA$$

The two quantities which comprise the second term of the right-hand side of Equation (3.5) therefore become

$$X_{Ct2} = \Delta t \int_{S_2} x\rho V_n \, dA$$

and

$$X_{At2} = -\Delta t \int_{S_1} x\rho V_n \, dA$$

The second limit expression thus becomes

$$\lim_{\Delta t \to 0} \frac{X_{Ct2} - X_{At2}}{\Delta t} = \lim_{\Delta t \to 0} \frac{\Delta t \int_{S_1} x\rho V_n \, dA + \Delta t \int_{S_1} x\rho V_n \, dA}{\Delta t} \tag{3.6}$$

$$= \int_{S_1+S_2} x\rho V_n \, dA = \int_{\substack{\text{control} \\ \text{surface}}} x\rho V_n \, dA$$

since $S_1 + S_2$ forms the entire surface enclosing the control volume.

Substituting Equations (3.5) and (3.6) into Equation (3.4), we obtain

$$\left. \frac{dX}{dt} \right|_{\substack{\text{system of} \\ \text{fluid particles}}} = \left. \frac{\partial X}{\partial t} \right|_{\substack{\text{control} \\ \text{volume}}} + \int_{\substack{\text{control} \\ \text{surface}}} x\rho V_n \, dA \tag{3.7}$$

We now have an equation relating the rate of change of X for the system of fluid particles as they move through the flow field to the changes of X as they take place within the fixed control volume.

Equation (3.4) can also be written in vector form. The student will recall that the scalar (dot) product of two vectors $\mathbf{A}$ and $\mathbf{B}$ is given by

$$\mathbf{A} \cdot \mathbf{B} = |A||B| \cos \theta$$

Figure 3.7   **Figure 3.8**

where θ is the angle between the two vectors as shown in Figure 3.7. We can define a vector **dA** with direction normal to the differential area dA (positive in the outward direction) and magnitude equal to the area dA as shown in Figure 3.8. Hence, for the scalar product of **V** and **dA**, we obtain

$$\mathbf{V} \cdot \mathbf{dA} = |V||dA| \cos \theta$$
$$= V_n \, dA$$

Therefore, Equation (3.7) becomes in vector form

$$\left. \frac{dX}{dt} \right|_{\text{system}} = \left. \frac{\partial X}{\partial t} \right|_{\substack{\text{control} \\ \text{volume}}} + \int_{\substack{\text{control} \\ \text{surface}}} x\rho \mathbf{V} \cdot \mathbf{dA} \qquad (3.8)$$

The choice of location and size of the control volume is dependent on the character of the problem to be analyzed. In some applications, the control volume will be of differential size. In others, a finite control volume will be more suitable. The student should note the choice of control volume in the examples of the next and subsequent chapters. In Chapter 4 we will utilize relationships (3.7) or (3.8) to derive equations of conservation of mass, momentum, and energy for fluid flow.

FUNDAMENTAL EQUATIONS OF FLUID MOTION

4.1 CONTINUITY EQUATION

Let us first apply the control volume–system relationship to the equation of conservation of mass, also called the continuity equation. In this case, X is the total mass M, with the law of conservation of mass stating that $dM/dt = 0$ for a system of fluid particles. Since the total mass M is, by definition, equal to $\int \rho \, d\Psi$ and $X = \int x\rho \, d\Psi$, then x is equal to 1. Applying Equation (3.7), we obtain

$$\left.\frac{dM}{dt}\right|_{\text{system}} = \left.\frac{\partial M}{\partial t}\right|_{\substack{\text{control} \\ \text{volume}}} + \int_{\substack{\text{control} \\ \text{surface}}} \rho V_n \, dA = 0 \qquad (4.1)$$

Or, applying Equation (3.8), we obtain the vector form

$$\left.\frac{dM}{dt}\right|_{\text{system}} = \left.\frac{\partial M}{\partial t}\right|_{\substack{\text{control} \\ \text{volume}}} + \int_{\substack{\text{control} \\ \text{surface}}} \rho \mathbf{V} \cdot \mathbf{dA} = 0 \qquad (4.2)$$

Again, the left-hand side represents the rate of change of total mass of the system of particles (equal to zero by conservation of mass). The first term of the right-hand side represents the rate at which mass is stored in the control volume, and the second term denotes the net rate of efflux of mass from the control volume.

The following examples will illustrate the application of this equation.

68

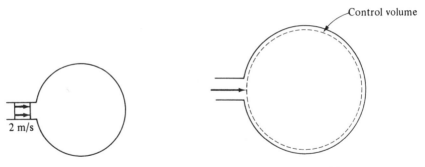

Figure 4.1   Figure 4.2

EXAMPLE 4.1

A gas flows into a rigid container initially evacuated. Assume that the inflow velocity is uniform at 2m/s, as shown in Figure 4.1. The tube inlet diameter is 10 cm with the volume of the tank equal to 2000 liters. The pressure and temperature in the inlet line are maintained constant at 400 kPa and 330 K, respectively. The gas can be assumed to obey the perfect gas law $p = \rho RT$, with R for the gas equal to 0.30 kJ/kg·K. Assume the tank to be noninsulated so that the temperature of the gas in the tank remains constant at a room temperature of 300 K. Determine the time required for the pressure in the tank to reach 300 kPa.

Solution   In this example the choice of the control volume is straightforward, as shown in Figure 4.2. Flow into the chosen control volume occurs only at the inlet pipe. From Equation (4.1) we obtain:

$$0 = \left.\frac{\partial M}{\partial t}\right|_{\substack{\text{control}\\\text{volume}}} + \int_{\substack{\text{inlet}\\\text{pipe}}} \rho V_n \, dA$$

Now V_n equals -2 m/s (since efflux is taken as positive) and

$$\rho = \frac{p}{RT} = \frac{400 \text{ kN/m}^2}{(0.30 \text{ kN} \cdot \text{m/kg} \cdot \text{K})(330 \text{ K})} = 4.04 \text{ kg/m}^3$$

Therefore, the second term of the right-hand side of Equation (4.1) becomes $-\int (4.04)2 \, dA$. Since density and velocity are constant across the inlet area, they can be taken outside the integral, yielding:

$$(-4.04 \text{ kg/m}^3)(2 \text{ m/s})\frac{\pi}{4}(0.10)^2 \text{ m}^2 = -0.0635 \text{ kg/s}$$

Therefore, $\partial M/\partial t = +0.0635$ kg/s. The total mass of gas inside the tank at any time is $\rho \forall$, with $\forall$ equal to 2 m³. It follows that

$$\frac{\partial M}{\partial t} = \forall \frac{\partial \rho}{\partial t} = \frac{\forall}{RT} \frac{\partial p}{\partial t}$$

We now have

$$\frac{\not V}{RT}\frac{\partial p}{\partial t} = 0.0635 \text{ kg/s}$$

Since pressure is a function only of time, we can write the total derivative:

$$\frac{\not V}{RT}\frac{dp}{dt} = 0.0635 \text{ kg/s}$$

Integrating, we obtain

$$\frac{\not V}{RT}\int_0^{300\text{kPa}} dp = 0.0635 \int_0^t dt$$

Substituting,

$$\frac{2 \text{ m}^3(300 \text{ kN/m}^2)}{(0.30 \text{ kN} \cdot \text{m/kg} \cdot \text{K})300 \text{ K}} = 0.0635t \text{ kg}$$

or

$$t = 105.0 \text{ s} \qquad \blacksquare$$

EXAMPLE 4.2

A circular swimming pool is 10 ft in diameter. It is to be filled to a uniform depth of 5 ft by means of a $\frac{1}{2}$-in-diameter hose, as shown in Figure 4.3. The velocity of the water in the hose is 5 ft/s. Determine the time in hours required to fill the pool.

Solution Select the control volume to include the entire volume to be filled, as indicated in Figure 4.3. Flow into the chosen control volume occurs only at the inlet hose. From Equation (4.1) we obtain

$$0 = \left.\frac{\partial M}{\partial t}\right|_{\substack{\text{control}\\\text{volume}}} + \int_{\substack{\text{inlet}\\\text{hose}}} \rho V_n \, dA$$

Since the density of water is relatively insensitive to pressure, we obtain

$$\frac{\partial M}{\partial t} = \rho V_n A$$

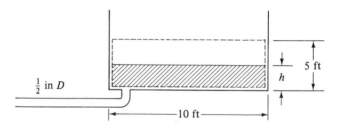

Figure 4.3

The mass of water contained in the pool at any given time can be written as

$$M = \rho \frac{\pi}{4}(10^2)h$$

Therefore,

$$\frac{\partial M}{\partial t} = \rho \frac{\pi}{4}(10^2)\frac{dh}{dt} = \rho V_n A$$

or

$$\frac{dh}{dt} = \frac{(5 \text{ ft/s})(\pi/4)(\frac{1}{24})^2 \text{ ft}^2}{(\pi/4)(10^2) \text{ ft}^2} = 8.68 \times 10^{-5} \text{ ft/s}$$

Integrating, we obtain

$$\int_0^{5 \text{ ft}} dh = 0.0000868 \int_0^t dt$$

$$t = \frac{5 \text{ ft}}{8.68 \times 10^{-5} \text{ ft/s}} = 57{,}600 \text{ s}$$

$$= \underline{16 \text{ h}} \qquad \blacksquare$$

For *steady flows,* the properties at a point do not vary with time. For this case, $\partial M/\partial t$ equals zero and the continuity equation simplifies to

$$\int_{\substack{\text{control} \\ \text{surface}}} \rho V_n \, dA = 0 \qquad (4.3)$$

A further simplification can be obtained if the density of the fluid does not vary within the flow field:

$$\int_{\substack{\text{control} \\ \text{surface}}} V_n \, dA = 0 \qquad (4.4)$$

Such flows, in which the density does not change as a consequence of the flow processes, are called *incompressible flows.* For example, imposing a pressure of 200 atmospheres on a volume of liquid water decreases the volume by only 1%, that is, raises the density by only 1%. Thus the flow of water can usually be treated as incompressible.

EXAMPLE 4.3

Water flows steadily through a pump (Figure 4.4). The velocity distribution in the circular inlet tube is parabolic, the velocity in m/s given by

$$V = 3\left(1 - \frac{r^2}{R^2}\right)$$

with R the radius of the inlet tube and r the radial distance from the centerline

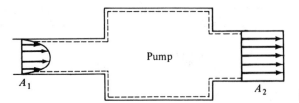

Figure 4.4

of the tube. The diameter of the inlet tube is 25 cm, while the outlet tube is 30 cm in diameter. The velocity in the outlet tube can be assumed to be uniform. Determine the magnitude of the outlet velocity.

Solution The control volume is chosen as indicated in Figure 4.4. Since efflux is taken as positive, V_n at section 1 is $-3(1 - r^2/R^2)$. At section 2, V_n is constant, so we obtain, from Equation (4.4),

$$-3 \int_{A_1} \left(1 - \frac{r^2}{R^2}\right) dA + V_2 \int_{A_2} dA = 0$$

Rewriting,

$$V_2 = \frac{3 \int_0^{25 \text{ cm}} (1 - r^2/R^2) \, 2\pi r \, dr}{\int_{A_2} dA} = \frac{3\left[\pi r^2 - \dfrac{\pi r^4}{2R^2}\right]_0^R}{\dfrac{\pi}{4}(0.30)^2}$$

$$= \frac{3\pi R^2/2}{(\pi/4)(0.30)^2}$$

$$= 4.167 \text{ m/s} \qquad \blacksquare$$

As has been explained previously, V_n represents the component of the velocity normal to the control surface. In many applications, the velocity vector itself is normal to the control surface. In many cases, it can be further approximated that the vector is constant across a section. For example, consider the flow in a pipe as shown in Figure 4.5. This type of flow, in which the flow at a given cross section is a function of only one space coordinate, is called *one-dimensional* flow. In the analysis of flow problems, the assumption of one-dimensionality affords great simplicity in obtaining

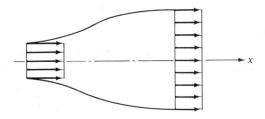

Figure 4.5

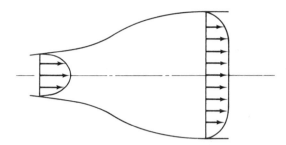

Figure 4.6

solutions, since the number of flow variables is reduced. In any actual case, due to viscosity, a layer of fluid particles at the wall sticks to the wall. Therefore, the velocity at a fixed wall is zero. The velocity profiles in an actual flow situation are as shown in Figure 4.6. One-dimensional flow, by its strictest definition, does not allow velocity components in the y or z directions. Therefore, in true one-dimensional flow, area changes like those shown in Figure 4.7 cannot occur. However, the one-dimensional approx-

Figure 4.7

imation becomes more exact the more gradual the area change. The real case illustrated in Figure 4.6 can be represented by one-dimensional flow, using an average velocity at each cross section. Note that one-dimensional analysis can yield information regarding variations in the x direction only; variations in a direction normal to the flow cannot be determined. The determination of average velocity for a given flow is provided next. By definition, the average velocity (V_{av}) is given by

$$V_{av} = \frac{Q}{A} = \frac{\int V_n \, dA}{A} \tag{4.5}$$

For example if the velocity distribution in a circular pipe is parabolic (see Figure 4.8a), that is, if

$$V = V_{max}\left(1 - \frac{r^2}{R^2}\right)$$

with V_{max} equal to the maximum velocity in the flow, then

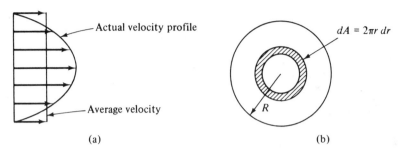

Figure 4.8

$$V_{av} = \frac{V_{max} \int_0^R \left(1 - \dfrac{r^2}{R^2}\right) 2\pi r \, dr}{\pi R^2}$$

$$= \tfrac{1}{2} V_{max}$$

Hence the average velocity in a circular pipe having parabolic velocity distribution is one-half the maximum velocity.

Next, let us examine the continuity equation for *steady one-dimensional* flow. For example, for one-dimensional flow and for the control volume shown in Figure 4.9, let the flow enter through section A_1 and exit through section A_2. We therefore obtain

$$\underset{\substack{\text{control}\\\text{surface}}}{\int} \rho V_n \, dA = \underset{\text{inflow}}{- \int_{A_1} \rho V_n \, dA} + \underset{\text{efflux}}{\int_{A_2} \rho V_n \, dA} = 0$$

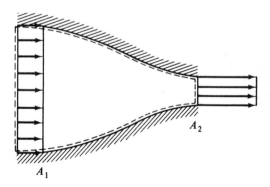

Figure 4.9

Since ρ and V_n do not vary across the sections but are different for each section, we can write

$$\rho_2 V_{n2} A_2 = \rho_1 V_{n1} A_1 \qquad (4.6)$$

Thus, for steady one-dimensional flow, the quantity $\rho V_n A$ is the mass flow rate $\dot{m}$ into or out of the control volume. With SI units, density is expressed in kg/m³, velocity in m/s, and area in m², so mass flow rate $\dot{m}$ is in kg/s. In the English system of units, density is expressed in slugs/ft³ or lbm/ft³, velocity in ft/s, area in ft²; hence the mass flow rate is in slugs/s or lbm/s.

EXAMPLE 4.4

In a rocket motor, 10 kg/s of liquid oxygen and 2 kg/s of liquid hydrocarbon fuel are fed into the combustion chamber, as shown in Figure 4.10. The gaseous products of combustion flow out of the exhaust nozzle at high velocity. The pressure and temperature of the gases at the nozzle exit plane are equal to 101 kPa and 800 K. The nozzle exit area is 400 cm². Assuming one-dimensional steady flow, with the exhaust gases behaving as a perfect gas with $R = 0.60$ kJ/kg · K, calculate the nozzle exit velocity.

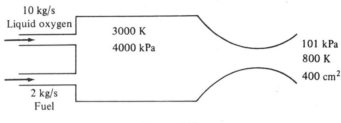

Figure 4.10

Solution Select a control volume as shown in Figure 4.11. The continuity equation for steady flow yields

$$\int_{\substack{\text{control} \\ \text{surface}}} \rho V_n \, dA = 0$$

For this case

$$\int_{\substack{\text{exit} \\ \text{plane}}} \rho V_n \, dA - 10 - 2 = 0$$

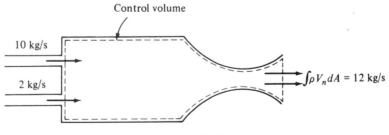

Figure 4.11

so that

$$\int \rho V_n \, dA = 12 \text{ kg/s}$$

For one-dimensional flow at the exit plane,

$$\rho_{\text{exit}} V_{\text{exit}} A_{\text{exit}} = 12 \text{ kg/s}$$

where

$$\rho_{\text{exit}} = \frac{p_{\text{exit}}}{RT_{\text{exit}}} = \frac{101 \text{ kN/m}^2}{(0.60 \text{ kNm/kg} \cdot \text{K})(800 \text{ K})} = 0.2104 \text{ kg/m}^3$$

Therefore,

$$V_{\text{exit}} = \frac{12 \text{ kg/s}}{(0.2104 \text{ kg/m}^3)(0.04 \text{ m}^2)} = \underline{1426 \text{ m/s}}$$

Note that for steady flow, it is not necessary to know the flow properties inside the control volume, only the flow quantities crossing the control surface.

■

4.2 MOMENTUM EQUATION

4.2(a) Linear Momentum Equation

Let us now apply the control volume–system relationship to the equation of conservation of *linear momentum*. In this case, X is the total linear momentum $M\mathbf{V}$, with the law of conservation of momentum in an inertial reference frame stating that

$$\Sigma \mathbf{F} = \frac{d(M\mathbf{V})}{dt}$$

for a system of fluid particles. It must be remembered that force and momentum are vector quantities; that is, they possess direction as well as magnitude. Therefore, there will be three scalar momentum equations; for example, in Cartesian coordinates, one each in the x, y, and z directions:

$$\Sigma F_x = \frac{d}{dt}(MV)_x$$

$$\Sigma F_y = \frac{d}{dt}(MV)_y$$

$$\Sigma F_z = \frac{d}{dt}(MV)_z$$

Applying Equation (3.7) to each equation, with $x = MV_n/M = V_x$, V_y, and V_z, respectively,

$$\Sigma F_x = \frac{d}{dt}(MV)_x\bigg|_{\text{system}} = \frac{\partial}{\partial t}(MV)_x\bigg|_{\substack{\text{control}\\\text{volume}}} + \int_{\substack{\text{control}\\\text{surface}}} V_x \rho V_n \, dA$$

$$\Sigma F_y = \frac{d}{dt}(MV)_y\bigg|_{\text{system}} = \frac{\partial}{\partial t}(MV)_y\bigg|_{\substack{\text{control}\\\text{volume}}} + \int_{\substack{\text{control}\\\text{surface}}} V_y \rho V_n \, dA \qquad (4.7)$$

$$\Sigma F_z = \frac{d}{dt}(MV)_z\bigg|_{\text{system}} = \frac{\partial}{\partial t}(MV)_z\bigg|_{\substack{\text{control}\\\text{volume}}} + \int_{\substack{\text{control}\\\text{surface}}} V_z \rho V_n \, dA$$

Applying Equation (3.8) to the vector momentum equation, with $x = MV/M = V$,

$$\Sigma \mathbf{F} = \frac{d}{dt}(M\mathbf{V})\bigg|_{\text{system}} = \frac{\partial}{\partial t}(M\mathbf{V})\bigg|_{\substack{\text{control}\\\text{volume}}} + \int_{\substack{\text{control}\\\text{surface}}} \mathbf{V}(\rho \mathbf{V} \cdot d\mathbf{A}) \qquad (4.8)$$

The left-hand side of Equation (4.7) or (4.8) represents the sum of all externally applied forces acting on the fluid within the control volume. These applied forces may involve pressure forces, viscous forces, gravity, magnetic forces, electric forces, surface tension, and so on. The first term of the right-hand side represents the rate at which linear momentum is stored in the control volume, and the second term denotes the net rate of efflux of linear momentum from the control volume. Note that the form of the equation of conservation of linear momentum (Newton's law) as given above applies only in an inertial reference frame. Therefore, the equations derived in this section apply only to a nonaccelerating control volume, that is, a nonmoving control volume or one moving with constant velocity. In a subsequent section we shall derive the momentum equation applicable in an accelerating coordinate system.

For steady flow, the first term on the right-hand side of Equation (4.7) equals zero. If the flow is also one-dimensional, then Equation (4.7) can be further reduced. For example, for one-dimensional flow in the x direction, let the flow enter and exit the control volume as shown in Figure 4.12. For this case, we obtain from Equation (4.7)

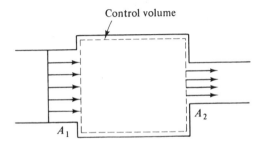

Figure 4.12

$$\Sigma F_x = \int_{\substack{\text{control} \\ \text{surface}}} V_x(\rho V_n \, dA) = -V_1\rho_1 V_1 A_1 + V_2\rho_2 V_2 A_2 = \dot{m}(V_2 - V_1) \qquad (4.9)$$

In SI units, $\dot{m}$ must be given in kg/s and velocity in m/s to be consistent with force in newtons. In the English system of units, $\dot{m}$ must be given in slugs/s and velocity in ft/s to be consistent with force in lbf.

The following examples will illustrate the application of the linear momentum equation in an inertial reference frame.

EXAMPLE 4.5

Let the rocket motor of Example 4.4 be mounted on a test stand, as shown in Figure 4.13. The liquid oxygen and hydrocarbon fuel are stored in the tanks as indicated. Ten kg/s of liquid oxygen and 2 kg/s of liquid hydrocarbon fuel are fed into the combustion chamber. The pressure and temperature of the combustion gases at the nozzle exit plane are equal to 101 kPa and 800 K. The nozzle exit area is 400 cm². Ambient pressure surrounding the test stand is 101 kPa. Assuming one-dimensional steady flow, calculate the rocket thrust, that is, the force necessary to hold the rocket in place.

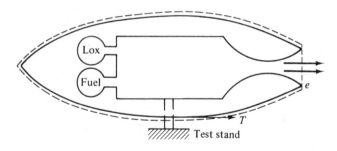

Figure 4.13

Solution   Select a control volume to include the entire rocket, including motor and tanks, so as to determine the entire force transmitted to the test stand. From Example 4.4, we have an exit mass flow of 12 kg/s with an exit velocity of 1426 m/s.

In this case, the only external force acting on the control volume is the force exerted on it by the test stand, as shown. Applying Equation (4.9), we obtain

$$T = \dot{m}(V_2 - V_1) = \dot{m}V_e$$

where V_e is the velocity at the exit plane of the nozzle and V_1 is zero. Therefore,

$$T = (12 \text{ kg/s})(1426 \text{ m/s})$$
$$= \underline{17.11 \text{ kN}}$$

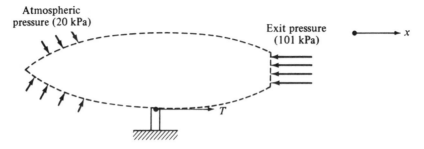

Figure 4.14

EXAMPLE 4.6

In another test of the rocket motor of Example 4.5, the ambient pressure surrounding the test stand is reduced to 20 kPa to simulate actual flight conditions. The conditions in the rocket motor, including exit plane pressure and velocity, are maintained as in Example 4.5. Determine the rocket thrust for this case.

Solution Here an additional external force is acting on the control volume, that is, a pressure force (see Figure 4.14). The unbalanced pressure force is due to the fact that the pressure at the nozzle exit plane is higher than the ambient pressure. It should be remembered that pressure always acts normal and inward to a surface, as indicated in the figure. The resulting pressure force equals $(p_a - p_e)A_e$, where p_e is 101 kPa and p_a is 20 kPa. Thus the pressure force acts in the negative x direction. Applying Equation (4.7) to this case, we obtain

$$T + (p_a - p_e)A_e = \dot{m}V_e$$

Therefore,

$$T = 17.11 \text{ kN} + [(101 - 20) \text{ kN/m}^2](0.04 \text{ m}^2)$$
$$= 17.11 \text{ kN} + 3.24 \text{ kN}$$
$$= \underline{20.35 \text{ kN}} \qquad \blacksquare$$

EXAMPLE 4.7

A vane, in the shape of a flat plate, is moving with velocity V_{vane} against a free jet, as indicated in Figure 4.15. Determine the force exerted on the vane. Assume that the jet has uniform velocity distribution.

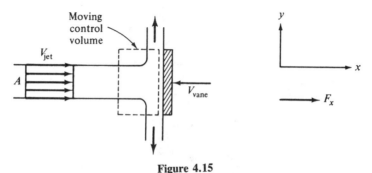

Figure 4.15

Solution Assume a control volume attached to the vane, that is, moving with the vane. Since this control volume is moving with constant speed, Equation (4.7) is applicable. All velocities in Equation (4.7) are measured relative to the moving coordinate system. Since the fluid is flowing as a free jet, the only external force acting on the fluid is the force exerted on it by the vane. Thus

$$F_x = \int_{\substack{\text{control} \\ \text{surface}}} V_x(\rho V_n \, dA) = \underbrace{-V_x \rho V_x A}_{\text{inflow}} + \underbrace{0}_{\text{efflux}}$$

where F_x is the force exerted on the fluid in the x direction and V_x is the velocity of the incoming jet relative to the moving coordinate system. In this case,

$$V_x = V_{\text{jet}} + V_{\text{vane}}$$

Hence we obtain

$$F_x = -\rho V_x^2 A = -\rho (V_{\text{jet}} + V_{\text{vane}})^2 A$$

That is, the force applied externally on the fluid (by the vane) is in the negative x direction. Alternatively, the fluid exerts an equal but opposite force (in the positive x direction) on the vane. Since the vane is inclined at a right angle to the direction of the jet, there is no force in the direction normal to the jet (the y direction).

If the vane were moving in the same direction as the jet, we would obtain

$$V_x = V_{\text{jet}} - V_{\text{vane}}$$

That is, the velocity of the fluid relative to the moving control volume would be smaller than the jet velocity. If the vane were moving away from the jet with a velocity equal to that of the jet, the jet particles would never reach the control surface; that is

$$V_x = V_{\text{jet}} - V_{\text{vane}} = 0$$

For the case of the vane moving away from the jet, the force exerted on the fluid by the vane is given by

$$F_x = -\rho (V_{\text{jet}} - V_{\text{vane}})^2 A$$

When the vane is held fixed ($V_{\text{vane}} = 0$), we obtain

$$F_x = -\rho V_{\text{jet}}^2 A$$

As a specific numerical example, examine the case in which the velocity of a 3-in-diameter jet striking a fixed plate is 30 fps and the density of the liquid is 62.0 lbm/ft^3:

$$F_x = -\left(\frac{62.0}{32.17} \text{ slugs/ft}^3\right)(900 \text{ ft}^2/\text{s}^2)\left(\frac{\pi}{4}\right)\left(\frac{3}{12}\right)^2 \text{ ft}^2$$

$$= \underline{85 \text{ lbf}}$$

■

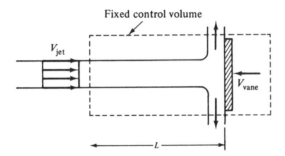

Fixed control volume

V_{jet}

V_{vane}

L

Figure 4.16

The solution of Example 4.7 has used a moving control volume. The problem of the moving vane can also be solved using a fixed control volume, as indicated in Figure 4.16. An instantaneous condition is depicted, for the volume of the free jet within the control volume decreases with time. Thus, at any given instant, the efflux from the control volume is not equal to the inflow due to the negative rate of storage of fluid mass within the control volume. To obtain the efflux, the unsteady continuity equation must be utilized:

$$\left.\frac{\partial M}{\partial t}\right|_{\substack{\text{control}\\\text{volume}}} + \int_{\substack{\text{control}\\\text{surface}}} \rho V_n \, dA = 0 \qquad (4.1)$$

or

$$\left.\frac{\partial M}{\partial t}\right|_{\substack{\text{control}\\\text{volume}}} - \int_{\text{inflow}} \rho V_n \, dA + \int_{\text{efflux}} \rho V_n \, dA = 0$$

With uniform velocity across the jet, the mass flow rate into the fixed control volume is $\rho V_{jet}A$. The volume of the jet within the fixed control volume changes due to the motion of the plate against the jet. The mass of the liquid will decrease at a rate

$$\left.\frac{\partial M}{\partial t}\right|_{\substack{\text{control}\\\text{volume}}} = \rho \frac{d(\text{volume})}{dt} = \rho A \frac{dL}{dt} = -\rho A V_{vane}$$

where L is the instantaneous length of the jet within the control volume. Thus, from Equation (4.1), we have

$$\int_{\text{efflux}} \rho V_n \, dA = \rho V_{jet}A - \left.\frac{\partial M}{\partial t}\right|_{\substack{\text{control}\\\text{volume}}}$$

$$= \rho V_{jet}A + \rho A V_{vane}$$

$$= \rho A(V_{jet} + V_{vane})$$

In order to obtain the force exerted on the fluid within the control volume, the unsteady momentum equation (4.7) must be used. Thus

$$F_x = \left.\frac{\partial(MV_x)}{\partial t}\right|_{\substack{\text{control} \\ \text{volume}}} - \int \underbrace{V_x(\rho V_n \, dA)}_{\substack{\text{momentum} \\ \text{inflow}}} + \int \underbrace{V_x(\rho V_n \, dA)}_{\substack{\text{momentum} \\ \text{efflux}}}$$

Since we have assumed uniform velocity across any section, the inflow of momentum is $V_{\text{jet}}\rho V_{\text{jet}}A$, while the efflux of momentum is

$$-V_{\text{vane}}\rho A(V_{\text{jet}} + V_{\text{vane}})$$

The latter expression is obtained since the fluid leaving the control volume has a velocity component in the x direction equal to V_{vane}. The rate of change of momentum within the control volume is

$$\left.\frac{\partial(MV_x)}{\partial t}\right|_{\substack{\text{control} \\ \text{volume}}} = -\rho A V_{\text{vane}} V_{\text{jet}}$$

$$F_x = -\rho V_{\text{vane}} V_{\text{jet}} A - \rho V_{\text{jet}}^2 A - \rho(V_{\text{jet}} + V_{\text{vane}}) V_{\text{vane}} A$$
$$= -\rho(V_{\text{jet}} + V_{\text{vane}})^2 A$$

Thus either choice of control volume (moving or fixed) leads to the same result, as it should. However, the choice of a moving control volume for this problem reduces the analysis to a steady-state analysis, which is simpler than the unsteady-state analysis when a fixed control volume is chosen.

4.2(b) Angular Momentum Equation

In treating the static equilibrium of fixed bodies or in studying the motion of rotating bodies, it was necessary to consider the moments relative to some axis, exerted on the body by the forces acting on it. In treating certain problems in fluid mechanics, it is sometimes required to evaluate the moments exerted on the fluid volume. An example occurs in the analysis of flow through rotating machinery, such as pumps or turbines.

When the axis about which the moments are taken is fixed, as in turbomachinery, we can utilize the linear momentum equation obtained in Section 4.2a to determine the moments acting on a fluid volume. Consider a control volume in a flow in which the flow properties are a function of only two space coordinates (called two-dimensional flow), as shown in Figure 4.17. The total momentum acting on the control volume will be determined by summing the moments acting on the differential volume elements making up the control volume. This procedure must be used here because the various parts of the control volume are located at varying distances from the origin. As indicated in Figure 4.17, a differential volume is located a distance of magnitude r from the origin. The moment dT_0 exerted on the differential volume equals the product of distance r times the component of the force dF_t exerted on the differential volume in a direction normal to r; that is,

$$dT_0 = (dF_t)r$$

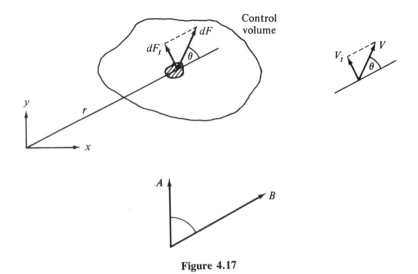

Figure 4.17

The differential force dF_t is obtained from the linear momentum equation (4.8) applied in the tangential direction:

$$dF_t = \frac{\partial}{\partial t}(V_t \, dM) + V_t(\rho V_n \, dA)$$

The differential moment dT_0 about the origin due to this force therefore becomes

$$dT_0 = r\frac{\partial}{\partial t}(V_t \, dM) + rV_t(\rho V_n \, dA)$$

To determine the total torque exerted on the entire control volume, let us integrate to obtain

$$T_0 = \frac{\partial}{\partial t} \int_{\substack{\text{control}\\\text{volume}}} rV_t \, dM + \int_{\substack{\text{control}\\\text{surface}}} rV_t(\rho V_n \, dA) \tag{4.10}$$

Integration of the differential moments exerted on the differential control volumes results in an expression for externally applied moments, since the moments exerted by the differential volumes on each other cancel (action-reaction). The left-hand side of Equation (4.10) thus represents the sum of all externally applied torques acting on the control volume. These applied torques may be due to pressure forces, viscous forces, gravity, magnetic forces, and so forth. The first term on the right-hand side represents the rate at which angular momentum is stored in the control volume, whereas the second term denotes the net rate of efflux of angular momentum from the control volume. Again, this result has been derived for an inertial reference system.

Referring to Figure 4.17, it can be seen that $V_t = V \sin \theta$. Therefore, Equation (4.10) can be written as

$$T_0 = \frac{\partial}{\partial t} \int_{\substack{\text{control} \\ \text{volume}}} rV \sin \theta \, dM + \int_{\substack{\text{control} \\ \text{surface}}} rV \sin \theta \, (\rho V_n \, dA) \qquad (4.11)$$

This result enables us to write Equation (4.10) in vector form. The student will recall that the cross product of two vectors **A** and **B** is given by

$$\mathbf{A} \times \mathbf{B} = |A||B| \sin \theta$$

where θ is the angle between the two vectors, as shown in Figure 4.17. The resulting vector is directed into the page. Therefore, Equation (4.10) becomes, in vector form,

$$\mathbf{T} = \frac{\partial}{\partial t} \int_{\substack{\text{control} \\ \text{volume}}} (\mathbf{r} \times \mathbf{V}) \, dM + \int_{\substack{\text{control} \\ \text{surface}}} (\mathbf{r} \times \mathbf{V})(\rho \mathbf{V} \cdot d\mathbf{A}) \qquad (4.12)$$

Although the foregoing vector equation has been derived for the two-dimensional case, it is actually applicable to the general three-dimensional case.

EXAMPLE 4.8

A lawn sprinkler discharges water in a horizontal direction through two identical, adjustable nozzles at opposite ends of the sprinkler rotor, which is pivoted at its center. After the water supply to the sprinkler has been turned on, the rotor eventually reaches a constant speed of rotation, this speed dependent on the magnitude of the opposing torque developed by bearing friction. Assume an adjustable nozzle with α the angle between the nozzle exit flow direction and the radial direction (see Figure 4.18). Determine the friction torque for rotor speeds ranging from zero (rotor held stationary) to maximum (zero torque, i.e., frictionless bearing).

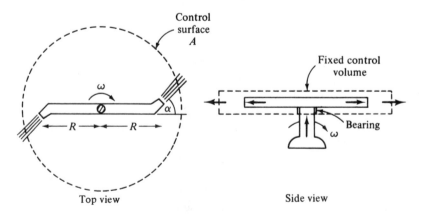

Top view　　　　　　　　　　　　Side view

Figure 4.18

Solution   Choose the fixed control volume to include the sprinkler rotor, as indicated in the figure. The size of the nozzle is usually very small in comparison to its distance from the pivot. Hence use a constant moment arm (R) to the center of the nozzle. The only externally applied torque is that due to bearing friction. Since the fluid enters the sprinkler in the direction of the axis of rotation of the sprinkler, it has no angular momentum. With uniform exit flow from the nozzle, Equation (4.11) becomes

$$T_0 = \underbrace{\int\limits_{\substack{\text{control}\\\text{surface}}} rV \sin\theta \,(\rho V_n \,dA)}_{} = \underbrace{-0}_{\text{inflow}} + \underbrace{\rho RV(\sin\theta)Q}_{\text{efflux}}$$

with Q the total volumetric discharge out of the control volume and V the absolute velocity of the water leaving the nozzles. Since we wish to relate torque to angular speed of rotation, it is now necessary to express the absolute fluid speed V in terms of rotational speed ω and fluid speed V_r relative to the moving nozzle. Thus from Figure 4.19 with V_t the tangential component of the absolute velocity V, we have

$$V_t = V \sin\theta$$

The rotor will move in the direction opposite to that of the tangential component of V_r. As can be seen from Figure 4.19, the tangential component of V_r is $V_r \sin\alpha$. It follows that

$$V_t = V_r \sin\alpha - \omega R$$

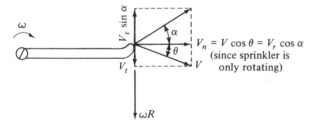

Figure 4.19

Hence

$$T_0 = \rho QR(V_r \sin\alpha - \omega R)$$

Using the continuity equation to relate nozzle discharge and discharge out of control volume, we obtain

$$Q = V_n S = 2V_r \cos\alpha \,\frac{(\pi/4)d^2}{\cos\alpha}$$

where V_n is the velocity normal to the control surface (taken here in a radial direction) and $(\pi/4)d^2$ is the cross-sectional area of each nozzle. Thus

$$V_r = \frac{Q}{(\pi/2)d^2}$$

Note that this result can also be obtained in a more direct manner by considering that mass must be conserved as the fluid passes through the sprinkler. That is, mass discharge out of the control volume is ρQ, which must equal the mass flow out of the nozzle $[\rho V_r(\pi/2)d^2]$, irrespective of the fact that it turns. Substituting into the expression for the sprinkler torque given above,

$$T_0 = \rho QR\left[\frac{Q}{(\pi/2)d^2}\sin\alpha - \omega R\right]$$

This equation shows that for a given nozzle area $(\pi/4)d^2$, nozzle setting α, and discharge Q, there exists a linear relation between friction torque and speed of rotation (Figure 4.20). The value of the T_0 intercept ($\omega = 0$) equals

$$T_0 = \rho Q^2\frac{R}{(\pi/2)d^2}\sin\alpha$$

and represents the torque required to hold the sprinkler stationary as water discharges through it. The value of the ω intercept ($T_0 = 0$) equals

$$\omega = \frac{Q}{(\pi/2)d^2R}\sin\alpha$$

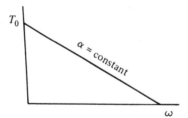

Figure 4.20

and represents the speed that the sprinkler reaches if no external torque is exerted. The maximum values of the intercepts are obtained when the nozzles are set at $\alpha = 90°$ ($\sin\alpha = 1$). For $\alpha = 0°$, the sprinkler will not rotate at all.

As a specific numerical example, consider the case of $\alpha = 90°$, $Q = 0.1$ gpm (gallons per minute), $R = \frac{1}{2}$ ft, and $(\pi/4)d^2 = 0.025$ in.^{2}. Hence

$$Q = 1.0 \text{ gpm} \times 0.00223 \text{ (ft}^3/\text{s)/gpm} = 0.00223 \text{ ft}^3/\text{s}$$

$$T_0 = (1.94 \text{ slugs/ft}^3)(0.00223 \text{ ft}^3/\text{s})\left(\frac{1}{2}\text{ ft}\right)\left(\frac{0.00223 \times 144 \text{ ft}^3/\text{s}}{2 \times 0.025 \text{ ft}^2} - \frac{1}{2}\omega\right)$$

$$= 0.0139 - 0.00108\,\omega$$

For $\omega = 0$,

$$T_0 = 0.0139 \text{ ft-lbf}$$

For $T_0 = 0$,

$$\omega = \frac{0.0139}{0.00108} = 12.9 \text{ rad/s} = 12.9\left(\frac{60}{2\pi}\right) = 123 \text{ rpm}$$

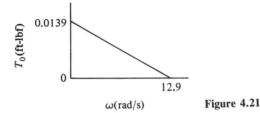
Figure 4.21

Results are plotted in Figure 4.21. ∎

4.3 ENERGY EQUATION

For a system of fluid particles, the law of conservation of energy states that the total energy E of the system increases, in going from state 1 to state 2, by an amount equal to the total heat energy $\tilde{Q}$ added to the system of particles less the work W' done by the system of fluid particles:

$$E_2 - E_1 = \tilde{Q} - W'$$

Here E represents the total energy possessed by the system in a given state and thereby includes the kinetic and potential energy of the entire system mass, the internal energy associated with the random motion of the molecules comprising the system, and other forms of storable energy, such as electrical energy (e.g., stored in a capacitor) and chemical energy. Heat and work are not properties of a system of particles but rather are forms of energy that are transferred across the system boundaries. Therefore, the heat and work transferred during a process are functions of the process itself, not just the end states.

The energy equation in differential form becomes

$$dE = d\tilde{Q} - dW'$$

We wish to apply the law of conservation of energy to a control volume in a fluid, in which mass is entering and leaving across the surface bounding the control volume. Let us apply Equations (3.7) and (3.8), respectively, with X equal to the total energy E of the fluid particles, and e the energy per unit mass, $e = E/M = x$. Substituting, we obtain

$$\left.\frac{dE}{dt}\right|_{\text{system}} = \left.\frac{\partial E}{\partial t}\right|_{\substack{\text{control} \\ \text{volume}}} + \int_{\substack{\text{control} \\ \text{surface}}} e\rho V_n \, dA \qquad (4.13)$$

$$\left.\frac{dE}{dt}\right|_{\text{system}} = \frac{d}{dt}(\tilde{Q} - W') = \left.\frac{\partial E}{\partial t}\right|_{\substack{\text{control} \\ \text{volume}}} + \int_{\substack{\text{control} \\ \text{surface}}} e\rho V_n \, dA \qquad (4.14)$$

$$\frac{d}{dt}(\tilde{Q} - W') = \left.\frac{\partial E}{\partial t}\right|_{\substack{\text{control} \\ \text{volume}}} + \int_{\substack{\text{control} \\ \text{surface}}} e\rho \mathbf{V} \cdot \mathbf{dA} \qquad (4.15)$$

The left-hand side represents the rate at which energy, in the form of heat and work, is transferred to the fluid in the control volume, with heat positive if it is added to the fluid and work positive if done by the fluid. The first term on the right represents the rate at which energy is stored in the control volume, the second term the net rate of efflux of energy from the control volume.

If the system possesses only internal, kinetic, and potential energies, then there results

$$E = U + \text{K.E.} + \text{P.E.} \qquad \text{or} \qquad e = u + \frac{V^2}{2} + gz$$

with U equal to total internal energy and u equal to internal energy per unit mass. Substituting into Equation (4.14):

$$\frac{d}{dt}(\tilde{Q} - W') = \frac{\partial E}{\partial t}\bigg|_{\substack{\text{control} \\ \text{volume}}} + \int_{\substack{\text{control} \\ \text{surface}}} \left(u + \frac{V^2}{2} + gz\right)\rho V_n\, dA \qquad (4.16)$$

For problems in fluid mechanics, with mass crossing the control volume boundaries, it is convenient to divide the work W' into two parts—the flow work necessary to push the fluid across the control surface, and all the other work W done by or on the fluid, such as shaft work, electric and magnetic work, and viscous shear work. In order to derive an expression for the former type of work, flow work, consider a mass Δm, as shown in Figure 4.22, and let us determine the work done by the fluid inside the control volume, acting against the external pressure, in pushing the mass out of the control volume. The force opposing the motion of the mass is $p\,\Delta A$, so that the work done is equal to $p\,\Delta A\,\Delta L$. The volume of the mass is $\Delta V\,(=\Delta A\,\Delta L)$, so that the work required is simply $p\,\Delta V$. Since the density ρ at a point in a fluid has been defined as the mass per unit volume (i.e., $\Delta m/\Delta V$), it follows that in the limit, the flow work required per unit mass of fluid is given by p/ρ. For mass flowing into the control volume, this expression is negative, for it represents work done by the fluid surrounding the control volume on the fluid within the control volume.

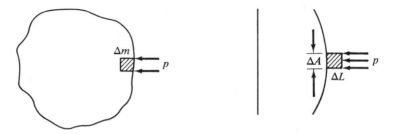

Figure 4.22

It is now convenient to combine the expressions for internal energy per unit mass u and flow work per unit mass p/ρ into a property called *enthalpy* h:

$$h = u + \frac{p}{\rho}$$

Rewriting the energy equation for a control volume in terms of enthalpy, we obtain

$$\frac{d}{dt}(\bar{Q} - W) = \left.\frac{\partial E}{\partial t}\right|_{\substack{\text{control}\\\text{volume}}} + \int_{\substack{\text{control}\\\text{surface}}} \left(h + \frac{V^2}{2} + gz\right)\rho V_n\, dA \qquad (4.17)$$

or, in vector form,

$$\frac{d}{dt}(\bar{Q} - W) = \left.\frac{\partial E}{\partial t}\right|_{\substack{\text{control}\\\text{volume}}} + \int_{\substack{\text{control}\\\text{surface}}} \left(h + \frac{V^2}{2} + gz\right)(\rho \mathbf{V} \cdot \mathbf{dA}) \qquad (4.18)$$

For steady flow, the first term on the right-hand side of Equation (4.18) is equal to zero. Furthermore, if the inlet and outlet flows to a control volume are one-dimensional with velocity vectors normal to the control surface, as shown in Figure 4.23, the energy equation reduces to

$$\frac{d}{dt}(\bar{Q} - W) = \left[\left(h_2 + \frac{V_2^2}{2} + gz_2\right) - \left(h_1 + \frac{V_1^2}{2} + gz_1\right)\right]\rho A V \qquad (4.19)$$

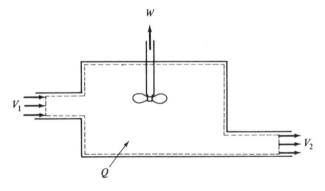

Figure 4.23

Each term of the energy equation as expressed by Equation (4.17), (4.18), or (4.19) is given as an energy rate, that is, energy per unit time (= power). Specifically, $d\bar{Q}/dt$ (heat addition rate) and dW/dt (work rate) must be given in joules per second (J/s) or watts (W), where $1\text{ W} = 1\text{ J/s}$. With velocity V in m/s, g in m/s^2, p in Pa or N/m^2, and ρ in kg/m^3, the units for h, $V^2/2$, and gz are N $\cdot$ m/kg or J/kg. Multiplication by $\rho A V$ (mass flow rate) in kg/s gives units of J/s for the right-hand side of Equation (4.19).

In English units, $d\bar{Q}/dt$ and dW/dt must be given in ft-lbf/s. With velocity in ft/s, g in ft/s², p in lbf/ft², and ρ in slugs/ft³, the units for h (enthalpy), $V^2/2$, and gz are ft-lbf/slug. Multiplication by ρAV (mass flow rate) in slugs/s gives units of ft-lbf/s for the right-hand side of Equation (4.19).

EXAMPLE 4.9

A pump is to be used to provide 1 cfs of water at atmospheric pressure through a 3-in-diameter pipe to a building 500 ft above sea level, the water coming from a reservoir at sea level (Figure 4.24). Determine the pump horsepower required. The density of water can be taken as constant at 62.4 lbm/ft³. Moreover, neglect heat transfer and assume negligible change in the internal energy of the water as it flows through the pipe.

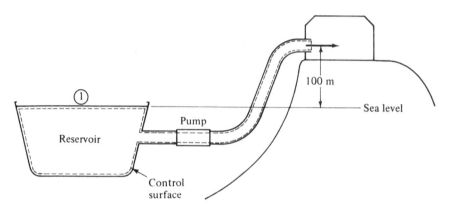

Figure 4.24

Solution Select a control volume with a surface that includes those cross sections at which properties are known. For example, as shown in the figure, the inlet to the control volume has been taken at the reservoir surface, where the pressure is atmospheric and the velocity negligible. The outlet from the control volume has been taken at the building, where again flow and pressure are known. For this problem, Equation (4.19) is applicable, with $\bar{Q} = 0$ and $h_2 - h_1 = (p_2 - p_1)/\rho$, since $u_2 - u_1 = 0$ and $\rho = $ constant. Substituting, we obtain

$$-\frac{dW}{dt} = \rho AV\left[\left(\frac{p_2}{\rho} + \frac{V_2^2}{2} + gz_2\right) - \left(\frac{p_1}{\rho} + \frac{V_1^2}{2} + gz_1\right)\right]$$

where $p_1 = p_2 = $ atmospheric pressure. The mass flow rate of water is equal to

$$\dot{m} = 62.4 \text{ lbm/ft}^3 \times 1 \text{ ft}^3/\text{s} = 62.4 \text{ lbm/s}$$

$$= \frac{62.4}{32.17} = 1.94 \text{ slugs/s}$$

with

$$V_2 = \frac{\dot{m}}{\rho A_2} = \frac{1.94 \text{ slugs/s}}{(1.94 \text{ slugs/ft}^3)[(\pi/4)(\frac{9}{144}) \text{ ft}^2]}$$
$$= 20.4 \text{ ft/s}$$

Therefore

$$-\frac{dW}{dt} = (1.94 \text{ slugs/s})\left(\frac{20.4^2}{2} \text{ ft}^2/\text{s}^2 + 32.17 \text{ ft/s}^2 \times 500 \text{ ft}\right)$$
$$= (1.94 \text{ slugs/s})(208.1 \text{ ft}^2/\text{s}^2 + 16{,}085 \text{ ft}^2/\text{s}^2)$$
$$= 31{,}608 \text{ ft-lbf/s} \qquad (\text{since } 1 \text{ ft}^2/\text{s}^2 = 1 \text{ ft-lbf/slug},$$
$$\text{as shown in Section 1.4})$$

$$\frac{dW}{dt} = -\frac{31{,}608 \text{ ft-lbf/s}}{550 \text{ (ft-lbf/s)/hp}} = \underline{-57.5 \text{ hp}}$$

where the negative sign denotes work done on the fluid.

■

EXAMPLE 4.10

In a turbojet engine, the gaseous products of combustion enter the turbine with a velocity of 20 m/s and temperature of 1000 K; at the turbine exit, the velocity and temperature of the gases are 80 m/s and 720 K (Figure 4.25). If the mass flow rate through the turbine is 40 kg/s, determine the power output for steady flow. Assume adiabatic flow (no heat transfer) as the gases flow through the turbine, and assume the gases behave as a perfect gas with constant specific heat ($c_p = 1.2 \text{ kJ/kg} \cdot \text{K}$) so that the enthalpy difference $h_2 - h_1$ can be written as $h_2 - h_1 = c_p(T_2 - T_1)$.

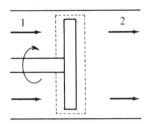

Figure 4.25

Solution   Applying Equation (4.17) to the control volume shown, we obtain for steady, adiabatic flow

$$-\frac{dW}{dt} = \int\limits_{\substack{\text{control} \\ \text{surface}}} \left(h + \frac{V^2}{2} + gz\right)\rho V_n \, dA$$

As is usual in problems dealing with gas flows, any possible change in potential energy can be neglected* in comparison with either the enthalpy change or the kinetic energy change, so that

$$
-\frac{dW}{dt} = \left[\left(h_2 + \frac{V_2^2}{2} \right) - \left(h_1 + \frac{V_1^2}{2} \right) \right] 40 \text{ kg/s}
$$

$$
= \left[c_p(T_2 - T_1) + \frac{V_2^2 - V_1^2}{2} \right] 40 \text{ kg/s}
$$

$$
= \Big[(1.2 \text{ kJ/kg} \cdot \text{K}) \times (720 - 1000) \text{ K}
$$

$$
+ \frac{80^2 \text{ m}^2/\text{s}^2 - 20^2 \text{ m}^2/\text{s}^2}{2} \Big] 40 \text{ kg/s}
$$

$$
= \left[-336 \text{ kJ/kg} + \frac{6400 - 400}{2} \text{ N} \cdot \text{m/kg} \right] 40 \text{ kg/s}
$$

$$
= [-336 \text{ kJ/kg} + 3 \text{ kJ/kg}]40 \text{ kg/s}
$$

$$
= [-333 \text{ kJ/kg}]40 \text{ kg/s}
$$

$$
= -13,320 \text{ kJ/s}
$$

or

$$
\text{Power out} = \underline{13,320 \text{ kW}}
$$

■

---

* For example, if the change in elevation were 3 m, then $g \, \Delta z$ would equal 9.81 m/s² × 3 m or 29.43 J/kg as compared to an enthalpy change of 336 kJ/kg and a kinetic energy change of 3 kJ/kg. Therefore, in subsequent examples and problems dealing with gas flows, the potential energy change in the energy equation will be neglected.

Similarly, in English units, if the change in elevation were 10 ft (about 3 m), then $g \, \Delta z$ would equal 32.17(10) ft²/s² = 321.7 ft-lbf/slug (as shown in Section 1.4). Now

$$
\frac{321.7 \text{ ft-lbf/slug}}{778 \text{ ft-lbf/Btu}} = 0.413 \text{ Btu/slug} = \frac{0.413 \text{ Btu/slug}}{32.17 \text{ lbm/slug}}
$$

$$
= 0.013 \text{ Btu/lbm}
$$

Compare this with an enthalpy change of

$$
336 \text{ kJ/kg} = \frac{(336 \text{ kJ/kg})(0.9478 \text{ Btu/kJ})}{2.2046 \text{ lbm/kg}} = 144.5 \text{ Btu/lbm}
$$

and a kinetic energy change of

$$
3 \text{ kJ/kg} = (3 \text{ kJ/kg}) \frac{0.4299 \text{ Btu/lbm}}{\text{kJ/kg}} = 1.3 \text{ Btu/lbm}
$$

4.4 BERNOULLI'S EQUATION

In order to solve problems in fluid flow, it is often necessary to determine the variation of pressure with velocity from point to point throughout the flow field. As a rule, such a relation between pressure and velocity may be extremely complex, involving changes in pressure and velocity in the x, y, and z directions, as well as other factors (such as friction) that have yet to be discussed. However, by making suitable simplifications or approximations, it is possible to obtain a relatively simple expression relating fluid pressure and velocity; this expression is called the Bernoulli equation. Because of its fundamental nature and simplicity, the equation has had wide usage in the field of fluid mechanics. Nevertheless, the student should carefully note the assumptions made in its derivation in order to be able to apply the equation to the proper physical situation.

Since we wish to consider the variation of pressure with velocity from point to point in the flow field, we shall consider flow along a streamline and shall examine the momentum equation in differential form. Then we shall integrate the momentum equation from one point to another along a streamline.

As stated in Chapter 1, a streamline is a continuous line drawn in the direction of the velocity vector at each point in the flow. By this definition, there can be no flow across a streamline. Select a differential control volume along a streamline in a flow field, as shown in Figure 4.26. The differential cross-sectional area normal to the streamline is δA with the length of the differential control volume along the streamline ds. Note that the symbol δA is used here for the differential element of cross-sectional area as distinguished from dA, which was used to denote the differential element of entire control surface. The flow enters the left-hand face of the control volume and leaves through the right-hand face. There is no flow through the sides of the control surface. The flow is assumed to be frictionless and steady, with pressure forces and gravity forces the only forces acting on the control volume. Let us write the linear momentum equation along the streamline (i.e., in the s direction):

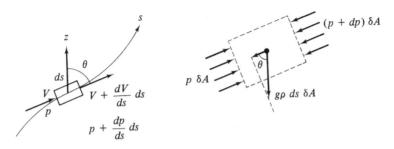

Figure 4.26

$$\Sigma F_s = \int_{\substack{\text{control} \\ \text{surface}}} V_s(\rho V_n \, dA)$$

In differential form, this equation is written as

$$\Sigma \, dF_s = (-V_{\text{in}} + V_{\text{out}})(\rho V_n \, \delta A) = (-V + V + dV)(\rho V \, \delta A)$$

since flow crosses only the areas that are normal to the streamline. The forces in the s direction are given by

$$\Sigma \, dF_s = p \, \delta A - (p + dp) \, \delta A - g\rho \, ds \, \delta A \cos \theta$$

with θ the angle between the vertical and the s direction. Therefore, the momentum equation becomes

$$-dp \, \delta A - g\rho \, ds \, \delta A \cos \theta = \rho \, \delta A \, V \, dV$$

or

$$dp + g\rho \, ds \cos \theta + \rho V \, dV = 0$$

$$\frac{dp}{\rho} + g \, dz + V \, dV = 0$$

Integrating between points designated 1 and 2 along the streamline, we obtain

$$\int_1^2 \frac{dp}{\rho} + \frac{V_2^2 - V_1^2}{2} + g(z_2 - z_1) = 0 \tag{4.20}$$

The first term on the left-hand side can only be integrated if some relationship exists between fluid pressure and density. For example, with incompressible constant-density flow, density is independent of pressure, so that Equation (4.20) can be integrated to yield

$$\frac{p_2 - p_1}{\rho} + \frac{V_2^2 - V_1^2}{2} + g(z_2 - z_1) = 0 \tag{4.21}$$

or

$$\frac{p}{\rho} + \frac{V^2}{2} + gz = \text{Constant}$$

The preceding equation is called the *Bernoulli equation*. This equation was derived here for the case of steady, frictionless flow along a streamline and with pressure and gravity the only forces acting on the control volume. For one-dimensional flow for which the flow properties do not vary in the direction normal to the streamline, the constant in the Bernoulli equation is the same for all streamlines.

Each term of Equation (4.21) is given in energy per unit mass. The term p/ρ is called *flow work* (energy/mass), as was shown in Section 4.3; the term $V^2/2$ is the kinetic energy per unit mass; and gz is the potential energy per unit mass. The statement of the equation gives the changes in energy levels between points along a streamline. Each term is given in J/kg.

In the English system of units, each term is given in ft-lbf/slug. To obtain these units, p is in lbf/ft², ρ in slugs/ft³, V in ft/s, g in ft/s², and z in ft. For kinetic and potential energy per unit mass, the units of ft²/s² are obtained, which was shown in Section 1.4 to equal ft-lbf/slug.

The following examples will illustrate the use of Bernoulli's equation.

EXAMPLE 4.11
Water flows through a tube called a Venturi meter (see Figure 4.27). From pressure measurements, the difference in pressure between sections 1 and 2 is known. The cross-sectional areas are specified. Assuming a horizontal position of the tube and one-dimensional flow through it, neglect frictional losses to find the volumetric flow Q through the Venturi tube as a function of the pressure difference $p_1 - p_2$.

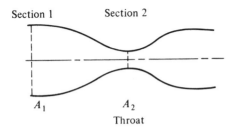

Section 1   Section 2

A_1 A_2
                 Throat   **Figure 4.27**

Solution   To obtain the desired expression for Q in terms of the pressure difference $p_1 - p_2$, use is made of the Bernoulli and continuity equations. First, the Bernoulli equation for one-dimensional frictionless flow between sections 1 and 2 is, from Equation (4.21),

$$\frac{p_1 - p_2}{\rho} + g(z_1 - z_2) = \frac{V_2^2 - V_1^2}{2}$$

where $z_1 - z_2 = 0$. From the continuity equation for constant-density flow, we obtain

$$Q = V_1 A_1 = V_2 A_2$$

or

$$V_1 = \frac{V_2 A_2}{A_1}$$

Substitute into Bernoulli's equation to obtain

$$\frac{V_2^2 - V_2^2 (A_2/A_1)^2}{2} = \frac{p_1 - p_2}{\rho}$$

Solving for V_2, we obtain

$$V_2 = \sqrt{\frac{2}{\rho} \frac{p_1 - p_2}{1 - A_2^2/A_1^2}}$$

Hence

$$Q = A_2 \sqrt{\frac{2}{\rho} \frac{p_1 - p_2}{1 - A_2^2/A_1^2}}$$ ■

EXAMPLE 4.12

1. Liquid flows from a large tank, open to the atmosphere, through a small, well-rounded aperture into the atmosphere, as shown in Figure 4.28. Neglecting losses, find an expression for the velocity of efflux from the tank. Assume the pressure at ②, aperture exit, to be atmospheric.

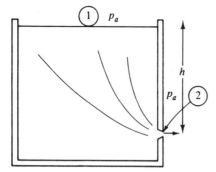

Figure 4.28

Solution   The expression for efflux velocity is obtained from Bernoulli's equation (4.21):

$$\frac{p_1 - p_2}{\rho} + \underbrace{g(z_1 - z_2)}_{= h} = \frac{V_2^2 - V_1^2}{2}$$

For this case, $p_1 = p_2 = p_a$ and V_1^2 is approximately zero, since the cross-sectional area of the tank (A_1) is much greater than the aperture area (A_2). Therefore,

$$V_2 = \sqrt{2gh}$$

2. Next, place a 90° elbow at the tank exit (Figure 4.29). Determine how high the water jet will reach. Again, assume no losses—take ② to be a distance h below the water surface.

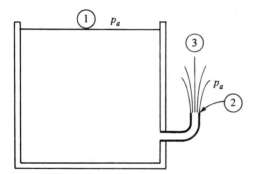

Figure 4.29

Solution   The velocity at the exit remains unchanged in magnitude and is given (for a lossless elbow) by

$$V_2 = \sqrt{2gh}$$

Next, use Bernoulli's equation between points 2 and 3. At point 3, the position of maximum elevation of the jet, the velocity of the jet V_3 is zero.

$$\frac{p_3 - p_2}{\rho} + \underbrace{g(z_3 - z_2)}_{z} = \frac{V_2^2 - V_3^2}{2}$$

Here $p_2 = p_3 = p_a$; therefore, $gz = V_2^2/2$. With $V_2^2 = 2gh$, it follows that $gz = gh$ and $z = h$. Thus if there are no losses, the water jet reaches the initial level of the water in the tank. With losses, the maximum elevation of the water jet will be below that of the water level in the tank.   ∎

We used the Bernoulli equation twice in the preceding example: once between points 1 and 2 and again between points 2 and 3. We could have solved this problem by taking Bernoulli's equation directly between points 1 and 3, namely (Figure 4.30),

$$\frac{p_1 - p_3}{\rho} + g(z_1 - z_3) = \frac{V_3^2 - V_1^2}{2}$$

where $p_1 = p_3 = p_a$ and $V_3 = 0$ and V_1 is negligibly small and taken as zero. Therefore, $z_1 = z_3$. Thus the jet reaches the initial water level in the tank, as was shown previously.

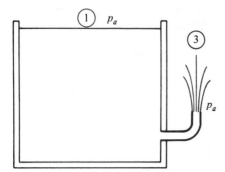

Figure 4.30

In the special case when no flow is taking place, the velocities are equal to zero, and Bernoulli's equation reduces to

$$\frac{p_2 - p_1}{\rho} = g(z_1 - z_2) \tag{4.22}$$

This equation can be recognized as the familiar *hydrostatic equation,* Equation (2.2).

It can be seen that the Bernoulli equation is similar in form to the one-dimensional energy equation shown in Section 4.3. We now wish to compare these two equations and show under what circumstances they are identical. The one-dimensional energy equation was shown to be [see Equation (4.19) with $\dot{m} = \rho V A$ and $h = u + p/\rho$]

$$\frac{1}{\dot{m}} \frac{d(\tilde{Q} - W)}{dt} = u_2 - u_1 + \frac{p_2}{\rho_2} - \frac{p_1}{\rho_1} + \frac{V_2^2 - V_1^2}{2} + g(z_2 - z_1)$$

For *incompressible* flow with no work, the energy equation reduces to

$$\frac{1}{\dot{m}} \frac{d\tilde{Q}}{dt} = u_2 - u_1 + \frac{p_2 - p_1}{\rho} + \frac{V_2^2 - V_1^2}{2} + g(z_2 - z_1)$$

Comparing this equation with Bernoulli's equation for frictionless flow [Equation (4.21)], we see that since both equations must be valid simultaneously, any heat that is added must go directly into an internal energy increase of the fluid. Furthermore, for frictionless flow with no work and no external heat addition, the energy equation and the Bernoulli equation (derived from the momentum equation) become identical. It is therefore clear that for incompressible flow with no work and no heat addition, only two basic equations are required to describe the flow. These are the continuity and Bernoulli equations.

Next, we will compare the energy and Bernoulli equations for the compressible-flow case. Recall, from thermodynamics, that for a reversible process

$$\tilde{q} = \int T \, ds$$

$\tilde{q}$ equals heat added per unit mass and s is the entropy per unit mass. It will also be recalled that for any thermodynamic process

$$T \, ds = du + p \, d\left(\frac{1}{\rho}\right)$$

Therefore,

$$\tilde{q} = \int du + \int p \, d\left(\frac{1}{\rho}\right)$$

Combining this result with the energy equation and utilizing the fact that

$$d\left(\frac{p}{\rho}\right) = \frac{dp}{\rho} + p \, d\left(\frac{1}{\rho}\right)$$

we obtain for a reversible process

$$0 = \frac{1}{\dot{m}} \frac{dW}{dt} + \int_1^2 \frac{dp}{\rho} + \frac{V_2^2 - V_1^2}{2} + g(z_2 - z_1)$$

Again, if there is no work done by the fluid, the energy equation reduces to the Bernoulli equation for frictionless flow as given in Equation (4.20).

4.5 EXAMPLES INVOLVING THREE FUNDAMENTAL EQUATIONS

We have derived the three basic equations of fluid mechanics, that is, the continuity, momentum, and energy equations. Up to this point we have used these equations in solving problems either singly or in conjunction with the continuity equation. There are, however, many applications in which it is necessary to use all three equations of fluid mechanics. The following examples will illustrate the use of the combination of several basic equations in solving fluid flow problems.

EXAMPLE 4.13

A liquid jet is issuing upward against a flat board holding an object, with total mass of board and object M; the jet supports the board as indicated in Figure 4.31. Determine the equilibrium height of the board above the nozzle exit as a function of nozzle area and nozzle exit velocity.

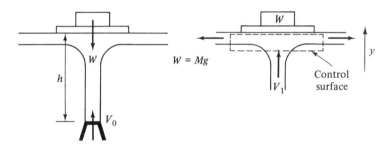

Figure 4.31

Solution   Since the jet is issuing into the atmosphere, the pressure acting on the surface of the jet is constant. Assuming one-dimensional flow, the pressure within the jet is also constant and equal to atmospheric pressure. Bernoulli's equation (4.21) thus gives us a relation between local jet velocity and distance above nozzle:

$$\frac{p_1 - p_0}{\rho} + \frac{V_1^2 - V_0^2}{2} + gh = 0$$

where

$$p_0 = p_1 = p_a$$

Therefore,

$$V_1^2 = -2gh + V_0^2$$

To determine the equilibrium height (h) at which the force exerted on the board by the jet having a local velocity V_1 is exactly balanced by the weight force of board and object Mg, use the momentum equation (4.9) or see Example 4.7:

$$F_y = -Mg = -\rho A_1 V_1^2$$

where F_y is the force exerted on the fluid within the control volume in the positive y direction. Substituting the expression for V_1 obtained above into the momentum equation, we get

$$\frac{Mg}{\rho A_1} = V_0^2 - 2gh$$

or

$$h = \frac{1}{2g}\left(V_0^2 - \frac{Mg}{\rho A_1}\right)$$

where A_1, the cross-sectional area of the jet at height h, is still to be determined. To determine A_1, use the continuity equation (4.6) to obtain

$$V_0 A_0 = V_1 A_1$$

where A_0 is the cross-sectional area of the jet at the nozzle exit. Substituting the expression for V_1 obtained previously, we have

$$V_0 A_0 = \sqrt{V_0^2 - 2gh} \, A_1$$

and

$$A_1 = \frac{A_0}{\sqrt{1 - 2gh/V_0^2}}$$

Combining with the expression for h obtained above, there results

$$h + \frac{1}{2g}\frac{Mg}{\rho A_0}\sqrt{1 - \frac{2gh}{V_0^2}} - \frac{V_0^2}{2g} = 0$$

Solving for h, we obtain

$$h = \frac{V_0^2}{2g} - \frac{1}{2g}\left(\frac{Mg}{\rho A_0 V_0}\right)^2$$

To illustrate with a numerical example, let the mass M be 1.0 slug with a nozzle exit diameter of 3 in and velocity of 30 fps. The density of the water in the jet is 1.93 slugs/ft^3. Substituting into the expression for h, we find the equilibrium height of the board:

$$h = \frac{30^2 \text{ ft}^2/\text{s}^2}{2 \times 32.17 \text{ ft/s}^2}$$

$$- \frac{1}{2 \times 32.17 \text{ ft/s}^2}\left[\frac{1.0 \text{ slug} \times 32.17 \text{ ft/s}^2}{1.93 \text{ slugs/ft}^3 \times (\pi/4)(\frac{3}{12})^2 \text{ ft}^2 \times 30 \text{ ft/s}}\right]^2$$

$$= 13.99 \text{ ft} - 1.99 \text{ ft} = \underline{12.0 \text{ ft}}$$

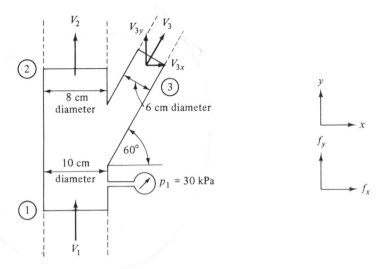

Figure 4.32

EXAMPLE 4.14
Water flows through the horizontal Y branch shown in Figure 4.32. For steady flow and neglecting losses, determine the force components required to hold the Y in place. Assume one-dimensional flow.

The gage pressure at station 1 is measured to be 30 kPa, with volumetric inflow 15.0 l/s (liters per second), volumetric outflow from station 2 equal to 10.0 l/s, and liquid density 1000 kg/m³.

Solution   From Equation (4.7) we have for one-dimensional flow at sections 1, 2, and 3 the two component momentum equations

$$\Sigma F_x = \int\limits_{\substack{\text{control} \\ \text{surface}}} V_x \rho V_n \, dA = \underset{\substack{\text{inflow}① \text{ efflux}②}}{-0 + 0} + \underset{\text{efflux}③}{(V_3)_x \rho V_3 A_3}$$

$$\Sigma F_y = \int\limits_{\substack{\text{control} \\ \text{surface}}} V_y \rho V_n \, dA = \underset{\text{inflow}①}{-V_1 \rho V_1 A_1} + \underset{\text{efflux}②}{V_2 \rho V_2 A_2} + \underset{\text{efflux}③}{(V_3)_y \rho V_3 A_3}$$

Here ΣF_x, the sum of all external forces exerted on the fluid in the x direction, is given by

$$\Sigma F_x = f_x - p_3 A_3 \cos 60°$$

with f_x the force in the x direction exerted on the fluid by the Y branch. And ΣF_y, the sum of all external forces exerted on the fluid in the y direction, is given by

$$\Sigma F_y = f_y + p_1 A_1 - p_2 A_2 - p_3 A_3 \sin 60°$$

with f_y the force in the y direction exerted on the fluid by the Y branch (Figure 4.33). The two component momentum equations therefore become

$$f_x - p_3 A_3 \cos 60° = V_3 \cos 60° \, \rho V_3 A_3$$

$$f_y + p_1 A_1 - p_2 A_2 - p_3 A_3 \sin 60° = -V_1 \rho V_1 A_1 + V_2 \rho V_2 A_2 + V_3 \sin 60° \, \rho V_3 A_3$$

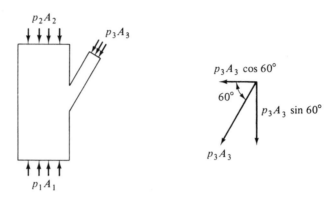

Figure 4.33

To obtain the force components, we must use the linear momentum equation. It can be seen that we must now find the unknown velocity components and pressures. These are obtained from the continuity and Bernoulli equations. From the continuity equation (4.1),

$$\int_{\substack{\text{control} \\ \text{surface}}} \rho V_n \, dA = -\rho V_1 A_1 + \rho V_2 A_2 + \rho V_3 A_3 = 0$$

or

$$Q_1 = Q_2 + Q_3$$

where $Q = AV$ and ρ is constant. Therefore,

$$Q_3 = 15 \, \text{l/s} - 10 \, \text{l/s} = 5 \, \text{l/s}$$

Also,

$$V_1 = \frac{Q_1}{A_1} = \frac{15 \times 10^{-3} \, \text{m}^3/\text{s}}{(\pi/4)(0.10)^2 \, \text{m}^2} = 1.910 \, \text{m/s}$$

$$V_2 = \frac{Q_2}{A_2} = \frac{10 \times 10^{-3} \, \text{m}^3/\text{s}}{(\pi/4)(0.08)^2 \, \text{m}^2} = 1.989 \, \text{m/s}$$

$$V_3 = \frac{Q_3}{A_3} = \frac{5 \times 10^{-3} \, \text{m}^3/\text{s}}{(\pi/4)(0.06)^2 \, \text{m}^2} = 1.768 \, \text{m/s}$$

From Bernoulli's equation (4.21),

$$\frac{p_3 - p_1}{\rho} = \frac{V_1^2 - V_3^2}{2} + \underbrace{g(z_1 - z_3)}_{= \, 0} = \frac{(1.910)^2 - (1.768)^2}{2} = 0.522 \, \text{N} \cdot \text{m/kg}$$

and

$$\frac{p_2 - p_1}{\rho} = \frac{V_1^2 - V_2^2}{2} + g\underbrace{(z_1 - z_2)}_{= \, 0} = \frac{(1.910)^2 - (1.989)^2}{2} = -0.308 \text{ N} \cdot \text{m/kg}$$

Hence

$$p_3 = 30 \text{ kPa} + 0.522 \text{ N} \cdot \text{m/kg} \times 1000 \text{ kg/m}^3 \times \frac{1}{1000} \text{ kPa/Pa} = 30.522 \text{ kPa}$$

and

$$p_2 = 30 \text{ kPa} - 0.308 \text{ N} \cdot \text{m/kg} \times 1000 \text{ kg/m}^3 \times \frac{1}{1000} \text{ kPa/Pa} = 29.692 \text{ kPa}$$

We can now substitute the values of pressure and velocity into the momentum equation to obtain f_x and f_y:

$$f_x = V_3 \cos 60° \, \rho V_3 A_3 + p_3 A_3 \cos 60°$$

$$= (1.768 \text{ m/s})(0.50)(1000 \text{ kg/m}^3)(1.768 \text{ m/s})\frac{\pi}{4}(0.06^2 \text{ m}^2)$$

$$+ (30.522 \text{ kN/m}^2)\frac{\pi}{4}(0.06^2 \text{ m}^2)(0.5)$$

$$= 4.419 \text{ N} + 0.0431 \text{ kN}$$

$$= \underline{0.0475 \text{ kN}}$$

$$f_y = -\rho V_1^2 A_1 + \rho V_2^2 A_2 + \rho V_3^2 A_3 \sin 60° + p_2 A_2 - p_1 A_1 + p_3 A_3 \sin 60°$$

$$= -(1000 \text{ kg/m}^3)(1.910^2 \text{ m}^2/\text{s}^2)\left(\frac{\pi}{4}0.10^2 \text{ m}^2\right)$$

$$+ (1000 \text{ kg/m}^3)(1.989^2 \text{ m}^2/\text{s}^2)\left(\frac{\pi}{4}0.08^2 \text{ m}^2\right)$$

$$+ (1000 \text{ kg/m}^3)(1.768^2 \text{ m}^2/\text{s}^2)\left(\frac{\pi}{4}0.06^2 \text{ m}^2\right)(0.8660)$$

$$+ (29.692 \text{ kN/m})\left(\frac{\pi}{4}0.08^2 \text{ m}^2\right) - (30 \text{ kN/m}^2)\left(\frac{\pi}{4}0.10^2 \text{ m}^2\right)$$

$$+ (30.522 \text{ kN/m}^2)\left(\frac{\pi}{4}0.06^2 \text{ m}^2\right)(0.8660)$$

$$= -28.65 \text{ N} + 19.89 \text{ N} + 7.65 \text{ N} + 0.1492 \text{ kN} - 0.2356 \text{ kN} + 0.0747 \text{ kN}$$

$$= \underline{-0.0128 \text{ kN}}$$

Since f_x and f_y are the forces exerted on the fluid by the Y and the fluid exerts an equal but opposite force on the Y, the force required to hold the

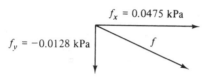

$f_y = -0.0128$ kPa

Figure 4.34

Y branch in place is the vector sum of its components, f_x, f_y, as shown in Figure 4.34. ■

EXAMPLE 4.15

A perfect gas flows without friction through a nozzle of varying cross-sectional area as shown in Figure 4.35. Heat is added through the nozzle walls to maintain the temperature constant. The temperature, pressure, and velocity at inlet (section 1) are specified. In addition, the inlet and exit cross-sectional areas are given. Assume one-dimensional, steady flow and compute the pressure and velocity at the exit (section 2) and the rate of heat addition.

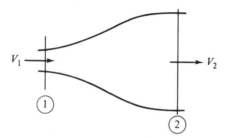

Figure 4.35

Solution   From the equation of state for a perfect gas, evaluate the densities at the two sections:

$$\rho = \frac{p}{RT_1}$$

From the steady flow continuity equation (4.6), we obtain a relationship between p_2 and V_2,

$$\rho_1 V_1 A_1 = \frac{p_1}{RT_1} V_1 A_1 = \rho_2 V_2 A_2 = \frac{p_2}{RT_1} V_2 A_2$$

or

$$p_2 = p_1 \frac{V_1 A_1}{V_2 A_2}$$

Similarly, the Bernoulli equation (4.20) also yields a relationship between the two unknowns p_2 and V_2. We can thus solve for p_2 and V_2:

$$\int_1^2 \frac{dp}{\rho} + \frac{V_2^2 - V_1^2}{2} + \underbrace{g(z_2 - z_1)}_{\cong 0} = 0$$

$$RT_1 \ln \frac{p_2}{p_1} = \frac{V_1^2 - V_2^2}{2}$$

Substituting the expression for p_2 obtained above, there results an equation for V_2:

$$RT_1 \ln \left(\frac{V_1 A_1}{V_2 A_2}\right) = \frac{V_1^2 - V_2^2}{2}$$

To compute the amount of heat addition to maintain the gas at constant temperature, use the energy equation (4.19):

$$\frac{d\bar{Q}}{dt} = \rho_1 A_1 V_1 \left[h_2 - h_1 + \frac{V_2^2 - V_1^2}{2} + g(z_2 - z_1) \right]$$

where $h_2 - h_1 = 0$ since the temperature is constant and $z_2 - z_1$ is negligibly small.

As a numerical example, take the case where A_1 is 0.5 ft^2 and A_2 equals 1 ft^2 with conditions at section 1 as follows: temperature 100°F, pressure 30 psia, and velocity 500 fps. Take the gas constant R to be that of air, 53.3 ft-lbf/lbm °R.

Substituting into the equation for V_2, we get

$$(53.3 \text{ ft-lbf/lbm °R})(32.17 \text{ lbm/slug}) \times (100 + 460) \text{ °R} \ln \left(\frac{500}{V_2} \frac{0.5}{1.0}\right)$$

$$= \frac{500^2 - V_2^2}{2} \text{ ft}^2/\text{s}^2$$

To solve the equation for V_2, use the graphical solution as shown in Figure 4.36. As we see from this figure, the intersection of the two curves gives a value of 225.6 fps for V_2.

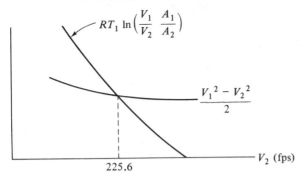

Figure 4.36

The pressure at section 2 is

$$p_2 = p_1 \frac{V_1 A_1}{V_2 A_2} = 30 \left(\frac{500}{225.6}\right)\left(\frac{0.5}{1.0}\right) = 33.2 \text{ lbf/in}^2$$

The rate of heat addition is

$$\frac{d\tilde{Q}}{dt} = \left(\frac{p_1}{RT_1}A_1V_1\right)\left(\frac{V_2^2 - V_1^2}{2}\right)$$

$$= \left[\frac{30 \times 144}{53.3 \times 32.17(560)} \text{ slugs/ft}^3\right](0.5 \text{ ft}^2)(500 \text{ ft/s})(-99,560 \text{ ft}^2/\text{s}^2)$$

$$= -\frac{111,980}{778} = \underline{-143.9 \text{ Btu/s}}$$

That is, the nozzle must be cooled to maintain the flow at constant temperature.

■

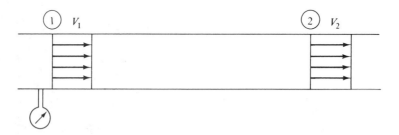

Figure 4.37

EXAMPLE 4.16

Water flows through the capillary tube shown in Figure 4.37. The tube has a constant diameter with steady flow through it and uniform velocity. From temperature measurements, the increase in internal energy ($u_2 - u_1$) between the two sections has been determined. Evaluate the frictional force exerted on the capillary tube by the fluid in terms of the internal energy change.

Solution Since the tube is of constant diameter, the continuity equation requires that $V_2 = V_1$. We can then find p_2 from the energy equation. Use of the momentum equation allows evaluation of the force exerted on the tube by the fluid. With friction, the fluid at the tube walls is stationary, so the friction force does no work on the fluid.

In the absence of external work and heat addition, the energy equation (4.19) becomes

$$\frac{p_1}{\rho} + \frac{V_1^2}{2} + gz = \frac{p_2}{\rho} + \frac{V_2^2}{2} + gz_2 + u_2 - u_1$$

Rewriting by using $V_2 = V_1$ and $z_2 - z_1 = 0$, we obtain

$$\frac{p_2}{\rho} = \frac{p_1}{\rho} - (u_2 - u_1)$$

The momentum equation (4.9) becomes

$$\Sigma F_x = \overset{\cdot}{f_x} + p_1 A - p_2 A = 0$$

with f_x the force exerted by the tube on the fluid due to friction. Therefore,

$$f_x = -p_1 A + [p_1 - \rho(u_2 - u_1)]A$$
$$\underline{f_x = -\rho(u_2 - u_1)A}$$

Hence the frictional force exerted on the tube is $-f_x$, that is, a force in the direction of flow, since there is an increase in internal energy in the direction of flow.   ■

4.6 APPLICATIONS INVOLVING CONTROL VOLUME IN NONINERTIAL COORDINATES

In the derivation of the linear momentum equation in Section 4.2, Newton's law of conservation of momentum for a system of particles was used in the form

$$\Sigma \mathbf{F} = \frac{d}{dt} M \mathbf{V}$$

It is to be remembered that this expression requires velocities to be measured relative to an inertial reference frame. In cases that we have considered, velocities were taken relative to the control volume, so that the control volume had to be either fixed or translating at constant velocity relative to a fixed reference frame. In some applications, however, it is desirable to take velocities relative to an accelerating control volume. For this reason, we wish to extend the linear momentum equation so as to be applicable to cases involving a noninertial control volume.

Consider first the simple case in which the control volume is translating along a straight line relative to an inertial, fixed reference. We shall denote velocities relative to the fixed frame of reference as $\mathbf{V}_{rf}$, with components V_{xrf}, V_{yrf}, and V_{zrf}, and velocities relative to the control volume by $\mathbf{V}_{rc}$, with components V_{xrc}, V_{yrc}, and V_{zrc}. Newton's law of motion is applicable to the fixed frame of reference, so that we can write, for a system of fluid particles of mass M,

$$\Sigma \mathbf{F} = \frac{d}{dt} M \mathbf{V}_{rf}$$

According to the law of conservation of mass, the rate of change of the total mass of the system of fluid particles is zero, so that

$$\Sigma \mathbf{F} = M \frac{d}{dt} \mathbf{V}_{rf} \qquad (4.23)$$

In component form, this becomes

$$\Sigma F_x = M \frac{d}{dt} V_{xrf}$$

$$\Sigma F_y = M \frac{d}{dt} V_{yrf}$$

$$\Sigma F_z = M \frac{d}{dt} V_{zrf}$$

The control volume–system relationship derived in Chapter 3 is applicable irrespective of the motion of the control volume. Note that velocities in this relationship are taken relative to the control volume. From Equation (3.8) we obtain

$$\frac{d}{dt}(M\mathbf{V}_{rc}) = \frac{\partial}{\partial t}(M\mathbf{V}_{rc})\bigg|_{\substack{\text{control}\\\text{volume}}} + \int_{\substack{\text{control}\\\text{surface}}} \mathbf{V}_{rc}(\rho\mathbf{V}_{rc} \cdot d\mathbf{A})$$

In scalar form this becomes

$$\frac{d}{dt}(M V_{xrc})_{\text{system}} = \frac{\partial}{\partial t}(M V_{xrc})\bigg|_{\substack{\text{control}\\\text{volume}}} + \int_{\substack{\text{control}\\\text{surface}}} V_{xrc}(\rho V_{nrc} \, dA)$$

$$\frac{d}{dt}(M V_{yrc})_{\text{system}} = \frac{\partial}{\partial t}(M V_{yrc})\bigg|_{\substack{\text{control}\\\text{volume}}} + \int_{\substack{\text{control}\\\text{surface}}} V_{yrc}(\rho V_{nrc} \, dA)$$

$$\frac{d}{dt}(M V_{zrc})_{\text{system}} = \frac{\partial}{\partial t}(M V_{zrc})\bigg|_{\substack{\text{control}\\\text{volume}}} + \int_{\substack{\text{control}\\\text{surface}}} V_{zrc}(\rho V_{nrc} \, dA)$$

For straight-line motion of the control volume with velocity $\mathbf{V}_{\text{C.V.}}$ relative to the fixed inertial frame, it follows that the velocity of a particle relative to the inertial reference is given by

$$\mathbf{V}_{rf} = \mathbf{V}_{\text{C.V.}} + \mathbf{V}_{rc}$$

Substituting into Equation (4.23), we obtain

$$\Sigma \mathbf{F} = M \frac{d}{dt} \mathbf{V}_{\text{C.V.}} + M \frac{d}{dt} V_{rc}$$

Applying the control volume–system relationship to this expression, there results

$$\Sigma \mathbf{F} - M \frac{d}{dt} \mathbf{V}_{\text{C.V.}} = \frac{\partial}{\partial t}(M\mathbf{V}_{rc})\bigg|_{\substack{\text{control}\\\text{volume}}} + \int_{\substack{\text{control}\\\text{surface}}} \mathbf{V}_{rc}(\rho\mathbf{V}_{rc} \cdot d\mathbf{A}) \qquad (4.24)$$

or, in component form,

$$\Sigma F_x - M\frac{d}{dt}V_{x\text{C.V.}} = \frac{\partial}{\partial t}(MV_{xrc})\bigg|_{\substack{\text{control}\\\text{volume}}} + \int_{\substack{\text{control}\\\text{surface}}} V_{xrc}(\rho V_{nrc}\, dA)$$

$$\Sigma F_y - M\frac{d}{dt}V_{y\text{C.V.}} = \frac{\partial}{\partial t}(MV_{yrc})\bigg|_{\substack{\text{control}\\\text{volume}}} + \int_{\substack{\text{control}\\\text{surface}}} V_{yrc}(\rho V_{nrc}\, dA) \qquad (4.25)$$

$$\Sigma F_z - M\frac{d}{dt}V_{z\text{C.V.}} = \frac{\partial}{\partial t}(MV_{zrc})\bigg|_{\substack{\text{control}\\\text{volume}}} + \int_{\substack{\text{control}\\\text{surface}}} V_{zrc}(\rho V_{nrc}\, dA)$$

The preceding equation thus presents the momentum equation for a noninertial control volume; it is applicable for the case in which the control volume is moving along a straight line and is in accelerative motion.

EXAMPLE 4.17

Develop the equation of motion for a rocket accelerating upward from the earth's surface. The rocket exhaust velocity V_e relative to the rocket, the rocket nozzle exit plane pressure p_e, and the rate of fuel consumption $\dot{m}_f$ are assumed constant throughout the flight. Let the air resistance (aerodynamic drag) be D, let g be the acceleration due to gravity, and M_R the rocket mass at any time t. Also solve for the absolute rocket velocity V_R versus time for the case in which D and g can be neglected (e.g., space travel).

Solution   In this example, take the control surface as shown in Figure 4.38; shown also are the forces acting on the control surface. For this case

$$\Sigma F_x = (p_e - p_a)A_e - D - M_R g$$

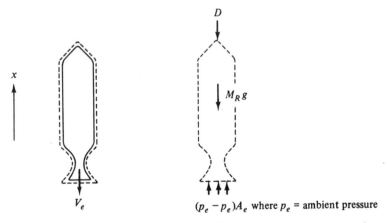

$(p_e - p_e)A_e$ where p_e = ambient pressure

Figure 4.38

Applying Equation (4.25), we obtain

$$(p_e - p_a)A_e - D - M_R g - M_R \frac{dV_R}{dt} = \frac{\partial}{\partial t}(M_R V_{xrc}) - V_e \dot{m}_f$$

The first term on the right-hand side refers to the rate of storage of linear momentum inside the control volume, with velocities expressed relative to the control volume. Certainly the structural mass of the rocket, the unburnt fuel, and the payload all travel at the rocket velocity, so their velocity relative to the control volume is zero. The burned gases possess a constant relative velocity V_e with respect to the control volume, so again there is no contribution to the time rate of change of momentum. Therefore, the first term on the right-hand side of the preceding equation is zero. The equation of motion of the rocket reduces to

$$(p_e - p_a)A_e - D - M_R g - M_R \frac{dV_R}{dt} = -V_e \dot{m}_f$$

If we neglect drag and gravity, the above simplifies to

$$(p_e - p_a)A_e = M_R \frac{dV_R}{dt} - V_e \dot{m}_f$$

But M_R is equal to $M_0 - \dot{m}_f t$, with M_0 the initial rocket mass, so that

$$(p_e - p_a)A_e = (M_0 - \dot{m}_f t)\frac{dV_R}{dt} - V_e \dot{m}_f$$

$$\frac{dV_R}{(p_e - p_a)A_e + \dot{m}_f V_e} = \frac{dt}{M_0 - \dot{m}_f t}$$

Integrating for $V_R = 0$ at $t = 0$, we obtain

$$V_R = -[(p_e - p_a)A_e + \dot{m}_f V_e]\frac{\ln(M_0 - \dot{m}_p t)}{\dot{m}_f}\Big|_0^t$$

$$= -\left[\frac{(p_e - p_a)A_e}{\dot{m}_f} + V_e\right]\ln\left(1 - \frac{\dot{m}_p t}{M_0}\right) \qquad (4.26)$$

∎

EXAMPLE 4.18

The Saturn 1B rocket, used for the initial Apollo launch, develops a sea level thrust of 7.295 MN, while consuming fuel at the rate of 2832 kg/s. If the initial mass of the rocket and its payload, including the Apollo capsule, is 586,000 kg, of which 400,000 kg is fuel, determine the velocity of the rocket after the fuel has been exhausted. Neglect any difference between the local ambient pressure and the rocket exit plane pressure, and neglect aerodynamic drag and gravity.

Solution   In the absence of pressure forces, the rocket thrust can be written as

$$T = \dot{m}_f V_e \quad \text{(see Example 4.5)}$$

or

$$V_e = \frac{(7.295 \times 10^6 \text{ N})}{2832 \text{ kg/s}} = 2576 \text{ m/s}$$

Applying Equation (4.26),

$$V_R = -V_e \ln \left(\frac{M_0 - \dot{m}_f t}{M_0} \right)$$

Assuming the mass flow rate of fuel to be constant, the term $\dot{m}_f t$ represents the total fuel consumed from time 0 to time t. If t is the burning time of the fuel, $\dot{m}_f t$ is the total fuel carried by the rocket, or 400,000 kg. Therefore, we have

$$V_R = -2576 \ln \left(1 - \frac{400{,}000}{586{,}000} \right)$$

$$= 2956 \text{ m/s} \qquad \blacksquare$$

When the fluid is at rest with respect to the control volume, the momentum equations (4.25) become

$$\Sigma F_x - M a_x = 0$$
$$\Sigma F_y - M a_y = 0$$
$$\Sigma F_z - M a_z = 0$$

where a_x, a_y, and a_z represent, respectively,

$$\frac{dV_{x\text{C.V.}}}{dt}, \qquad \frac{dV_{y\text{C.V.}}}{dt}, \qquad \text{and} \qquad \frac{dV_{z\text{C.V.}}}{dt}$$

In differential form, these equations are

$$\Sigma \, dF_x - (dM)a_x = 0$$
$$\Sigma \, dF_y - (dM)a_y = 0 \qquad\qquad (4.27)$$
$$\Sigma \, dF_z - (dM)a_z = 0$$

We shall now consider the incremental volume $dx \, dy \, dz$ (Figure 4.39) in a fluid at rest with respect to the control volume, but with the control volume undergoing constant linear acceleration. With pressure and gravity the only forces, and with $dM = \rho \, dx \, dy \, dz$, we obtain, in the x direction,

$$p \, dy \, dz - \left(p + \frac{\partial p}{\partial x} dx \right) dy \, dz - \rho \, dx \, dy \, dz \, a_x = 0$$

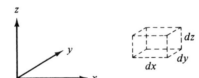

Figure 4.39

or

$$\frac{\partial p}{\partial x} = -\rho a_x$$

Similarly, in the y direction,

$$\frac{\partial p}{\partial y} = -\rho a_y$$

However, in the z direction we also have the gravity force, so that

$$p \, dx \, dy - \left(p + \frac{\partial p}{\partial z} \, dz\right) dx \, dy - \rho g \, dx \, dy \, dz - \rho \, dx \, dy \, dz \, a_z = 0$$

or

$$\frac{\partial p}{\partial z} = -\rho(g + a_z)$$

The total change in pressure (dp) is, from differential calculus,

$$dp = \frac{\partial p}{\partial x} \, dx + \frac{\partial p}{\partial y} \, dy + \frac{\partial p}{\partial z} \, dz$$

Combining,

$$dp = -\rho[a_x \, dx + a_y \, dy + (a_z + g) \, dz] \tag{4.28}$$

For cases in which the density of the fluid is constant, Equation (4.28) can be integrated to obtain

$$p = -\rho[a_x x + a_y y + (a_z + g)z] + C$$

To evaluate the constant of integration C, take $p = p_0$ at the origin, hence

$$p = -\rho[a_x x + a_y y + (a_z + g)z] + p_0 \tag{4.29}$$

EXAMPLE 4.19

A pail of water, filled with water to a depth of 0.5 ft, is placed on the floor of an elevator. Find the equilibrium pressure at the bottom of the pail when the elevator is moving upward at a constant acceleration of 15 ft/s², when the elevator is moving downward at a constant acceleration of 15 ft/s² and 32.17 ft/s², and when the elevator is standing still.

Solution In this case, the expression for the pressure of a fluid at rest with respect to a coordinate system which moves along a straight line [Equation (4.29)] becomes

$$p = p_0 - \rho(a_z + g)z$$

At the bottom of the pail, $z = 0.5$ ft. (Figure 4.40). For an upward acceleration of 15 ft/s², $a_z = 15$ ft/s². Thus

$$p - p_0 = 1.94 \text{ slugs/ft}^3 \, (15 + 32.17) \text{ ft/s}^2 \, (-0.5) \text{ ft}$$

$$= 45.75 \text{ slugs/ft} \cdot \text{s}^2 = 45.75 \text{ lbf/ft}^2 = \underline{0.318 \text{ psi}}$$

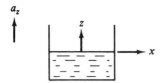

Figure 4.40

For a downward acceleration of 15 ft/s^2, $a_z = -15$ ft/s^2 and

$p - p_0 = -1.94$ slugs/ft^3 $(-15 + 32.17)$ ft/s^2 (-0.5) ft

$$= 16.65 \text{ lbf/ft}^2 = \underline{0.116 \text{ psi}}$$

For a downward acceleration of 32.17 ft/s^2, $a_z = -32.17$ ft/s^2. Hence $p - p_0 = 0$. Thus when the water in the pail moves downward with an acceleration equal to that of gravity (free fall), there is no hydrostatic pressure increase due to submergence.

When the elevator is standing still (or moving at constant velocity),

$p - p_0 = -1.94$ slugs/ft^3 $(0 + 32.17)$ ft/s^2 (-0.5)

$$= 31.20 \text{ lbf/ft}^2 = \underline{0.217 \text{ psi}}$$

For this special case of nonacceleration of the elevator, the hydrostatic equation can be used to give

$$p - p_0 = \rho g(0 - z)$$

or

$p - p_0 = 1.94$ slugs/ft^3 (32.17) ft/s^2 $[-(-0.5)]$ ft

$$= 31.20 \text{ lbf/ft}^2 = \underline{0.217 \text{ psi}}$$

■

EXAMPLE 4.20

Find the equation of the free surface of water in a tank 1.0 m long which moves in a horizontal plane with constant acceleration 0.8 m/s^2 (Figure 4.41).

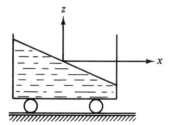

Figure 4.41

Solution   In this case, $a_x = 0.8$ m/s^2, $a_y = 0$, $a_z = 0$, and the expression for the pressure [Equation (4.29)] becomes

$$p - p_0 = -\rho(0.8x + gz)$$

Along the free surface, the pressure is constant and equal to that of the atmosphere p_0; hence

$$0.8x = -gz$$

or

$$z = \frac{-0.8x}{9.81} = -0.0815x \qquad \text{(with } x \text{ in meters)}$$

the equation of a straight line. ∎

Another simple application of the control volume in noninertial coordinates is the case of a body of liquid rotating at constant angular speed about a fixed axis. When the particle velocity with respect to the rotating coordinate system is zero—i.e., $V_{rc} = 0$—the time rate of change of the velocity vector (absolute velocity $\overline{V}_{rf}$) of a particle in a fixed inertial coordinate system is the centrifugal acceleration due to the angular velocity and is given by

$$\frac{d\overline{V}_{rf}}{dt} = \overline{\omega} \times (\overline{\omega} \times \overline{r})$$

where $\overline{\omega}$ is the angular rotation vector and $\overline{\omega} \times \overline{r}$ is the angular velocity of a point in the moving coordinate system due to rotation, with $\overline{r}$ the distance of the point from the origin (Figure 4.42).

Figure 4.42

The momentum equation becomes

$$\Sigma \overline{F} = M\overline{\omega} \times (\overline{\omega} \times \overline{r}) \qquad (4.30)$$

or, in component form,

$$\Sigma F_x = -M\omega^2 x$$
$$\Sigma F_y = -M\omega^2 y \qquad (4.31)$$
$$\Sigma F_z = -M\omega^2 z$$

The forces in the x direction on an incremental volume (Figure 4.39) are

$$p \, dy \, dz - \left(p + \frac{\partial p}{\partial x} dx \right) dy \, dz$$

Substituting into Equation (4.31):

$$\frac{\partial p}{\partial x} = \rho\omega^2 x$$

Similarly,

$$\frac{\partial p}{\partial y} = \rho\omega^2 y$$

However, in the z direction we have the gravity force; thus

$$\frac{\partial p}{\partial z} = -\rho g$$

The total change in pressure is

$$dp = \rho(\omega^2 x\, dx + \omega^2 y\, dy - g\, dz).$$

Integrating, we obtain for constant density ρ

$$p = \rho\left(\frac{\omega^2 x^2}{2} + \frac{\omega^2 y^2}{2} - gz\right) + C$$

or with $x^2 + y^2 = r^2$

$$p = \rho\left(\frac{\omega^2 r^2}{2} - gz\right) + C$$

For the evaluation of the constant of integration C, take $p = p_0$ at $r = 0$ and $z = h_0$ (Figure 4.43):

$$p_0 = \rho(-gh_0) + C \qquad \text{or} \qquad C = p_0 + \rho g h_0$$

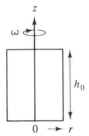

Figure 4.43

Hence

$$p = p_0 - \rho g(z - h_0) + \rho\frac{\omega^2 r^2}{2} \tag{4.32}$$

EXAMPLE 4.21

A sealed can of radius r and height H, completely filled with liquid, is rotating about its central axis at a constant angular speed ω. Find the equilibrium pressure distribution in the liquid.

Solution   From Equation (4.32) we get the pressure distribution with $h_0 = H$:

$$p = p_0 + \rho[(H - z)g + \tfrac{1}{2}\omega^2 r^2]$$

When the can is at rest, the lines of the constant pressure (isobars) are horizontal straight lines (Figure 4.44a), the pressure (hydrostatic pressure) increasing with increase in submergence $(H - z)$, i.e., distance from $z = H$. If the can were rotating in outer space where gravity is negligibly small, the pressure distribution (centrifugal pressure) would be $p - p_0 = \rho\tfrac{1}{2}\omega^2 r^2$

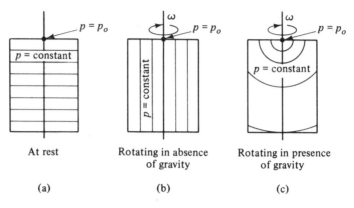

Figure 4.44

and the lines of constant pressure would be vertical straight lines (Figure 4.44b). Finally, when the rotation takes place in the presence of gravity, the constant-pressure lines (actually surfaces) are parabolas (Figure 4.44c). The intercepts of the isobars with the z axis are obtained by letting $r = 0$ in Equation (4.32), namely,

$$z = H - \frac{p - p_0}{\rho g}$$ ∎

PROBLEMS

In the following problems, unless otherwise indicated, assume atmospheric pressure $= 101$ kPa or 14.7 psi, density of water 1000 kg/m^3 or 62.4 lbm/ft^3 (1.94 slugs/ft^3). Also, unless specified as a gage reading, take pressures to be absolute pressures.

4.1. Fluid with specific gravity 0.07 enters a Y, as shown in Figure P4.1, with velocity $V_1 = 5$ m/s. The diameter at 1 is 10 cm; diameter at 2, 7 cm; diameter at 3, 6 cm. If equal mass flows are to occur at 2 and 3, find V_2 and V_3.

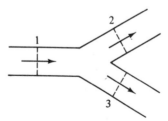

Figure P4.1

4.2. A 1-in-diameter pipe carries a liquid with specific gravity equal to 0.86. The average flow velocity at a pipe cross section is 20 fps. Find the mass flow rate at this cross section.

4.3. The average velocity of a liquid at a section of pipe having a diameter of 4 cm is 3 m/s. What will be the average velocity of the liquid at another section where the diameter is 6 cm? Assume steady, incompressible flow.

4.4. The velocity distribution of two-dimensional flow between two parallel plates is given by $V = V_m(1 - y^2/h^2)$ as shown in Figure P4.4. Find the mass flow rate and average flow velocity, with water the working fluid.

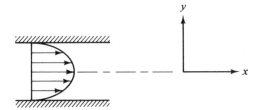

Figure P4.4

4.5. A water jet emerges from a tank through a nozzle as shown in Figure P4.5. The diameter of the nozzle constriction is 1.25 in, while that at the jet exit is 1.5 in. Find the water velocity and pressure in the nozzle constriction. Neglect frictional effects. If the water level in the tank is increased to 10 ft above the centerline of the nozzle, find the water velocity and pressure in the constriction.

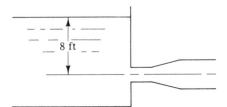

8 ft

Figure P4.5

4.6. Air enters a centrifugal compressor at a pressure of 100 kPa, temperature 20°C, flow rate 0.3 m³/s. Compressed air is discharged through an outlet pipe at a pressure of 1000 kPa, temperature 80°C, velocity 30 m/s. Find the diameter of the discharge pipe assuming the air to obey the perfect gas law.

4.7. A two-dimensional jet of water emerges from a 1-in horizontal slot with an average velocity of 20 fps. Assume that the horizontal component of the water velocity remains constant (i.e., neglect frictional effects of the air). Calculate the absolute velocity and thickness of the water jet at a distance of 1 ft from the slot for initial jet slopes of 30°, 45°, and 60° with the horizontal.

4.8. Liquid oxygen flows through a 10-mm-diameter tube. At a given section, the density of the oxygen is 20 kg/m³. Find the average velocity at the section if the mass flow rate is 0.03 kg/s.

4.9. Under certain conditions, the velocity distribution for fluid flow in a circular pipe is parabolic, with $V = C(a^2 - r^2)$, as shown in Figure P4.9. If the total flow through a pipe of internal diameter 2 in is 0.01 cfs, determine the value of C. Plot to scale V versus r for the above velocity distribution. Determine the relation between maximum and average velocities.

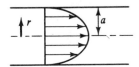

Figure P4.9

4.10. Water flows radially toward a bathtub drain, as shown in Figure P4.10. At a radius of 0.05 m, the radial velocity of the water is uniform at 10 cm/s, with a water depth of 2 cm. Determine the average velocity of the water in the 2-cm drainpipe.

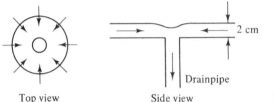

Top view   Side view   **Figure P4.10**

4.11. A rectangular pool, 10 m × 30 m, is to be filled to a depth of 3 m. For a filling time of 2 hours, determine the inlet flow required in m^3/s. If 6-cm hoses are available and the water velocity in each hose is not to exceed 3 m/s, determine the number of hoses required.

4.12. A tank is to be filled with air to a pressure of 150 psia. If the initial tank pressure is 14.7 psia, determine the time required if the velocity of air in the 1-in-diameter inlet line to the tank is 4 fps, the pressure of air in the inlet line is 175 psia, and the temperature of air in line and tank is 85°F. Tank volume = 350 ft^3.

4.13. A 2-cm-diameter horizontal pipe contains a 90° elbow (Figure P4.13). The pressure at the elbow inlet is 500 kPa. The volume flow rate of the water is 1.5 l/s. Neglecting losses, find the force exerted by the flowing water on the pipe system.

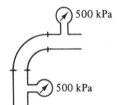

500 kPa

500 kPa   **Figure P4.13**

4.14. Water flows past one side of a flat plate as shown in Figure P4.14. The flow is parallel to the plate with uniform velocity at the leading edge, while at the trailing edge the velocity increases linearly with y as shown; that is, $v = cy$ for $y \leqslant h$. For y greater than h, the velocity is V. For two-dimensional flow, find the x component of the force exerted on the plate by the fluid.

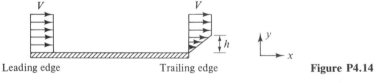

Leading edge   Trailing edge   **Figure P4.14**

4.15. Two horizontal water jets strike a vertical plate as shown in Figure P4.15. Calculate the force and moment required to hold the plate stationary.

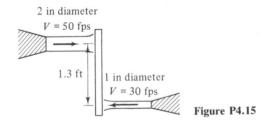

Figure P4.15

4.16. In Figure P4.16 a conical paper cup, initially filled with water, develops a hole in its bottom 2.5 mm in diameter. Determine the time required for the water to drain out of the cup. The cup has a base of 6 cm diameter and a height of 8 cm. Assume quasi-steady flow.

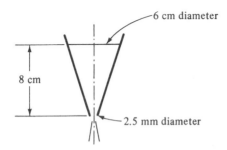

Figure P4.16

4.17. Find the horsepower required to pump 5 gpm of water from the bottom of a crater on the moon to the crater rim, the difference in elevation being 100 ft. The acceleration due to gravity (g) equals one-sixth of that on earth.

4.18. Hot flue gases are flowing in a constant-area duct (6 cm in diameter and 5 m long). The inlet temperature of the gas stream is 600 K, while the inlet pressure is 110 kPa. There is a heat loss from the duct of 10 W/m. Assume that the flue gases behave as a perfect gas ($p = \rho RT$) with $h = 1.0T$, where h is the enthalpy per unit mass (kJ/kg) and T is the absolute temperature. Compute the exit temperature of the gases for an inlet velocity of 10 m/s. The gas constant R is 0.30 kJ/kg · K.

4.19. A free jet of water is being discharged from a hydrant as shown in Figure P4.19. The water pressure within the hydrant is measured to be 120 psia. Determine the location of impact of the water jet on the ground.

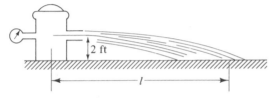

Figure P4.19

4.20. Water flows through a horizontal reducing 90° elbow. The inlet diameter is 25 mm, with an exit diameter of 20 mm. The pressure at the elbow inlet is 300 kPa. Neglect flow losses and calculate the force exerted on the elbow for a flow rate of 0.001 m³/s.

4.21. Two pipes are connected to the machine as shown in Figure P4.21. Water flows through the machine as indicated with pressure at point A (elevation 220 ft) equal to 30 psig and at point B (elevation 100 ft) equal to 100 psig. The volume flow is 5 cfs. Is the machine shown a turbine or a pump? Calculate its horsepower. (*Hint:* A turbine produces work; a pump requires work input.)

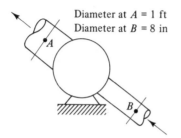

Diameter at A = 1 ft
Diameter at B = 8 in

Figure P4.21

4.22. A nozzle is fastened to the end of a U tube with dimensions as shown in Figure P4.22. The nozzle exhausts into atmospheric pressure; neglect friction and compute the force exerted on the U tube by the water.

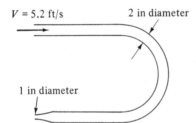

V = 5.2 ft/s   2 in diameter

1 in diameter

Figure P4.22

4.23. A two-dimensional jet of incompressible fluid strikes a flat plate as shown in Figure P4.23. Neglecting friction and gravity, find V_2, V_3, d_2, and d_3 in terms of ρ, V_1, d_1, and α. Determine the magnitude and direction of the force required to hold the plate stationary.

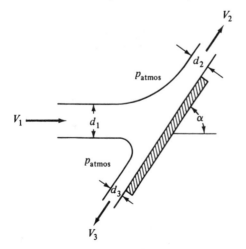

V_2

d_2

p_{atmos}

V_1 d_1

p_{atmos}

α

d_3

V_3

Figure P4.23

4.24. Water collects on the bottom of a gasoline tank of an automobile as indicated in Figure P4.24. To drain the water from the tank, the 1-in-diameter plug is removed. Compute how long it takes to rid a 2 ft by 4 ft tank of accumulated water. Assume quasi-steady flow; that is, assume that at any instant the flow can be treated as steady.

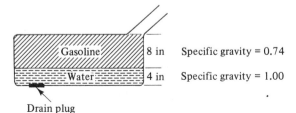

Specific gravity = 0.74

Specific gravity = 1.00

Drain plug   **Figure P4.24**

4.25. A water nozzle is directed vertically downward against a flat metal plate as shown in Figure P4.25. Find the force exerted on the plate by the water jet. Neglect friction.

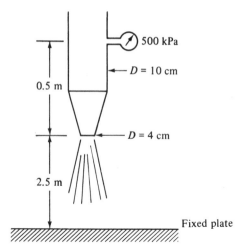

Figure P4.25

4.26. A jet of water 6 cm in diameter has a velocity of 30 fps. The jet of water strikes a stationary blade which diverts the water through an angle of 130°. Find the force of the jet on the curved blade. (See Figure P4.26.)

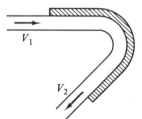

Figure P4.26

4.27. Water flows over the dam as shown in Figure P4.27. Upstream the flow has an elevation of 12 m and an average speed of 0.1 m/s; downstream the water elevation is 1.0 m. What is the horizontal force on the 10-m-wide dam? Assume frictionless flow with water density = 1000 kg/m³.

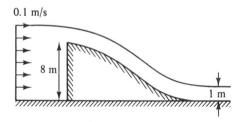

Figure P4.27

4.28. A water siphon consisting of 2-in-diameter pipe is arranged as shown in Figure P4.28. Assuming frictionless flow, find the rate of discharge in cfs and the pressure at point *A*.

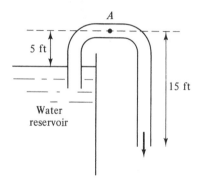

Figure P4.28

4.29. Water flows through a pipe with a U tube (trap) as shown in Figure P4.29. The flow rate is 0.1 cfs. The pressure at inlet is 70 psia. Calculate the water force on the bend. Neglect losses.

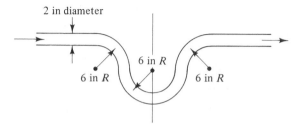

Figure P4.29

4.30. Suppose that the U tube of Problem 4.29 is placed in a vertical position as shown in Figure P4.30. Determine the fluid force on the bend.

Figure P4.30

4.31. Water is drawn from a reservoir through a vertical 300-mm-diameter pipe by
a pump which discharges it into a horizontal 150-mm-diameter pipe (Figure
P4.31). The inlet pressure gage reads a gage pressure of -20 kPa, while the
discharge pressure gage shows an absolute pressure of 150 kPa for a discharge
rate of 0.05 m³/s. Calculate the kW input to the pump.

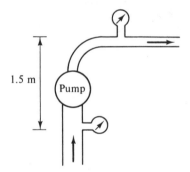

Figure P4.31

4.32. A circular open tank is filled with water to a depth of 7 ft. The diameter of
the tank is 3 ft. An opening at the bottom of the tank has a diameter of $\frac{1}{2}$ in.
How much water will flow out of the hole during an interval of 30 minutes?

4.33. A circular tank has two openings in its vertical wall as indicated in Figure
P4.33. Find the time it takes to completely empty the tank if $H = 0.5$ m,
$D = 0.2$ m, and the openings are 10 mm in diameter.

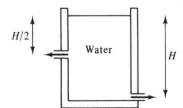

Figure P4.33

4.34. In problem 4.33, the left-side opening is closed when the water level reaches 75% of *H*. Find the total time it takes to empty the tank completely.

4.35. In order to double the flow of water through the tube and nozzle in Figure P4.35, a water pump is to be installed in the discharge line. Determine the pump horsepower required. Does the horsepower requirement depend on the location of the pump in the discharge line? Neglect friction and losses.

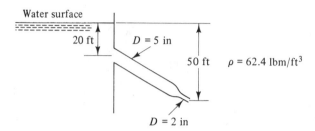

Figure P4.35

4.36. Two firemen are holding the nozzle 1 m off the ground as indicated in Figure P4.36. The nozzle has an outlet diameter of 5 cm and is connected to a 10-cm-diameter fire hose. The pressure in the hose is 1000 kPa. What is the maximum horizontal distance from the building the firemen can stand and still get the water on the roof 30 m high?

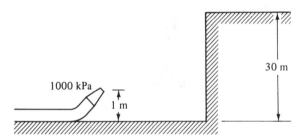

Figure P4.36

4.37. A large tank is attached to the cart as shown in Figure P4.37. A water jet issues from the tank through a 3-in-diameter hole with a velocity of 20 fps. The jet is turned through an angle of 45° by a trough on the cart. What is the tension in the cable holding the cart?

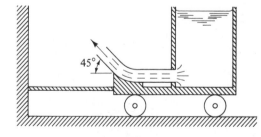

Figure P4.37

4.38. A jet of water is discharged from a nozzle as shown in Figure P4.38, with the jet inclined at $\pi/6$ rad to the horizontal. How high will the jet go? Neglect friction.

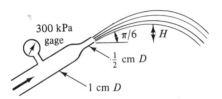

Figure P4.38

4.39. Neglecting losses, determine the pressures at 2 and 3 and the x and y components of force needed to hold the fitting in place in Figure P4.39. Since the plane of the fitting is horizontal, neglect the weight of water in the bend.

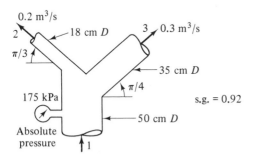

Figure P4.39

4.40. Water in a reservoir is under a pressure of 20 psia at the free surface (Figure P4.40). What is the pump power required? Assume frictionless flow.

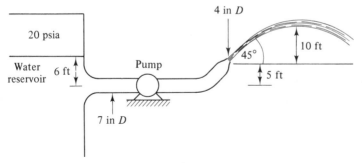

Figure P4.40

4.41. A horizontal jet of water 10 cm² in area moving at 20 m/s is divided in half by a splitter on a stationary flat plate inclined $\pi/4$ rad from the jet direction (Figure P4.41). What is the magnitude and direction of the resultant force on the plate?

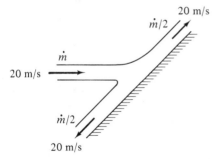

Figure P4.41

4.42. Find the flow rate of water through the system shown in Figure P4.42 for the turbine to produce 30 hp. Neglect friction in your calculations.

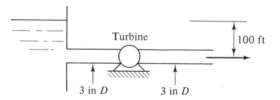

Turbine

100 ft

3 in D 3 in D **Figure P4.42**

4.43. A jet of water is shot vertically upward as shown in Figure P4.43. The jet impinges on a flat plate of mass M. The diameter of the jet at the nozzle exhaust is 5 cm. Neglecting losses,
  (a) Find the jet velocity at the nozzle exit.
  (b) Find the jet velocity at height h.
  (c) Compute the mass M that can be held in place at elevation h.

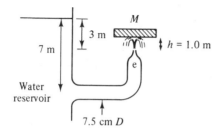

7 m

3 m

M

$h = 1.0$ m

e

Water
reservoir

7.5 cm D **Figure P4.43**

4.44. Find the x and y components of the force necessary to hold the elbow AB in place in Figure P4.44. The gage pressure at A is 60 psi, and the velocity at A is 10 fps. Neglect friction and neglect gravitational forces. The pipe has an inner diameter of 2 in and is carrying water of density 1.94 slugs/ft^3.

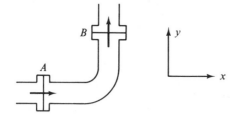

B

y

A

x

Figure P4.44

4.45. In Figure P4.45 exhaust gases are expanded isentropically in a rocket nozzle from a chamber pressure and temperature of 3000 kPa and 2000 K, respectively, to an exhaust pressure of 30 kPa. Assume that the exhaust gases behave as a perfect gas with constant specific heats so that for this isentropic process in the nozzle, $p/\rho^\gamma =$ constant. Assume that the gases have $R = 0.40$ kJ/kg $\cdot$ K with γ equal to $c_p/c_v = 1.3$. Determine the rocket thrust.

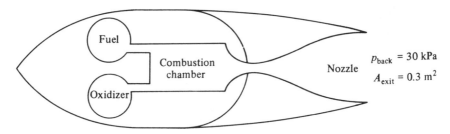

Figure P4.45

4.46. An open container of water slides down a plane which inclines 30° with the horizontal. Determine the angle of inclination of the water level if
(a) The container velocity is constant.
(b) The container slides without frictional resistance.
(c) The container slides with a frictional resistance so that its acceleration equals one-half of the accelerative component of gravity in the direction of container motion.

4.47. A U tube filled with water is used to measure the acceleration of a vehicle (Figure P4.47). Devise a scale which will indicate values of acceleration directly.

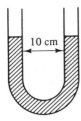

Figure P4.47

4.48. A 55-gallon drum of oil is rotated about a vertical center axis at a constant speed of 300 rpm. Determine the forces due to angular acceleration at radial distances of 1, 6, and 12 in from the axis of rotation. The density of the oil is 1.8 slugs/ft^3.

4.49. A cylindrical container with height $H = 0.6$ m and radius $R = 0.1$ m is filled with water to a height $h_1 = 0.4$ m. It is then rotated about its vertical axis with constant angular speed. Determine the angular speed required for the surface of the liquid to reach the upper rim of the container.

4.50. Determine the slope of the water surface in a small tank mounted on a turntable 3 ft from the axis of turntable rotation. The rotative speed of the turntable is 20 and 300 rpm, respectively.

5

DIMENSIONAL ANALYSIS AND SIMILITUDE

5.1 INTRODUCTION

In the preceding chapter we obtained solutions to flow problems by analytic means—that is, we started from a mathematical description of the basic laws governing fluid flow and solved these equations by suitable assumptions and simplifications, to obtain an algebraic description (solution) of the given problem. The solution was in the form of a relationship between the various quantities describing the flow. Another way of obtaining this information is by means of actual measurements; that is, by using suitable instrumentation we can measure the flow quantities of interest. Often, because some flows are too complicated to solve analytically, experimental evaluation is the only means available to us. Such experimental studies can be carried out on the full-scale machinery or flow setup, but more often the studies are conducted on smaller models in order to reduce the cost of the evaluation. For model investigations, certain conditions of similarity must be met between actual flow and model flow. In either case, a technique called dimensional analysis becomes useful and at times necessary in the conduct of the experimental study.

5.2 DIMENSIONAL ANALYSIS

Dimensional analysis is a procedure that allows us to formulate a functional relationship between a set of nondimensional groups composed of physical variables, the groups numbering less than the variables. As a consequence, considerably fewer experimental measurements are required to establish a mathematical relationship between the physical variables of the flow over the desired range. To carry out the process of dimensional analysis the physical variables describing the phenomenon must be known or assumed.

The basis of dimensional analysis is the fact that an equation expressing a relationship between flow quantities must be dimensionally homogeneous. Thus the terms on both sides of an equality must possess the same dimensions, for dissimilar quantities cannot be added or subtracted. For example, recall from the study of mechanics that the distance $s - s_0$ traveled by a body falling in a vacuum in time t after being dropped with initial velocity V_0 is given by

$$s - s_0 = \tfrac{1}{2}gt^2 + V_0 t$$

Here the terms $\tfrac{1}{2}gt^2$ and $V_0 t$ must have the same dimension of length to correspond to the dimension of $s - s_0$ on the left-hand side of the equality. The term $V_0 t$ is composed of dimensions of velocity (length/time) and time; hence it has the dimension of length. Similarly, the $\tfrac{1}{2}gt^2$ term is made up of the product of acceleration (length/time^2) and time^2; hence it too has the dimension of length.

The primary dimensions consist of mass (M), length (L), time (T), and temperature (θ) or force (F), length (L), time (T), and temperature (θ). Remember that force and mass are not independent; they are related by the equation $F = ma$, and so force and mass cannot both be primary dimensions in a system. Generally, in fluid mechanics, we are not concerned with temperature as a primary dimension; physical quantities describing fluid flow problems in this text can be expressed in terms of the fundamental dimensions of M, L, and T or F, L, and T. Thus with length as the primary dimension of space, units of area and volume can be obtained as secondary (derived) units. With length represented by the symbol L, area is given by L^2 and volume by L^3. With time as the other primary dimension, linear velocity, for example, is given by L/T. Similarly, linear acceleration becomes L/T^2. In the M, L, T system, force has dimensions of ML/T^2; in the F, L, T system, mass has dimensions of FT^2/L.

Table 5.1 presents the dimensions of various physical quantities useful in fluid mechanics based on the M, L, T system and on the F, L, T system.

To illustrate how the process of dimensional analysis is carried out, consider the problem of determining the frictional force exerted by a flowing fluid on a smooth pipe. Assume that the significant physical quantities which are to be functionally related by dimensional analysis are the fluid density ρ, the average fluid velocity V, the interior surface area of pipe S, the dynamic viscosity μ, the internal diameter of pipe D, and the frictional force F exerted on the pipe by the fluid. Expressed in a formal mathematical manner, this statement takes the form

$$F = F(\rho, V, S, \mu, D) \tag{5.1}$$

TABLE 5.1   Dimensions of Various Physical Quantities

Quantity	Symbol	Dimensions (M, L, T)	Dimensions (F, L, T)
Length	l	L	L
Time	t	T	T
Mass	m	M	FT^2/L
Force	F	ML/T^2	F
Velocity (linear)	V	L/T	L/T
Acceleration (linear)	a	L/T^2	L/T^2
Area	A	L^2	L^2
Volume	$\rlap{—}V$	L^3	L^3
Pressure	p	M/LT^2	F/L^2
Density	ρ	M/L^3	FT^2/L^4
Acceleration due to gravity	g	L/T^2	L/T^2
Dynamic viscosity	μ	M/LT	FT/L^2
Kinematic viscosity	v	L^2/T	L^2/T
Surface tension	σ	M/T^2	F/L
Angle (radians)	θ	No dimensions	No dimensions
Velocity (angular)	ω	$1/T$	$1/T$
Acceleration (angular)	α	$1/T^2$	$1/T^2$
Torque or moment	T_0	ML^2/T^2	FL
Work, energy	W	ML^2/T^2	FL
Momentum (linear)	mV	ML/T	FT
Volume flow rate	Q	L^3/T	L^3/T
Mass flow rate	$\dot{m}$	M/T	FT/L
Power	$\mathcal{P}$	ML^2/T^3	FL/T
Moment of inertia	I	ML^2	FLT^2
Momentum (angular)	$I\omega$	ML^2/T	FLT

The relationship (5.1) merely states that the physical quantity F, the frictional force exerted by the fluid, depends in some unspecified way on the set of physical variables, ρ, V, S, μ, and D. Generally it is possible to expand this expression into a power series (which may contain a single term, multiple terms, or an infinite number of terms) in the form

$$F = c_1\rho^{\alpha_1}V^{\alpha_2}S^{\alpha_3}\mu^{\alpha_4}D^{\alpha_5} + c_2\rho^{\beta_1}V^{\beta_2}S^{\beta_3}\mu^{\beta_4}D^{\beta_5} + \cdots \qquad (5.2)$$

where the α, β, etc., are numerical exponents that depend on the nature of the function F and the c's are numerical coefficients which do not contribute to the dimensions of the expression, that is, are dimensionless. As indicated, the terms on the right-hand side of Equation (5.2) are to be added; therefore every term must have the same dimensions—in this case, the dimension of force. The dimensions of the variables in terms of M, L, and T are as follows (see Table 5.1):

$$F: \qquad ML/T^2$$
$$\rho: \qquad M/L^3$$
$$V: \qquad L/T$$

$$S: \quad L^2$$
$$\mu: \quad M/LT$$
$$D: \quad L$$

Next, substitute the physical dimensions into Equation (5.2) to obtain

$$\frac{ML}{T^2} = \left(\frac{M}{L^3}\right)^{\alpha_1}\left(\frac{L}{T}\right)^{\alpha_2}(L^2)^{\alpha_3}\left(\frac{M}{LT}\right)^{\alpha_4}L^{\alpha_5} + \left(\frac{M}{L^3}\right)^{\beta_1}\left(\frac{L}{T}\right)^{\beta_2}(L^2)^{\beta_3}\left(\frac{M}{LT}\right)^{\beta_4}L^{\beta_5} + \cdots$$

For each term, collect like exponents for the primary dimensions M, L, and T to obtain three algebraic equations with five unknowns. For example, for the first term, we obtain

$$\text{for } M: \qquad 1 = \alpha_1 + \alpha_4$$
$$\text{for } L: \qquad 1 = -3\alpha_1 + \alpha_2 + 2\alpha_3 - \alpha_4 + \alpha_5$$
$$\text{for } T: \qquad -2 = -\alpha_2 - \alpha_4$$

while for the second term we obtain

$$\text{for } M: \qquad 1 = \beta_1 + \beta_4$$
$$\text{for } L: \qquad 1 = -3\beta_1 + \beta_2 + 2\beta_3 - \beta_4 + \beta_5$$
$$\text{for } T: \qquad -2 = -\beta_2 - \beta_4$$

and so on for all the other terms. Solving these equations in terms of two unknowns, say, α_4, α_5 and β_4, β_5, we obtain

$$\alpha_1 = 1 - \alpha_4$$
$$\alpha_2 = 2 - \alpha_4$$
$$\alpha_3 = 1 - \frac{\alpha_4}{2} - \frac{\alpha_5}{2}$$

and

$$\beta_1 = 1 - \beta_4$$
$$\beta_2 = 2 - \beta_4$$
$$\beta_3 = 1 - \frac{\beta_4}{2} - \frac{\beta_5}{2}$$

Returning to Equation (5.2), if F, L, and T were taken as primary dimensions, we would have for the variables:

$$F: \quad F$$
$$P: \quad FT^2/L^4$$
$$V: \quad L/T$$
$$S: \quad L^2$$
$$\mu: \quad FT/L^2$$
$$D: \quad L$$

Substituting into Equation (5.2):

$$F = \left(\frac{FT^2}{L^4}\right)^{\alpha_1}\left(\frac{L}{T}\right)^{\alpha_2}(L^2)^{\alpha_3}\left(\frac{FT}{L^2}\right)^{\alpha_4}L^{\alpha_5}$$

there results

for F: $1 = \alpha_1 + \alpha_4$

for L: $0 = -4\alpha_1 + \alpha_2 + 2\alpha_3 - 2\alpha_4 + \alpha_5$

for T: $0 = 2\alpha_1 - \alpha_2 + \alpha_4$

Solving, $\alpha_1 = 1 - \alpha_4$, $\alpha_2 = 2 - \alpha_4$, $\alpha_3 = 1 - \alpha_4/2 - \alpha_5/2$; and, in similar fashion, $\beta_1 = 1 - \beta_4$, $\beta_2 = 2 - \beta_4$, $\beta_3 = 1 - \beta_4/2 - \beta_5/2$. Thus, the result agrees with the same analysis in the M, L, T dimension system.

Substituting these values into Equation (5.2), it follows that

$$F = \rho V^2 S\left[c_1\left(\frac{\mu}{\rho DV}\right)^{\alpha_4}\left(\frac{D^2}{S}\right)^{(\alpha_4+\alpha_5)/2} + c_2\left(\frac{\mu}{\rho DV}\right)^{\beta_4}\left(\frac{D^2}{S}\right)^{(\beta_4+\beta_5)/2} + \cdots\right]$$

The series on the right-hand side is merely an unspecified function of $\mu/\rho DV$ and D^2/S (since we do not know the values of the constants c and the exponents) and it can be written symbolically as $f(\mu/\rho DV, D^2/S)$. Hence

$$F = \rho V^2 Sf\left(\frac{\mu}{\rho DV}, \frac{D^2}{S}\right) \qquad (5.3)$$

From the above derivation it becomes clear that it is necessary to consider only one term of the series in Equation (5.2) to obtain the functional relation between the physical quantities. Further, the functional relation between F, ρ, V, S, μ, and D is equivalent to a relation between the three nondimensional products, which is expressed symbolically as

$$\frac{F}{\frac{1}{2}\rho V^2 S} = f\left(\frac{\rho VD}{\mu}, \frac{D^2}{S}\right) \qquad (5.4)$$

where the product $F/\rho V^2 S$ has been multiplied by 2 to conform to usual practice. Therefore, in an experimental investigation of the relationship between force F and the other physical parameters, one has to find only the relation between the nondimensional parameters

$$\frac{F}{\frac{1}{2}\rho V^2 S} \qquad \text{(called the force coefficient)}$$

$$\frac{\rho VD}{\mu} \qquad \text{(called the Reynolds number)}$$

and

$$\frac{D^2}{S} \qquad \text{(a ratio of cross-sectional area and surface area)}$$

For a circular pipe, $S = \pi Dl$, where l is the length of pipe, D^2/S becomes

$D/\pi l$. Thus for a circular pipe, the relation between the nondimensional parameters is

$$\frac{F}{\frac{1}{2}\rho V^2 S} = f\left(\frac{\rho VD}{\mu}, \frac{D}{l}\right) \tag{5.5}$$

When the inner surface of the pipe is rough instead of smooth, as assumed above, we must add at least one additional physical parameter to describe the roughness of the pipe; this parameter is usually designated ϵ, having the dimension of length. Thus, carrying out the dimensional analysis with seven significant quantities instead of six as above, we obtain an additional nondimensional ratio, namely, ϵ/D. Hence the nondimensional relationship now reads

$$\frac{F}{\frac{1}{2}\rho V^2 S} = f\left(\frac{\rho VD}{\mu}, \frac{D^2}{S}, \frac{\epsilon}{D}\right) \tag{5.6}$$

The significance of using nondimensional parameters in lieu of dimensional ones lies in the reduction in the number of parameters required to describe the phenomenon. Here we have shown that the total number of parameters required for description was reduced by 3. The number 3 arose from the fact that we utilized three primary dimensions (M, L, T). The choice of appropriate physical parameters in a given flow situation is based on experience or good guessing; or it can be taken from the differential equation describing the given flow phenomenon, without solving the differential equation.

The number of dimensionless parameters will, in general, be equal to the total number of variables n (includes the physical quantity of interest) minus the number of dimensions d. In the example presented, $n - d = 6 - 3 = 3$ dimensionless parameters. (In some cases, the dimensional equations are not independent. This would have the effect of reducing the number of independent dimensions d.)

EXAMPLE 5.1

As a first example of the application of dimensional analysis, consider a problem studied in dynamics: the motion of a mass suspended from a spring executing simple harmonic motion. It is desired to find an expression for the period of oscillation.

Solution   Assume that the significant physical parameters describing the oscillations are: P, the period of oscillation; a, the amplitude of oscillation; m, the mass; and k, the stiffness of the spring. The spring stiffness is defined as the constant of proportionality relating the restitutive force of the spring to its displacement, with units of force per unit distance.

Expressing the above dependency in a formal manner, we write

$$P = P(m, k, a)$$

The dimensions of the four variables in the F, L, T system are:

$$P: \quad T$$
$$m: \quad FT^2/L$$
$$k: \quad F/L$$
$$a: \quad L$$

Writing the functional relation as a single term of a series, we obtain

$$P = cm^{\alpha_1}k^{\alpha_2}a^{\alpha_3}$$

Substitute the physical dimensions into the equation to obtain

$$T = \left(\frac{FT^2}{L}\right)^{\alpha_1}\left(\frac{F}{L}\right)^{\alpha_2}L^{\alpha_3}$$

Since this equation must be homogeneous in the primary dimensions F, L, T, we obtain

$$\text{for } F: \quad 0 = \quad \alpha_1 + \alpha_2$$
$$\text{for } L: \quad 0 = -\alpha_1 - \alpha_2 + \alpha_3$$
$$\text{for } T: \quad 1 = \quad 2\alpha_1$$

Solving the three equations, we obtain

$$\alpha_1 = \tfrac{1}{2}$$
$$\alpha_2 = -\tfrac{1}{2}$$
$$\alpha_3 = 0$$

Therefore,

$$P = c\sqrt{\frac{m}{k}}$$

Thus dimensional analysis shows that the period of the oscillating mass does not depend on amplitude but is proportional to the square root of the ratio m/k. Recall from dynamics that the solution of the equation of motion of a simple mass spring system gave the period as

$$P = 2\pi\sqrt{\frac{m}{k}}$$

Thus dimensional analysis has resulted in the correct form of the expression in this case but cannot supply the value of the constant. The value of the constant can be found by a single experiment in which we measure the period (P) for a spring with known stiffness k and supporting a known mass m. On the other hand, without the result of dimensional analysis, we would have to vary the values of m, k, and a. This would result in a large number of tests as well as additional work in correlating the data obtained and would have required more time and effort than a test guided by dimensional analysis.

In the above example, we chose to use the F, L, T system; if we had used the M, L, T system, we would have obtained exactly the same result.

EXAMPLE 5.2

Consider the problem of steady flow of water in an open channel (i.e., exposed to the atmosphere) of constant cross section and constant slope (Figure 5.1). Determine the functional relationship between appropriate nondimensional parameters.

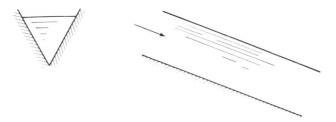

Figure 5.1. Flow in an open channel.

Solution   In order to establish a set of significant flow variables for carrying out the dimensional analysis, it is reasonable to consider that two effects are involved for steady flow having a constant depth: the gravitational component causing flow and the flow-resisting force. The gravitational component will be a function of the density of the fluid ρ, the cross-sectional area A, the acceleration due to gravity g, the length of the channel l, and the slope of the channel s. The force resisting the flow will be assumed to be a function of the flow velocity V and the wetted surface area S. Expressed formally, this statement becomes

$$V = V(\rho, g, s, S, A, l)$$

The single-term expression of the series expansion of the preceding statement is

$$V = c\rho^{\alpha_1} g^{\alpha_2} s^{\alpha_3} S^{\alpha_4} A^{\alpha_5} l^{\alpha_6}$$

Selecting the M, L, T system, we obtain

$$\frac{L}{T} = \left(\frac{M}{L^3}\right)^{\alpha_1} \left(\frac{L}{T^2}\right)^{\alpha_2} (0)^{\alpha_3} (L^2)^{\alpha_4} (L^2)^{\alpha_5} (L)^{\alpha_6}$$

From the homogeneity in the primary dimensions, we get

for the exponents of M: $0 = \alpha_1$

for the exponents of L: $1 = -3\alpha_1 + \alpha_2 + 2\alpha_4 + 2\alpha_5 + \alpha_6$

for the exponents of T: $-1 = -2\alpha_2$

$$\alpha_1 = 0$$

$$\alpha_2 = \tfrac{1}{2}$$

$$\alpha_4 = \tfrac{1}{4} - \alpha_5 - \tfrac{1}{2}\alpha_6$$

Therefore,

$$V = c \sqrt{g} \, S^{1/4} s^{\alpha_3} \left(\frac{A}{S}\right)^{\alpha_5} \left(\frac{l}{\sqrt{S}}\right)^{\alpha_6} = c \sqrt{g} \, A^{1/4} s^{\alpha_3} \left(\frac{A}{S}\right)^{\alpha_5 - (1/4)} \left(\frac{l}{\sqrt{S}}\right)^{\alpha_6}$$

which can be rewritten in nondimensional form as

$$\frac{V^2}{g \sqrt{A}} = f\left(s, \frac{A}{S}, \frac{l}{\sqrt{S}}\right)$$

Thus the results of dimensional analysis show that the velocity in an open channel of constant cross section and depth is a function of slope, area ratio, length ratio, acceleration due to gravity, and cross-sectional area, but it is independent of density of the fluid.

The nondimensional parameter V^2/gL is called the Froude number. Dimensional analysis has shown that the Froude number for the open-channel flow under consideration is functionally dependent on the channel slope, area ratio, and length ratio.

In order to obtain the functional form of the relationship between the dimensionless ratios given above, we need to conduct a series of tests on a channel of one cross section only, over a range of slopes and wetted surfaces (channel lengths). The test results obtained can then be used for different values of the cross section, slope, and length; again a considerable saving in experimental work results. The results of actual tests show that the Froude number is a function of only two nondimensional parameters, in that the area and length ratios can be combined to form the single parameter $(A/S \cdot l^2/S)$, where S can be expressed as $P \cdot l$ with P the wetted perimeter of the cross section. Thus we obtain

$$\frac{V^2}{g \sqrt{A}} = f\left(s, \frac{A}{P^2}\right)$$ ∎

5.3 CORRELATION OF EXPERIMENTAL DATA

One of the uses of dimensional analysis is in correlating test results in terms of nondimensional parameters. This application will be discussed here, and two examples will be presented to illustrate how empirical relations are obtained.

First, consider the problem of flow of water through a diffuser consisting of a diverging circular pipe (Figure 5.2). A diffuser is a device used to slow down a flow and bring about an increase of pressure. By use of suitable

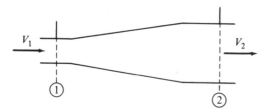

Figure 5.2. Flow through a diffuser.

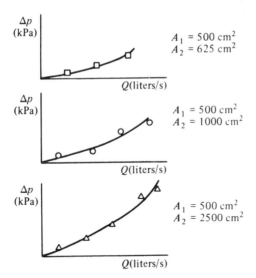

Figure 5.3.   Results of tests on a diffuser.

instrumentation, measurements are made of pressure changes (Δp) across the length of the diffuser over a range of volume flow rates (Q) for one inlet area (A_1) and three different outlet areas (A_2). The results of these tests are shown in Figure 5.3. Based on the results of these measurements, it is desired to predict the pressure changes for a different inlet area over a range of flow rates and outlet areas. By replotting the test results in terms of nondimensional parameters obtainable from dimensional analysis, it is possible to establish an empirical relation for pressure drop as a function of flow and geometric factors of the diffuser. For this purpose, assume that p, ρ, Q, A_1, A_2 are the significant physical parameters. Thus

$$\Delta p = f(Q, A_1, A_2, \rho)$$

and

$$\Delta p = cQ^{\alpha_1}A_1^{\alpha_2}A_2^{\alpha_3}\rho^{\alpha_4}$$

Substitute the primary physical dimensions into the equation to obtain

$$\frac{M}{LT^2} = \left(\frac{L^3}{T}\right)^{\alpha_1}(L^2)^{\alpha_2}(L^2)^{\alpha_3}\left(\frac{M}{L^3}\right)^{\alpha_4}$$

Dimensional homogeneity in the primary dimensions requires that

$$\text{for } M: \qquad 1 = \alpha_4$$
$$\text{for } L: \qquad -1 = 3\alpha_1 + 2\alpha_2 + 2\alpha_3 - 3\alpha_4$$
$$\text{for } T: \qquad -2 = -\alpha_1$$

from which we obtain

$$\alpha_1 = 2, \qquad \alpha_2 = -2 - \alpha_3, \qquad \text{and} \qquad \alpha_4 = 1$$

so that

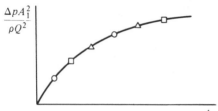

$$\frac{\Delta p A_1^2}{\rho Q^2}$$

$$\frac{A_2}{A_1}$$ **Figure 5.4.**   Nondimensional plot of diffuser test results given in Figure 5.2.

$$\Delta p = cQ^2 A_1^{-2-\alpha_3} A_2^{\alpha_3} \rho$$

or

$$\frac{\Delta p\, A_1^2}{\rho Q^2} = f\left(\frac{A_2}{A_1}\right) \tag{5.7}$$

From dimensional analysis there results a functional relationship between the two nondimensional parameters indicated, where $f(A_2/A_1)$ is an unspecified function of A_2/A_1. Next plot the test results shown in Figure 5.3 in terms of the parameters suggested by Equation (5.7). Thus we have shown that the use of nondimensional parameters allowed us to reduce the results of three series of tests on diffuser flow as shown in Figure 5.3 into a single curve (Figure 5.4). It can also be seen that dimensional analysis prior to actual testing would have reduced the amount of data to be taken. Furthermore, the curve of Figure 5.4 allows the prediction of performance for diffuser sizes and test conditions not previously tested. For example, consider the following case: It is desired to determine the water flow rate through a diffuser with inlet area 0.15 m² and outlet area 0.225 m² such that the pressure change does not exceed 2.3 kPa. From experimental data plotted as shown in Figure 5.4, we might obtain, for example, for $A_2/A_1 = 0.225/0.15 = 1.5$, the quantity $\Delta p\, A_1^2/\rho Q^2 = 0.27$. Hence

$$Q^2 = \frac{A_1^2 \Delta p}{0.27\rho} = \frac{(0.15^2\ \text{m}^2)\Delta p}{(0.27)(1000\ \text{kg/m}^3)}$$

For $\Delta p = 2.3 \times 10^3$ kg · m/s², $Q^2 = 0.1917$ m⁶/s². Therefore, $Q = 0.4378$ m³/s or less.

As a second illustration, consider the problem of the propulsion of small bacteria by use of a tail in the shape of a coiled spring, as shown in Figure 5.5. It has been speculated that the bacterium propels itself by rotating its spiral-like flagellum (tail). Direct observations of the flagellum of living, motile bacteria are as of this date not experimentally possible, since they lie below the limit of resolution of the optical microscope. It is therefore desired to support the speculation by observation on a larger-size model of the

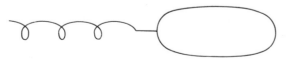

Flagellum (tail)   Bacterial body   **Figure 5.5.**   Schematic of bacterium.

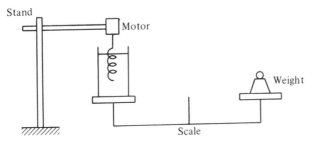

Stand

Motor

Weight

Figure 5.6. Test setup to model bacteria propulsion.

Scale

bacterium. For this purpose, a fine wire is attached to a small electric motor and submerged into a beaker of high-viscosity glycerine (Figure 5.6). The results of measurements of thrust force with this test setup over a range of speeds of rotation are shown in Figure 5.7. To obtain applicable nondimensional

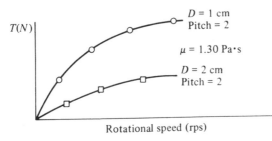

$T(N)$

$D = 1$ cm
Pitch = 2

$\mu = 1.30$ Pa·s

$D = 2$ cm
Pitch = 2

Rotational speed (rps)

Figure 5.7. Results of tests on model of bacterium.

parameters for correlation of test results, a dimensional analysis assuming the following significant parameters is performed:

$$T = T(N, \mu, \rho, D)$$

where T is the thrust force generated by the flagellum, N its rotational speed, D the diameter of the coiled flagellum, μ the viscosity of the liquid, and ρ the density of the liquid. Performing the analysis, we obtain the following functional relation:

$$\frac{T}{\rho N^2 D^4} = f\left(\frac{\rho N D^2}{\mu}\right) \tag{5.8}$$

The parameter $T/\rho N^2 D^4$ is called the thrust coefficient K_T, $\rho N D^2/\mu$ the rotational Reynolds number. We can now replot the test results of Figure 5.7 into nondimensional form, as shown in Figure 5.8. By calculating the Reynolds number for the real flagellum (based on its tiny diameter), we then use Figure 5.8 to find K_T and hence thrust.

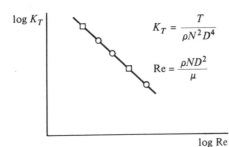

$\log K_T$

$$K_T = \frac{T}{\rho N^2 D^4}$$

$$\text{Re} = \frac{\rho N D^2}{\mu}$$

$\log$ Re

Figure 5.8. Nondimensional test results on bacteria model.

5.4 MODELING AND SIMILITUDE

In obtaining information by experimental means, it is often more convenient and less costly to conduct the actual measurements on a model rather than on the original. The model is usually smaller in size than the original (which is called the prototype) but can occasionally be larger. Examples of model testing involving fluid flow are those conducted on pumps, turbines, airplanes, ships, pipes, canals, and so on. In conducting such tests on models, certain conditions of similarity must be observed to ensure that the model test data are applicable to the prototype. Two kinds of conditions must be satisfied to ensure similarity between model and prototype: (1) geometric similarity of the physical boundaries and (2) dynamic similarity of the flow fields.

The most familiar kind of similarity is the geometric one, particularly relating to triangles. Figure 5.9 shows two similar triangles, ABC and $A_m B_m C_m$. From the laws of similarity as given in geometry, the following ratios between sides of the triangles must be satisfied:

$$\frac{a}{a_m} = \frac{b}{b_m} = \frac{c}{c_m} = \lambda$$

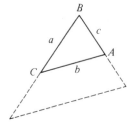 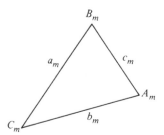

Figure 5.9. Similar triangles.

The ratio λ is called the scale ratio, that is, $a = \lambda a_m$, $b = \lambda b_m$, and $c = \lambda c_m$, which states that the three sides of the triangle with subscript m differ from the reference triangle (no subscript) by a constant scale ratio. In a similar fashion, we can find similar configurations for any geometric shape. For example, the equation for the circle with radius R shown in Figure 5.10 is $x^2 + y^2 = R^2$. The geometrically similar circle with scale ratio such that $R = \lambda R_m$ has the equation $x_m^2 + y_m^2 = R_m^2$. Hence every point on the circle with radius R_m has a corresponding point on the original circle with radius R such that $x = \lambda x_m$ and $y = \lambda y_m$. An example of such corresponding points is point A and point A_m as shown.

We have thus seen that geometric similarity between two configurations is achieved if a constant scale ratio exists between corresponding linear distances in the two configurations (e.g., $x/x_m = \lambda = $ constant). Note that this also implies that the ratio between two distances in the original must be equal to the ratio between the two corresponding distances in the model.

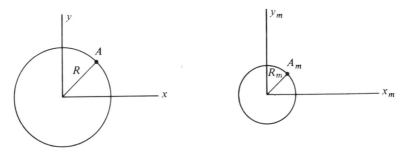

Figure 5.10.   Similar circles.

For example, the ratio of the distances x_1 and x_2 in the original is x_1/x_2, but from the preceding discussion we have $x_1 = \lambda x_{1m}$ and $x_2 = \lambda x_{2m}$. Therefore,

$$\frac{x_1}{x_2} = \frac{\lambda x_{1m}}{\lambda x_{2m}} = \frac{x_{1m}}{x_{2m}}$$

which confirms the preceding statement.

Thus far we have discussed geometric similarity of the physical boundaries of model and prototype. Examples of such physical boundaries are the external outline of a body (e.g., a circular cylinder) over which flow takes place or the internal walls and blades of a pump that passes the fluid from inlet to outlet. In order to achieve the geometric similarity of the boundaries, we merely need to construct the model in a constant scale ratio of the prototype. However, we must also ensure that the flow patterns in model and prototype are geometrically similar. To achieve these geometrically similar flow patterns, we need to have similarity in the velocities, accelerations, and forces of the two flow fields. We speak of dynamic similarity when the kinematic quantities that describe motion (displacement, velocity, and acceleration) and the forces causing the motion are related in linear ratios.

We have seen that the concept of similarity in geometric quantities exists. In an analogous way, there can exist similarity between other physical quantities, such as velocity, density, and pressure; that is, there exist ratios between the various quantities such that

$$\text{for velocity:} \quad V = \nu V_m$$

$$\text{for density:} \quad \rho = \delta p_m$$

$$\text{for pressure:} \quad p = \sigma p_m$$

In examining the dynamic similarity of flow fields, consider first the scaling of the kinematic quantities, displacement (s) and time (t):

$$s = \lambda s_m$$

$$t = \tau t_m$$

Since the velocity of a particle is the limit of the ratio of displacement change and time change, the scaling of velocities becomes

$$V = \frac{\lambda}{\tau}V_m = \nu V_m$$

where $\nu = \lambda/\tau$. Similarly, the scaling of acceleration is

$$a = \frac{\lambda}{\tau^2}a_m = \frac{\lambda^2}{\tau^2}\frac{1}{\lambda}a_m = \frac{\nu^2}{\lambda}a_m$$

Finally, let the scale ratio between forces be

$$F = \eta F_{\text{model}}$$

All flows must satisfy Newton's second law (the momentum equation), which states that a relation exists between external forces and produced acceleration. We therefore have for the prototype flow

$$F = ma$$

Substituting the scaling relations given above into this equation, we get

$$\eta F_{\text{model}} = m\frac{\nu^2}{\lambda}a_m$$

To achieve scaling of fluid masses, we express the fluid mass in terms of a fluid volume and density such that $m = \rho l^3$. Thus

$$F_{\text{model}} = \frac{\rho l^3}{\eta}\frac{\nu^2}{\lambda}a_m = \frac{\delta\rho_m\lambda^3 l_m^3}{\eta}\frac{\nu^2}{\lambda}a_m$$

where $\rho = \delta\rho_m$. Comparing this equation with the momentum equation for the model flow,

$$F_{\text{model}} = \rho_m l_m^3 a_m$$

we see that

$$\rho_m l_m^3\frac{\delta\lambda^3\nu^2}{\lambda\eta}a_m = \rho_m l_m^3 a_m$$

or

$$\frac{\delta\lambda^2\nu^2}{\eta} = 1$$

Since the various scale ratios are $\lambda = s/s_m = l/l_m$, $\nu = V/V_m$, $\eta = F/F_m$, and $\delta = \rho/\rho_m$, we obtain for dynamic similarity the statement that

$$\frac{(\rho/\rho_m)(l^2/l_m^2)(V^2/V_m^2)}{F/F_{\text{model}}} = 1$$

or

$$\frac{F}{\rho V^2 l^2} = \left(\frac{F}{\rho V^2 l^2}\right)_{\text{model}} \tag{5.9}$$

As expressed by Equation (5.9), dynamic similarity will be achieved when the value of the nondimensional parameter $F/\rho V^2 l^2$ is the same at geometrically similar locations. Note that the ratio of flow velocities of corresponding fluid particles is the same in model and prototype as well as the ratio of all forces acting on corresponding fluid particles. When dynamic similarity of two flow fields with geometrically similar boundaries is achieved, the flow fields exhibit geometrically similar flow patterns.

Next, let us examine, in further detail, the forces acting on the fluid particles; specifically, let us examine the forces F of Equation (5.9). As indicated in Section 4.2, the forces F may involve pressure forces, viscous (friction) forces, gravity forces, magnetic forces, electric forces, surface tension forces, and others. Restricting ourselves to the first three types of forces, Equation (5.9) becomes

$$\frac{F_{\text{pressure}}}{\rho V^2 l^2} + \frac{F_{\text{viscous}}}{\rho V^2 l^2} + \frac{F_{\text{gravity}}}{\rho V^2 l^2}$$
$$= \left(\frac{F_{\text{pressure}}}{\rho V^2 / l^2}\right)_{\text{model}} + \left(\frac{F_{\text{viscous}}}{\rho V^2 l^2}\right)_{\text{model}} + \left(\frac{F_{\text{gravity}}}{\rho V^2 l^2}\right)_{\text{model}} \tag{5.10}$$

Since each type of force can vary in a different manner in the flow field, dynamic similarity will certainly be attained if each term for the prototype is identical to the corresponding term for the model; for example, the second term on the left-hand side of Equation (5.10) should equal the second term on the right-hand side. Thus, for dynamic similarity, we require that

$$\frac{F_{\text{viscous}}}{\rho V^2 l^2} = \left(\frac{F_{\text{viscous}}}{\rho V^2 l^2}\right)_{\text{model}} \tag{5.11}$$

$$\frac{F_{\text{gravity}}}{\rho V^2 l^2} = \left(\frac{F_{\text{gravity}}}{\rho V^2 l^2}\right)_{\text{model}} \tag{5.12}$$

Using Equations (5.11) and (5.12), it follows from Equation (5.10) that

$$\frac{F_{\text{pressure}}}{\rho V^2 l^2} = \left(\frac{F_{\text{pressure}}}{\rho V^2 l^2}\right)_{\text{model}} \tag{5.13}$$

It should be noted here that for flow in which only pressure forces are involved, the geometry of the boundaries determines the geometry of the flow pattern. That is, complete dynamic similarity between two frictionless flow fields exists if the boundaries are geometrically similar. This result follows from the fact that Equations (5.9) and (5.13) are then identical.

In order to transform the modeling criteria obtained in Equations (5.11), (5.12), and (5.13) into more useful forms, let us express the various types of forces in terms of characteristic quantities associated with a given flow. First examine the case in which friction forces play a significant role. In Section 1.7 the frictional force exerted on the surface area S was shown to be equal to

$$F_{\text{friction}} = \tau S = \mu \frac{du}{dy} S$$

The term du/dy expresses the rate of change of flow velocity with respect to distance. Thus it can be represented as being proportional to velocity V per unit length, that is, V/l, where V is a characteristic velocity associated with the flow. The surface area S can be represented as being proportional to the square of a characteristic length l. The friction force F_{friction} can then be written as being proportional to $\mu l^2 V/l = \mu l V$. As an example of characteristic quantities, take the case of uniform flow about a circular cylinder, where a characteristic length would be the diameter of the cylinder and a characteristic velocity would be the uniform velocity at large distances from the cylinder. As shown earlier, dynamic similarity in the case in which friction forces are significant will be obtained when Equation (5.11) is satisfied. Therefore, substituting the expression obtained for F_{friction} into Equation (5.11),

$$\frac{\mu l V}{\rho V^2 l^2} = \left(\frac{\mu l V}{\rho V^2 l^2}\right)_{\text{model}}$$

or inverting,

$$\frac{\rho V l}{\mu} = \left(\frac{\rho V l}{\mu}\right)_{\text{model}} \tag{5.14}$$

The parameter $\rho V l/\mu$, called the Reynolds number, was previously obtained in Section 5.2 by using dimensional analysis. It represents the ratio of inertia forces to viscous (friction) forces and thus expresses the relative importance of inertia versus friction forces in a given flow situation. The term *inertia force* refers to the mass times acceleration term in Newton's second law. The inertia force can thus be represented as being proportional to $(\rho l^3 a)$, with l a characteristic length associated with the flow. We know from particle dynamics that the acceleration (a) can be expressed as the rate of change of velocity with time (dV/dt) or as the product $V(dV/dl)$ [since $V = dl/dt$ or $dt = dl/V$, therefore $dV/dt = V(dV/dl)$], where V is a characteristic velocity associated with the flow. As shown before, dV/dl can be represented as being proportional to the characteristic velocity per unit length, V/l. Thus acceleration is proportional to V^2/l; whence we obtain that the so-called inertia force is proportional to $\rho l^3 V^2/l$ or $\rho l^2 V^2$. The term $\rho l^2 V^2$ is the denominator appearing in Equation (5.11).

Next, consider the criterion for dynamic similarity of the pressure forces. Since the pressure force exerted on a fluid element can be considered to be proportional to the characteristic pressure difference Δp times l^2, where the characteristic pressure difference is taken as the difference between local pressure and a reference pressure—say, the pressure at large distance from a body in uniform flow—we obtain from Equation (5.13)

$$\frac{\Delta p l^2}{\rho V^2 l^2} = \frac{\Delta p}{\rho V^2} = \left(\frac{\Delta p}{\rho V^2}\right)_{\text{model}} \tag{5.15}$$

The nondimensional parameter $\Delta p/\frac{1}{2}\rho V^2$ is called the *pressure coefficient*.

Summarizing, we conclude that for two dynamically similar flows in which friction forces are important, their Reynolds numbers must be equal. Examples of systems in which friction forces play an important role are the flow of an incompressible fluid in a closed pipe and the drag on bodies (e.g., submarines, airplanes) moving through an incompressible fluid.

Finally, examine the flow case in which gravity forces play a significant role. Examples are the flow caused by the motion of a ship on the water's surface or the propagation of waves along the surface of the water. The gravity force on a body of mass m is equal to mg; it can therefore be taken to be proportional to $\rho l^3 g$. Using Equation (5.12), we obtain

$$\frac{\rho l^3 g}{\rho V^2 l^2} = \left(\frac{\rho l^3 g}{\rho V^2 l^2}\right)_{model}$$

or inverting,

$$\frac{V^2}{lg} = \left(\frac{V^2}{lg}\right)_{model} \tag{5.16}$$

In this case, dynamic similarity is achieved when the two parameters V^2/lg are equal. This nondimensional parameter is called the *Froude number*. For the case in which friction and gravity forces play a vital role, dynamic similarity is attained when both Reynolds and Froude numbers are equal.

In summary, we have seen that for proper modeling of flow systems, the model and prototype must be of the same shape although they can differ in size (geometric similarity) and they must be dynamically similar. The significance of certain nondimensional parameters, such as Reynolds and Froude numbers, in dynamic similarity has been shown.

EXAMPLE 5.3

Air at 1 atmosphere, 20°C flows with an average velocity of 10 m/s through a pipe having a diameter of 25 cm. A model of this flow is to be constructed using water as flow medium. What must the average water velocity be in a 6-cm pipe if the flow is dynamically similar to the prototype? If the pressure drop in the model is 200 kPa, find the corresponding pressure drop in the prototype.

Solution To achieve dynamic similarity in completely enclosed flows, the Reynolds numbers must be equal; that is, using Equation (5.14) and taking the pipe diameter as characteristic length,

$$\left(\frac{\rho_m V_m D_m}{\mu_m}\right)_{model} = \left(\frac{\rho V D}{\mu}\right)_{prototype}$$

Remember that the ratio μ/ρ is called kinematic viscosity ν. From Appendix A, the kinematic viscosities of water and air at 20°C are, respectively, 1.00×10^{-6} m^2/s and 1.51×10^{-5} m^2/s. Using these values, the average water velocity becomes

$$V_m = \frac{D}{D_m}\frac{\nu_m}{\nu}V$$

$$= \left(\frac{25}{6}\right)\left(\frac{1.00 \times 10^{-6}}{1.51 \times 10^{-5}}\right)(10\ \text{m/s})$$

$$= 2.76\ \text{m/s}$$

Thus to obtain dynamic similarity, the average water velocity must be 2.76 m/s.

To find the pressure drop in the prototype, use Equation (5.15):

$$\frac{\Delta p}{\rho V^2} = \left(\frac{\Delta p}{\rho V^2}\right)_{\text{model}}$$

Rewrite the equation in terms of Δp:

$$\Delta p = \frac{\rho V^2}{\rho_m V_m^2}\Delta p_m$$

$$= \left(\frac{1.204\ \text{kg/m}^3}{998.3\ \text{kg/m}^3}\right)\left(\frac{10\ \text{m/s}}{2.76\ \text{m/s}}\right)^2(200\ \text{kPa})$$

$$= 3.166\ \text{kPa}$$

Thus for a measured pressure in the model (using water as fluid), we can compute a corresponding pressure drop for the prototype (using air) amounting to 3.166 kPa.

■

EXAMPLE 5.4

Water discharges into the atmosphere from an orifice on the side of an open tank. A small model of the tank is to be built, also using water as flow medium. If the scale ratio of prototype to model is 10:1, what are the ratios of volume discharge and force exerted on the tanks at dynamically similar conditions?

Solution Since this flow involves a free surface, dynamic similarity will be achieved for equal Froude number. Thus from Equation (5.16),

$$\frac{V^2}{h} = \left(\frac{V^2}{gh}\right)_{\text{model}}$$

where V is the velocity of discharge and h is the depth of water above the orifice. Since the magnitude of the gravity is identical for model and prototype, the foregoing equation becomes

$$V^2 = \frac{h}{h_{\text{model}}}V_{\text{model}}^2 = \lambda V_m^2$$

Thus the velocities are related as follows:
$$V = V_m\sqrt{\lambda}$$
where the scale ratio $\lambda = h/h_m$.

The discharges are given by
$$Q = VA \qquad \text{and} \qquad Q_m = V_m A_m$$
where A is the orifice area. Since we also have complete geometric similarity, the orifice areas scale as λ^2 and we obtain
$$\frac{Q}{Q_m} = \frac{VA}{V_m A_m} = \sqrt{\lambda}\,\lambda^2 = \lambda^{2.5}$$
For a scale ratio of 10, this results in
$$\frac{Q}{Q_m} = 10^{2.5} = \underline{316}$$
The relationship between the forces exerted on the tanks is, using Equation (5.13),
$$\frac{F}{F_m} = \frac{\rho V^2 A}{\rho V_m^2 A_m} = \lambda\lambda^2 = \lambda^3$$
For a scale ratio of 10, we obtain
$$\frac{F}{F_m} = 10^3 = \underline{1000}$$
Thus in modeling that involves free liquid surfaces, such as occurs in spillways of dams, weirs, and harbor breakwaters, the discharge ratio varies as the $\frac{5}{2}$ power of the geometric scale ratio, while the force ratio varies as the cube of the geometric scale ratio. ■

5.5 NONDIMENSIONAL PRODUCTS BY INSPECTION

In Section 5.2 we presented an analytical method of finding nondimensional products. This method involved the solution of a set of linear algebraic equations for the unknown exponents associated with each physical variable. In this section we will show that it is also possible to find nondimensional products by less mathematical means, namely by inspection. In the inspection method, we arrange all assumed physical (dimensional) variables, by trial and error, in such a way that all dimensions cancel and thereby produce nondimensional variables. The number of independent nondimensional products obtainable is given by the difference between the number of primary physical dimensions and the number of physical variables assumed. Usually, we try to form "standard" nondimensional products (Reynolds number, Froude number, pressure coefficient, etc). Some of the more common dimensionless products used in fluid flow problems are listed in Table 5.2.

TABLE 5.2 Some Dimensionless Parameters Used in Fluid Flow

Name	Symbol	Expression
Pressure coefficient	C_p	$(p - p_\infty)/\frac{1}{2}\rho V^2$
Drag coefficient	C_D	$D/\frac{1}{2}\rho V^2 A$
Force coefficient	C_F	$F/\frac{1}{2}\rho V^2 S$
Friction coefficient	f	$4\tau w/\frac{1}{2}\rho V^2$
Thrust coefficient	K_T	$T/\rho\omega^2 D^4$
Power coefficient	C_P	$\mathscr{P}/\rho\omega^3 D^5$
Froude number	Fr	V^2/gL
Reynolds number	Re	$\rho VD/\mu$
Relative roughness		ϵ/D
Mach number	M	V/a^*
Weber number	We	$\rho V^2 L/\sigma$

* a is velocity of sound.

As an example of the method, let us reconsider the problem of flow through a diffuser examined in Section 5.3. There we assumed that the pressure drop is a function of density, areas, and flow rate; that is,

$$\Delta p = f(Q, A_1, A_2, \rho)$$

Next, place the appropriate primary physical dimensions underneath each assumed physical variable. We will use the M, L, T system.

$$\frac{M}{LT^2} \qquad \frac{L^3}{T}, L^2, L^2, \frac{M}{L^3}$$

Since the number of assumed physical variables is five, with three primary dimensions, we expect $5 - 3 = 2$ nondimensional products. As a first step in obtaining a nondimensional product, try the combination $\Delta p/\rho$, which has dimensions of

$$\frac{M}{LT^2}\frac{L^3}{M} \qquad \text{or} \qquad \frac{L^2}{T^2}$$

Next, add the variable Q to the combination:

$$\frac{\Delta p}{\rho Q^2}$$

where Q^2 has been selected to cancel the T^2 of $\Delta p/\rho$. We now have

$$\frac{M/LT^2}{(M/L^3)(L^6/T^2)} \qquad \text{or} \qquad \frac{1}{L^4}$$

Add A_1^2 to obtain the dimensionless parameter

$$\frac{\Delta p\, A_1^2}{\rho Q^2}$$

Of the five physical variables, only A_2 has not been utilized as yet; hence another nondimensional product is A_2/A_1. These results are identical to those of Section 5.2. In the next two examples, additional illustrations will be given of the inspection method for finding nondimensional parameters.

EXAMPLE 5.5

Air flowing past the blades of a turbomachine, such as a compressor or fan, generates noise. Assume that the rate of generation of noise energy $\mathscr{P}$ is a function of the rotative speed ω, the diameter of the rotating blades D, air density ρ, and speed of sound a. Determine the appropriate dimensionless parameters describing the generation of noise by the machine.

Solution   First list the physical variables and their primary dimensions:

$$\mathscr{P} = f(\omega, D, \rho, a)$$

In the F, L, T system, these variables have the dimensions:

$$\frac{FL}{T} \quad \frac{1}{T}, L, \frac{FT^2}{L^4}, \frac{L}{T}$$

The number of nondimensional parameters required to describe this phenomenon will be $5 - 3 = 2$. Start with $\mathscr{P}/\rho$, which has dimensions

$$\frac{FL/T}{FT^2/L^4} \quad \text{or} \quad \frac{L^5}{T^3}$$

The variable containing time as primary dimension is ω; hence take $\mathscr{P}/\rho\omega^3$ with dimensions now of L^5. The variable D contains L as primary dimension, so one dimensionless parameter is $\mathscr{P}/\rho\omega^3D^5$. The remaining variable not used is a, with dimensions L/T. But L/T can be obtained from the product $D\omega$. Thus our second dimensionless parameter is $D\omega/a$, a rotational Mach number. Thus two dimensionless parameters for this phenomenon are the power coefficient $\mathscr{P}/\rho\omega^3D^5$ and the rotational Mach number $D\omega/a$. Other pairs of dimensionless parameters can be formed by combination of these two or by taking different combinations of physical variables from the start. ∎

EXAMPLE 5.6

A body falls freely in a vacuum. The distance traveled Δs by the body is assumed to be a function of the elapsed time t, the initial velocity V_0, and the acceleration due to gravity g. Find a set of applicable nondimensional parameters.

Solution   Here the functional relationship is, with dimensions:

$$\Delta s = f(g, \quad t, \quad V_0)$$

$$L \quad \frac{L}{T^2}, T, \frac{L}{T}$$

Note that there are only two primary physical dimensions involved here, L

and T. The dimension L can also be obtained from the product $tV_0(T \cdot L/T)$; hence one nondimensional product is

$$\frac{\Delta s}{tV_0}$$

The dimensions of the remaining variable g (L/T^2) are obtainable from the product

$$\frac{V_0}{t}\left(\frac{L}{T} \cdot \frac{1}{T}\right)$$

Hence another nondimensional product is

$$\frac{gt}{V_0}$$

In lieu of gt/V_0 we could also use $gt^2/\Delta s$, which when multiplied by the first product $\Delta s/tV_0$ gives

$$\frac{gt^2}{\Delta s} \cdot \frac{\Delta s}{tV_0} = \frac{gt}{V_0}$$

Dimensional analysis thus gives a possible relationship in the form

$$\frac{\Delta s}{V_0 t} = f\left(\frac{gt}{V_0}\right)$$

The analytic solution for this problem was given in Section 5.2 as

$$\Delta s = \tfrac{1}{2}gt^2 + V_0 t$$

or, dividing by $V_0 t$:

$$\frac{\Delta s}{V_0 t} = \frac{1}{2}\frac{gt}{V_0} + 1$$

Thus the function $f(gt/V_0)$ is shown to be

$$\frac{1}{2}\frac{gt}{V_0} + 1$$

This functional relation could also be obtained from experimental data of the phenomenon.

If in addition to the physical variables specified we had also included the mass of the body, we would have seen that the "mass" dimension is contained in no other physical variable and hence should not have been included. ■

5.6 FURTHER COMMENTS ON MODELING AND DIMENSIONAL ANALYSIS

Only one nondimensional parameter was required to achieve dynamic modeling in Examples 5.3 and 5.4. In other cases, two or more dimensionless groups would be necessary to achieve dynamic modeling. In some instances of model testing, it is not feasible or desirable to carry out tests on the model to satisfy even two nondimensional parameters. An example of this occurs in the testing of models of surface ships. Then two parameters (the Reynolds number and the Froude number) must be equal for complete scaling of the physical quantities, specifically,

$$\frac{V^2}{gL} = \frac{V_m^2}{gL_m} \qquad \text{for Froude scaling}$$

and

$$\frac{VL}{\nu} = \frac{V_m L_m}{\nu_m} \qquad \text{for Reynolds scaling}$$

where the subscript m signifies the model. From the Froude scaling, we obtain the ratio between speed and length in the form

$$\frac{V}{V_m} = \sqrt{\frac{L}{L_m}}$$

while from the Reynolds scaling we have the ratio

$$\frac{V}{V_m} = \frac{\nu}{\nu_m} \frac{L_m}{L}$$

Thus

$$\frac{\nu}{\nu_m} = \frac{L^{3/2}}{L_m^{3/2}}$$

The latter relationship indicates that the viscosity of liquid in which the model tests are to be conducted must be different from that of the prototype liquid, if the ship model is of different length from that of the prototype ship. However, in actual practice, ship models are tested in water—the liquid in which the prototype operates. In order to deduce the ship drag from model tests in water, two separate tests are required based on the assumption that the total drag is composed of the sum of wave drag and friction drag. First, in order to obtain the wave drag—a free-surface phenomenon—a model of the ship is built and tested according to Froude scaling. For example, if the model is one-tenth of the full-scale ship, the model speed is 0.316 times the full-scale speed. For this first test, Reynolds scaling is not adhered to. The total drag of the model is measured. The wave drag of the model is then

$$(\text{Wave drag})_m = (\text{Total drag})_m - (\text{Friction drag})_m$$

To obtain the friction drag of the model, a separate test is conducted on a fully submerged flat plate to eliminate the free-surface effect. To scale up the wave drag of the model to the wave drag of the full-scale ship, we use Equation (5.9), namely,

$$\frac{D_{\text{wave}}}{\rho V^2 L^2} = \frac{(D_{\text{wave}})_m}{\rho V_m^2 L_m^2}$$

Substituting the relation between V and L obtained from Froude scaling, we have

$$D_{\text{wave}} = (D_{\text{wave}})_m \frac{L^3}{L_m^3}$$

That is, the wave drag of the prototype is increased as the cube of the length ratio of prototype to model ship. The frictional drag is scaled according to Reynolds scaling with

$$D_{\text{friction}} = (D_{\text{friction}})_m \frac{V^2}{V_m^2} \frac{L^2}{L_m^2}$$

Substituting the relation between V and L obtained from Reynolds scaling, we have

$$D_{\text{friction}} = (D_{\text{friction}})_m$$

That is, the friction drag of the prototype is equal to that of the model, if the model is tested in water at the same temperature as that experienced by the prototype.

A different problem arises in the modeling of flows of rivers, harbors, and estuaries. We cannot geometrically scale down the vertical distances as much as the horizontal distances. Otherwise we would obtain strong effects of viscosity and surface tension which are absent in the prototype. This dual geometric scaling results in an approximate modeling of the flow.

We have shown in this chapter that dimensional analysis permits us to obtain a functional dependence of an unknown physical quantity on the physical variables involved in the phenomenon under consideration. As pointed out previously, the result obtained from any dimensional analysis can be no better than the validity of the significant physical variables assumed. To illustrate this point, if in Example 5.1, which deals with the oscillation of a mass-spring system, we had assumed that the significant variables were the spring constant k, the amplitude of oscillation a, and the average velocity of oscillation V, that is

$$P = P(k, a, V)$$

then we would have obtained the single nondimensional group

$$\frac{PV}{a} = \text{Constant}$$

Subsequent experimental tests would show no correlation of results based on this nondimensional parameter. Hence the experimenter must look for additional or different significant physical variables. Here the change would have been found to be the replacement of V by mass m.

PROBLEMS

5.1. Which of the following ratios are dimensionless?

$$\frac{\rho V}{L\sigma}, \frac{p}{\rho V^2}, \frac{L^3 p}{\mu V}, \frac{c_p \Delta T}{\rho V^2}, \frac{\mathscr{P}}{\rho V^3 L^2}$$

5.2. Repeat Example 5.1 using the M, L, T system.

5.3. Assume that the speed of sound a in a gas (i.e., the speed at which small disturbances propagate through the gas) depends on the gas density ρ, the pressure p, and the dynamic viscosity μ. Find a functional relationship for the speed of sound. (Use the F, L, T system).

5.4. Repeat Problem 5.3 using the M, L, T system.

5.5. Consider an airplane flying at high speeds. Assume that the significant physical parameters affecting drag force on the plane are the speed of the plane V, a characteristic length l of the plane, the fluid density ρ, and the dynamic viscosity μ. At high speeds, an additional significant physical parameter is the sonic speed a of the undisturbed air. Show by use of dimensional analysis that the drag coefficient (see Table 5.2) will now be a function of the Reynolds number $\rho V l / \mu$ and the nondimensional ratio V/a, called the Mach number.

5.6. Repeat Example 5.2 using the F, L, T system.

5.7. Assume that the thrust T of a propeller depends on its diameter D, the fluid density ρ, dynamic viscosity μ, the revolutions per unit time N, and the velocity of advance V with respect to the undisturbed fluid. By means of dimensional analysis, determine the appropriate nondimensional parameters.

5.8. Find the dimensions of the following in the M, L, T system:

$$\frac{\omega}{\rho V}, \frac{T_0}{I\omega}, \frac{\sigma}{\rho L}, \frac{\mathscr{P}}{\rho V^2}$$

5.9. Find the dimensions of the ratios in Problem 5.8 in the F, L, T system.

5.10. A shaft rotates in a lubricated bearing. Assume that the tangential frictional resistance F_T depends on the load normal to the shaft F_n, the shaft revolutions per unit time N, the viscosity of the lubricant μ, and the diameter of the shaft d. By use of dimensional analysis, establish the nondimensional parameters.

5.11. A jet of water issues from an orifice in the side of an open tank into the atmosphere. Use dimensional analysis to find a functional relation for the jet velocity. Assume the velocity V to be a function of water depth h, water density ρ, orifice diameter d, and acceleration due to gravity g.

5.12. Assume that the power input to a liquid pump depends on its discharge (volume flow rate) Q, the pressure increase (Δp) between pump inlet and outlet, the density of the liquid ρ, its characteristic diameter D, and its efficiency η. Find the functional relation for the power input by use of dimensional analysis.

5.13. Use dimensional analysis to establish a functional equation for the force exerted by a free jet on a moving flat plate. The plate is moving in a direction parallel to the jet and is inclined at a right angle to the jet.

5.14. A model of an airplane is tested in a wind tunnel using air at standard conditions. If the tunnel air speed is 150 ft/s, calculate the water speed if the same model is to be tested in a water tunnel at dynamically similar conditions.

5.15. A rotary mixer with propellerlike blades 1 ft in diameter is to be used to stir oil at a temperature of 80°F. Tests on a $\frac{1}{5}$-size model in water at 60°F show the optimum operating speed to be 30 rpm. Estimate the optimum angular speed for the prototype mixer.

5.16. An ocean liner 200 m long has a maximum speed of 20 m/s. At what speed should a model i0 m long be towed in a test basin to establish surface wave resistance?

5.17. An ocean liner 600 ft long has a maximum speed of 30 knots (1 knot = 1689 fps). At what speed should a model 30 ft long be towed in a test basin to establish surface wave resistance?

5.18. A model of a low-speed air compressor is tested at the same speed as the prototype compressor. The model compressor is one-fifth of the full-scale compressor. Compare the power requirements for model and prototype.

5.19. In an underwater explosion, the explosive is almost immediately converted into gas, the initial pressure depending on the type of explosive. The explosion causes a spherical shock to be propagated through the water. Assume that the pressure at the shock front at a given instant is a function of the radius of the shock front R, the initial explosion pressure p_i, the mass of the explosive M, the density of the water ρ, and the bulk modulus E of the water. Establish appropriate nondimensional parameters.

In a model test in a test pond it is determined that 0.5 kg of TNT causes a shock pressure of 20 MPa at a distance of 2 m from the center of explosion. What would be the corresponding conditions if 200 kg of TNT were exploded?

5.20. Under certain conditions, the flow behind a body immersed in a fluid oscillates with period P. Assume that the period of oscillation is a function of the upstream flow velocity V, the density of the fluid ρ, the viscosity μ, and the length of the body L. Determine nondimensional parameters describing this flow phenomenon. Two geometrically similar bodies having a scale ratio of 3 are tested separately in the same water tunnel. What will be the ratio of the period of oscillation and the ratio of the velocity at dynamically similar points?

5.21. It is desired to study the motion of water waves in a horizontal channel by using a geometrically similar model having a $\frac{1}{20}$ scale ratio. Determine the scale ratios (prototype-to-model) for the (a) time, (b) velocity, (c) force, and (d) acceleration. In the prototype, a wave travels a certain distance in 30 seconds. How long does it take the model wave to traverse the corresponding distance in the model tank?

5.22. Show that for hydraulic machinery operating in the turbulent flow regime, a nondimensional grouping is

$$\frac{\mathscr{P}}{\rho N^3 D^5}, \qquad \frac{Q}{ND^3}, \qquad \text{and} \qquad \frac{gh}{N^2 D^2}$$

if we assume the following significant physical parameters: power $\mathscr{P}$, flow rate Q, speed of rotation of machinery N, diameter of impeller d, liquid density ρ, liquid head h, and acceleration due to gravity g.

5.23. The formation of waves about bridge piers is to be determined by the use of model tests. What must be the speed of water in the model tank to correspond to a river speed of 5 m/s? A $\frac{1}{20}$ scale model is to be used.

The model discharge rate between two adjacent piers is measured to be 0.1 m^3/s. Find the corresponding prototype discharge rate.

5.24. Oil discharges into an open container from a large open tank through a horizontal pipe. The discharge rate is to be determined from a $\frac{1}{10}$ scale model. What must be the ratio between velocity, discharge rate, and kinematic viscosity for prototype and model? Note that in this problem both friction and gravity forces are of significance.

5.25. An airplane propeller with a diameter of 1 m rotates at a speed of 100.0 rad/s and moves with a forward speed of 50 m/s. Model tests are to be conducted in an atmospheric wind tunnel on a $\frac{1}{5}$ scale model. What will be the ratio of propeller thrust and turning torque between prototype and model? How will these results change if a pressurized wind tunnel ($p = 1000$ kPa) is utilized?

5.26. The aerodynamic drag force D on an automobile is found to be a function of air density and viscosity, automobile velocity, frontal area A, and a characteristic length L of the body. Determine the appropriate dimensionless parameters to be used for model testing.

5.27. Air at 1 atmosphere, 70°F flows through a 3-in-diameter pipe with a velocity of 3 fps. What must the average velocity of water be in a 1-in tube if its flow is to be dynamically similar to the air flow?

5.28. The noise power generated by a fan is found to be dependent on fan rotational speed, fan diameter, air density, and speed of sound a. Determine the appropriate dimensionless parameters.

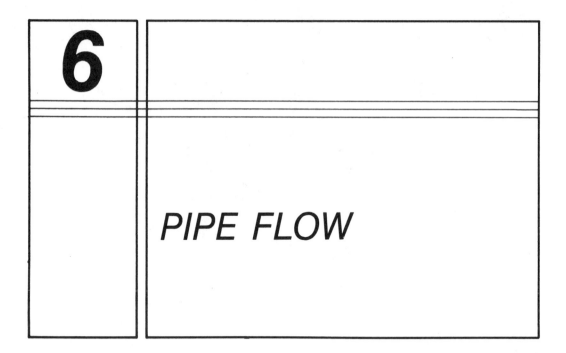

PIPE FLOW

6.1 INTRODUCTION

In this chapter, we will consider internal flow through ducts or pipes in which the presence of frictional forces acting on the fluid must be taken into account. An analysis of such flows is important in the many situations in which fluid must be transported from one place to another. For example, a calculation of the pressure drop due to friction in a pipe must be made to determine the pumping requirements for a water supply system; a similar analysis is employed in a determination of the outflow from a reservoir through a pipe or pipe network.

We know from the discussion of viscosity in Section 1.7 that when a fluid flows through a pipe, the layer of fluid at the wall has zero velocity; layers of fluid at progressively greater distances from the pipe surface have higher velocities, with the maximum velocity occurring at the pipe centerline. A velocity distribution in the pipe is built up somewhat as shown in Figure 6.1. The actual velocity distribution in the pipe is dependent on the type of flow existent in the pipe. Similarly, the type of flow will play an important part in the determination of the magnitude of the frictional forces acting on the fluid. There are two basic types of flow, each possessing fundamentally different characteristics. The first type is called laminar flow: the fluid flows in smooth layers, or laminae. In this type of flow, a particle of fluid in a given layer stays in that layer. For example, suppose that fluid moves steadily through a horizontal glass tube in laminar flow. If dye of neutral buoyancy is injected at a certain distance from the tube wall, as shown in Figure 6.2,

156

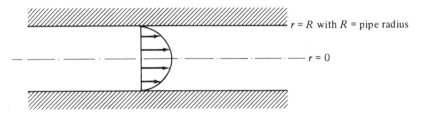

Figure 6.1.   Velocity distribution in pipe flow.

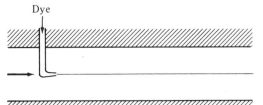

Figure 6.2.   Laminar flow through horizontal tube.

the dye streak will appear as a straight line. In other words, the dye will remain in the same fluid layer. For laminar flow, shear stress in the fluid is caused by the sliding of one layer of fluid over another. As the velocity of the flow through the glass tube of Figure 6.2 is increased, the dye streak is observed to become wavy. Above a certain velocity, the line breaks down entirely a short distance from the point of injection and is dispersed through the fluid. The flow is then said to be turbulent (Figure 6.3). As opposed to the smooth motion of the fluid in laminar flow, turbulent flow is characterized by an irregular, random motion of fluid particles in time and space. Velocity fluctuations occur both in the flow direction and normal to the flow direction.

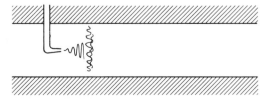

Figure 6.3.   Turbulent flow through horizontal tube.

For example, in a turbulent flow, if the axial component of velocity V_x at a point is plotted versus time, an irregular pattern results, as shown in Figure 6.4, with random fluctuations about a mean value. Generally the velocity

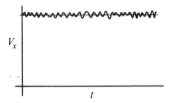

Figure 6.4.   Variation of velocity with time.

fluctuations are small in comparison to the mean. As long as the mean velocity does not vary with time, we can still represent the flow as turbulent, steady flow. However, even though the velocity fluctuations are small, they have a great effect on the flow characteristics. For example, the transverse movement of a particle of fluid from a faster-moving layer to a slower-moving layer will have the effect of increasing the velocity in the slower-moving layer. Similarly, the movement of a slower-moving particle to a layer of faster-moving particles will decrease the fluid velocity of the faster-moving layer. The motion of particles in a direction normal to the flow direction thus acts as an equivalent shear stress; this turbulent shear stress may be hundreds of times greater than the laminar stress due to the sliding of one layer over another. Furthermore, with the large number of random particle fluctuations present in a turbulent flow, there is a tendency toward mixing of the fluid and a more uniform velocity profile. The interchange of momentum between faster- and slower-moving particles tends to "even out" the velocity profile. Thus, for turbulent pipe flow, the mean velocity plotted versus radius might appear as shown in Figure 6.5, whereas, for laminar flow, the profile is parabolic.

Turbulent flow   Laminar flow

Figure 6.5.   Velocity distributions in pipe flow.

Everyone is familiar with some of the examples of the occurrence of laminar and turbulent flows. When smoke leaves a cigarette, it travels upward initially in a smooth, regular pattern; at a certain distance above the cigarette, however, the smoke breaks down into an irregular pattern (Figure 6.6). In other words, with the smoke, a transition has occurred from a laminar to a turbulent pattern. When a water faucet is turned on very slowly (Figure 6.7), with just a small amount of flow, the flow pattern observed is a very smooth, laminar one; when the faucet is opened wide, the flow breaks up and becomes turbulent.

Another way of looking at the difference between laminar and turbulent flows is to consider what happens when a small disturbance is introduced into a flow. If the flow is laminar, a small disturbance is damped out by viscous forces. If the disturbance cannot be damped out but continues to grow and affect the entire stream, we have turbulent flow. Even in turbulent pipe flow, with the great majority of the flow characterized by rough, irregular motions, there will always be a thin layer of smooth laminar flow near a

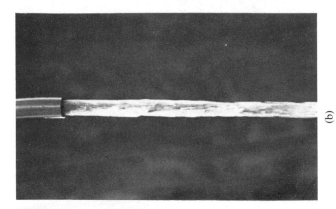

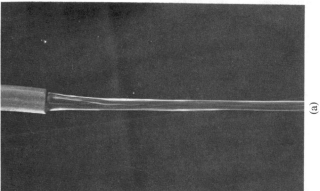

(a)

(b)

Figure 6.7. Flow patterns from water faucet: (a) laminar flow; (b) turbulent flow.

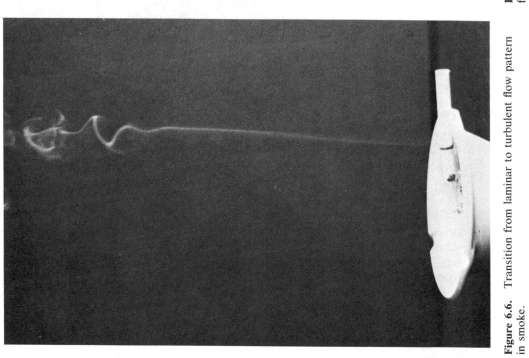

Figure 6.6. Transition from laminar to turbulent flow pattern in smoke.

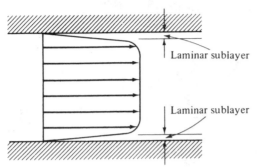

Figure 6.8. Laminar sublayer in pipe flow.

wall, for the particle fluctuations must die out near a boundary. This thin layer is called the laminar sublayer (Figure 6.8). The thickness of the laminar sublayer depends on the degree of turbulence of the main stream—the more turbulent the flow, the thinner the sublayer. In any case, the thickness of the sublayer is only a very small fraction of the pipe diameter.

In an analysis of flow through a pipe, we are interested in the type of flow, whether laminar or turbulent, since the shear stress and resultant frictional forces acting on the fluid vary greatly for the two types. It is important, then, to establish a criterion for the type of flow existent in a pipe. From the discussion of flow through a glass tube, we have established that velocity is one of the important determining variables. Moreover, the magnitude of fluid viscosity would be expected to play a role, since it is the viscous forces that damp out disturbances introduced into the flow. In the late nineteenth century, a series of experimental tests was conducted by Osborne Reynolds using dye injected into fluids flowing through glass tubes of varying diameter. He established that the proper parameter to use as a criterion for describing the type of flow in a circular pipe is $\rho \overline{V} D / \mu$, with $\overline{V}$ the mean velocity. This dimensionless ratio, called the Reynolds number, has already been derived (Chapter 5) and is of extreme importance in pipe flow. It has been shown experimentally that a critical Reynolds number exists for a given apparatus, below which flow in the pipe is laminar and above which the flow is turbulent. It is not possible to specify a universal value of critical Reynolds number that would hold for all systems. The actual value must depend to a certain extent on the magnitude of the disturbances introduced into the flow by such factors as the pipe inlet, pipe bends, and extraneous vibrations due to the proximity of pumping machinery. Under ordinary conditions, the critical Reynolds number is between 2000 and 3000. It has been found that for Reynolds numbers less than 2000, the flow is always laminar. It is interesting to note that by taking extreme care to minimize disturbances, laminar flow has been observed at a Reynolds number of 40,000. However, this is an unusual example; for most industrial applications, the critical Reynolds number lies between 2000 and 2300. In our work, unless otherwise specified, we shall take the critical Reynolds number for flow in pipes to be 2200. To gain an idea of the magnitude of flow velocities involved with this value of Reynolds number, consider the flow of water at 20°C (68°F) through a circular pipe of 5 cm (2 in) inner diameter. At 20°C, the kinematic viscosity of water, defined as $\nu = \mu/\rho$, is equal to 1.00×10^{-6} m²/s (1.08×10^{-5} ft²/s), so that the mean flow velocity for a critical Reynolds number of 2200 is

$$\overline{V} = \frac{2200\nu}{D} = \frac{2200 \times 1.00 \times 10^{-6} \text{ m}^2/\text{s}}{5 \times 10^{-2} \text{ m}} = 0.044 \text{ m/s } (0.144 \text{ fps})$$

In other words, under normal conditions, if the flow velocity is less than 0.044 m/s, the flow in the 5-cm pipe will be laminar; if the velocity is greater than 0.044 m/s, the flow will be turbulent. Again, it is emphasized that the critical Reynolds number for pipe flow is not a fixed number but varies slightly, depending on the history of the flow. Also note that if a light oil (which might have a kinematic viscosity of 1.00×10^{-4} m²/s) were flowing through the 5-cm pipe with a critical Reynolds number of 2200, laminar flow would exist in the pipe until a velocity of

$$\overline{V} = \frac{2200 \times 10^{-4} \text{ m}^2/\text{s}}{5 \times 10^{-2}} = 4.4 \text{ m/s } (14.4 \text{ fps})$$

was attained.

We have assumed up to this point that viscous flow, whether laminar or turbulent, completely fills the entire cross section of the pipe. Near a pipe inlet, however, this is not the case. For example, suppose that uniform flow enters a straight pipe, as shown in Figure 6.9. Again, due to viscosity, the fluid velocity at the pipe wall must be zero. The fluid of zero velocity gradually slows down the fluid next to it; this fluid in turn gradually slows down the faster-moving fluid next to it, and so on, until the effects of viscosity are felt across the entire pipe. In other words, near an inlet, the effects of wall friction or viscosity are confined to an ever-increasing layer which eventually fills the pipe. The central or core region of the flow is a region of uniform flow in which viscous effects do not exist. Further downstream, the core region disappears, and the flow is termed fully developed viscous flow. Again, the nature of the fully developed flow, that is, laminar or turbulent, will depend on the Reynolds number.

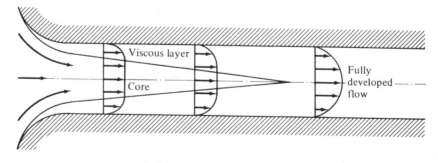

Figure 6.9. Development of flow at pipe inlet.

The inlet length required to attain fully developed flow is dependent on the type of flow. For instance, for laminar flow in circular tubes, the required inlet length is given approximately by

$$\frac{L}{D} = (0.057)\text{Re}_D,\dagger \qquad \text{with} \qquad \text{Re}_D = \frac{\rho \overline{V} D}{\mu}$$

so that the inlet length to attain fully developed laminar flow is at most 125.4 pipe diameters for a critical Reynolds number of 2200. For turbulent flow, the inlet length to achieve fully developed flow is between 25 and 50 pipe diameters, the value depending on the wall roughness and inlet shape. For instance, the inlet length required for a square-edged opening is less than that required for a rounded-edge opening. It can be seen that in an overall piping system, the inlet region is relatively small compared to the region of fully developed flow. If necessary, inlet effects can generally be handled by applying a small correction factor to values obtained for fully developed flow.

6.2 EQUATIONS OF MOTION

First we shall derive the equations of motion for fully developed constant-area steady flow of an incompressible fluid. To simplify the analysis, the flow will be approximated as one-dimensional. For fully developed turbulent flow with an almost uniform velocity profile, the assumption of one-dimensionality is a good one. Laminar flow, with a substantial velocity variation with radial distance from the centerline, will be represented as one-dimensional, using a mean velocity at each cross section.

Select a differential control volume, as shown in Figure 6.10, with ends normal to the flow direction. The continuity equation (4.3) for steady flow yields

$$\int_{\substack{\text{control} \\ \text{surface}}} \rho V_n \, dA = 0$$

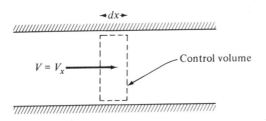

Figure 6.10

---

† H. L. Langhaar, "Steady Flow in the Transition Length of a Straight Tube," *Journal of Applied Mechanics,* Vol. 64, A55–A58 (1942).

For constant-area incompressible flow, with velocity in the x direction only, this reduces to

$$V_x = \text{Constant}$$

The x-momentum equation for one-dimensional steady flow [Equation (4.7)] is

$$\Sigma F_x = \int_{\substack{\text{control} \\ \text{surface}}} V_x(\rho V_n \, dA)$$

But from the continuity equation, V_x is constant, so that

$$\Sigma F_x = 0$$

Figure 6.11 depicts the forces acting on the control surface for a horizontal pipe with no gravity forces; τ_w is the wall shear stress due to friction acting on the fluid in the control volume. Summing forces, we obtain

$$pA - (p + dp)A - \tau_w dA_s = 0$$

where A is the cross-sectional area of the pipe and dA_s is the peripheral differential surface area over which the shear stress acts. Canceling terms, there results

$$A \, dp + \tau_w \, dA_x = 0$$

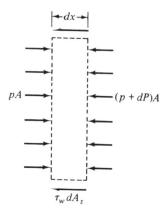

$$\tau_w \, dA_s \qquad \textbf{Figure 6.11}$$

For a circular duct, $A = \pi D^2/4$ and $dA_s = \pi D \, dx$, so that

$$dp + \tau_w \frac{4dx}{D} = 0$$

If the duct or pipe is not circular, an equivalent hydraulic diameter D_h is used, defined as

$$D_h = \frac{4 \times \text{Cross-sectional area of flow}}{\text{Perimeter wetted by fluid}} \tag{6.1}$$

Note that for a circular pipe flowing full of fluid,

$$D_h = \frac{4(\pi D^2/4)}{\pi D} = D$$

as might be expected. For a circular duct flowing half full of fluid,

$$D_h = \frac{4 \times \frac{1}{2} \times (\pi D^2/4)}{\pi D/2} = D$$

For a square duct of side s flowing full,

$$D_h = \frac{4s^2}{4s} = s$$

For all subsequent analysis in this chapter, we shall assume the duct or pipe to be flowing full; a later chapter on open channel flow (Chapter 10) will be concerned with ducts and pipes not flowing full, such as sewer pipes, drains, and streams.

In general, then, we can write

$$dp + \tau_w \frac{4dx}{D_h} = 0 \tag{6.2}$$

In order to solve Equation (6.2), we need an expression for τ_w in terms of the usual flow variables. It is customary to express τ_w in terms of a dimensionless friction factor f, with f defined such that

$$\tau_w = \frac{f}{4} \frac{1}{2} \rho \overline{V}^2 \tag{6.3}$$

with $\overline{V}$ the mean or average velocity at a flow cross section (See equation 4.5). Substituting Equation (6.3) into Equation (6.2), we obtain

$$dp + \frac{1}{2}\rho \overline{V}^2 \frac{f \, dx}{D_h} = 0 \tag{6.4}$$

The foregoing equation can be integrated between cross section 1 and cross section 2 of a pipe a distance L apart (Figure 6.12):

$$p_2 - p_1 = -\frac{1}{2}\rho \overline{V}^2 \frac{fL}{D_h} \tag{6.5}$$

If there is a difference in elevation $z_2 - z_1$ between sections 1 and 2, it is necessary to include a term in the momentum equation to account for the gravity force, as was done in Section 4.4. It follows that

$$p_2 - p_1 + \frac{1}{2}\rho \overline{V}^2 \frac{fL}{D_h} + (z_2 - z_1)\rho g = 0 \tag{6.6}$$

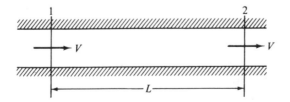

Figure 6.12

It is seen that Equation (6.6) is similar to the Bernoulli equation (4.21) for constant-area flow with a term added to include friction.

For frictional flow in which a change of cross-sectional area occurs between sections 1 and 2, a term expressing the change in kinetic energy must be added to Equation (6.6):

$$\frac{p_2}{\rho} + \frac{V_2^2}{2} + gz_2 + \frac{fL}{D}\frac{\overline{V}^2}{2} = \frac{p_1}{\rho} + \frac{V_1^2}{2} + gz_1 \tag{6.7}$$

Equation (6.7) is sometimes called the modified Bernoulli equation. The application of Equation (6.7) to problems involving area change will be shown in Section 6.4.

EXAMPLE 6.1

Determine the pressure drop between two cross sections 1 ft apart in a horizontal 2-in-diameter tube (Figure 6.12) for laminar flow of water when the average velocity $\overline{V}$ is 0.15 fps, the friction factor f is 0.030 and the density of the water is 62.35 lbm/ft³.

Solution   Substituting into Equation (6.5) for a horizontal pipe, we get,

$$p_2 - p_1 = -\frac{1}{2}\rho V^2 \frac{fL}{D_h}$$

$$= -\frac{1}{2}\left(\frac{62.35 \text{ lbm/ft}^3}{32.17 \text{ lbm/slug}}\right)(0.15^2 \text{ ft}^2/\text{s}^2)\left[\frac{0.030(1) \text{ ft}}{\frac{2}{12} \text{ ft}}\right]$$

$$= -0.00392 \text{ lbf/ft}^2 = \underline{-2.73 \times 10^{-5} \text{ psi}}$$ ∎

EXAMPLE 6.2

Five liters per second of water flows through a horizontal circular pipe of inner diameter 6 cm, 20 m long. If the friction coefficient for the flow is 0.02 and the density of water is 998 kg/m³, determine the pressure drop in the pipe.

Repeat the above for the case in which the pipe is aligned at a positive angle of 0.1 rad with the horizontal.

Solution   The flow velocity is

$$V = \frac{Q}{A} = \frac{5 \times 10^{-3} \text{ m}^3/\text{s}}{(\pi/4)(0.06)^2 \text{ m}^2} = 1.768 \text{ m/s}$$

Substituting into Equation (6.5) for a horizontal pipe, we obtain

$$p_2 - p_1 = -\frac{1}{2}\rho V^2 \frac{fL}{D}$$

$$= -\frac{1}{2}(998 \text{ kg/m}^3)(1.768^2 \text{ m}^2/\text{s}^2)(0.02)\left(\frac{20}{0.06}\right)$$

$$= \underline{-10.40 \text{ kPa}}$$

For the pipe rising at an angle of 5 deg (0.1 rad) with the horizontal, we must also consider the gravity force term in the momentum equation. For this case, $z_2 - z_1$, the difference in elevation between cross sections 1 and 2, is equal to $20 \sin 0.1 = 1.997$ m; therefore, for inclined pipe we have, from Equation (6.6):

$$p_2 - p_1 = -\frac{1}{2}\rho V^2 \frac{fL}{D} - (z_2 - z_1)\rho g$$
$$= -10.40 \text{ kPa} - (1.997 \text{ m})(998 \text{ kg/m}^3)(9.81 \text{ m/s}^2)$$
$$= -10.40 - 19.55$$
$$= -29.95 \text{ kPa} \qquad \blacksquare$$

EXAMPLE 6.3

The pump in Figure 6.13 is to supply 0.05 m³/s of water from a reservoir to a hilltop 150 m above the reservoir surface. The supply pipe has an inner diameter of 10 cm with a total pipe length of 500 m. The friction coefficient f is equal to 0.023 and the density of water is 1000 kg/m³. Neglecting losses in the inlet pipe (i to 1) and pump, determine the pump power required.

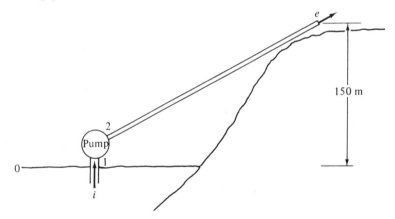

Figure 6.13

Solution   Neglecting losses, we can write Bernoulli's equation (4.21) between the reservoir surface and pump inlet as

$$gz_0 + \frac{p_0}{\rho} = gz_1 + \frac{V_1^2}{2} + \frac{p_1}{\rho}$$

Now apply the energy equation (4.19) for the pump, assuming that any change of internal energy as the water flows through the pump is lost as heat [$u_2 - u_1 = (1/\dot{m})(d\overline{Q}/dt)$]:

$$-\frac{1}{\dot{m}}\frac{dW_p}{dt} = \frac{V_2^2 - V_1^2}{2} + (z_2 - z_1)g + \frac{p_2 - p_1}{\rho}$$

where $V_2 = V_1$ for $A_2 = A_1$ from continuity for incompressible flow. From 2 to e, apply the modified Bernoulli equation with friction term [Equation (6.7)]:

$$\frac{p_2}{\rho} + gz_2 = \frac{p_e}{\rho} + gz_e + \frac{fL}{D}\frac{V^2}{2}$$

where $V_2 = V_e$ since $A_2 = A_e$. Combining these equations so as to express the pump work in terms of reservoir and outlet conditions, we have

$$gz_0 + \frac{p_0}{\rho} - \frac{1}{\dot{m}}\frac{dW_p}{dt} = \frac{fL}{D}\frac{V^2}{2} + gz_e + \frac{p_e}{\rho} + \frac{V_e^2}{2}$$

or

$$-\frac{1}{\dot{m}}\frac{dW_p}{dt} = \frac{p_e - p_0}{\rho} + g(z_e - z_0) + \frac{fL}{D}\frac{V^2}{2} + \frac{V_e^2}{2} \qquad (6.8)$$

For our case,

$$\overline{V} = \frac{Q}{A} = \frac{0.05 \text{ m}^3/\text{s}}{(\pi/4)(0.10)^2 \text{ m}^2} = 6.366 \text{ m/s}$$

and $p_0 = p_e$, so that

$$-\frac{1}{\dot{m}}\frac{dW_p}{dt} = 0 + (9.81 \text{ m/s}^2)150 \text{ m} + (0.023)\left(\frac{500}{0.10}\right)\left(\frac{6.366^2}{2}\right)\text{m}^2/\text{s}^2$$

$$+ \frac{6.366^2}{2} \text{ m}^2/\text{s}^2$$

$$= 0 + 1.472 \text{ kN} \cdot \text{m/kg} + 2.330 \text{ kN} \cdot \text{m/kg} + 0.020 \text{ kN} \cdot \text{m/kg}$$

$$= 3.822 \text{ kJ/kg}$$

In order to calculate pumping power required, we must now determine $\dot{m}$.

$$\dot{m} = \rho A V$$
$$= \rho Q$$
$$= (1000 \text{ kg/m}^3)(0.05 \text{ m}^3/\text{s})$$
$$= 50 \text{ kg/s}$$

and

$$\frac{dW_p}{dt} = (-3.822 \text{ kJ/kg})(50 \text{ kg/s})$$

$$= -191.1 \text{ kW}$$

(*Note:* The minus sign denotes work done on the fluid.)   ■

As was done in Section 4.4, we now wish to compare the modified Bernoulli equation (6.7) with the energy equation for flow with friction. The energy equation for flow with friction through a constant-area pipe with no work takes the form (4.19):

$$\frac{1}{\dot{m}}\frac{d\overline{Q}}{dt} = u_2 - u_1 + \frac{p_2 - p_1}{\rho} + g(z_2 - z_1)$$

The modified Bernoulli equation with friction [Equation (6.7)] was found to be

$$\frac{fL}{D}\frac{V^2}{2} + \frac{p_2 - p_1}{\rho} + g(z_2 - z_1) = 0$$

Comparing the two equations, we see that the effect of friction is to cause either a heat loss from the fluid or a gain in internal energy of the fluid. In both instances there is a loss in the mechanical energy per unit mass of fluid.

The Bernoulli equation is sometimes expressed in terms of *head*, given in meters. By dividing Equation (6.7) by g, we obtain

$$\frac{fL}{D}\frac{V^2}{2g} + \frac{p_2 - p_1}{\rho g} + (z_2 - z_1) + \frac{V_2^2 - V_1^2}{2g} = 0 \qquad (6.9)$$

with each term now having units of meters. The term $V^2/2g$ is called *velocity head* h_v, $p/\rho g$ the *pressure head* h_p, and z the *gravity head* h_z. The *total head* h_T at the entrance to the pipe of Example 6.2 is thus equal to

$$h_{T_i} = h_{v_i} + h_{p_i} + h_{z_i}$$

with the total head at exit equal to

$$h_{T_e} = h_{v_e} + h_{p_e} + h_{z_e}$$

and the term $(fL/D)V^2/2g$ representing a loss in head h_f due to friction. A pump, which adds mechanical energy to the flow, serves to increase the head. In terms of head, then, Equation (6.8) can be written as

$$h_{T_i} + h_{\text{pump}} - h_f = h_{T_e}$$

with h_{pump} the magnitude of the head added by the pump. In other words, the initial total head, plus the head added by a pump, less the head loss due to friction, is equal to the final total head.

6.3 FRICTION FACTOR *f*

In the preceding section we established the method of including a friction term in the equations of motion. This process permitted the solution of some flow problems for the simple case in which f is given. In an actual situation, however, the friction coefficient is not given but must be found as part of the overall problem.

The value of the coefficient of friction between the fluid and the wall will be greatly influenced by the velocity profile in the pipe, since wall shear stress τ_w is equal to $\mu(dV/dy)_{\text{wall}}$. Note that this expression is true for both laminar and turbulent pipe flows, for even in fully developed turbulent pipe flow, the fluid near the wall is in the laminar sublayer. For well-ordered laminar flow, with shear stress throughout the flow given by $\mu(dV/dy)$, the velocity profile is calculable, and an expression for friction coefficient can be derived. In turbulent flow, however, the fluid is in chaotic, disordered motion, with random fluctuations of fluid from one layer to another. For turbulent flow, no theoretical analysis has yet been able to describe the situation completely. For example, the shear stress is no longer due merely to the sliding of one fluid layer over another, so it cannot be represented as $\mu(dV/dy)$. Rather, turbulent shear stress is due primarily to the random transverse motion of fluid particles; a discrete expression for turbulent shear is not available. Therefore, in order be to able to calculate *f* for fully developed turbulent pipe flow, it is necessary to resort to experimental data. With this point in mind, we shall first derive expressions for *f* for fully developed incompressible laminar flow in a constant-area circular pipe and in a constant-area rectangular channel. Then we shall present experimental data arranged in dimensionless form that will allow a determination of *f* for fully developed turbulent flow.

Consider laminar flow in a circular pipe with velocity V in the x direction. For fully developed flow, we shall assume that velocity is a function of radial distance r measured from the pipe centerline but not of axial distance x; in other words, the velocity profiles are no longer changing with x. It can be assumed further that pressure is a function of x but not of r. This flow, shown in Figure 6.14, is called *Poiseuille flow*.

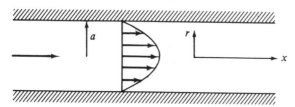

Figure 6.14

Consider the momentum equation for a differential control volume in the shape of a circular cylinder, with cylinder radius less than *a*, as shown in Figure 6.15. There is no velocity change in the axial direction, so the momentum equation yields $\Sigma\, F_x = 0$. Let τ be the shear stress acting on the surface area of the cylindrical element, as shown in the figure. Taking a force balance, we obtain

$$+\tau 2\pi r\, dx + pA - (p + dp)A = 0$$

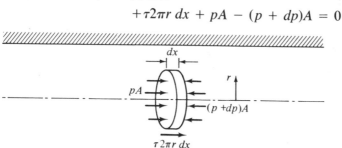

Figure 6.15

or

$$+\tau 2\pi r \, dx - (dp)\pi r^2 = 0$$

so that

$$\frac{dp}{dx} = +\frac{2\tau}{r}$$

We can write τ as $\mu(dV/dr)$; note that, since dV/dr is negative, the shear force is directed in the negative x direction, as might be anticipated. Substituting, we obtain

$$\frac{dp}{dx} = \frac{2}{r}\,\mu\,\frac{dV}{dr}$$

The left-hand side of the preceding equation is a function of x alone, and the right-hand side is a function of r alone. Since x and r are independent variables, it follows that each side must be equal to a constant. We now wish to integrate to obtain V as a function of r. First multiply by $r/2$:

$$\mu\,\frac{dV}{dr} = \frac{r}{2}\,\frac{dp}{dx}$$

Integrating for dp/dx constant, we obtain

$$\mu V = \frac{r^2}{4}\,\frac{dp}{dx} + c_1 \qquad \text{with } c_1 \text{ a constant}$$

Now apply the boundary condition $V = 0$ at $r = a$:

$$0 = \frac{a^2}{4}\,\frac{dp}{dx} + c_1$$

or

$$c_1 = -\frac{a^2}{4}\,\frac{dp}{dx}$$

Substituting, we obtain a parabolic velocity profile:

$$V = -\frac{1}{\mu}\frac{dp}{dx}\left(\frac{a^2}{4} - \frac{r^2}{4}\right) \tag{6.10}$$

The average velocity for this flow is (see Section 4.1)

$$\overline{V} = \frac{\int_0^a V2\pi r \, dr}{\pi a^2} = -\frac{dp}{dx}\frac{a^2}{8\mu} \tag{6.11}$$

We have expressed the pressure gradient in terms of friction coefficient as

$$\frac{dp}{dx} = -\frac{f}{D}\frac{1}{2}\rho\overline{V}^2 = -\frac{f}{4a}\rho\overline{V}^2$$

so that

$$f = \frac{32\mu}{\rho\overline{V}a} = \frac{64\mu}{\rho\overline{V}D} \tag{6.12}$$

The expression $\rho \overline{V} D / \mu$ can be recognized as the Reynolds number based on pipe diameter, so that $f = 64/\mathrm{Re}_D$ for laminar flow in a circular pipe.

As another example of the calculation of friction factor for fully developed laminar flow, consider flow in a rectangular channel with width w much greater than depth d (as shown in Figure 6.16), so that the flow is two-dimensional. Take the velocity V in the x direction, a function of y only, with pressure a function of x but not y. The momentum equation for the control volume shown is

$$\Sigma F_x = -2y \, dp + 2\tau \, dx = 0$$

or

$$\mu \frac{dV}{dy} = y \frac{dp}{dx}$$

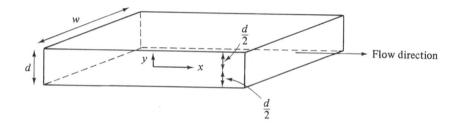

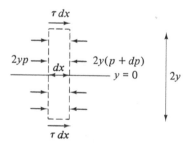

Control volume

Figure 6.16.   Flow through rectangular channel.

We can now integrate the above, with $V = 0$ at $y = \pm d/2$, to obtain

$$V = \frac{1}{2\mu} \frac{dp}{dx} \left(y^2 - \frac{d^2}{4} \right)$$

and

$$\overline{V} = \frac{\displaystyle\int_{-d/2}^{+d/2} V \, dy}{d} = -\frac{1}{\mu} \frac{dp}{dx} \frac{d^2}{12}$$

For the cross section of Figure 6.16, the hydraulic diameter [Equation (6.1)] is given by

$$D_h = \frac{4(d \times w)}{2d + 2w}$$

For $d \ll w$, this reduces to $D_h = 2d$. According to Equation (6.4),

$$f = -2\frac{d}{dx}\frac{D_h}{\overline{V}^2}\frac{1}{\rho}$$

Substituting our expressions for $\overline{V}$ and D_h, we find

$$f = \frac{96}{\mathrm{Re}_{D_h}}$$

where

$$\mathrm{Re}_{D_h} = \frac{\rho\overline{V}D_h}{\mu}$$

Results are available for any rectangular channel with laminar flow.* These show that $f \cdot \mathrm{Re}_{D_h}$ varies from 96 (infinitely wide rectangle) to 56.91 (square). Figure 6.17 shows values of $f \cdot \mathrm{Re}_{D_h}$ over a range of width-to-height ratios of rectangular ducts.

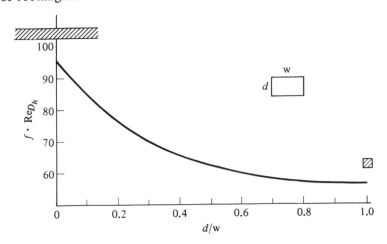

Figure 6.17. Friction factors for laminar flow in rectangular pipes.

EXAMPLE 6.4

The laminar flow described in Example 6.1 takes place in a $1\frac{1}{4}$ in by $2\frac{1}{2}$ in rectangular duct. Determine the pressure drop in a 1-ft-long duct and compare

* See H. L. Dryden, F. D. Murnaghan, and H. Bateman, *Hydrodynamics* (New York: Dover, 1956), p. 197.

it with the results for the circular tube. The kinematic viscosity of the water is $\nu = 1.20 \times 10^{-5}$ ft²/s.

Solution   The hydraulic diameter of the rectangular cross section is obtained from Equation (6.1)

$$D_h = \frac{4\,dw}{2(d + w)} = 2\frac{1.25(2.5)}{(1.25 + 2.5)} = 1.667 \text{ in}$$

From Figure 6.17, the friction factor product for $d/w = 0.5$ is $f \cdot \text{Re}_{D_h} = 63$. The Reynolds number becomes

$$\text{Re}_{D_h} = \frac{\overline{V}D_h}{\nu} = \frac{(0.15\text{ft/s})(1.667/12)\text{ft}}{1.20 \times 10^{-5}\text{ft}^2/\text{s}} = 2083$$

Hence

$$f = 63/2083 = 0.030$$

The pressure drop per foot length of horizontal pipe is, from Equation (6.5),

$$p_2 - p_1 = -\frac{1}{2}\rho\overline{V}^2\frac{fL}{D_h}$$

$$= -\frac{1}{2}\left(\frac{62.35 \text{ lbm/ft}^3}{32.17 \text{ lbm/slug}}\right)(0.15^2 \text{ ft}^2/\text{s}^2)\left[\frac{0.030 \times 1 \text{ ft}}{(1.667/12) \text{ ft}}\right]$$

$$= -0.00471 \text{ lbf/ft}^2 = \underline{3.27 \times 10^{-5} \text{ psi}}$$

as compared with a pressure drop of 2.73×10^{-5} psi in the circular tube of the same cross-sectional area.   ∎

For fully developed laminar flow, therefore, *f* depends on the Reynolds number and channel shape alone. It is interesting to note that the roughness of the pipe wall does not enter into the expression for *f*. The reason is that, for the parabolic laminar flow velocity profile, very little of the flow comes in contact with the roughness elements or protuberances of the wall surface; the velocities in the vicinity of the wall surface are quite low. This situation is in contrast to turbulent flow, where the almost uniform profile across the entire pipe provides relatively high velocities near the wall. In fact, for turbulent flow, the roughness elements may protrude through the laminar sublayer into the main, higher-velocity portions of the flow and create a large effect on the flow. Consequently, for turbulent flow, we might expect *f* to depend on a characteristic roughness dimension as well as on the Reynolds number. Once again, whereas for laminar flow we were able to develop an analytical solution for *f* using an expression for shear stress, for turbulent flow, because of its randomness, a discrete expression for τ is not available and we must resort to experimental data to obtain *f*.

Over the years a great many investigators have run large numbers of experiments in an attempt to express the variation of *f* for turbulent flow in circular pipes with such variables as flow velocity *V*, pipe diameter *D*, pipe roughness (measured by a characteristic roughness height ϵ), and fluid properties μ and ρ. From our study of dimensional analysis in Chapter 5, it seems advisable to arrange all the experimental data in the form of plots of dimensionless parameters. For this case, f is already dimensionless, and we know that the Reynolds number based on pipe diameter is significant. With the variables $(\rho, V, D, \mu, f, \epsilon)$, we require another dimensionless parameter that will involve ϵ.

It seems logical to scale roughness with pipe diameter, so that the third dimensionless parameter is relative roughness ϵ/D. From experimental data, then, curves have been plotted of f as a function of Re_D for various ϵ/D. The resultant set of curves is called the Moody diagram (Figure 6.18a).

Several interesting features can be observed. First, as we have shown, in the laminar region there is no dependence of f on roughness. For Reynolds numbers greater than critical, there are two regions of the Moody diagram—the transition region and the region of complete turbulence. In the latter region, f is independent of the Reynolds number and is dependent solely on pipe roughness (see Figure 6.18b). Here a large number of the roughness elements project through the laminar sublayer and into the turbulent, uniform region of the flow. The resistance to flow is caused primarily by the presence of the protuberances in the high-velocity portion of the stream. In the transition region, only a small number or possibly none of the roughness elements project through the laminar sublayer. The resistance to flow in this region is provided by laminar shear at the wall, as well as the protuberances in the flow. Therefore, in the transition region, f is dependent on the Reynolds number as well as on ϵ/D.

Note that for large values of relative roughness, the dividing line between the two regions occurs at a lower Reynolds number than for small values of relative roughness. This is because the thickness of the laminar sublayer decreases as the Reynolds number, and hence the turbulence of the flow, increases. So, at low Reynolds numbers, with a relatively thick laminar sublayer, only very rough pipes will have protuberances that will project through the sublayer.

In order to determine a value of the friction factor f from the Moody diagram, a knowledge of relative roughness is necessary. Typical values of roughness for various types of pipe are shown in Table 6.1.

It should be noted that after pipes have been in service for a time, deposits build up on the pipe walls which may substantially increase the preceding values of ϵ.

For noncircular pipes, good correlation for fully developed turbulent pipe flow is obtained by using Figure 6.18, with Re_{D_h} substituted for Re_D and ϵ/D_h for ϵ/D. The values of critical Reynolds number from Section 6.1 can also be used, with Re_{D_h} in place of Re_D.

For laminar flow in noncircular pipes, however, an appreciable error can result from using the value of f for circular pipes, as was shown for flow

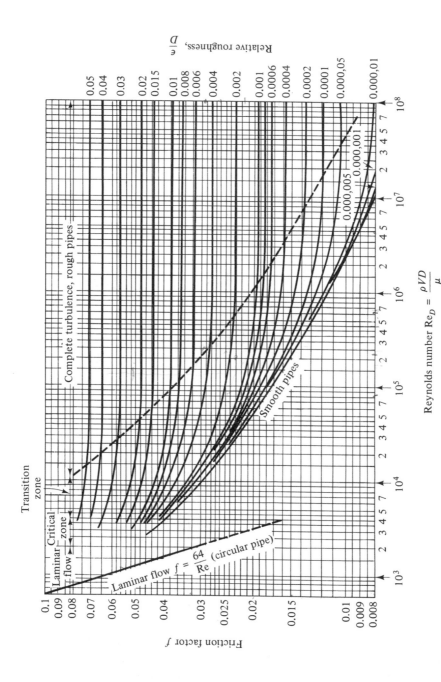

Figure 6.18(a).   Friction factors for pipe flow. (Adapted from L. F. Moody, "Friction Factors for Pipe Flow," *Trans. ASME*, Vol. 68, 1944, p. 672.)

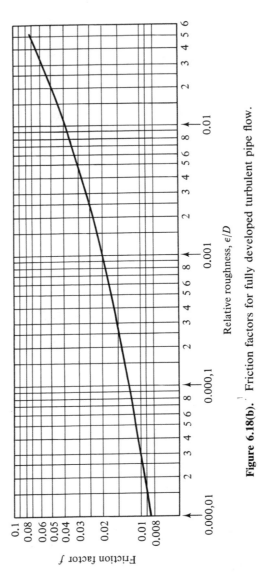

Figure 6.18(b). Friction factors for fully developed turbulent pipe flow.

TABLE 6.1 Roughness of Pipe Materials

Type	ϵ (mm)	ϵ (ft)
Glass	Smooth	Smooth
Asphalted cast iron	0.12	0.0004
Galvanized iron	0.15	0.0005
Cast iron	0.26	0.00085
Wood stave	0.18–0.90	0.0006–0.003
Concrete	0.30–3.0	0.001–0.01
Riveted steel	1.0–10	0.003–0.03
Drawn tubing	0.0015	0.000005

through the rectangular channel worked out previously in this section. The magnitude of the error is dependent on the shape of the cross section.

The following example illustrates the use of the Moody diagram.

EXAMPLE 6.5

Ten liters per second of water at 25°C is to flow through a horizontal pipe 100 m long. Compare the pressure drop in a cast iron pipe of circular cross section of 10 cm diameter with the pressure drop in a wood pipe ($\epsilon = 0.30$ mm) of square cross section 10 cm on a side.

Solution The water velocity in the circular pipe, $\overline{V}$, is given by

$$\overline{V} = \frac{10 \times 10^{-3}\ \text{m}^3/\text{s}}{(\pi/4)(0.10)^2\ \text{m}^2} = 1.273\ \text{m/s}$$

At 25°C, $\nu = 0.898 \times 10^{-6}$ m²/s, so that

$$\text{Re}_D = \frac{(1.273\ \text{m/s})(0.10)\text{m}}{0.898 \times 10^{-6}\ \text{m}^2/\text{s}} = 1.418 \times 10^5$$

This is well over the critical value of 2200 given in Section 6.1, so we are in the region of fully developed turbulent pipe flow. In order to find f from Figure 6.18, we need relative roughness:

$$\frac{\epsilon}{D} = \frac{0.30 \times 10^{-3}\ \text{m}}{0.10\ \text{m}} = 0.0030 \quad \text{for the circular pipe}$$

With relative roughness and Reynolds number, we now go to the Moody diagram to find $f = 0.027$. From Equation (6.5),

$$p_2 - p_1 = -\frac{1}{2}\rho \overline{V}^2 \frac{fL}{D}$$

$$= -\frac{1}{2}(997\ \text{kg/m}^3)(1.273)^2\ \text{m}^2/\text{s}^2\left(\frac{0.027 \times 100\ \text{m}}{0.10\ \text{m}}\right)$$

$$= \underline{-21.81\ \text{kPa}}$$

For the square wood pipe,

$$\overline{V} = \frac{10 \times 10^{-3} \text{ m}^3/\text{s}}{(0.10)^2 \text{ m}^2} = 1.0 \text{ m/s}$$

$$D_h = 10 \text{ cm} \quad (\text{see Section 6.2})$$

$$\text{Re}_{D_h} = \frac{(1.0 \text{ m/s})(0.10 \text{ m})}{0.898 \times 10^{-6} \text{ m}^2/\text{s}} = 1.114 \times 10^5$$

$$\frac{\epsilon}{D_h} = \frac{0.30 \times 10^{-3} \text{ m}}{0.10 \text{ m}} = 0.003$$

so that, from Figure 6.18,

$$f = 0.027$$

$$p_2 - p_1 = -\frac{1}{2}\rho\overline{V}^2\frac{fL}{D_h}$$

$$= -\frac{1}{2}(997 \text{ kg/m}^3)(1.0)^2 \text{ m}^2/\text{s}^2\left(\frac{0.027 \times 100 \text{ m}}{0.10 \text{ m}}\right)$$

$$= \underline{-13.46 \text{ kPa}} \qquad ■$$

The discussion of the friction factor has dealt solely with fully developed flow. At a pipe inlet, however, the flow is not fully developed. As can be seen in Figure 6.9, the velocity gradient at the wall and resultant shear stress are actually greater near the inlet. The problem of trying to predict an equivalent friction coefficient for flow near an inlet and its variation with axial distance is extremely complex. Fortunately, it has been shown experimentally that the local friction factor attains its fully developed value roughly 10 diameters from the inlet for turbulent flow; a somewhat greater length is required for laminar flow. It can be seen, then, that any inlet effect can be neglected except for very short pipes. Throughout this chapter we shall neglect any increase in frictional loss due to the inlet effect.

6.4 LOSSES IN PIPE FITTINGS AND VALVES

In addition to losses due to wall friction in a piping system, there are also losses of mechanical energy and pressure due to flow through valves or fittings (called minor losses). For example, in flow through a 90° elbow (Figure 6.19), there is a tendency for the flow to separate from the curved walls, giving rise to an eddying motion, this motion taking some energy away from the flow. The minor losses can be treated as an equivalent frictional loss by expressing the pressure drop due to the fitting as

$$\Delta p = -K\tfrac{1}{2}\rho\overline{V}^2$$

or in terms of head loss h:

$$\Delta h = -K\frac{\overline{V}^2}{2g}$$

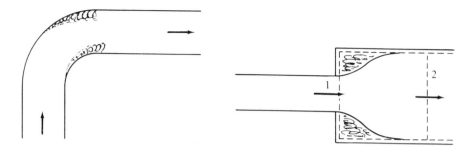

Figure 6.19   **Figure 6.20**

In a very few cases, K can be determined analytically; generally we must resort to experimental data. One case that can be treated analytically is flow through a sudden enlargement, as shown in Figure 6.20. The flow at section 1 must expand to a larger area at 2. Since the flow is unable to negotiate a sudden enlargement, there will be a "dead water" or stagnant region, as shown in the figure, with an eddying motion taking place in this region. Select a control volume as indicated, with section 2 far enough downstream so that the flow is uniform and parallel at this section. The pressure at section 1 and in the stagnant region can be assumed equal to p_1. From the momentum equation for incompressible, steady flow, we obtain

$$\Sigma F_x = \int\limits_{\substack{\text{control}\\\text{surface}}} V_x \rho V \, dA$$

In summing forces, we shall neglect the viscous shear at the wall, for the fluid in the vicinity of the wall is in the stagnant region. The only forces acting on the fluid bounded by the control surface are pressure forces, and we have

$$p_1 A_2 - p_2 A_2 = \rho A_2 V_2 (V_2 - V_1)$$

or

$$p_1 - p_2 = \rho V_2 (V_2 - V_1)$$

From continuity, $A_1 V_1 = A_2 V_2$. Substituting in the above,

$$p_1 - p_2 = \rho \frac{A_1}{A_2} V_1^2 \left(\frac{A_1}{A_2} - 1 \right) \tag{6.13}$$

Now apply the modified Bernoulli equation between 1 and 2, with a term $K V_1^2 / 2$ accounting for the loss from the expansion:

$$\frac{p_2 - p_1}{\rho} + \frac{K V_1^2}{2} + \frac{V_2^2 - V_1^2}{2} = 0 \tag{6.14}$$

Substituting Equation (6.13) into Equation (6.14) gives

$$\frac{K V_1^2}{2} = \frac{V_1^2}{2} \left[1 - \left(\frac{A_1}{A_2} \right)^2 \right] + V_1^2 \frac{A_1}{A_2} \left(\frac{A_1}{A_2} - 1 \right)$$

or

$$K = \left(1 - \frac{A_1}{A_2}\right)^2$$

For flow into a reservoir from a square-edged pipe, with A_1/A_2 very small, K approaches 1 (see Figure 6.21).

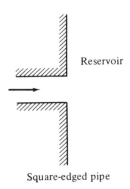

Reservoir

Square-edged pipe

$K = 1$ **Figure 6.21**

For a sudden contraction, the situation is somewhat more complex. Uniform flow at 1 is to pass through a contraction, as shown in Figure 6.22. The flow necks down to a minimum area at 0, where the flow is parallel and uniform; this area is called the *vena contracta*. The flow then expands to fill the pipe at 2. It is valid to assume the major portion of the losses occurs between 0 and 2, so that the same analysis could be used here as was used for the expansion. However, the minimum area A_0 cannot be predicted and

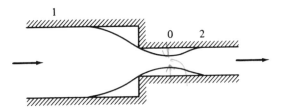

Figure 6.22

must be found experimentally, so that the value of K is essentially an empirical value. Typical values of K for sudden contractions are given in Table 6.2.

For flow from a reservoir into a pipe through a square-edged inlet, the value for K is approximately 0.5. If the inlet is rounded, the amount of eddying and the extent of the stagnant region are reduced. For a well-rounded entrance, K can be reduced to 0.05 or less (Figure 6.23).

For reentrant pipe openings, the loss coefficient K is generally taken as 1.0. Here the eddying region is much larger and more pronounced (Figure 6.24).

TABLE 6.2 Loss Coefficients
for Sudden Contractions*

A_2/A_1	K
0.1	0.37
0.2	0.35
0.3	0.32
0.4	0.27
0.5	0.22
0.6	0.17
0.7	0.10
0.8	0.06
0.9	0.02
1.0	0

* S. Whitaker, *Introduction to Fluid Mechanics* (Englewood Cliffs, N. J.: Prentice-Hall, 1968), p. 314.
Note: $\Delta p = -K\frac{1}{2}\rho V_2^2$.

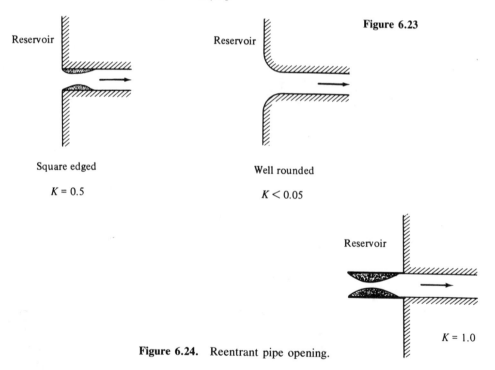

Reservoir

Reservoir

Figure 6.23

Square edged

$K = 0.5$

Well rounded

$K < 0.05$

Reservoir

$K = 1.0$

Figure 6.24.   Reentrant pipe opening.

Values of K for other fittings, as determined from experimental data, are shown in Table 6.3. These values are relatively insensitive to the Reynolds number.

Pipe fittings such as couplings, unions, and tees are used to join pipes by threading or soldering (Figures 6.25 to 6.27). Elbow fittings are used to

TABLE 6.3 Loss Coefficients for Valves and Fittings*

Fitting or Valve		K
Standard 45° elbow		0.35
Standard 90° elbow		0.75
Long-radius 90° elbow		0.45
Coupling		0.04
Union		0.04
Gate valve	Open	0.20
	$\frac{3}{4}$ Open	0.90
	$\frac{1}{2}$ Open	4.51
	$\frac{1}{4}$ Open	24.0
Globe valve	Open	6.4
	$\frac{1}{2}$ Open	9.5
Tee (along run, line flow)		0.4
Tee (branch flow)		1.5

* S. Whitaker, *Introduction to Fluid Mechanics* (Englewood Cliffs, N.J.: Prentice-Hall, 1968), p. 316.

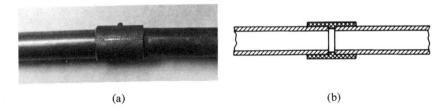

(a)   (b)

Figure 6.25. Coupling.

Figure 6.26. Union.

Figure 6.27. Tee connection.

change the direction of a pipe. A 90° elbow was shown in Figure 6.19, while a 45° elbow is shown in Figure 6.28.

Valves are used to control the amount of flow in a piping system. The desired amount of flow is obtained by adjusting the opening within the valve, resulting in pressure losses. Several types of valves are utilized. For example, a gate valve (Figure 6.29) uses a gate which moves in its seat in a direction perpendicular to the flow. A globe valve (Figure 6.30) uses a circular disk and seat. The flow experiences two 90° turns in its passage through the

Figure 6.28.   45° elbow.

(a)   (b)   (c)

Figure 6.29.   Gate valve.

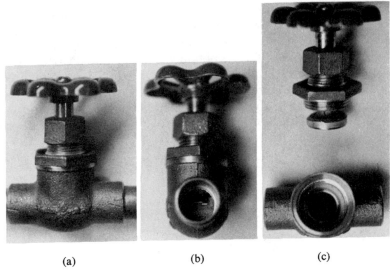

(a)   (b)   (c)

Figure 6.30. Globe valve.

valve. Hence the pressure loss through a fully opened globe valve is greater than through a fully opened gate valve.

EXAMPLE 6.6

A flow of 0.1 cfs of water enters a large chamber through a 1-in-diameter tube (Figure 6.31). It emerges from the chamber through a $\frac{1}{2}$-in-diameter tube. The chamber consists of a 6-in-diameter, 6-in-long tube. Calculate the pressure drop across the chamber assuming the density of water to be 1.94 slugs/ft^3.

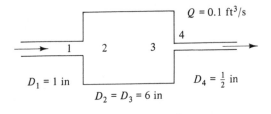

Figure 6.31

Solution   The entrance loss coefficient K_e is given by

$$K_e = \left(1 - \frac{A_1}{A_2}\right)^2 = [1 - (\tfrac{1}{6})^2]^2 = 0.945$$

Since $A_4/A_3 = (\tfrac{1}{2}/6)^2 = 0.00694$, the contraction loss coefficient $K_c = 0.50$. The pressure drop inside the chamber is negligible; that is, $p_2 \cong p_3$. Therefore

$$p_1 - p_4 = \rho\frac{V_4^2}{2} - \rho\frac{V_1^2}{2} + K_e\rho\frac{V_1^2}{2} + K_c\rho\frac{V_4^2}{2}$$

where

$$V_1 = \frac{Q}{A_1} = \frac{0.1 \text{ ft}^{3/s}}{\pi/4(\frac{1}{12})^2 \text{ft}^2} = 18.33 \text{ ft/s}$$

$$V_4 = 4V_1 = 73.34 \text{ ft/s}$$

$$p_1 - p_4 = \rho_2(1.5V_4^2 - .055\ V_1^2)$$

$$= \frac{1.94}{2} \text{ slugs/ft}^3\ (8068 \text{ ft}^2/\text{s}^2 - 18 \text{ ft}^2/\text{s}^2)$$

$$= 7809 \text{ lbf/ft}^2 = \underline{54.23 \text{ lbf/in}^2}$$

■

6.5 PIPING SYSTEMS

We are now in a position to work through several examples of complete piping systems. Three basic types of problems are encountered. First, for a given piping system and flow rate, it might be required to find the pressure drop. Second, for a given system and pressure drop, the flow rate might be required. Third, for a given flow and pressure drop, it might be required to design the system, that is, to determine the necessary pipe diameter or pipe length. In the following material, an example of each type of problem is presented.

EXAMPLE 6.7

For the piping system shown in Figure 6.32, determine the pressure p_2. There are 100 m of 15-cm-I.D. cast iron pipe and 30 m of 7.5-cm-I.D. cast iron pipe. Water flow rate is 0.01 m^3/s, the water temperature is 20°C, and the gage pressure at 1 is 250 kPa.

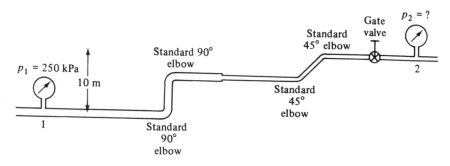

Figure 6.32

Solution First write the modified Bernoulli equation (6.7) between 1 and 2, including minor losses:

$$\frac{p_1}{\rho} + \frac{V_1^2}{2} + gz_1 = \left(\frac{fL}{D}\frac{V^2}{2}\right)_{\substack{15\text{-cm}\\ \text{pipe}}} + \left(\Sigma K\frac{V^2}{2}\right)_{\substack{15\text{-cm}\\ \text{pipe}}} + \left(\frac{fL}{D}\frac{V^2}{2}\right)_{\substack{7.5\text{-cm}\\ \text{pipe}}}$$

$$+ \left(\Sigma K\frac{V^2}{2}\right)_{\substack{7.5\text{-cm}\\ \text{pipe}}} + \frac{V_2^2}{2} + gz_2 + \frac{p_2}{\rho}$$

The velocity in the 15-cm pipe is

$$V_{15} = \frac{Q}{A} = \frac{0.01\ \text{m}^3/\text{s}}{(\pi/4)(0.15)^2\ \text{m}^2} = 0.5659\ \text{m/s}$$

and in the 7.5-cm pipe it is

$$V_{7.5} = \frac{Q}{A}\frac{0.01}{(\pi/4)(0.075)^2\ \text{m}^2} = 2.264\ \text{m/s}$$

At 20°C, for water, $\nu = 1.00 \times 10^{-6}\ \text{m}^2/\text{s}$, so that

$$\text{Re}_{15} = \frac{(0.5659\ \text{m/s})(0.15\ \text{m})}{1.00 \times 10^{-6}\ \text{m}^2/\text{s}} = 8.489 \times 10^4$$

$$\text{Re}_{7.5} = \frac{(2.264\ \text{m/s})(0.075\ \text{m})}{1.00 \times 10^{-6}\ \text{m}^2/\text{s}} = 1.698 \times 10^5$$

Also,

$$\left(\frac{\epsilon}{D}\right)_{15} = \frac{0.26 \times 10^{-3}\ \text{m}}{0.15\ \text{m}} = 0.00173$$

$$\left(\frac{\epsilon}{D}\right)_{7.5} = \frac{0.26 \times 10^{-3}\ \text{m}}{0.075\ \text{m}} = 0.00346$$

From the Moody diagram, we obtain

$$f_{15} = 0.025$$

$$f_{7.5} = 0.027$$

The minor losses in the 15-cm pipe are given by

$$(\Sigma K)_{15} = \underset{\substack{\text{standard}\\ \text{elbow}}}{0.75} + \underset{\substack{\text{standard}\\ \text{elbow}}}{0.75} = 1.50$$

For the 7.5-cm pipe,

$$(\Sigma K)_{7.5} = \underset{\text{contraction}}{0.33} + \underset{\text{elbow}}{0.35} + \underset{\text{elbow}}{0.35} + \underset{\text{gate valve}}{0.20} = 1.23$$

Substituting into the Bernoulli equation, we have

$$\frac{p_2 - p_1}{\rho} = g(z_1 - z_2) + \frac{V_{15}^2}{2}\left(1 - \frac{fL}{D} - -\Sigma K\right)_{15} - \frac{V_{7.5}^2}{2}\left(1 + \frac{fL}{D} + \Sigma K\right)_{7.5}$$

$$= 9.81 \text{ m/s}^2(-10 \text{ m}) + \left(\frac{0.5659^2}{2} \text{ m}^2/\text{s}^2\right)\left(1 - \frac{0.025 \times 100}{0.15} - 1.50\right)$$

$$- \left(\frac{2.264^2}{2} \text{ m}^2/\text{s}^2\right)\left(1 + \frac{0.027 \times 30}{0.075} + 1.23\right)$$

$$= -98.1 \text{ m}^2/\text{s}^2 + (0.1601 \text{ m}^2/\text{s}^2)(-17.17) - (2.563 \text{ m}^2/\text{s}^2)(13.03)$$

$$= -98.1 \text{ m}^2/\text{s}^2 - 2.749 \text{ m}^2/\text{s}^2 - 33.40 \text{ m}^2/\text{s}^2$$

$$= -134.2 \text{ m}^2/\text{s}^2$$

or

$$p_2 - p_1 = -(998.3 \text{ kg/m}^3)(134.2 \text{ m}^2/\text{s}^2)$$
$$= -134 \text{ kPa}$$

or

$$p_2 = 250 - 134$$
$$= \underline{116 \text{ kPa (gage)}} \qquad \blacksquare$$

EXAMPLE 6.8
For the piping system shown in Figure 6.33, determine the water flow rate. The water temperature is 70°F, with all pipe asphalted cast iron. There are 1000 ft of 3-in-I.D. pipe.

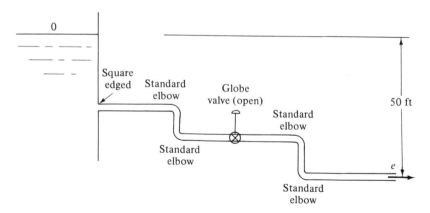

Figure 6.33

Solution Write the modified Bernoulli equation between the reservoir surface and pipe outlet:

$$\frac{p_0}{\rho} + gz_0 + \frac{V_0^2}{2} = \frac{fL}{D}\frac{V^2}{2} + \Sigma K\frac{V^2}{2} + \frac{p_e}{\rho} + gz_e + \frac{V_e^2}{2}$$

where $p_0 = p_e$ = atmospheric pressure; $z_0 - z_e = 50$ ft; and $V_0 = 0$. For this case, with flow velocity the unknown, Reynolds numbers cannot be directly determined; a trial-and-error solution is called for. As a first trial, assume $f = 0.02$.

$$(32.17 \text{ ft/s}^2)(50 \text{ ft}) = \left[\frac{0.02 \times 1000}{3/12} + \left(\underset{\substack{\text{square-} \\ \text{edged} \\ \text{inlet}}}{0.5} + \underset{\text{4 elbows}}{3.0} + \underset{\substack{\text{globe} \\ \text{valve}}}{6.4}\right) + 1\right]\frac{V_e^2}{2}$$

$$\frac{V_e^2}{2} = \frac{32.17(50)}{80 + 9.9 + 1} \text{ ft}^2/\text{s}^2$$

or

$$V_e = 5.95 \text{ ft/s}$$

For this velocity,

$$\text{Re} = \frac{VD}{\nu} = \frac{(5.95 \text{ ft/s})(\frac{3}{12}\text{ ft})}{1.04 \times 10^{-5} \text{ ft}^2/\text{s}} = 1.41 \times 10^5$$

$$\frac{\epsilon}{D} = \frac{0.0004 \text{ ft}}{3/12 \text{ ft}} = 0.0016$$

From the Moody diagram, we obtain $f = 0.023$. Since this value does not agree with our initial trial, we shall now make a second trial, starting with $f = 0.023$.

$$\frac{fL}{D} = \frac{0.023 \times 1000}{3/12} = 92$$

$$\frac{V_e^2}{2} = \frac{32.17(50)}{92 + 9.9 + 1} \text{ ft}^2/\text{s}^2$$

or

$$V_e = 5.59 \text{ fps}$$

For this velocity,

$$\text{Re} = \frac{VD}{\nu} = \frac{5.59 \text{ ft/s} \times \frac{3}{12}\text{ ft}}{1.04 \times 10^{-5} \text{ ft}^2/\text{s}} = 1.34 \times 10^5$$

From the Moody diagram, we find $f = 0.023$. It can be seen that generally we are able to converge on the correct answer quite rapidly; no more than two trials are required, as a rule. The water flow rate is

$$Q = AV$$

$$= \left[\frac{\pi}{4}\left(\frac{3}{12}\right)^3 \text{ft}^2 \right](5.59 \text{ ft/s})$$

$$= \underline{0.27 \text{ cfs}} \qquad \text{or} \qquad \underline{16.5 \text{ cfm}} \qquad \blacksquare$$

EXAMPLE 6.9

A pump is to be used to supply 5 liters per second of water from a reservoir to a point 400 m from the reservoir at the same level as the reservoir surface. Determine the minimum-size cast iron pipe required. The water temperature is 15°C; assume that minor losses can be neglected. (See Figure 6.34). The pump supplies 50 kW of power to the water flow.

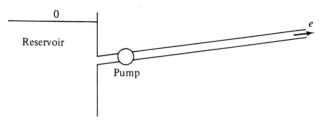

Figure 6.34

Solution   From Equation (6.8) we can express the pump work in terms of reservoir surface and outlet conditions:

$$-\frac{1}{\dot{m}}\frac{dW_p}{dt} = \frac{p_e - p_0}{\rho} + g(z_e - z_0) + \left(\frac{fL}{D}\frac{V^2}{2}\right) + \frac{V_e^2}{2}$$

For our case,

$$\dot{m} = \rho Q = (999.2 \text{ kg/m}^3)(5 \times 10^{-3} \text{ m}^3/\text{s}) = 4.996 \text{ kg/s}$$

$$-\frac{1}{\dot{m}}\frac{dW_p}{dt} = -\frac{1}{4.996 \text{ kg/s}}(-50 \times 10^3 \text{ N} \cdot \text{m/s}) = +10,010 \text{ m}^2/\text{s}^2$$

(*Note: dW_p/dt is negative work in a thermodynamic sense, for it represents work done on the fluid.*)

We now have

$$10,010 = \left(\frac{fL}{D} + 1\right)\frac{V^2}{2}$$

where

$$V = \frac{Q}{A} = \frac{5 \times 10^{-3} \text{ m}^3/\text{s}}{(\pi/4)D^2 \text{ m}^2}$$

$$= \frac{6.366 \times 10^{-3}}{D^2} \text{ m/s} \qquad \text{with} \qquad D \text{ in m}$$

$$10,010 = \left(\frac{400f}{D} + 1\right)\frac{4.053 \times 10^{-5}}{2D^4}$$

For this case, with D the unknown, the Reynolds number and f cannot be immediately determined. A trial-and-error procedure is required. As a first trial, assume that $f = 0.025$.

$$10{,}010 = \left[\frac{400(0.025)}{D} + 1\right]\frac{2.026 \times 10^{-5}}{D^4}$$

$$= \frac{20.26 \times 10^{-5}}{D^5} + \frac{2.026 \times 10^{-5}}{D^4}$$

The second term on the right is small compared to the first, so, to a first approximation,

$$10{,}010 \cong \frac{2.026 \times 10^{-4}}{D^5} \quad \text{and} \quad D = 0.0289 \text{ m}$$

For this first trial,

$$V = \frac{6.366 \times 10^{-3}}{(0.0289)^2} = 7.622 \text{ m/s}$$

$$\text{Re} = \frac{7.622 \times 0.0289}{1.15 \times 10^{-6}} = 1.915 \times 10^5$$

$$\frac{\epsilon}{D} = \frac{0.26 \times 10^{-3}}{0.0289} = 0.0090$$

From the Moody diagram, we obtain $f = 0.036$. To start a second iteration, let $f = 0.036$. Therefore,

$$10{,}010 = \left[\frac{400(0.036)}{D} + 1\right]\frac{2.026 \times 10^{-5}}{D^4}$$

or

$$10{,}010 \sim \frac{0.0002917}{D^5} \quad \text{and} \quad D = 0.0311 \text{ m}$$

For this second trial,

$$V = \frac{6.366 \times 10^{-3}}{(0.0311)^2} = 6.582 \text{ m/s}$$

$$\text{Re} = \frac{6.582 \times 0.0311}{1.15 \times 10^{-6}} = 1.78 \times 10^5$$

$$\frac{\epsilon}{D} = \frac{0.26 \times 10^{-3}}{0.0311} = 0.0084$$

From the Moody diagram, we obtain $f = 0.036$. The agreement with the assumed value is good enough so that $D = 0.0311$ m can be taken as the required diameter.

■

EXAMPLE 6.10

The supply line for a small air-driven hand tool (stapler, nailer, drill, etc.) consists of a $\frac{1}{4}$-in-diameter nylon hose. The tool requires an air pressure of 90 psig. Supply air is available at 92.5 psig and a temperature of 80°F. The rated air consumption of the tool is 4 scfm (cubic feet per minute at standard conditions). Determine the maximum allowable length of hose that may be used.

Solution   At standard conditions (60°F and 1 atmosphere), the density of air is 0.0763 lbm/ft³; hence the mass flow rate of the tool is

$$\dot{m} = \rho Q = (0.0763 \text{ lbm/ft}^3)(4 \text{ ft}^3/\text{min}) = 0.305 \text{ lbm/min} = 0.0051 \text{ lbm/s}$$

From Equation (6.5), we get

$$p_2 - p_1 = -\frac{\rho}{2}\overline{V}^2 \frac{fL}{D}$$

or

$$L = \frac{p_1 - p_2}{\rho} \frac{2D}{f\overline{V}^2}$$

As a first approximation, use the air density at the hose inlet; that is, assume incompressible flow. From the equation of state, we get the density ρ_1.

$$\rho_1 = \frac{p_1}{RT_1} = \frac{107.2(144) \text{ lbf/ft}^2}{(53.35 \text{ lbf/lbm°R})(32.17 \text{ lbm/slug})(460 + 80)°R}$$

$$= 0.01666 \text{ slugs/ft}^3$$

From the continuity equation, we obtain the average velocity V in the hose:

$$V = \frac{Q}{A} = \frac{\dot{m}}{\rho A} = \frac{0.0051 \text{ lbm/s}}{0.01666 \text{ slugs/ft}^3(32.17 \text{ lbm/slug})(\pi/4)[1/4(12)]^2\text{ft}^2}$$

$$= 27.9 \text{ ft/s}$$

The Reynolds number is

$$\text{Re} = \frac{\rho D\overline{V}}{\mu} = \frac{0.01666 \text{ slugs/ft}^3[1/4(12)] \text{ ft} (27.9 \text{ ft/s})}{3.85 \times 10^{-7} \text{ lbf-s/ft}^2} = 25{,}150$$

where 3.85×10^{-7} lbf - s/ft² is the dynamic viscosity of air at 80°F.
From the Moody diagram, for smooth tubes we get

$$f = 0.0245$$

Hence

$$L = \frac{(2.5(144)\text{lbf/ft}^2)(2)(\frac{1}{48}) \text{ ft}}{(0.01666 \text{ slugs/ft}^3)(0.0245)(32.3)^2\text{ft}^2/\text{s}^2} = \underline{35.2 \text{ ft}}$$

Next, check the air density at the tool inlet, assuming constant temperature flow:

$$\rho_2 = \frac{p_2}{RT_1} = \frac{(144)\text{lbf/ft}^2}{53.35\text{ft-lbf/lbm°R}(32.17)\text{lbm/slug}(540)°R} = 0.01627 \text{ slugs/ft}^3$$

which represents a change in density of less than 3%. Hence, the assumption of incompressibility is valid.   ■

6.6 PIPES IN PARALLEL

The previous section has dealt with systems in which the resistances (friction and minor losses) were placed in series. In many cases, however, a system involves pipes in parallel, as shown in Figure 6.35. For example, it might be required to find the flows through 2 and 3 for given pipe diameters D_1, D_2, D_3, and D_4, given $p_1 - p_4$, and given flow at 1. In this case, the flow resistances are in parallel. There is a direct analog to a dc electrical circuit. In an electrical circuit, electrical current (amperes) flows through an electrical resistance measured in ohms. The driving force for such a system is voltage drop. In the circuit shown (Figure 6.36) resistances R_2 and R_3 are in parallel. In order to find the currents I_2 and I_3, we use the fact that the voltage drop across each resistor must be the same:

$$I_2 = \frac{V_1 - V_4}{R_2}$$

$$I_3 = \frac{V_1 - V_4}{R_3}$$

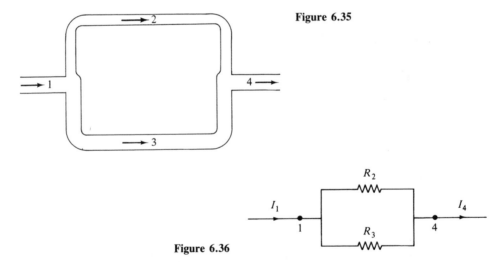

Figure 6.35

Figure 6.36

Similarly, in the flow circuit, with pressure drop the driving force analogous to voltage drop, the pressure drop $p_1 - p_4$ in Figure 6.35 is the same regardless of the path taken. Neglecting minor losses and elevation differences, we have for $A_1 = A_4$ (and hence for $V_1 = V_4$):

$$\frac{p_1 - p_4}{\rho} = \left(\frac{fL}{D}\right)_2 \frac{V_2^2}{2}$$

$$\frac{p_1 - p_4}{\rho} = \left(\frac{fL}{D}\right)_3 \frac{V_3^2}{2}$$

For a given pressure drop, therefore, the flow through each branch can be determined as in Example 6.9.

EXAMPLE 6.11

Branch 3 of a parallel pipe system as shown in Figure 6.35 is 50 ft long with a smooth 1-in-diameter tube. Branch 2 consists of a smooth ½-in-diameter tube. Water at 60°F enters the system at a flow rate of 0.15 cfs. Determine the effect of changing the length of branch 2 on the flow rate in each branch and on the pressure drop across the branches. Neglect minor losses and assume cross-sectional areas A_1 and A_4 to be equal.

Solution   The pressure drops across the two branches will be equal; hence

$$\Delta p_2 = \Delta p_3$$

or

$$\rho \frac{f_2 L_2}{D_2} \frac{V_2^2}{2} = \rho \frac{f_3 L_3}{D_3} \frac{V_3^2}{2}$$

$$\frac{f_2 L_2}{\frac{1}{2}/12} V_2^2 = \frac{f_3 (50)}{1/12} V_3^2$$

$$L_2 = 25 \frac{f_3}{f_2} \left(\frac{V_3}{V_2}\right)^2$$

Next, we assume a partitioning of volume flow rates between the two branches and determine the resulting length of branch 2; therefore let

$$Q_2 = m Q_3$$

$$\frac{\pi}{4} D_2^2 V_2 = m \frac{\pi}{4} D_3^2 V_3$$

Hence

$$\frac{V_3}{V_2} = \frac{1}{4m}$$

Therefore

$$L_2 = 25 \frac{f_3}{f_2} \frac{1}{16m^2} = 1.5625 \frac{1}{m^2} \frac{f_3}{f_2}$$

Next, we determine the average velocity and Reynolds number in each branch to obtain friction factor values from the Moody diagram. Now

$$Q_1 = Q_2 + Q_3 = 0.15 \text{ ft}^3/\text{s} = \frac{\pi}{4} D_2^2 V_2 + \frac{\pi}{4} D_3^2 V_3$$

$$= \frac{\pi}{4}\left(\frac{1}{24}\right)^2 V_2 + \frac{\pi}{4}\left(\frac{1}{12}\right)^2 \frac{V_2}{4m} = 0.0013635\left(1 + \frac{1}{m}\right)V_2$$

Hence

$$V_2 = \frac{110}{1 + 1/m} \quad \text{and} \quad V_3 = \frac{110}{4(m + 1)}$$

$$\text{Re}_2 = \frac{D_2 V_2}{\nu} = \frac{(\frac{1}{2}/12) \text{ ft } V_2 \text{ ft/s}}{1.22 \times 10^{-5} \text{ ft}^2/\text{s}} = 3415 V_2$$

$$\text{Re}_3 = \frac{D_3 V_3}{\nu} = \frac{\frac{1}{12} \text{ ft } V_3 \text{ ft/s}}{1.22 \times 10^{-5} \text{ ft}^2/\text{s}} = 6831 V_3$$

The pressure drop across both branches is given by

$$\Delta p = \frac{\rho V_3^2 f_3 L_3}{2 D_3} = f_3 V_3^2 \left(\frac{1.94}{2} \text{ slugs/ft}^3\right)\frac{50}{1/12} = 582 f_3 V_3^2$$

As an example, take $Q_2 = \frac{1}{2}Q_3$; hence, $V_2 = \frac{110}{3} = 36.67$ fps and $V_3 = 110/4(\frac{3}{2}) = 18.33$ fps; $\text{Re}_2 = 3415(36.67) = 125{,}230$ and $\text{Re}_3 = 6831(18.33) = 125{,}210$. From the Moody diagram, we find $f_2 = 0.0165$ and $f_3 = 0.0165$. The length of branch 2, L_2, is $1.5625(4)(0.0165)/0.0165 = 6.25$ ft. The pressure drop becomes $\Delta p = 582(0.0165)(18.33)^2 = 3226$ psf $= 22.4$ psi.

The results of calculations for a range of flow rate ratios m are shown in the following table:

m	V_2 (fps)	V_3 (fps)	Re_2	Re_3	f_2	f_3	L_2 (ft)	Q_2 (cfs)	Q_3 (cfs)
1	55	13.75	187,825	93,925	.015	.0175	1.82	0.075	0.075
$\frac{1}{2}$	36.67	18.33	125,230	125,210	.0165	.0165	6.25	0.05	0.10
$\frac{1}{4}$	22	22	75,130	150,280	.0185	.016	21.62	0.03	0.12
$\frac{1}{5}$	18.33	22.91	62,600	155,500	.019	.0157	32.28	0.025	0.125
$\frac{1}{6}$	15.71	23.57	53,650	161,010	.020	.0155	43.59	0.0215	0.1285
$\frac{1}{8}$	12.22	24.44	41,730	166,950	.0215	.015	69.77	0.0167	0.1333

■

EXAMPLE 6.12

A horizontal heating duct supplies air at 60°C to two branches as shown in Figure 6.37. The two branches lead to two registers (duct outlets which direct the air into the room). Branch B contains two well-rounded 90° duct turns with a loss coefficient K of 0.2. The supply duct is 350 mm by 80 mm. At the juncture, each branch has a width of 175 mm. The depth of all ducts is maintained at 80 mm. The pressure drop from the supply duct to each register outlet is the same. Supply flow is 10 m³/min.

Assume that any contracting or expanding section, if required in branch B, is sufficiently gradual so that the pressure drop due to this section can

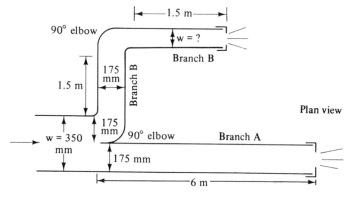

Figure 6.37

be neglected. In the design of heating ducts, the friction factor is usually taken at 0.02.

Determine the width of branch B at the register.

Solution From Table 2 of Appendix A we get at a temperature of 60°C, $\rho = 1.06 \text{ kg/m}^3$ and $\nu = 1.89 \times 10^{-5} \text{ m}^2/\text{s}$.

In the supply duct, the average velocity is

$$\overline{V} = \frac{Q}{A} = \frac{10 \text{ m}^3/\text{min}}{(60 \text{ s/min})(0.35 \text{ m})(0.08 \text{ m})} = 5.95 \text{ m/s}$$

Since the depth of all ducts remains constant, we obtain from the continuity condition that the average velocity in the two ducts will be the same, that is, $Q = \overline{V}A = \overline{V}wd$; for example, $Q_A = Q/2 = \overline{V}_A d(w/2)$, and hence $\overline{V} = \overline{V}_A$.

Branch A:

From Equation (6.5),

$$\frac{\Delta p}{L} = -\frac{1}{2}\rho \overline{V}^2 \frac{f}{D_h} = -\frac{1}{2}\frac{1.06}{1}(5.95)^2 \frac{0.02}{D_h}$$

with

$$D_h = 2\frac{wd}{(w+d)} = 2\frac{(0.175)(0.08)}{0.255} = 0.110 \text{ m}$$

Hence

$$\frac{\Delta p}{L} = -3.41 \text{ Pa/m}$$

or, with $L = 6$ m,

$$\Delta p = -20.47 \text{ Pa}$$

(As a check on the friction factor, we get from Figure 6.18, for

$$\text{Re}_{D_h} = \frac{\overline{V}D_h}{\nu} = \frac{5.95(0.110)}{1.89 \times 10^{-5}} = 0.356 \times 10^5$$

$f = 0.022$ for a smooth sheet metal duct.)

Branch B:

$$\Delta p \text{ (2 elbows)} = -2K \frac{1}{2} \rho \overline{V}^2 = -2(0.2)\frac{1}{2}\frac{1.06}{1}(5.95)^2$$
$$= 7.51 \text{ Pa}$$

For the first 1.5-m length, $\Delta p = -3.41(1.5) = -5.11$ Pa. Hence in the remainder of branch B, the pressure drop is $20.47 - (7.51 + 5.11) = 7.85$ Pa.

Hence, in the section of unknown width in branch B,

$$\frac{\Delta p}{L} = -\frac{7.85}{1.5} = -5.23 \text{ Pa/m} = \frac{1}{2}\frac{1.06}{1}\frac{\overline{V}^2}{D_h}f$$

or, with $f = 0.02$,

$$\frac{\overline{V}^2}{D_h} = \frac{5.23(2)}{1.06(0.02)} = 493.7 \text{ m/s}^2$$

Now $Q = \overline{V}A = \overline{V}w(0.08)$.

$$\overline{V} = \frac{5}{60(0.08)w} = \frac{1.042}{w}$$

and

$$D_h = \frac{2dw}{(d + w)} = 2\frac{0.08w}{0.08 + w}$$

Hence

$$\frac{\overline{V}^2}{D_h} = \frac{(1.042)^2(w + 0.08)}{w^3 0.16} = 6.782\frac{w + 0.08}{w^3}$$

Thus

$$493.7 = 6.782\frac{w + 0.08}{w^3} \quad \text{or} \quad 72.75w^3 = w + 0.08$$

Using a trial-and-error solution for w,

w	$72.75w^3$	$w + 0.08$
0.1	0.073	0.18
0.15	0.246	0.23
0.14	0.200	0.22
0.145	0.222	0.225
0.146	0.226	0.226

we obtain $w = \underline{146 \text{ mm}}$.   ■

EXAMPLE 6.13

Twelve cubic meters per minute of water flow into a horizontal pipe network consisting of two cast iron pipe branches (Figure 6.38). The first branch is 100 m long with a 200-mm pipe diameter, and the second branch is 200 m long with a 250-mm pipe diameter and a half-open gate valve. Determine the flow through each pipe and the pressure drop across the branches. Water temperature is 20°C.

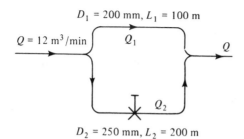

$D_1 = 200$ mm, $L_1 = 100$ m

$Q = 12$ m³/min

Q_1

Q

Q_2

$D_2 = 250$ mm, $L_2 = 200$ m

Figure 6.38

Solution   From Equation (6.7) and Section 6.5, the pressure drop in each branch is given by:

$$\Delta p = \rho f \frac{L}{D} \frac{V^2}{2} + \rho \Sigma K \frac{V^2}{2}$$

The velocity in each branch can be expressed as $V = Q/(\pi/4)D^2$, so that, for each branch:

$$\Delta p = \frac{8}{\pi^2} \rho \left(\frac{fL}{D} + \Sigma K \right) \frac{Q^2}{D^4}$$

But $\Delta p_1 = \Delta p_2$ and $\Sigma K_1 = 0$; therefore, the ratio of volume flow rates is

$$\frac{Q_1}{Q_2} = \left(\frac{D_1}{D_2} \right)^2 \sqrt{\frac{f_2 L_2/D_2 + \Sigma K_2}{f_1 L_1/D_1}}$$

As a first trial, assume $f_1 = f_2 = 0.020$. From Table 6.3, $K_2 = 4.5$ for a half-open gate valve; hence

$$\frac{f_2 L_2}{D_2} + \Sigma K_2 = 0.020 \frac{200}{0.250} + 4.5 = 20.5$$

and

$$\frac{f_1 L_1}{D_1} = 0.020 \frac{100}{0.2} = 10$$

Therefore,

$$\frac{Q_1}{Q_2} = \left(\frac{0.200}{0.250} \right)^2 \sqrt{\frac{20.5}{10}} = 0.916$$

Total flow $Q = Q_1 + Q_2$, or

$$\frac{Q}{Q_2} = \frac{Q_1}{Q_2} + 1 = 1.916$$

and

$$Q_2 = \frac{\frac{12}{60}}{1.916} = 0.1044 \text{ m}^3/\text{s}$$

with $Q_1 = 0.2 - 0.1044 = 0.0956 \text{ m}^3/\text{s}$
  Then

$$V_2 = \frac{0.1044}{(\pi/4)(0.250)^2} = 2.13 \text{ m/s}$$

From Appendix A, $\nu = 1.00 \times 10^{-6} \text{ m}^2/\text{s}$ at 20°C for water, so

$$\text{Re}_2 = \frac{V_2 D_2}{\nu} = \frac{(2.13 \text{ m/s})(0.250 \text{ m})}{1.00 \times 10^{-6} \text{ m}^2/\text{s}} = 5.32 \times 10^5$$

From Table 6.1, $\epsilon = 0.26$ mm for cast iron pipe; hence,

$$\frac{\epsilon}{D_2} = \frac{0.00026}{0.250} = 0.001$$

Hence, from Figure 6.18, we find $f_2 = 0.02$, which was our initial assumption.
  Now,

$$\Delta p_2 = \frac{\rho V_2^2}{2}\left(\frac{f_2 L_2}{D} + K_2\right) = \frac{(998.3 \text{ kg/m}^3)(2.13^2 \text{ m}^2/\text{s}^2)}{2} 20.5$$

$$= 46.5 \text{ kPa}$$

where $\rho = 998.3 \text{ kg/m}^3$ at 20°C. Further,

$$V_1 = \frac{0.0956 \text{ m}^3/\text{s}}{(\pi/4)(0.200)^2 \text{ m}^2} = 3.04 \text{ m/s}$$

$$\text{Re}_1 = \frac{(3.04 \text{ m/s})(0.2 \text{ m})}{1 \times 10^{-6} \text{ m}^2/\text{s}} = 6.09 \times 10^5$$

$$\frac{\epsilon}{D_1} = \frac{0.00026 \text{ m}}{0.200 \text{ m}} = 0.0013$$

From Figure 6.17, $f_1 = 0.0208$, and hence

$$\Delta p_1 = \frac{\rho V_1^2}{2}\frac{f L_1}{D_1} = \frac{(998.3 \text{ kg/m}^3)(3.04^2 \text{ m}^2/\text{s}^2)}{2}\frac{0.0208 (100 \text{ m})}{0.2 \text{ m}}$$

$$= 48.0 \text{ kPa}$$

as compared to $\Delta p_2 = 46.5$ kPa.
  For our second trial, we increase Q_2 and decrease Q_1 to balance the pressure drops. Let $Q_2 = 0.105 \text{ m}^3/\text{s}$ and $Q_1 = 0.095 \text{ m}^3/\text{s}$. Then

$$V_2 = \frac{0.105 \text{ m}^3/\text{s}}{(\pi/4)(0.25)^2} = 2.14 \text{ m/s}$$

$$\text{Re}_2 = \frac{(2.14 \text{ m/s})(0.25 \text{ m})}{1 \times 10^{-6} \text{ m}^2/\text{s}} = 5.34 \times 10^5$$

Therefore,

$$f_2 = 0.02 \quad \text{and} \quad \Delta p_2 = 47.0 \text{ kPa}$$

$$V_1 = \frac{0.095 \text{ m}^3/\text{s}}{(\pi/4)(0.200)^2 \text{ m}^2} = 3.02 \text{ m/s}$$

$$\text{Re}_1 = 6.05 \times 10^5$$

$$f_1 = 0.0208, \qquad \Delta p_1 = 47.3 \text{ kPa}$$

Thus the solution is

$$Q_1 = 0.095 \text{ m}^3/\text{s}, \qquad Q_2 = 0.105 \text{ m}^3/\text{s}$$

with a pressure drop Δp of 47 kPa. Note that, although branch 2 has a larger diameter than branch 1, it carries less flow due to the presence of a valve and due to its greater length. ∎

PROBLEMS

6.1. Water at 10°C flows steadily through a 5-cm-diameter pipe. The pressure drop in a 3-m length of pipe is found to be 750 Pa. Compute the wall shear stress.

6.2. Air at 80°F flows steadily through a rectangular duct, 2 in by 6 in with a velocity of 5 fps. Is the flow laminar or turbulent?

6.3. For Problem 6.2, compute the pressure drop per foot of duct, with $f = 0.02$. Express your answer in inches of water.

6.4. Water flows through an annular section 15 m long. The inner diameter of the annulus is 10 cm, outer diameter 12 cm. For a water velocity of 10 cm/s, is the flow laminar or turbulent? Determine the friction factor and pressure drop per meter. Water temperature is 20°C; assume smooth surface.

6.5. Find the hydraulic diameter of a 30°-60° triangular section flowing full of fluid. Express your answer in terms of a, the smallest side of the triangle.

6.6. A crude oil has a kinematic viscosity of 1×10^{-5} m²/s at 10°C. The oil is flowing in a smooth circular tube, 1 cm in diameter, at a velocity of 3.0 cm/s. Is the flow laminar or turbulent? Find the friction factor.

6.7. A 1000-gallon swimming pool is to be filled with a $\frac{1}{2}$-in-diameter garden hose. If the supply pressure is 50 psi, compute the time required to fill the pool. The hose is 50 ft long, with $f = 0.023$.

6.8. The pressure difference across the windowsill of Figure P6.8 is 50 Pa. The gap is 0.1 mm, and the width of the window is 1 m. What is the average velocity and the volume flow rate of air through the gap? Take an air temperature of 28°C.

2 cm

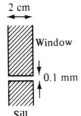

Window

0.1 mm

Sill   **Figure P6.8**

6.9. A container has a diameter of 4 in. A horizontal tube (d = 0.2 in, 4 in long) is attached to the container as shown in Figure P6.9. Initially, the liquid depth in the container is 4 in. During a period of 10 minutes, a quantity of 0.5 in^3 leaves the container. Find the kinematic viscosity of the liquid.

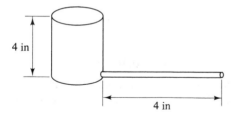

4 in

4 in   **Figure P6.9**

6.10. A liquid flows through a circular pipe resulting in a pressure drop of 2 kPa in a pipe length of 10 m. The Reynolds number of the flow is 100. What will be the pressure drop if the flow rate is tripled, assuming the same liquid properties? Repeat for a Reynolds number of 10^7.

6.11. Air flows through a $\frac{1}{2}$-in-diameter smooth tube at 450°R and absolute pressure of 3 psi. Determine the range of air velocities over which laminar flow will occur in the tube.

6.12. From tests made on a 75-mm-diameter circular pipe, a friction factor of 0.015 is calculated for an average water velocity of 10 m/s. The water temperature is 20°C. Find the average roughness height of the pipe.

6.13. Mercury at 68°F flows in a horizontal 1-in-diameter steel pipe (ϵ = 0.003 ft). The pressure drop is 14.5 psi over a distance of 50 ft. The density of the mercury is 26.2 slug/ft^3, and its kinematic viscosity is 1.23×10^{-6} ft^2/s. What is the volume and mass flow rate of the mercury?

6.14. Air at atmospheric pressure and 20°C flows through a horizontal duct 100 m long. The volume flow rate is 10 m^3/s; the pressure drop in the duct is not to exceed 300 Pa. Using riveted sheet metal with ϵ = 0.005 m, would a circular or square cross section require less sheet metal?

6.15. SAE 10 oil flows through a $\frac{1}{2}$-in-diameter tube, 150 ft long. The pressure drop is measured to be 10 psi. For a kinematic viscosity of 0.001 ft^2/s, find the volumetric flow through the tube.

6.16. Water at 5°C flows through a smooth tube at a rate of 50 liters per second. The pressure drop in 1000 m of tubing is 376 Pa. What is the tube diameter?

6.17. Show that for a constant rate of discharge and a constant value of friction factor, the friction loss in a pipe varies inversely as the fifth power of the diameter.

6.18. A 5-ft-diameter pipe has a length of 2000 ft. The vertical drop from a reservoir to the end of the pipe is 120 ft. Assume a friction factor of 0.02 in the pipe and calculate the maximum amount of power the water flowing through the pipe could deliver.

6.19. Determine the power which must be added to pump 0.5 m^3/min of a liquid through a horizontal smooth pipe 300 m long and 50 mm in diameter. Density of the liquid is 2000 kg/m^3, kinematic viscosity 2.5×10^{-8} m^2/s.

6.20. A $\frac{1}{4}$-horsepower fan is to be used to supply air through an air conditioning duct 4 in by 8 in, 30 ft long. Determine the volume flow rate of air through the duct and the pressure just downstream of the fan, for an air temperature of 60°F.

6.21. Repeat Problem 6.20 with six elbows (90°) installed in the duct.

6.22. A water drainage pipe is to be installed so that 2 m³/min flow through it solely by gravity. The pipe is made of smooth concrete ($\epsilon = 0.3$ mm) with a diameter of 150 mm. What must be the slope of the pipe to obtain the desired flow? Assume the pipe flows full.

6.23. Calculate the rate of water flow through the pipe shown in Figure P6.23. Assume a power input to the pump of 3 horsepower, with a pump efficiency (output power/input power) of 80%. Compute the pressure at the pump inlet.

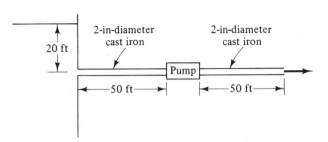

Figure P6.23

6.24. If the flow through the system described in Problem 6.23 is to be doubled, what size pump would be required?

6.25. A pump delivers 0.2 m³/s of water at 0°C through a 75-m-long cast iron pipe ($D = 0.1$ m) to a tank with water level 15 m above the centerline of the pump. Calculate the pump input power required if the pump efficiency is 80%.

6.26. A concrete culvert passes underneath a highway as shown in Figure P6.26. The horizontal culvert flows full and is rectangular, 3 ft × 1.5 ft. Determine the volume flow through the culvert for $\epsilon = 0.003$ ft.

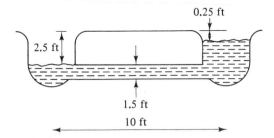

Figure P6.26

6.27. Chilled water at 5°C is flowing in a horizontal 250-mm-diameter cast iron pipe at a rate of 2 m³/min. The pipe is 20 m long with a standard 90° elbow and a half-open globe valve. Calculate the total pressure drop.

6.28. Water at 77°F is siphoned between the two large tanks shown in Figure P6.28. The connecting tube is a 1½-in-diameter smooth plastic hose 15 ft long. Determine the volumetric flow between the two reservoirs and the pressure at A.

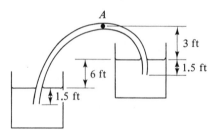

Figure P6.28

6.29. Water at 30°C discharges through a 250-mm-diameter cast iron pipe, as shown in Figure P6.29. Calculate the flow rate for the globe valve half and fully open; $p_1 - p_2 = 300$ kPa, and the pipe makes an angle of 0.05 rad with the horizontal.

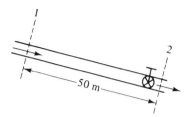

Figure P6.29

6.30. Determine the flow rate of water at 60°F through the siphon shown in Figure P6.30. Also calculate the minimum pressure in the siphon. The siphon will no longer operate when the minimum pressure becomes less than the vapor pressures of the liquid, with resultant vaporization. Give dimensions for the siphon shown at which vaporization would occur.

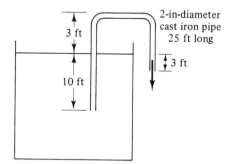

Figure P6.30

6.31. Due to corrosion and scaling, the roughness of a pipe increases with its years of service, varying approximately as

$$\epsilon = \epsilon_0 + kt$$

with ϵ_0 the roughness of the new surface given in Table 6.1, t the time in years, and k a constant. Assuming a value of k for cast iron of 0.00001 m per year, determine the discharge of water through a 20-cm-diameter cast iron pipe

500 m long as a function of time in years over a 20-year period. Assume the pressure drop across the pipe is maintained at 15 kPa.

6.32. A garden hose ($\frac{1}{2}$ in I.D.) is 100 ft long and is connected to a water tap where the pressure is 45 psig. Calculate the velocity at the hose exit. What will the exit velocity be if a nozzle ($\frac{1}{4}$ in diameter) is attached to the garden hose? Assume $\epsilon = 0.00075$ ft, water temperature $= 68°F$.

6.33. Determine the power output of the water turbine shown in Figure P6.33. Water temperature is 10°C, all pipe cast iron; neglect minor losses. Velocity in the 30-cm pipe is 2 m/s.

26.64KW

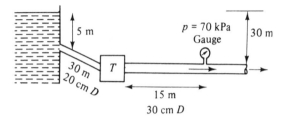

5 m $p = 70$ kPa   30 m
           Gauge

30 m
20 cm D T

15 m

30 cm D   Figure P6.33

6.34. Determine the horizontal component of the water force on pipe AB in Figure P6.34; assume water temperature $= 60°F$, pipe is cast iron, 1 ft diameter, and pipe exit from reservoir at A is well-rounded ($K = 0.05$).

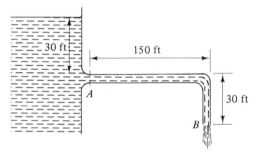

30 ft   150 ft

A 30 ft

B Figure P6.34

6.35. Find the pressure drop from section A to section B in Figure P6.35 for water at 30°C flowing at a rate of 0.01 m³/s. There are 5 m of 40-mm-diameter tube, and 0.5 m of 75-mm tube between A and B.

A B

Figure P6.35

6.36. Calculate the pressure p_1 such that 2.0 cfs of water will flow through the system shown in Figure P6.36. Assume a water temperature of 60°F, with the pipe asphalted cast iron.

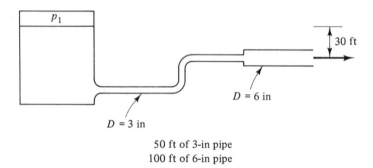

p_1

30 ft

$D = 6$ in

$D = 3$ in

50 ft of 3-in pipe
100 ft of 6-in pipe

Figure P6.36

6.37. Find the flow rate through the system shown in Figure P6.37. Assume cast iron pipe, with a water temperature of 20°C. The valve is a gate valve, fully open.

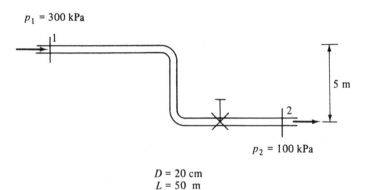

$p_1 = 300$ kPa

5 m

$p_2 = 100$ kPa

$D = 20$ cm
$L = 50$ m

Figure P6.37

6.38. The gate valve of Problem 6.37 is to be used to control the flow rate through the system. Calculate the flow for the valve $\frac{1}{4}$ open, $\frac{1}{2}$ open, and $\frac{3}{4}$ open.

6.39. What size pump will be necessary in the 3-in line so that 5 cfs will flow through each pipe? Neglect losses in bends and elbows. Assume that water is flowing at 70°F, with smooth pipe (see Figure P6.39).

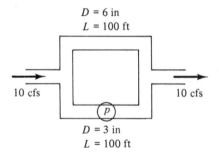

$D = 6$ in
$L = 100$ ft

10 cfs

10 cfs

p

$D = 3$ in
$L = 100$ ft

Figure P6.39

6.40. A factory draws its water supply from a mountain lake located 75 m above
the factory as shown in Figure P6.40. The factory is at present supplied via
300 m of 15-cm-diameter cast iron pipe. For control purposes, there are 5 gate
valves and 1 globe valve in the line, as well as three standard 90° elbows and
two 45° elbows. The water is supplied at atmospheric pressure. The factory
manager reports that even without a pump the present water supply system
as described is adequate to meet the maximum requirements of the factory,
although he doesn't know the maximum flow. However, the manager envisions
doubling the size of the factory in the near future, which would mean doubling
the water requirements. The manager would like to determine whether it would
be more economical to do this by increasing the pipe diameter size or by
installing a pump in the line. The cheapest available pipe is commercial steel.
Assist the manager by determining the following:
a. What size pump would be required to double the flow?
b. What diameter pipe would be required to double the flow? (Assume water
temperature to be 5°C.)

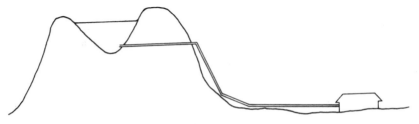

Figure P6.40

6.41. One hundred cfm of water is to be supplied from a reservoir to a building
located 100 ft below the reservoir surface. The water is to be supplied at
atmospheric pressure, the pipe to be cast iron, 1000 ft long, with two globe
valves and five standard elbows in the line. Calculate the minimum pipe size
that could be used.

6.42. Water flows from two reservoirs as shown in Figure P6.42, with a pipe from
each reservoir joined to form a single pipe. Calculate the total discharge through
the single pipe.

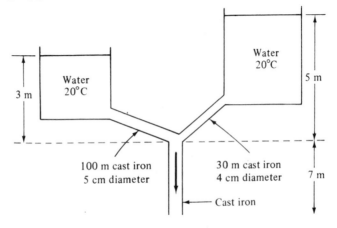

Figure P6.42

6.43. Water enters a horizontal pipe network from a large standpipe. The network consists of three pipes as shown in Figure P6.43. Assume a friction factor f of 0.025 for each pipe and determine the height of the water in the standpipe.

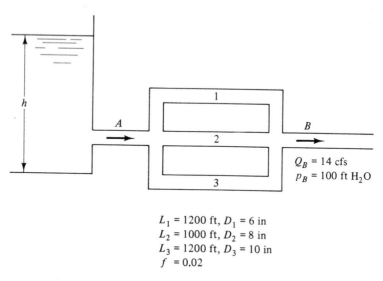

$L_1 = 1200$ ft, $D_1 = 6$ in
$L_2 = 1000$ ft, $D_2 = 8$ in
$L_3 = 1200$ ft, $D_3 = 10$ in
$f = 0.02$

Figure P6.43

6.44. Water flows from a reservoir through the piping system shown in Figure P6.44. Determine the flow through each pipe and the pressures at points 2 and 3 for the friction factors shown.

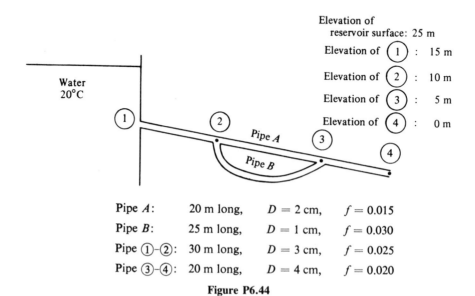

Pipe A:	20 m long,	$D = 2$ cm,	$f = 0.015$
Pipe B:	25 m long,	$D = 1$ cm,	$f = 0.030$
Pipe ①-②:	30 m long,	$D = 3$ cm,	$f = 0.025$
Pipe ③-④:	20 m long,	$D = 4$ cm,	$f = 0.020$

Figure P6.44

6.45. Two sites, both at sea level, are to be provided with water from a large storage reservoir, with surface 65 ft above sea level (Figure P6.45). Since one site is 300 ft farther from the reservoir than the other, a pump is to be installed in *AC* so as to provide the same flow to both sites. The pipe leading from the reservoir is 3-in pipe, *AB* and *AC* are 2-in pipe. Determine the pump power to balance the flows, assuming all pipes cast iron, water temperature 68°F. Neglect minor losses.

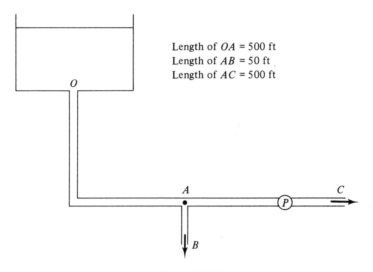

Length of *OA* = 500 ft
Length of *AB* = 50 ft
Length of *AC* = 500 ft

Figure P6.45

6.46. Calculate the flow rates in the system of pipes shown in Figure P6.46.

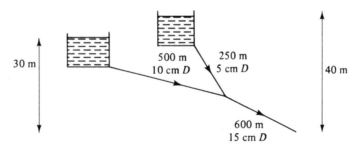

30 m

500 m
10 cm *D*

250 m
5 cm *D*

40 m

600 m
15 cm *D*

Figure P6.46

6.47. The water level in the open tank of Figure P6.47 is 7 ft above the entrance of the cast iron pipe. The pipe is 25 ft long and has a diameter of 4 in over the first 15 ft and 2 in over the remaining 10 ft. The pipe is inclined at 15° with the horizontal. Determine the initial discharge rate for a water temperature of 45°F.

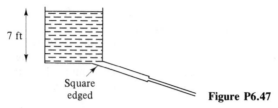

7 ft

Square
edged

Figure P6.47

6.48. A horizontal pipeline 10 km long carries petroleum oil. The diameter of the pipeline is 15 cm with the volume rate of flow 10 1/s. Calculate the pressure drop in the pipe. The viscosity of the oil is 0.045 Pa · s; the density of the oil is 890 kg/m^3, with the pipe made of cast iron.

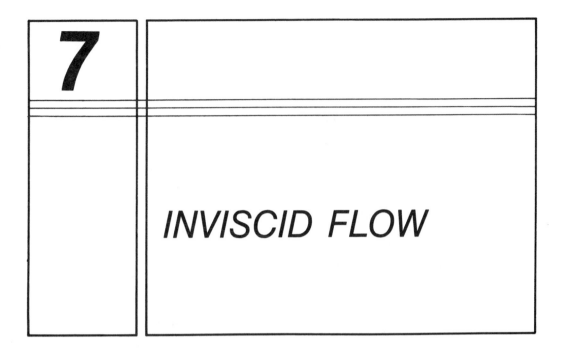

INVISCID FLOW

7.1 INTRODUCTION

In this and the following two chapters, we shall study external fluid flow over body surfaces. Such a study is important, for example, in analyzing the forces exerted on a body by a fluid passing over it. The resistance to the motion of a vehicle passing through a fluid—called drag—determines the propulsive requirements of such a vehicle. Equally important is the force normal to the direction of body motion—called lift—generated, for example, by the motion of an airfoil through a fluid.

An analysis of flow about bodies involves a solution of the equations of motion in two or three dimensions. Unfortunately, the complete equations of viscous fluid motion represent a set of complicated partial differential equations that can only be solved for certain simplified cases. For this reason, it is necessary to make certain approximations in analysis of the flow to enable the engineer to effect a workable solution for many important problems.

The fluids that we commonly deal with—water and air—possess relatively low viscosities. Consequently, over most of the flow field, the fluid can be treated as nonviscous. At the surface of a body, however, the fluid sticks to that surface, and no matter how small the value of μ, the velocity of a real fluid must be zero relative to the surface. Therefore, it follows that viscous fluid forces can generally be neglected, except in the immediate vicinity of a body surface. In terms of the dimensionless Reynolds number $\rho VL/\mu$ discussed in Chapter 5, as long as the fluid velocity V and characteristic body dimension L are not very small, the Reynolds number for a fluid of low viscosity will be large, implying that the viscous fluid forces are small compared to the inertial forces.

It is useful, then, to divide the flow over a body into two parts: a viscous region, where the fluid is either in the immediate vicinity of the body or has been in the immediate vicinity of the body, and a nonviscous region, where the effects of viscosity can be neglected and viscous forces need not be included in the equations of motion.

For example, for flow over the streamlined shape shown in Figure 7.1, viscous forces can be confined to a thin layer, called the *boundary layer,* next to the body surface; to a separated region near the rear of the body, where the boundary layer flow can no longer follow the body contour; and to the wake region, downstream of the body, which consists of fluid that has been affected by or has been in the boundary layer. The large majority of the flow field can be treated as inviscid.

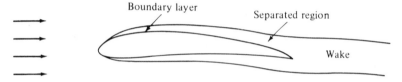

Figure 7.1.   Flow over streamline shape.

In this chapter, then, we shall look at the equations of motion for nonviscous flow. Boundary layer flow will be studied in Chapter 8, and, finally, in Chapter 9 we shall combine the two and analyze the resultant flow about a body. As a result of a study of inviscid flow, we hope to be able to determine, for example, the shape of the streamlines for flow around some simple geometrical shapes and the pressure distribution about these shapes when immersed in a nonviscous fluid flow. The latter is of more than academic interest, in that the boundary layer can often be considered thin enough so that there is no pressure variation through the boundary layer in a direction normal to the body surface. Thus the inviscid pressure distribution can be used for flow of a real, viscous fluid over a large portion of the body surface.

7.2 EQUATIONS OF TWO-DIMENSIONAL MOTION

In Chapter 4 we derived the fundamental equations of fluid flow and solved a number of one-dimensional flow problems—that is, those flow problems in which it was reasonable to assume that flow at a given cross section was a function of only one space variable. Here we shall utilize the results of Chapter 4 and treat two-dimensional steady flow, but shall restrict ourselves to the inviscid, incompressible case. Specifically, we shall derive the equations of continuity and momentum, using a differential element as the control volume. Several techniques for solving the equations will be demonstrated.

7.2(a) Continuity Equation

First let us derive the continuity equation of steady incompressible two-dimensional flow. In Section 4.1 the continuity equation for steady flow was obtained for a control volume in the form

$$\int_{\substack{\text{control}\\ \text{surface}}} \rho \mathbf{V} \cdot \mathbf{dA} = 0$$

or

$$\int_{\substack{\text{control}\\ \text{surface}}} \rho V_n \, dA = 0 \qquad\qquad (4.3)$$

where V_n is the local velocity component normal to the control surface element. For one-dimensional flow, where the velocity does not vary over a given cross section or in which the flow can be represented by an average velocity at each cross section, the foregoing equation can be integrated quite simply, as was done in Chapter 4. However, for two-dimensional flow, in which we have variations of flow quantities in two directions, we shall have to establish the continuity equation for a differential control volume, and the result will be a partial differential equation.

We shall derive the two-dimensional continuity equation for steady incompressible flow in the Cartesian (rectangular) coordinate system. Select a differential control volume in the flow field, as shown in Figure 7.2, with sides dx and dy and unit dimension in the z direction. Let u represent the velocity component in the x direction, v the velocity component in the y direction. Consider a mass balance for the control volume. If the x component of velocity at the left-hand face is u, the mass flow crossing this face and entering the control volume is

$$\underbrace{\rho u}\ \underbrace{dy}$$
$$\rho V_n \ dA$$

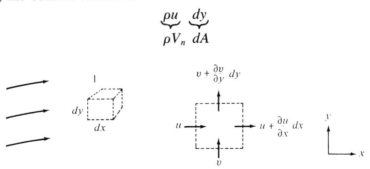

Figure 7.2

The velocity components u and v vary throughout the flow field, so from calculus, the x component of velocity at the right-hand face is $u + (\partial u/\partial x)$ dx. Thus the mass flow crossing the right-hand face and leaving the control volume is

$$\rho \underbrace{\left(u + \frac{\partial u}{\partial x} dx \right)}_{V_n} \underbrace{dy}_{dA}$$

Similarly, if the velocity in the y direction at the bottom face is v, the mass flow entering the control volume across the bottom face is $\rho v\, dx$. The velocity in the y direction at the top face is $v + (\partial v/\partial y)\, dx$, so that the mass flow rate leaving the control volume across the top face is

$$\rho \left(v + \frac{\partial v}{\partial y} dy \right) dx$$

For steady flow, the rate at which mass enters the control volume is equal to the rate at which mass leaves the control volume. Summing, we obtain

$$\rho u\, dy + \rho v\, dx = \rho \left(u + \frac{\partial u}{\partial x} dx \right) dy + \rho \left(v + \frac{\partial v}{\partial y} dy \right) dx$$

Simplifying, we have the continuity equation for steady two-dimensional incompressible (ρ = constant) flow:

$$\frac{\partial u}{\partial x} + \frac{\partial v}{\partial y} = 0 \tag{7.1}$$

7.2(b) Momentum Equation

Next let us derive the momentum equation for two-dimensional inviscid flow. For inviscid flow, there are no frictional (viscous) forces acting on the control volume. As in the case of the continuity equation, we shall utilize a differential volume element (Figure 7.2) in Cartesian coordinates. The momentum equation for steady flow is

$$\Sigma \mathbf{F} = \int_{\substack{control \\ surface}} \mathbf{V}(\rho \mathbf{V} \cdot \mathbf{dA}) \tag{4.8}$$

or, in differential form,

$$\Sigma\, d\mathbf{F} = \mathbf{V}(\rho \mathbf{V} \cdot \mathbf{dA})$$

Momentum is a vector quantity, so for this two-dimensional flow we will have one momentum equation in the x direction and one in the y direction.

The x-momentum equation (4.7) becomes, for steady flow,

$$\Sigma F_x = \int_{\substack{control \\ surface}} V_x \rho V_n\, dA \tag{7.2}$$

First consider the momentum flux in the x direction. The rate at which x momentum enters the control volume of Figure 7.2 across the left-hand face is

$$\underbrace{u}_{V_x} \underbrace{\rho u}_{\rho V_n} \underbrace{dy}_{dA}$$

The rate at which x momentum leaves the control volume across the right-hand face is

$$\underbrace{\left(u + \frac{\partial u}{\partial x} dx\right)}_{V_x} \underbrace{\rho\left(u + \frac{\partial u}{\partial x} dx\right)}_{\rho V_n} \underbrace{dy}_{dA}$$

Also, the fluid flow crossing the bottom face, $\rho v \, dx$, has a velocity component in the x direction and hence brings into the control volume an x-momentum flux $\rho u v \, dx$, as shown in Figure 7.3.

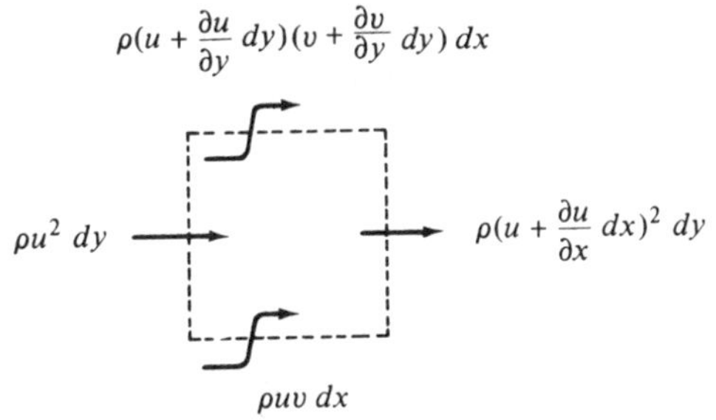

$$\rho(u + \frac{\partial u}{\partial y} dy)(v + \frac{\partial v}{\partial y} dy) \, dx$$

$\rho u^2 \, dy$ →   → $\rho(u + \frac{\partial u}{\partial x} dx)^2 \, dy$

$\rho u v \, dx$

Figure 7.3. x-momentum flux.

From calculus, the x-momentum flux leaving the control volume across the top face is

$$\rho\left(u + \frac{\partial u}{\partial y} dy\right)\left(v + \frac{\partial v}{\partial y} dy\right) dx$$

According to the sign convention adopted in Chapter 4, the rate of efflux of momentum is taken as positive, the rate of influx as negative. Therefore, the net x-momentum flux becomes

$$\rho\left(u + \frac{\partial u}{\partial x} dx\right)\left(u + \frac{\partial u}{\partial x} dx\right) dy + \rho\left(u + \frac{\partial u}{\partial y} dy\right)\left(v + \frac{\partial v}{\partial y} dy\right) dx$$
$$- \rho u^2 \, dy - \rho u v \, dx$$

Simplifying and dropping second-order terms, we obtain

$$\rho u \frac{\partial u}{\partial x} dx \, dy + \rho u \frac{\partial u}{\partial x} dx \, dy + \rho u \frac{\partial v}{\partial y} dx \, dy + \rho v \frac{\partial u}{\partial y} dx \, dy$$

But, from the continuity equation, $\partial u/\partial x + \partial v/\partial y = 0$, so that we finally have for the net x-momentum flux

$$\rho u \frac{\partial u}{\partial x} \, dx \, dy + \rho v \frac{\partial u}{\partial y} \, dx \, dy$$

Next, we shall evaluate the external forces acting on the differential control volume. Since we restrict ourselves to inviscid flow, we need include only pressure forces and gravity. With gravity acting in the y direction, we need only include pressure forces in the x-momentum equation. Pressure forces acting on the control volume in the x direction are shown in Figure 7.4.

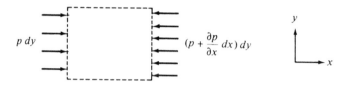

Figure 7.4. Forces acting on the control volume in the x direction.

The external forces acting in the x direction are given by

$$\Sigma F_x = p \, dy - \left(p + \frac{\partial p}{\partial x} \, dx \right) dy$$

$$= -\frac{\partial p}{\partial x} \, dx \, dy$$

Substituting the expressions obtained for the net x-momentum flux and for the external forces into Equation (7.2), we obtain the x-momentum equation

$$-\frac{\partial p}{\partial x} \, dx \, dy = \left(\rho u \frac{\partial u}{\partial x} + \rho v \frac{\partial u}{\partial y} \right) dx \, dy$$

or

$$-\frac{1}{\rho} \frac{\partial p}{\partial x} = u \frac{\partial u}{\partial x} + v \frac{\partial u}{\partial y} \qquad (7.3)$$

The y-momentum equation becomes, for steady flow,

$$\Sigma F_y = \int_{\substack{\text{control} \\ \text{surface}}} V_y (\rho V_n \, dA) \qquad (7.4)$$

First, let us consider the rate of inflow and efflux of y momentum across the boundaries of the control volume shown in Figure 7.5. The rate at which y momentum enters the differential control volume across the bottom face is

$$\underbrace{v}_{V_y} \; \underbrace{\rho v}_{\rho V_n} \; \underbrace{dx}_{dA}$$

The rate at which y momentum leaves the control volume across the top face is

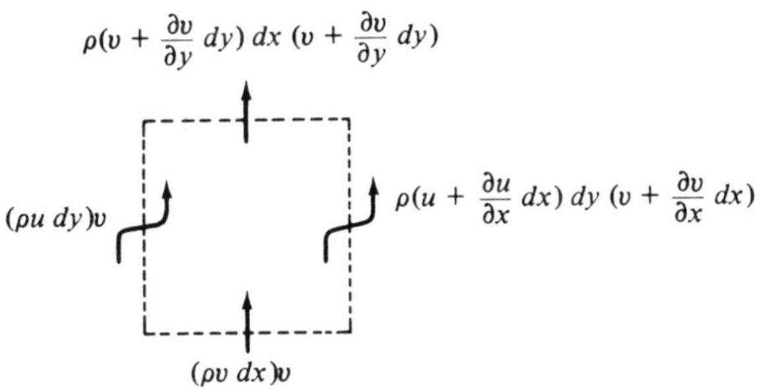

$$\rho(v + \frac{\partial v}{\partial y} dy) dx \, (v + \frac{\partial v}{\partial y} dy)$$

$(\rho u \, dy)v$

$\rho(u + \frac{\partial u}{\partial x} dx) dy \, (v + \frac{\partial v}{\partial x} dx)$

$(\rho v \, dx)v$

Figure 7.5.   y-momentum flux.

$$\underbrace{\left(v + \frac{\partial v}{\partial y} dy\right)}_{V_y} \underbrace{\rho\left(v + \frac{\partial v}{\partial y} dy\right)}_{\rho V_n} \underset{dA}{dx}$$

Also, the flow crossing the left face, $\rho u \, dy$, has a velocity component in the y direction and hence brings into the control volume a y-momentum flux $\rho u v$ dy. From calculus, the y-momentum flux leaving the control volume across the right face is

$$\rho\left(u + \frac{\partial u}{\partial x} dx\right)\left(v + \frac{\partial v}{dx} dx\right) dy$$

The net y-momentum flux becomes

$$\rho\left(u + \frac{\partial u}{\partial x} dx\right)\left(v + \frac{\partial v}{\partial x} dx\right) dy + \rho\left(v + \frac{\partial v}{\partial y} dy\right)\left(v + \frac{\partial v}{\partial y} dy\right) dx$$
$$- \rho u v \, dy - \rho v^2 \, dx$$

Simplifying, dropping second-order terms, and utilizing the continuity equation, we obtain for the net y-momentum flux

$$\rho u \frac{\partial v}{\partial x} dx \, dy + \rho v \frac{\partial v}{\partial y} dx \, dy$$

As shown in Figure 7.6, the external forces acting on the control volume in the y direction are pressure forces and gravity. Summing, we obtain

$$\Sigma F_y = -\frac{\partial p}{\partial y} dx \, dy - \rho g \, dx \, dy$$

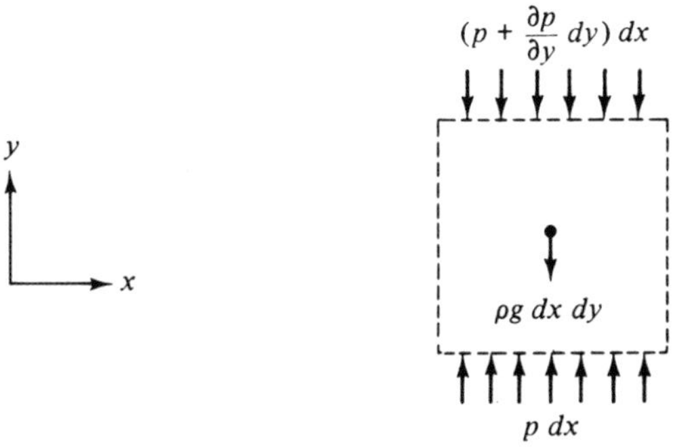

$(p + \frac{\partial p}{\partial y} dy) dx$

$\rho g \, dx \, dy$

$p \, dx$

Figure 7.6.   Forces acting on the control volume in the y direction.

Substituting the expressions obtained for the net y-momentum flux and for the external forces into Equation (7.4), we have the y-momentum equation

$$-\frac{\partial p}{\partial y}\,dx\,dy - \rho g\,dx\,dy = \rho\left(u\,\frac{\partial v}{\partial x} + v\,\frac{\partial v}{\partial y}\right)dx\,dy$$

or

$$-\frac{1}{\rho}\frac{\partial p}{\partial y} - g = u\,\frac{\partial v}{\partial x} + v\,\frac{\partial v}{\partial y} \tag{7.5}$$

The momentum equations (7.3) and (7.5), given in differential form and valid for frictionless flow, are called *Euler's equations*.

$$-\frac{1}{\rho}\frac{\partial p}{\partial x} = u\,\frac{\partial u}{\partial x} + v\,\frac{\partial u}{\partial y} \tag{7.3}$$

$$-\frac{1}{\rho}\frac{\partial p}{\partial y} - g = u\,\frac{\partial v}{\partial x} + v\,\frac{\partial v}{\partial y} \tag{7.5}$$

Together with the continuity equation (7.1), they constitute the equations of motion for steady frictionless two-dimensional flow.

7.3 STREAM FUNCTION AND VELOCITY POTENTIAL

In order to solve problems of steady incompressible two-dimensional fluid motion, using the continuity and Euler's equations, we must obtain solutions of three partial differential equations for the three unknowns (the velocity components u and v and the fluid pressure p) subject to certain boundary conditions. It was stated earlier that, for viscous flow, the fluid velocity at a body surface is zero relative to the surface. On the other hand, for flow that is assumed inviscid, the fluid velocity at the body surface is not zero. However, since the fluid cannot penetrate the solid surface or move away from the surface and leave a vacuum behind, the component of velocity normal to the body surface must be zero. It therefore follows that, for inviscid flow, the component of fluid velocity tangential to the body surface is nonzero.

The complexity of solving three simultaneous partial differential equations can be reduced by the introduction of a single function capable of describing the two velocity components of the flow field. Two such functions are available for this purpose: the stream function and the velocity potential function.

The stream function, designated $\psi(x, y)$, is a function of the space coordinates x, y and is related to the velocity components such that

$$u = \frac{\partial \psi}{\partial y}$$

$$v = -\frac{\partial \psi}{\partial x} \tag{7.6}$$

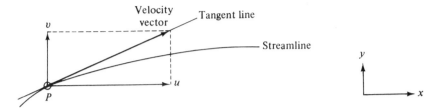

Figure 7.7.   Velocity vector and streamline.

where the velocity component u is given by the partial derivative with respect to y and the velocity component v is given by the partial derivative with respect to x.

From a knowledge of the stream function of a given flow field, we can obtain the velocity distribution and the streamlines of the flow. Before discussing the stream function further, let us relate the concept of stream function to the streamlines of the flow pattern. A streamline was defined (Section 1.3) as a continuous line drawn in the direction of the velocity vector at each point in the flow. In Figure 7.7 a tangent is drawn through the point P on the streamline. The slope of the tangent at point P is dy/dx, while the slope of the velocity vector is v/u. By definition of a streamline, the local velocity vector is in the direction of the tangent of the streamline. Therefore, the two slopes must be equal, and we obtain the differential equation of the streamlines:

$$\frac{dy}{dx} = \frac{v}{u} \tag{7.7}$$

Substituting the expressions for u and v (7.6) into the differential equation of the streamlines (7.7), we obtain

$$\frac{\partial \psi}{\partial y}\, dy + \frac{\partial \psi}{\partial x}\, dx = 0$$

As defined in calculus, the total differential $(d\psi)$ of the stream function ψ is given by

$$d\psi = \frac{\partial \psi}{\partial x}\, dx + \frac{\partial \psi}{\partial y}\, dy$$

It therefore follows that $d\psi = 0$ along a streamline: hence the stream function ψ remains constant along a streamline. In other words, the equation for the streamline in a flow field can be obtained from the expression for the stream function by setting the stream function equal to a constant. Different values of the constant will define different streamlines.

As an example of a stream function and streamlines, consider uniform parallel flow along the x axis, as shown in Figure 7.8, with constant uniform velocity U from left to right. The stream function in this case is

$$\psi = Uy$$

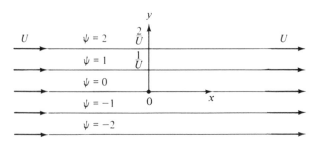

Figure 7.8. Streamlines in uniform flow.

This result can easily be verified by taking partial derivatives as indicated:

$$u = \frac{\partial \psi}{\partial y} = U$$

$$v = -\frac{\partial \psi}{\partial x} = 0$$

The streamlines are obtained by assigning constant values to the stream function, for example, letting $\psi = 0$; therefore, $Uy = 0$ corresponds to $y = 0$, that is, a horizontal line through the origin. Next let $\psi = 1$; therefore, $Uy = 1$ and $y = 1/U$ again correspond to a horizontal line with y intercept of $1/U$, as indicated in Figure 7.8. For $\psi = 2$ we obtain $y = 2/U$ as intercept. Similarly, we obtain the streamlines for negative y's.

The stream function serves as a means of establishing the streamlines of the flow. Another useful function in solving flow problems is the velocity potential. The velocity potential $\phi(x, y)$ is a function of the space coordinates x, y and is related to the velocity components as follows:

$$u = \frac{\partial \phi}{\partial x}$$

$$v = \frac{\partial \phi}{\partial y}$$

$\qquad(7.8)$

That is, the velocity component u is given by the partial derivative of ϕ in the x direction, and the velocity component v by the partial derivative of ϕ in the y direction. A flow for which the potential function ϕ exists, as defined above, at each point in the flow field, is called potential flow.

As an example of a velocity potential, again consider the case of uniform flow along the x axis (Figure 7.9). In this case, the velocity potential will be

$$\phi = Ux$$

This is verified by taking the partial derivatives as indicated:

$$u = \frac{\partial \phi}{\partial x} = U$$

$$v = \frac{\partial \phi}{\partial y} = 0.$$

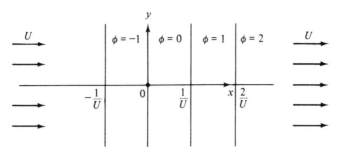

Figure 7.9. Equipotential lines for uniform flow.

The lines of equal value of velocity potential (called *equipotential lines*) are obtained by assigning constant values to the velocity potential. For example, let $\phi = 0$; therefore, $0 = Ux$ and $x = 0$ corresponds to the vertical line through the origin. Similarly, $x = 1/U$ is the x intercept for the $\phi = 1$ equipotential line (Figure 7.9).

EXAMPLE 7.1

Find the velocity potential of a uniform stream with velocity V inclined at an angle α to the x axis (Figure 7.10).

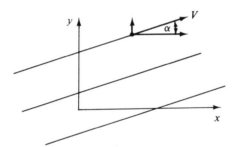

Figure 7.10. Inclined uniform flow.

Solution   With $u = V \cos \alpha$ and $v = V \sin \alpha$ in Equation (7.8), we obtain

$$\frac{\partial \phi}{\partial x} = V \cos \alpha$$

and

$$\frac{\partial \phi}{\partial y} = V \sin \alpha$$

Integrating, we obtain

$$\phi = V \cos \alpha \cdot x + V \sin \alpha \cdot y + C$$

where C is the constant of integration, which can be set equal to zero, since the velocity components are related to the spatial derivatives of the potential and not to the value of the potential. ■

From the example shown in Figures 7.8 and 7.9, we can see that the streamlines and equipotentials are perpendicular to each other. This statement is true in general, as the following material shows. For any equipotential line, which is a line along which the velocity potential is constant, the expression for the total differential $d\phi$ becomes zero; that is,

$$d\phi = \frac{\partial \phi}{\partial x} dx + \frac{\partial \phi}{\partial y} dy = 0$$

Similarly, we have along a streamline, which is a line along which the stream function is constant,

$$d\psi = \frac{\partial \psi}{\partial x} dx + \frac{\partial \psi}{\partial y} dy = 0$$

Rewriting the first expression using the definition of velocity potential given in Equation (7.8), we obtain

$$\frac{dy}{dx}\bigg|_{\phi = \text{constant}} = -\frac{\partial \phi / \partial x}{\partial \phi / \partial y} = -\frac{u}{v}$$

while the second expression, after appropriate substituting of Equation (7.6), results in

$$\frac{dy}{dx}\bigg|_{\psi = \text{constant}} = -\frac{\partial \psi / \partial x}{\partial \psi / \partial y} = \frac{v}{u}$$

Thus we conclude that

$$\frac{dy}{dx}\bigg|_{\psi = \text{constant}} \times \frac{dy}{dx}\bigg|_{\phi = \text{constant}} = -1$$

which will be recognized from analytic geometry as stating that the lines of constant stream function and velocity potential are perpendicular to each other.

Next, let us examine a special characteristic of potential flow. If we differentiate the first equation of equation pair (7.8) with respect to y, we obtain

$$\frac{\partial u}{\partial y} = \frac{\partial^2 \phi}{\partial y \, \partial x}$$

and similarly, if we differentiate the second equation of equation pair (7.8) with respect to x, we obtain

$$\frac{\partial v}{\partial x} = \frac{\partial^2 \phi}{\partial x \, \partial y}$$

Since the order of differentiation can be interchanged, the following relation between the partial derivatives of the velocity components results:

$$\frac{\partial u}{\partial y} = \frac{\partial v}{\partial x}$$

or

$$\frac{\partial u}{\partial y} - \frac{\partial v}{\partial x} = 0 \tag{7.9}$$

This equation expresses mathematically the condition that the fluid elements do not rotate during the course of their movement in the flow field; hence Equation (7.9) is called the *irrotationality condition* for two-dimensional flow. The absence of rotation of a fluid element is demonstrated in Figure 7.11, where a simple example of irrotational flow is shown. The fluid elements move in circular paths without changing their orientation. The orientation of the fluid element at two different instances is indicated by the two diagonals shown in the figure. It is seen that the direction of the diagonals is preserved. To achieve this, the inner side of the element must move faster than the outer side. Indeed, we will see in a subsequent flow example that the tangential velocity must vary as the inverse of the radial distance from the center of the circles to achieve irrotationality.

An example of rotational flow occurs when the fluid elements move in circular trajectories, as in rigid body rotation. In this case, the tangential velocity is proportional to the radial distance from the center of the circle. The orientation of the diagonals is shown in Figure 7.12.

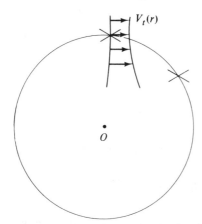

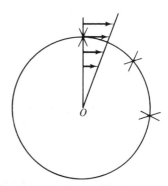

Figure 7.11. Preservation of orientation of fluid element in irrotational flow.

Figure 7.12. Change in orientation of fluid element in rotational flow.

A more general case of irrotational motion than that given in Figure 7.11 occurs when the change in orientation of one diagonal is balanced by an equal but opposite change in orientation of the other diagonal. An example of this is shown for the case when only elongation of the fluid element occurs (Figure 7.13). Thus irrotational flow can be described as flow in which the net rotation of the moving fluid element does not change as the element moves from location to location in the flow field.

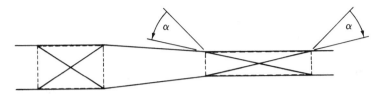

Figure 7.13. Irrotational flow with elongation of fluid element.

To derive the condition of irrotationality directly from consideration of the motion of a fluid element, consider a rectangular element as shown in Figure 7.14. Rotation in the counterclockwise direction will be taken as positive, as seen from Figure 7.14b. The elemental length dx has rotated with respect to point 0 at a rate equal to

$$\frac{(\partial v/\partial x)\, dx}{dx} = \frac{\partial v}{\partial x}$$

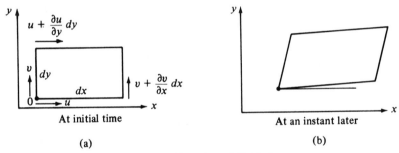

At initial time

(a)

At an instant later

(b)

Figure 7.14. Rotation of fluid element.

Similarly, the elemental length dy has rotated with respect to point 0, during the same time interval, at a rate equal to

$$\frac{-(\partial u/\partial y)\, dy}{dy} = -\frac{\partial u}{\partial y}$$

The negative sign accounts for the fact that its rotation is clockwise. Defining the rotation of the fluid element as the average of the rotation of the two sides, the rotation ω is given by

$$\omega = \frac{1}{2}\left(\frac{\partial v}{\partial x} - \frac{\partial u}{\partial y}\right) \tag{7.10}$$

For irrotational motion, the rotation ω of the fluid element must equal zero, and Equation (7.10) reduces to Equation (7.9).

EXAMPLE 7.2
Determine whether the specified flows are rotational or irrotational:

1. $u = y$
   $v = -\frac{3}{2}x$

2. $u = xy^2$
   $v = x^2y$

Solution If $\partial v/\partial x - \partial u/\partial y = 0$, the flow is irrotational; otherwise it will be rotational flow.

1. $\dfrac{\partial v}{\partial x} = -\dfrac{3}{2}$

   $\dfrac{\partial u}{\partial y} = 1$

   $\overline{\dfrac{\partial v}{\partial x} - \dfrac{\partial u}{\partial y} = -\dfrac{5}{2} \neq 0}$

Hence the flow is *rotational*.

2. $\dfrac{\partial v}{\partial x} = 2xy$

   $\dfrac{\partial u}{\partial y} = 2xy$

   $\overline{\dfrac{\partial v}{\partial x} - \dfrac{\partial u}{\partial y} = 2xy - 2xy = 0}$

Hence the flow is *irrotational*.

In this instance, we can also determine the expression for the velocity potential. Since

$$u = \frac{\partial \phi}{\partial x} = xy^2$$

and

$$v = \frac{\partial \phi}{\partial y} = x^2y$$

we obtain by integration that the velocity potential ϕ is

$$\phi = \tfrac{1}{2}x^2y^2 + C$$ ■

   When the flow is irrotational, it becomes possible to integrate Euler's equation, and the result is a simple relationship between velocity of flow and pressure. To show this, substitute the irrotationality relation (7.9) into Euler's equations (7.3) and (7.4), thereby obtaining

$$u\frac{\partial u}{\partial x} + v\frac{\partial v}{\partial x} = -\frac{1}{\rho}\frac{\partial p}{\partial x}$$

and

$$u \frac{\partial u}{\partial y} + v \frac{\partial v}{\partial y} = -\frac{1}{\rho} \frac{\partial p}{\partial y} - g$$

These two equations can be simplified as follows $\left(\text{since } \dfrac{\partial u^2}{\partial x} = 2u \dfrac{\partial u}{\partial x} \text{ etc.} \right)$:

$$\frac{1}{2} \frac{\partial u^2}{\partial x} + \frac{1}{2} \frac{\partial v^2}{\partial x} = -\frac{1}{\rho} \frac{\partial p}{\partial x}$$

$$\frac{1}{2} \frac{\partial u^2}{\partial y} + \frac{1}{2} \frac{\partial v^2}{\partial y} = -\frac{1}{\rho} \frac{\partial p}{\partial y} - g$$

If the density ρ remains constant throughout the flow field, we can integrate each equation to obtain

$$\frac{1}{2}(u^2 + v^2) = -\frac{1}{\rho} p + F_1(y) \tag{7.11a}$$

and

$$\frac{1}{2}(u^2 + v^2) = -\frac{1}{\rho} p - gy + F_2(x) \tag{7.11b}$$

The functions $F_1(y)$ and $F_2(x)$, each a function of a single variable y or x, respectively, must be included, for we integrated partial derivatives. Since Equations (7.11a) and (7.11b) must be identical, we obtain

$$F_1(y) + gy = F_2(x)$$

Since x and y are independent variables, and since the left-hand side is a function of y only while the right-hand side is a function of x only, we conclude that

$$F_1(y) = -gy + \text{constant}$$

and

$$F_2(x) = \text{constant}$$

Thus we write a single equation

$$\frac{1}{2} V^2 + \frac{p}{\rho} + gy = C \tag{7.12}$$

where $V^2 = u^2 + v^2$ is the total velocity with components u and v, and C is constant throughout the region of irrotational flow.

Equation (7.12) will be recognized as the Bernoulli equation, which relates the total velocity of flow to the local pressure. Thus, having obtained the distribution of velocity in the flow field from a knowledge of the velocity potential, the Bernoulli equation allows us to obtain the pressure distribution in the flow field.

In Section 4.4 we derived Bernoulli's equation for the movement of a fluid element along a streamline and integrated along the streamline without requiring irrotationality. In that case, the constant C would have different values for different streamlines unless the motion is irrotational.

7.4 LAPLACE'S EQUATION AND SOME SIMPLE SOLUTIONS

We shall now use the velocity potential to solve some simple problems in two-dimensional fluid dynamics. To do this, we must first derive the differential equation that the velocity potential satisfies. The velocity components u and v expressed in terms of the velocity potential as given in Equation (7.8) can be substituted into the continuity equation (7.1)

$$\frac{\partial}{\partial x}\left(\frac{\partial \phi}{\partial x}\right) + \frac{\partial}{\partial y}\left(\frac{\partial \phi}{\partial y}\right) = 0$$

to obtain the differential equation of the velocity potential, namely,

$$\frac{\partial^2 \phi}{\partial x^2} + \frac{\partial^2 \phi}{\partial y^2} = 0 \tag{7.13}$$

This equation is called *Laplace's equation* and is given here in rectangular coordinates for the two-dimensional case.

In the following material, we shall present several flow examples: source flow, free vortex flow, and flow about bodies. The expressions for the velocity potential and stream function for these examples can be written in a more compact form if we utilize polar coordinates instead of rectangular coordinates. It will be recalled that polar coordinates are related to rectangular coordinates as shown in Figure 7.15. It can be shown, using the chain rule of partial differentiation, that Laplace's equation in polar coordinates is

$$\frac{\partial^2 \phi}{\partial r^2} + \frac{1}{r}\frac{\partial \phi}{\partial r} + \frac{1}{r^2}\frac{\partial^2 \phi}{\partial \theta^2} = 0 \tag{7.14}$$

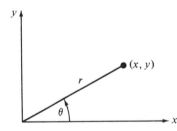

Figure 7.15.   Definition of polar coordinates.

7.4(a) Source Flow

A simple solution of Laplace's equation is given by

$$\phi = m \ln r \tag{7.15}$$

where the potential is a function of the radial distance from the origin only and m is a constant. To verify that $m \ln r$ is indeed a solution of Laplace's equation, substitute Equation (7.15) into Equation (7.14). We obtain

$$r^2\left(-\frac{m}{r^2}\right) + r\frac{m}{r} = 0$$

$$0 = 0$$

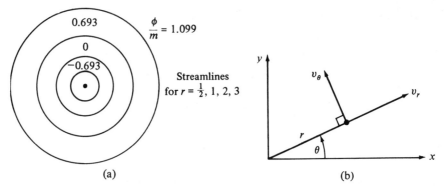

Figure 7.16. Equipotentials for source flow.

The equipotential lines for this flow are obtained by assigning a constant value to the potential and solving for the radius. The equipotentials are lines of constant radius or concentric circles, as shown in Figure 7.16. To establish the streamlines of this flow, we need to determine the stream function. In polar coordinates, the velocity components relate to the velocity potential as follows:

$$v_r = \frac{\partial \phi}{\partial r}$$

$$v_\theta = \frac{1}{r}\frac{\partial \phi}{\partial \theta}$$

The relationship between velocity components and stream function is

$$v_r = \frac{1}{r}\frac{\partial \psi}{\partial \theta}$$

$$v_\theta = -\frac{\partial \psi}{\partial r}$$

Thus we obtain

$$v_r = \frac{1}{r}\frac{\partial \psi}{\partial \theta} = \frac{\partial \phi}{\partial r}$$

and

$$v_\theta = -\frac{\partial \psi}{\partial r} = \frac{1}{r}\frac{\partial \phi}{\partial \theta}$$

Therefore,

$$\frac{1}{r}\frac{\partial \psi}{\partial \theta} = \frac{m}{r}$$

and

$$\frac{\partial \psi}{\partial r} = 0$$

Integrating the first equation with respect to θ, we obtain

$$\psi = m\theta + f(r) + \text{constant}$$

where $f(r)$ is a function of r only and must be included in addition to the constant, since we integrated a partial derivative with respect to θ. Integrating the second equation with respect to r, we obtain

$$\psi = F(\theta) + \text{constant}$$

Since the two expressions obtained for the stream function must be identical, we conclude that $f(r) = 0$ and $F(\theta) = m$. Therefore, the expression for the stream function of the source flow becomes

$$\psi = m\theta + \text{constant} \tag{7.16}$$

The constant can be set equal to zero. The streamlines of this flow are obtained by setting ψ equal to different constant values, as shown in Figure 7.17. The streamline pattern obtained consists of straight lines emanating from the origin; hence this kind of flow is called *source flow*. The flow velocities are in a radial direction only and decrease inversely with distance from the origin according to the expression

$$v_r = \frac{\partial \phi}{\partial r} = \frac{m}{r}$$

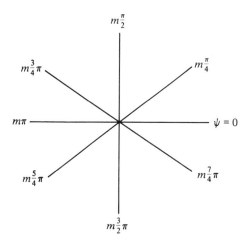

Figure 7.17. Streamlines for source flow.

The constant m is called the *source strength*, for it is related to the volume flow out of the source, as can be shown. The volume flow Q is determined by evaluating the flow past a circle of radius R (Figure 7.18). The flow across the circle is

$$\int v_r \, dA = \int_0^{2\pi} v_r R \, d\theta$$

Since v_r has the same magnitude at a given distance R from the origin and is equal to m/R, the flow is $2\pi R m/R$. From continuity, the volume flow of the incompressible fluid out of the source located at the origin must be equal

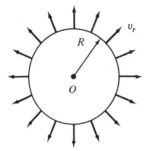

Figure 7.18. Flow from source.

to the flow across the circle R. Thus $Q = 2\pi m$ and m (the source strength) is related to Q (the volume flow) by

$$m = \frac{Q}{2\pi}$$

When the value of m is negative, we have flow into the origin. Such flows are called *sink flows*.

7.4(b) Free Vortex Flow

Next, examine another simple solution of Laplace's equation (7.13). This time take the potential to be a function of θ, only with

$$\phi = -C\theta$$

Again, it can easily be verified that the preceding expression is a solution of Laplace's equation. In this case, the equipotential lines are straight lines emanating from the origin. The stream function is obtained from the velocity potential as follows:

$$\frac{\partial \psi}{\partial \theta} = r\frac{\partial \phi}{\partial r} = 0$$

and

$$\frac{\partial \psi}{\partial r} = -\frac{1}{r}\frac{\partial \phi}{\partial \theta} = \frac{C}{r}$$

Integration of the two equations results in

$$\psi = F(r) + \text{constant}$$

and

$$\psi = C\ln r + f(\theta) + \text{constant}$$

Comparison of the two expressions obtained for the stream function ψ gives $F(r) = C\ln r$ and $f(\theta) = 0$. The constant can be set equal to zero. Thus the expression for the stream function is

$$\psi = C\ln r$$

The streamlines are therefore concentric circles about the origin, as shown in Figure 7.19, and the flow continuously circles the origin. This type of

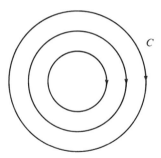

Figure 7.19. Streamlines for free vortex flow.

flow is called *free vortex flow*. In free vortex flow, we have the velocity components

$$v_r = \frac{\partial \phi}{\partial r} = 0$$

and

$$v_\theta = \frac{1}{r}\frac{\partial \phi}{\partial \theta} = -\frac{C}{r}$$

The flow velocity is thus in the tangential direction only and varies inversely as the distance from the origin. For positive values of the constant C, the flow is in the clockwise direction, since v_θ was taken as positive in the counterclockwise direction.

EXAMPLE 7.3
The eye of a tornado has a radius of 20 m. Find the variation of pressure in the flow field of the tornado if the maximum wind velocity is 50 m/s.

Solution The flow field due to a tornado can be represented by a free vortex and solid-body rotation. The core or so-called eye of the tornado behaves closely to that of solid-body rotation, while the flow field outside the eye is well represented by a free vortex field. Thus the tangential velocity distribution (there are no radial velocities in this case) will be a function of the radial distance from the center of the tornado, as shown in Figure 7.20. In the analysis, we neglect any translational motion of the tornado and consider motion only with respect to the center of the tornado. The streamlines of flow are circular, as shown in the figure. In the eye ($r \leq R$), we have rigid body rotation with velocity distribution

$$v_\theta = \omega r$$

Outside the eye ($r \geq R$), the flow is represented by a free vortex, and hence the velocity distribution is

$$v_\theta = \frac{C}{r}$$

Since the tangential velocity as given by the two equations must be equal at the edge of the eye ($r = R$), we can evaluate the constant C:

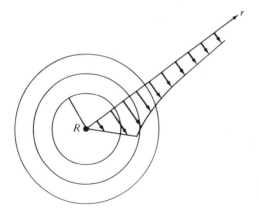

Figure 7.20. Velocity distribution in a tornado.

$$v_\theta = \omega R = \frac{C}{R}$$

Hence

$$C = \omega R^2$$

Therefore, the velocity distribution is

$$v_\theta = \omega r \qquad r \leqslant R$$

and

$$v_\theta = \frac{\omega R^2}{r} \qquad r \geqslant R$$

The maximum wind velocity occurs at the edge of the eye. If the maximum velocity is 50 m/s at a radius of 20 m, we obtain

$$\omega = \frac{v_\theta}{R} = \frac{50 \text{ m/s}}{20 \text{ m}} = 2.5 \text{ rad/s}$$

and the velocity variation is given by

$$v_\theta = 2.5r \text{ m/s} \qquad r \leqslant 20 \text{ m}$$

and

$$v_\theta = \frac{1000}{r} \text{ m/s} \qquad r \geqslant 20 \text{ m}$$

Since the flow outside the eye is potential, we can obtain its pressure distribution from Bernoulli's equation (7.12)—namely,

$$\frac{1}{2}v_\theta^2 + \frac{p}{\rho} = \frac{p_a}{\rho}$$

where p_a is the atmospheric pressure far from the eye and where the flow velocity vanishes. Rewrite to obtain the pressure variation external to the eye:

$$(p_a - p)\frac{1}{\rho} = \frac{1}{2}v_\theta^2 = \frac{1}{2}\frac{\omega^2 R^4}{r} = \frac{1}{2}\frac{V_\theta^2 R^2}{r^2} \qquad (7.17)$$

where V_θ is the magnitude of the velocity at $r = R$, the edge of the eye. Since v_θ^2 is always positive, the pressure external to the eye will always be less than atmospheric pressure. This underpressure is in part responsible for the damage caused by tornadoes.

To evaluate the pressure distribution within the eye, we must refer to Euler's equations, for the flow is rotational there. In steady two-dimensional flow and in the absence of gravity effects, Euler's equations in cylindrical coordinates can be shown to be

$$-\frac{1}{\rho}\frac{\partial p}{\partial r} = v_r\frac{\partial v_r}{\partial r} + \frac{v_\theta}{r}\frac{\partial v_r}{\partial \theta} - \frac{v_\theta^2}{r}$$

and

$$-\frac{1}{\rho r}\frac{\partial p}{\partial \theta} = v_r\frac{\partial v_\theta}{\partial r} + \frac{v_\theta}{r}\frac{\partial v_\theta}{\partial \theta} + \frac{v_\theta v_r}{r}$$

In the case under consideration here, these equations simplify, since $v_r = 0$ and v_θ is independent of the variable θ, to

$$-\frac{1}{\rho}\frac{\partial p}{\partial r} = -\frac{v_\theta^2}{r}$$

and

$$-\frac{1}{\rho r}\frac{\partial p}{\partial \theta} = 0$$

The second equation indicates that, within the eye, the pressure p is a function of the radial distance r only, and the first equation can thus be written as an ordinary differential equation:

$$\frac{1}{\rho}\frac{dp}{dr} = \frac{\omega^2 r^2}{r} = \omega^2 r$$

Integrating, we obtain

$$p = \frac{\rho}{2}\omega^2 r^2 + \text{constant}$$

The constant is evaluated from the previously obtained pressure relation in the potential region of the flow by equating the two pressures at the edge of the eye ($r = R$). Thus

$$p = \tfrac{1}{2}\rho\omega^2 R^2 + \text{constant} = p_a - \tfrac{1}{2}\rho V_\theta^2$$

Hence the constant is

$$\text{constant} = p_a - \frac{1}{2}\rho V_\theta^2 - \frac{1}{2}\frac{\rho V_\theta^2 R^2}{R^2}$$

$$= p_a - \rho V_\theta^2$$

and

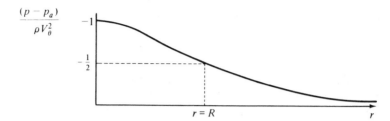

Figure 7.21.   Pressure variation in a tornado.

$$p = \frac{\rho}{2}V_\theta^2\frac{r^2}{R^2} - \rho V_\theta^2 + p_a$$

Simplifying, we obtain for the pressure distribution in the eye:

$$\frac{p_a - p}{\rho} = V_\theta^2\left(1 - \frac{1}{2}\frac{r^2}{R^2}\right) \tag{7.18}$$

We can now combine Equations (7.17) and (7.18) and sketch the variation of pressure in the entire flow field as in Figure 7.21. From the figure it is seen that the minimum pressure occurs at the center of the eye and that the entire pressure field of the tornado is below atmospheric. For example, if the maximum wind velocity is 50 m/s, as given in this example, the minimum pressure or the maximum underpressure is

$$p - p_a = -\rho(V_\theta^2) = -(1.225 \text{ kg/m}^3)(2500 \text{ m}^2/\text{s}^2)$$
$$= -3.063 \text{ kPa}$$

where ρ is taken at 15°C, 1 atmosphere.   ■

   Let us now return to rectangular coordinates to establish the form of the differential equation which the stream function for two-dimensional potential flow must satisfy. To do this, substitute the definition of the stream function in terms of the velocity components, that is, $u = \partial\psi/\partial y$, $v = -\partial\psi/\partial x$, as given in Equation (7.6), into the irrotationality condition as given in Equation (7.9), namely, $\partial u/\partial y - \partial v/\partial x = 0$. We then obtain

$$\frac{\partial}{\partial y}\left(\frac{\partial\psi}{\partial y}\right) - \frac{\partial}{\partial x}\left(-\frac{\partial\psi}{\partial x}\right) = 0$$

Thus the stream function for two-dimensional potential flow also satisfies Laplace's equation; that is,

$$\frac{\partial^2\psi}{\partial x^2} + \frac{\partial^2\psi}{\partial y^2} = 0 \tag{7.19}$$

Since the stream function ψ and the velocity potential ϕ satisfy the same differential equation in two-dimensional flow, their roles may be interchanged to obtain different flows. This was shown to be the case when we considered source flow and free vortex flow.

7.5 SUPERPOSITION OF POTENTIAL FLOWS

An interesting property of potential flows is noteworthy here—the property of superposition. It is useful to establish more complicated flows by combining simple potential flows. Specifically, superposition of potential flows—that is, the combining of several different potential flows to obtain a new flow that is also potential—is possible since the differential equation that the velocity potential satisfies (i.e., Laplace's equation) is a linear differential equation.

To illustrate this property, consider two potential flows with velocity potentials ϕ_1 and ϕ_2. Each flow satisfies Laplace's equation; that is,

$$\frac{\partial^2 \phi_1}{\partial x^2} + \frac{\partial^2 \phi_1}{\partial y^2} = 0$$

and

$$\frac{\partial^2 \phi_2}{\partial x^2} + \frac{\partial^2 \phi_2}{\partial y^2} = 0$$

Adding these two equations, we obtain

$$\frac{\partial^2 (\phi_1 + \phi_2)}{\partial x^2} + \frac{\partial^2 (\phi_1 + \phi_2)}{\partial y^2} = 0$$

This shows that the combined flow with velocity potential $\phi = \phi_1 + \phi_2$ also satisfies the differential equation and is therefore also a potential flow.

7.5(a) Flow about a Half-body

As an example of the application of the superposition principle, consider the flow obtained by combining a source with a uniform stream; that is, add the expression for the velocity potential of uniform flow (Ux) to that of a source as given in Equation (7.15) to obtain

$$\phi = Ux - m \ln r$$
$$= Ur \cos \theta + m \ln r$$

The stream function for this combined flow is similarly obtainable by adding the expression for the stream function of uniform flow (Uy) to that of the source given in Equation (7.16):

$$\psi = Uy + m\theta$$
$$= Ur \sin \theta + m\theta$$

The velocity components of the resulting flow field are obtainable by appropriate differentiation as follows:

$$v_r = \frac{1}{r} \frac{\partial \psi}{\partial \theta} = \frac{1}{r}(Ur \cos \theta + m)$$

$$v_\theta = -\frac{\partial \psi}{\partial r} = -U \sin \theta$$

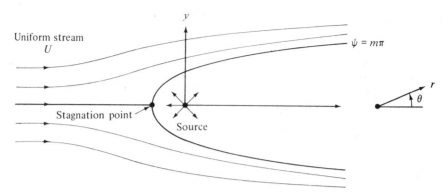

Figure 7.22. Flow about a two-dimensional half-body.

Figure 7.22 indicates the location of the source at the coordinate origin and shows several streamlines of the resulting flow field. Next, determine the location of the stagnation point in the flow field, that is, the point where the velocity of the fluid is zero. Using the tangential component of flow velocity, we obtain

$$v_\theta = 0: \qquad U \sin \theta = 0 \quad \text{or} \quad \theta = 0, \pi$$

From the radial component of flow velocity, we obtain

$$v_r = 0: \qquad Ur \cos \theta + m = 0$$

Hence the angle θ must be equal to π, so that $-Ur + m = 0$; therefore,

$$r = \frac{m}{U}$$

which states that the distance of the stagnation point from the location of the source is a function of the ratio of source strength m and uniform flow velocity U.

Since the flow velocity is always tangential to a streamline, any of the streamlines in Figure 7.22 can represent the contour of a body in inviscid flow. The contours of two such bodies are shown in Figure 7.23, where it

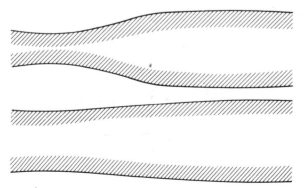

Figure 7.23. Contour of infinitely long bodies.

is seen that the bodies are infinitely long. Since we desire to obtain at least a semi-infinite body, however, the streamline passing through the stagnation point is of particular interest here. The value of the stream function at the stagnation point is

$$\psi = U\frac{m}{U}\sin \pi + m\pi = m\pi$$

The equation of the streamline passing through the stagnation point is obtained by substituting the value $m\pi$ into the expression of the stream function:

$$m\pi = Ur \sin \theta + m\theta$$

This streamline is indicated in Figure 7.22. It outlines the contour of a body that extends on the right to infinity. Such a body is called a *half-body*.

7.5(b) Flow about a Rankine Body

If we combine a source and sink of equal strength with uniform flow, we obtain a closed body called a *Rankine body*. As indicated in Figure 7.24, the source is located at point $(-a, 0)$, while the sink is located at point $(a, 0)$. The resulting velocity potential is

$$\phi = \underbrace{Ux}_{\substack{\text{uniform}\\\text{flow}}} + m[\underbrace{\ln \sqrt{(x + a)^2 + y^2}}_{\text{source}} - \underbrace{\ln \sqrt{(x - a)^2 - y^2}}_{\text{sink}}]$$

In this case, the source and sink are not located at the coordinate origin but are displaced by a distance a; hence the expressions for source and sink take the form shown. The stream function for the combined flow becomes

$$\psi = Uy - m \tan^{-1} \frac{2ay}{x^2 + y^2 - a^2}$$

The velocity components of the flow field are

$$u = \frac{\partial\phi}{\partial x} = U + m\left[\frac{x + a}{(x + a)^2 + y^2} - \frac{x - a}{(x - a)^2 + y^2}\right]$$

$$v = \frac{\partial\phi}{\partial y} = m\left[\frac{y}{(x + a)^2 + y^2} - \frac{y}{(x - a)^2 + y^2}\right]$$

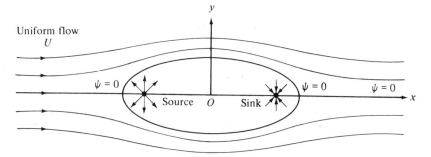

Figure 7.24.   Flow about a Rankine body.

Next, locate the stagnation points in the flow field, that is, points at which u and v vanish. From the preceding expression for the velocity component v, it is easily seen that it will vanish for $y = 0$, that is, along the x axis. This is as expected, since the flow is symmetric about the x axis. Thus we have to find points on the x axis for which u vanishes; that is, values of x where $u = 0$ for $y = 0$:

$$U + m\left[\frac{x + a}{(x + a)^2} - \frac{x - a}{(x - a)^2}\right] = 0$$

or

$$U = \frac{2ma}{x^2 - a^2}$$

and

$$x = \pm \sqrt{a^2 + \frac{2ma}{U}}$$

which states that the distance of the stagnation points (there are two in this flow case) from the coordinate origin is a function of source (sink) strength, magnitude of uniform flow velocity, and distance between source and sink.

The value of the stream function at the stagnation points, since $y = 0$, is

$$\psi = 0 - m\tan^{-1}\frac{0}{x^2 - a^2} = 0$$

The equation of the body streamline shown in Figure 7.24 therefore becomes

$$0 = Uy - m\tan^{-1}\frac{2ay}{x^2 + y^2 - a^2}$$

or

$$\tan\frac{Uy}{m} = \frac{2ay}{x^2 + y^2 - a^2}$$

By combining a number of sources and sinks with uniform flow, we can obtain a variety of bodies with different shapes. However, in order to obtain a closed body, the sum of the source strengths must equal the sum of the sink strengths. Thus far we have discussed only two-dimensional point sources and sinks. For the purpose of generating different bodies, we could also use sources and sinks distributed over a surface or even over a volume. When combined with uniform flow, a further variation in the body shapes obtainable results. Examples of different body shapes are shown in Figure 7.25, where several two-dimensional source and sink distributions have been combined with a uniform stream.

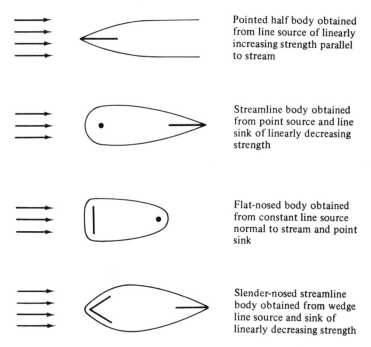

Pointed half body obtained from line source of linearly increasing strength parallel to stream

Streamline body obtained from point source and line sink of linearly decreasing strength

Flat-nosed body obtained from constant line source normal to stream and point sink

Slender-nosed streamline body obtained from wedge line source and sink of linearly decreasing strength

Figure 7.25.   Examples of body shapes obtainable.

7.6 POTENTIAL FLOW ABOUT A CIRCULAR CYLINDER

When the distance between a point source and sink in uniform flow is made small, the resulting body obtained approaches that of a circular cylinder. This fact is illustrated in Figure 7.26, where several Rankine bodies with different distances between source and sink are shown. When the distance is made zero, the body becomes a circular cylinder.

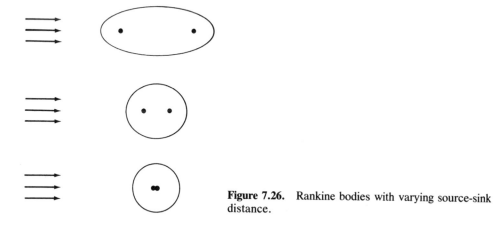

Figure 7.26.   Rankine bodies with varying source-sink distance.

The velocity potential of uniform flow about a circular cylinder can be shown to be given by

$$\phi = Ur \cos \theta + U\frac{R^2}{r} \cos \theta$$

where the first term is the potential due to uniform flow and the second term is due to the doublet flow (the term applied to the case when source and sink are located at the same point). The corresponding stream function is

$$\psi = Ur \sin \theta - U\frac{R^2}{r} \sin \theta$$

The streamlines of the flow are shown in Figure 7.27. On the cylinder—that is, for $r = R$—the value of the stream function is zero. Furthermore, the stream function $\psi = 0$ extends on either side of the cylinder to infinity, since for $\theta = 0$ and π the stream function vanishes for all r.

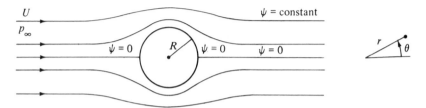

Figure 7.27. Flow about a circular cylinder.

From the expressions of the velocity potential (or, alternately, from the stream function), we can obtain the velocity distribution of the flow field. Specifically,

$$v_r = \frac{\partial \phi}{\partial r} = U \cos \theta \left(1 - \frac{R^2}{r^2} \right)$$

$$v_\theta = \frac{1}{r} \frac{\partial \phi}{\partial \theta} = -U \sin \theta \left(1 + \frac{R^2}{r^2} \right)$$

If we set $r = R$ in the above equations, we obtain the velocity distribution on the cylinder:

$$v_r(R) = 0$$

$$v_\theta(R) = -2U \sin \theta$$

The first equation shows that there is no flow in the direction normal to the surface of the cylinder since the cylinder surface constitutes an impermeable wall. As shown in Figure 7.28, the velocity tangential to the surface varies from zero at $\theta = \pi$ (180°) through a maximum of $2U$ at $\theta = \pi/2$ (90°) to zero at $\theta = 0$. The stagnation points of the flow are thus at $\theta = 0$ and π.

From Bernoulli's equation (7.12), we can determine the pressure dis-

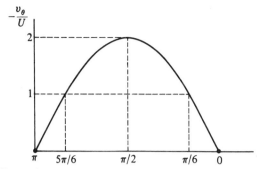

Forward stagnation point

Rear stagnation point

Figure 7.28. Velocity distribution on a circular cylinder.

tribution in the potential flow. In particular, let us evaluate the pressure distribution on the surface of the cylinder. From Equation (7.12),

$$\frac{1}{2}V^2 + \frac{p}{\rho} + gy = C = \frac{1}{2}U^2 + \frac{p_\infty}{\rho} + gy$$

Here V is the total velocity with components u and v, and U and p_∞ are velocity and pressure, respectively, at large distances from the cylinder. Rewriting the foregoing equation, we obtain

$$p = p_\infty + \frac{\rho}{2}(U^2 - V^2)$$

The pressure distribution is usually presented in nondimensional form, using the pressure coefficient C_p as defined in Chapter 5, Equation (5.15), with the flow velocity at infinity (U) taken as the reference velocity. Thus

$$C_p = \frac{(p - p_\infty)}{\frac{1}{2}\rho U^2} = \frac{(\rho/2)(U^2 - V^2)}{\frac{1}{2}\rho U^2} = 1 - \frac{V^2}{U^2} = 1 - 4\sin^2\theta$$

The resulting pressure distribution is shown graphically in Figure 7.29, where it is seen that the pressure coefficient equals unity at the two stagnation points. The minimum pressure coefficient of -3 occurs at $\theta = \pi/2$ (90°) and $3\pi/2$ (270°). At angles of $\pi/6$ (30°) with respect to the horizontal, the pressure coefficient becomes zero, signifying that the pressure on the cylinder equals the value at infinity.

Integration of the pressure distribution over the surface of the cylinder yields the force exerted on the cylinder by the fluid. Figure 7.30 indicates the direction of the pressure force exerted on the surface of the cylinder. Let F_x be the force exerted in the x direction and F_y be the force exerted in the y direction. Therefore,

$$F_x = -\int p\cos\theta\, dA = -R\int_0^{2\pi} p\cos\theta\, d\theta$$

and

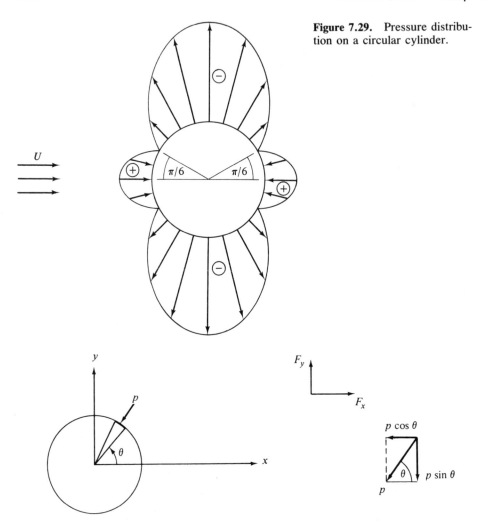

Figure 7.29. Pressure distribution on a circular cylinder.

Figure 7.30. Pressure force on a circular cylinder.

$$F_y = -\int p \sin \theta \, dA = -R \int_0^{2\pi} p \sin \theta \, d\theta$$

Substituting the expression for the pressure given above, that is,

$$p = p_\infty + \frac{\rho}{2}U^2 - \rho 2U^2 \sin^2 \theta$$

there results

$$F_x = -R\left(p_\infty + \frac{\rho}{2}U^2\right) \underbrace{\int_0^{2\pi} \cos \theta \, d\theta}_{= \, 0} + \rho 2U^2 R \underbrace{\int_0^{2\pi} \sin^2 \theta \cos \theta \, d\theta}_{= \, 0}$$

and

$$F_y = -R\left(p_\infty + \frac{\rho}{2}U^2\right)\underbrace{\int_0^{2\pi} \sin\theta \, d\theta}_{= \, 0} + \rho 2U^2 R \underbrace{\int_0^{2\pi} \sin^3\theta \, d\theta}_{= \, 0}$$

Therefore the total force exerted on the circular cylinder by the fluid is zero, since both F_x and F_y are zero. Although shown here only for a circular cylinder, this result—that is, the absence of a net force exerted by the fluid—is typical for potential flow about a body when the fluid stream surrounds the entire body.

7.7 POTENTIAL FLOW ABOUT A CIRCULAR CYLINDER WITH CIRCULATION

If we further combine uniform flow about a circular cylinder discussed in the preceding section with a free vortex flow, we can examine the effect of circulation on the flow field about a cylinder. This effect results in a force being exerted on the cylinder in a direction perpendicular to the direction of the uniform stream, as will be shown below.

The velocity potential of this combined flow is given by

$$\phi = Ur\cos\theta + U\frac{R^2}{r}\cos\theta - C\theta$$

while the corresponding stream function is

$$\psi = Ur\sin\theta - U\frac{R^2}{r}\sin\theta + C\ln r$$

As before, the first term represents the uniform flow, while the second term is the doublet flow. The third term represents free vortex flow. The various flow patterns are shown in Figure 7.31.

The velocity components of this combined flow field are

$$v_r = \frac{\partial\phi}{\partial r} = U\cos\theta\left(1 - \frac{R^2}{r^2}\right)$$

$$v_\theta = \frac{1}{r}\frac{\partial\phi}{\partial\theta} = -U\sin\theta\left(1 + \frac{R^2}{r^2}\right) - \frac{C}{r}$$

The velocity distribution on the circular cylinder is (setting $r = R$):

$$v_r(R) = 0$$

$$v_\theta(R) = -2U\sin\theta - \frac{C}{R}$$

From Bernoulli's equation, the pressure distribution is obtained as

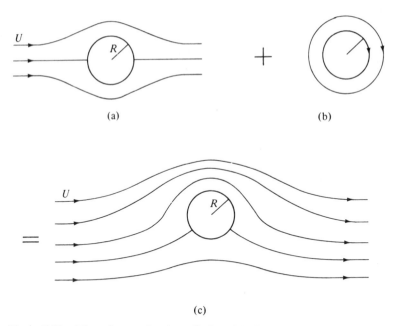

Figure 7.31. Flow about a circular cylinder with circulation: (a) flow pattern about a cylinder in uniform flow; (b) circulatory flow pattern; (c) combined flow pattern.

$$\frac{(p - p_\infty)}{\frac{1}{2}\rho U^2} = \left[1 - \left(2 \sin \theta + \frac{C}{RU}\right)^2\right]$$

The force F_x exerted in the x direction on the cylinder is

$$F_x = -\int p \cos \theta \, dA = 0$$

The force F_y exerted in the y direction on the cylinder is

$$F_y = -\int p \sin \theta \, dA = \frac{1}{2}\rho U^2 4 \frac{C}{RU} \underbrace{\int_0^{2\pi} \sin^2 \theta \, R \, d\theta}_{R\pi}$$

$$= \rho U(2\pi C)$$

Thus, in uniform flow with circulation, a force perpendicular to the direction of the uniform flow is obtained. The quantity $2\pi C$ is called the *circulation* and is designated by Γ; F_y is called the *lift* and is usually designated by L. Thus the preceding equation is written as

$$L = \rho U \Gamma \qquad (7.20)$$

Here L is the lift per unit length of cylinder. In SI units, L is given in N/m, ρ in kg/m^3, U in m/s, and Γ in m^2/s. In English units, L is in lbf/ft, ρ in slugs/ft^3, U in ft/s and Γ in ft^2/s.

The magnitude of the circulation can be varied independently of the magnitude of the velocity of the incoming uniform flow. When Γ is zero, no force is exerted by the fluid on the cylinder. As the circulation Γ about the cylinder is increased, the lift force also increases; specifically, it increases linearly with circulation. Since we are considering only ideal fluid flow, no force will be exerted by the fluid on the cylinder in the x direction.

Next, let us determine the location of the stagnation points indicated in Figure 7.28. At the stagnation points, both velocity components will vanish. Thus on the surface of the cylinder we have

$$v_\theta(R) = -2U \sin \theta - \frac{C}{R} = 0$$

or

$$\sin \theta = -\frac{C}{2UR} = -\frac{\Gamma}{4\pi UR}$$

For $\Gamma < 4\pi UR$, two stagnation points will occur on the cylinder. When the circulation is equal to $4\pi UR$, the two stagnation points merge at $\theta = 3\pi/4$, as shown in Figure 7.32. For values of circulation greater than $4\pi UR$, a stagnation point external to the cylinder occurs. In this case, the cylinder is engulfed in circulatory flow and no fluid particles from the external uniform flow reach the cylinder (Figure 7.33).

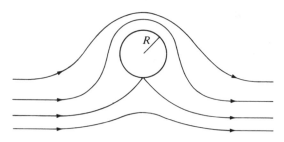

Figure 7.32. Flow about a circular cylinder with circulation $4\pi UR$.

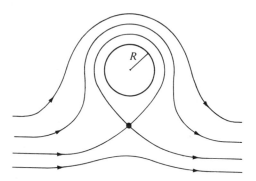

Figure 7.33. Flow about a circular cylinder with circulation greater than $4\pi UR$.

An illustration of the effect of the various terms of Equation (7.20) can be shown by the behavior of a curve ball. When a baseball pitcher applies a spin to the ball at release from his fingers, the trajectory is curved as shown in Figure 7.34. Here U represents the velocity of the air relative to the ball.

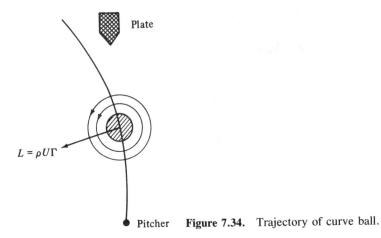

Figure 7.34.   Trajectory of curve ball.

The curvature of the ball's trajectory depends on the magnitude of the lift L generated. As seen from Equation (7.20), this force is dependent on the density of the air, the relative velocity U, and Γ. It is well known that at high altitudes (e.g., Denver or Mexico City), where the air density is somewhat less than at sea level, curve ball pitchers are not as effective as in sea level locations. Furthermore, when the wind is blowing from center field toward home plate, again the curve ball pitcher is not as effective, for the relative velocity between ball and air is decreased. Finally, the greater the spin imparted to the ball, the greater is the curvature of the trajectory. The spin of the ball creates a circulatory flow about the ball with the circulation Γ increasing with spin.

7.8 LIFT ON AIRFOILS

The lift generated by the flow about circular cylinders with circulation has few practical applications. However, the characteristics of the results obtained for the cylinder are generally applicable and will be used here to discuss the flow about two-dimensional bodies especially shaped to achieve high lift forces. In particular, let us examine the flow about the two-dimensional body with cross-sectional shape as shown in Figure 7.35. The cross section of this body, called *airfoil* when the flow medium is air or *hydrofoil* when used in water, has a rounded nose and a sharp trailing edge. As we shall see, the latter feature gives the foil its unique ability to achieve the desired high lift forces. If we immerse the foil in a uniform stream of fluid, we would expect a flow pattern as indicated in Figure 7.35. The location of the stagnation points A and B is a function of the cross-sectional shape of the foil and the relative attitude of the foil with respect to the flow. Note that the rear stagnation point B in Figure 7.35 is not at the trailing edge. Since we have assumed ideal (i.e., inviscid) flow without circulation, no forces are exerted

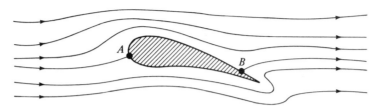

Figure 7.35.   Flow about airfoil without circulation.

on the body. A closer examination of the flow field, however, shows that at the trailing edge (not being a stagnation point) the flow must make an impossibly sharp turn with infinite velocities occurring. Since the occurrence of such infinite velocities is irreconcilable with physical reality even in an ideal fluid, the flow must adjust itself in such a manner that the stagnation point B moves to the trailing edge. This process can be accomplished by superposing a circulatory flow on the foil, as shown in Figure 7.36. The resulting lift on the foil can be obtained by integrating the pressure distribution over the surface of the foil or by using the results of Section 7.7, which stated that if a lifting body is immersed in a uniform flow field with velocity U, the lift force normal to the direction of U will be proportional to the circulation of the flow, namely,

$$L = \rho U \Gamma \qquad (7.20)$$

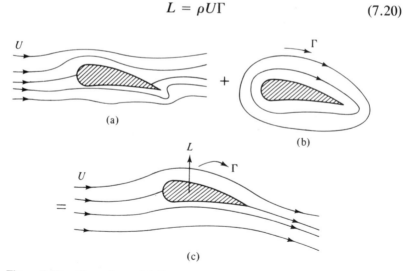

Figure 7.36.   Flow about airfoil: (a) flow pattern about foil in uniform flow; (b) circulatory flow pattern; (c) combined flow pattern.

It should be noted that, unlike the circular cylinder where the magnitude of circulation can be varied arbitrarily, the presence of the sharp trailing edge prescribes a unique value of circulation for the foil. Since for different foil

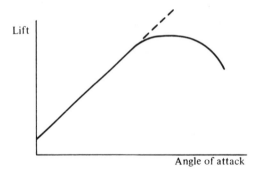

Figure 7.37. Lift of foil versus angle of attack.

shapes and inclination to the flow the value of circulation required to move the aft stagnation point to the trailing edge will vary, it follows that the magnitude of the resulting circulation will depend on the shape of the foil and its inclination to the flow. Equation (7.20) states that the lift on the foil is proportional to the circulation about it; hence the lift force will be a function of the foil cross section and its inclination to the flow. In general, the lift increases with increase in angle of flow incidence (also called angle of attack). For small angles of incidence, the increase is linear, as shown in Figure 7.37. This relationship between lift and angle of attack is typical for most foil sections. Foils are used extensively in machinery and devices requiring lift forces. They are used as wings on airplanes and hydrofoil boats, as blades in turbomachinery, such as fans and pumps, and as blades in thrust devices, such as airplane and marine propellers.

As will be discussed in Chapter 8, viscous effects of the flow must also be considered when evaluating the forces exerted on a body by a fluid. These effects account for the force exerted in the direction of flow (drag force) and for the deviation of the lift from linear behavior shown in Figure 7.37 due to flow separation. In spite of the presence of these viscous effects, the lift on airfoils at small angles of attack (say, less than 10°) can be estimated with reasonable accuracy by using ideal flow with circulation.

PROBLEMS

7.1. A uniform steady incompressible flow field has a velocity component u of 2 m/s and a velocity component v of 3 m/s. Determine expressions for the velocity potential and stream function. Sketch the streamlines.

7.2. A two-dimensional incompressible flow field has the x component of velocity given by the expression

$$u = \frac{x}{(x^2 + y^2)^{3/2}}$$

Using the continuity equation, determine the expression for the y component of velocity v. Is this flow irrotational?

7.3. The velocity components of a possible incompressible fluid field are $u = 2cxy$ and $v = c(a^2 + x^2 - y^2)$.

(a) Do the above components satisfy the continuity equation?

(b) Is the flow rotational or irrotational?

(c) Find the stream function.

(d) Find the velocity potential.

(e) Sketch the streamlines.

7.4. The velocity potential for a certain flow is $x^2 - y^2$. Find the equation for the streamlines in terms of x and y.

7.5. Consider the potential flow between two planes inclined at an angle of $\pi/2$ to each other as shown in Figure P7.5. In this case, the velocity potential is given by the expression

$$\phi = Cr^2 \cos 2\theta$$

where C is a constant to be determined by the value of velocity specified at a given point. Show that the expression given above represents flow in a right-angle corner. The velocity at a point on the horizontal wall a distance of 0.5 ft from the origin is -5 fps. Evaluate the velocity at the point $r = 1$ ft, $\theta = 45°$.

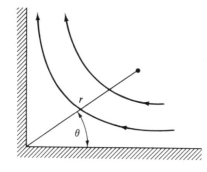

Figure P7.5

7.6. Consider the parallel two-dimensional flow shown in Figure P7.6. Find the stream function. Is the flow irrotational? ($u = 2$ m/s at $y = 0$; $u = 4$ m/s at $y = 0.8$ m.)

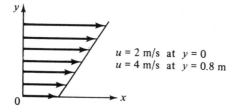

$u = 2$ m/s at $y = 0$
$u = 4$ m/s at $y = 0.8$ m

Figure P7.6

7.7. A whirlpool occurs near the water surface, as shown in Figure P7.7. Assume that the flow in the whirlpool can be represented by free vortex flow and deduce the equation of the free surface of the whirlpool. Remember that along the free surface, pressure is constant.

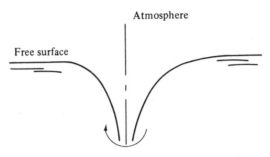

Atmosphere

Free surface

Figure P7.7

7.8. Air (density equal to 0.076 lbm/ft³) flows between two concentric circular streamlines. The inner streamline has a radius of 2 ft, while the radius of the outer streamline is 2.5 ft. The pressure difference between the two streamlines is 0.02 psi. Find the value of the circulation.

7.9. Examine the two-dimensional potential flow field which is obtained by combining a source of strength m with a free vortex with circulation Γ. Both are located at the origin of the coordinate system. Sketch the resulting streamline pattern.

7.10. A source of strength m is located at point $(a, 0)$, while a sink of equal strength is located at point $(-a, 0)$. Draw the equipotentials and the streamlines of this flow.

7.11. Two sources of equal strength m are located a distance $2d$ apart. Show that this is equivalent to having a source located near a flat wall a distance d from the center of the source. Evaluate the velocity distribution along the flat wall.

7.12. A two-dimensional free vortex is located near the plane, as indicated in Figure P7.12. The velocity at large distance from the vortex is U, and the pressure there is p_∞. The free vortex has a circulation of Γ, and the flow is inviscid and incompressible. Find the force exerted on the plane.

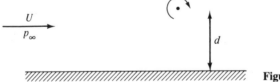

U

p_∞

d

Figure P7.12

7.13. Two sources of strength m each are located at points $(2, 0)$ and $(\frac{1}{2}, 0)$, and two sinks of equal strength are located at $(-2, 0)$ and $(-\frac{1}{2}, 0)$. Show that the circle with center at the coordinate origin having a radius of unity is a streamline.

7.14. A hut in the shape of a semicircular cylinder experiences a lift force as the wind blows over it. What is the lift per unit length if the wind velocity U is 20 m/s and the air temperature is 30°C? The radius R of the hut is 3 m.

7.15. A circulation of 10 ft²/s is superposed about a cylinder having a radius of 1 ft. The velocity of the water stream is 20 fps, while its temperature is 60°F. Find the location of the stagnation points and determine the lift force per unit length of cylinder.

7.16. A cylinder having a diameter of 1 m is immersed in air ($\rho = 1.22$ kg/m^3) having a uniform velocity of 3 m/s. A circulatory flow is added such that the two stagnation points coincide on the surface of the cylinder. Calculate the lift on the cylinder.

7.17. The pressure distribution on a probe (see Figure P7.17) can be approximated as that on the front half of a circular cylinder in a uniform flow. Where should a pressure tap be located so as to indicate the pressure in the uniform flow a long distance in front of the probe?

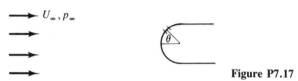

Figure P7.17

7.18. A circular cylinder of 1.5 ft diameter is immersed in a uniform airflow of velocity 15 fps. For potential steady incompressible flow about the cylinder, determine the difference between the maximum and minimum pressures on the cylinder surface. Air temperature is 32°F, and pressure in the uniform flow far from the cylinder is 14.5 psia.

7.19. Repeat Problem 7.18 for water flow at 68°F.

7.20. Two half-cylinders are joined together and placed in a uniform potential flow (see Figure P7.20). A hole is to be drilled at angle θ such that there will be no net force between the half-cylinders at the joints. Detemine angle θ, assuming the internal pressure p_{int} to be equal to the static pressure on the external surface of the cylinder at which the hole is drilled.

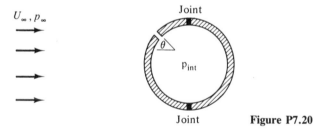

Joint   **Figure P7.20**

7.21. Water flow about a half-body is obtained by combining a two-dimensional source with uniform flow of velocity 1.0 m/s in the x direction. The flow emanating from the source is 0.1 (m^3/s)/m.
  (a) Sketch the half-body.
  (b) Determine the half-width of the body.
  (c) Find the minimum pressure on the surface at the half-body (take pressure at infinity to be 100 kPa). Assume a water temperature of 20°C.

7.22. A circular cylinder moves with a constant velocity through a water reservoir at a depth of 10 m. At what cylinder velocity will cavitation be incipient on the cylinder surface? Assume a water temperature of 5°C. (*Note*: Cavitation occurs when the pressure on the body surface falls below the saturation pressure; for water at 5°C, the saturation pressure is 0.87 kPa.)

7.23. Flow about a Rankine body is obtained by combining a two-dimensional uniform flow of velocity 3 fps in the x direction with a source and sink. Flows out of the source and into the sink are each 0.1 (ft^3/s)/ft (see Figure 7.24) with source at $(-0.5$ m, 0), sink at (0.5 m, 0). Determine the length and width of the body. Where does the minimum pressure occur? Find the maximum velocity on the body surface.

7.24. What effect would an increase of uniform flow velocity have on the dimensions of the Rankine body of Problem 7.23? (Assume source and sink strength and location to remain the same.)

7.25. What effect would an increase of source and sink strength have on the dimensions of the Rankine body of Problem 7.23? (Assume uniform flow velocity and source and sink location to remain the same.)

7.26. Consider uniform flow over a cylinder of radius 1.0 ft, with $p_\infty = 10$ psia. Plot pressure in the flow field versus radial distance for $\theta = \pi/2$ and for $\theta = 0$. Assume steady two-dimensional flow of water at 60°F, with a uniform flow velocity $U_\infty = 10$ ft/s. (See Figure P7.26.)

p_∞, U_∞

Figure P7.26

7.27. A circulation of 5 ft^2/s is imposed on the flow of Problem 7.26. Plot pressure in the flow field versus radial distance for $\theta = \pi/2$ and for $\theta = 0$. Repeat for $\theta = -\pi/2$. Calculate the lift force on the cylinder.

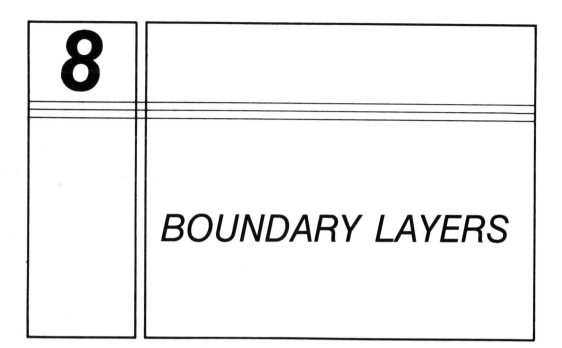

8.1 INTRODUCTION

We shall now examine the steady flow of a real, viscous fluid over a fixed body. Most of the fluids that we are accustomed to dealing with, such as water or air, possess relatively low viscosities, so that the Reynolds number $\rho U_\infty L/\mu$ of such flows, based on the approach velocity U_∞, will generally be quite high unless either the characteristic length L is very small or the flow velocity U_∞ is very low. Indeed, for the frictionless, nonviscous flow described in the preceding chapter, it can be seen that the Reynolds number is infinite.

Flow at high Reynolds number implies that, at least over most of the flow field, inertial fluid forces are predominant over viscous forces; in other words, that the effect of viscosity is small over most of the flow field. However, at the fixed wall, the viscosity of the fluid will cause the layer of fluid at the wall surface to stick to the wall; this layer of fluid will have zero velocity relative to the surface. This boundary condition imposed on the flow, that is, that the fluid velocity be equal to zero at the fixed surface, holds true for real fluids no matter how small the value of viscosity. Due to the shearing action of one fluid layer on the layer adjacent moving at a higher velocity, a velocity distribution is built up near the surface, as shown in Figure 8.1. It can be seen that large velocity gradients occur at the body surface; a small distance away from the surface, the flow velocity approaches asymptotically a free stream value.

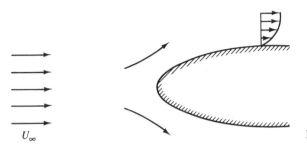

Figure 8.1

From the discussion of viscosity in Section 1.7, we found that viscous shear force is proportional to the product of μ and velocity gradient. Thus, even though a fluid might have a small coefficient of viscosity, the large velocity gradients occurring near a surface indicate that, in this region, viscous shear cannot be neglected. Away from the wall, the velocity gradients are small, and consequently the viscous forces are negligible in comparison with the inertial forces. In the region away from the wall, the flow can be treated with the equations of motion for frictionless flow developed in Chapter 7.

It is thus possible, for flow of low-viscosity fluids at high Reynolds numbers, to divide the flow field over a surface into two parts. The effects of viscosity can be confined to a thin layer in the vicinity of the surface, called the *boundary layer*. Outside the boundary layer, the flow can be treated as nonviscous. This method of dividing the flow was first proposed by Prandtl in 1904; it has proven to be extremely useful in analyzing the complex behavior of real fluid flows. According to the method, the boundary layer is assumed thin enough so as not to affect the frictionless region of the flow. In fact, it can be assumed that there is no pressure variation in the direction normal to the surface in the boundary layer, so that the pressure variation throughout the boundary layer can be obtained from the frictionless flow solution outside the boundary layer.

As an example, consider the flow of a viscous fluid over a flat plate aligned with the flow direction, as shown in Figure 8.2. If the entire flow were frictionless, the velocity and pressure throughout the flow field would be uniform, with a velocity distribution as shown in Figure 8.3.

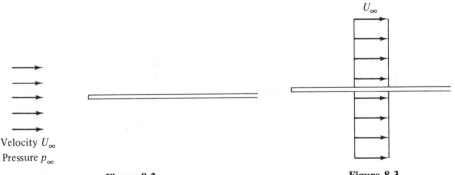

Velocity U_∞
Pressure p_∞

Figure 8.2   **Figure 8.3**

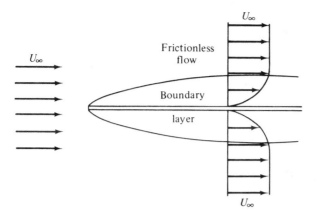

Figure 8.4

For flow of a viscous fluid over a flat plate, a boundary layer grows from the leading edge of the plate (Figure 8.4). Everywhere outside the boundary layer, the velocity is U_∞ and the pressure p_∞, as was the case for frictionless flow. Inside the boundary layer, the velocity varies from zero at the surface to U_∞ at the outer edge of the boundary layer. The thickness of the boundary layer increases in the downstream direction as more and more of the flow comes under the action of viscous shear. Again, the boundary layer is very thin, perhaps amounting to a thickness of only a few thousandths of an inch. Yet it occurs at a surface, in a region of the flow that is most important to the engineer. If we wish to calculate the resistance to motion of a body moving through a fluid, or a heat transfer coefficient at a body surface, we must have a means for analyzing the flow in the immediate vicinity of a body surface. It is the purpose of this chapter to describe qualitatively the behavior of viscous boundary layer flow, to present the equations of motion for this flow, and to provide methods of solution that will be used here to analyze the boundary layer on a flat plate. In Chapter 9 we shall discuss boundary layer flows on bodies of arbitrary shape.

8.2 LAMINAR AND TURBULENT BOUNDARY LAYERS

Near the leading edge of the flat plate shown in Figure 8.4, flow in the boundary layer is in smooth layers, a particle of fluid in a given layer tending to stay in that layer. Flow in this portion of the boundary layer is laminar. Any slight disturbance present in the flow is damped out by the action of viscous forces. Farther along the plate, however, the laminar boundary layer breaks down and becomes unstable, just like the cigarette smoke of Figure 6.6 or the water jet coming out of the faucet in Figure 6.7, and transition to a turbulent boundary layer occurs. The turbulent portion of the boundary layer is characterized by random motions or fluctuations of fluid particles or groups of particles from one layer to another. The mixing action incurred by this random movement of slower-moving particles into faster-moving layers

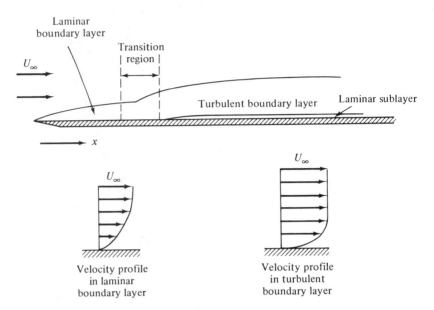

Figure 8.5

and vice versa gives rise to a much more uniform velocity profile in the turbulent boundary layer, as shown in Figure 8.5.

Even after transition to the turbulent boundary layer, a region exists in the vicinity of the wall in which the turbulent fluctuations are damped out by the presence of the wall. This very thin layer, which we also encountered in our discussion of pipe flow, is called the *laminar sublayer*.

Shear stress in the laminar boundary layer is due to the sliding of one fluid layer over another. In this region of the flow, according to the definition of viscosity presented in Chapter 1, shear stress τ_l is given by

$$\tau_l = \mu \frac{du}{dy} \tag{8.1}$$

In the turbulent portion of the boundary layer, the momentum exchange caused by the random motion of fluid particles from one layer to another causes an additional apparent shear stress, τ_t, as was explained in Section 6.1. It is possible to define a turbulent kinematic viscosity, or eddy viscosity ϵ, analogous to kinematic viscosity ν, such that

$$\tau_t = \rho \epsilon \frac{du}{dy} \tag{8.2}$$

In the turbulent boundary layer, the eddy viscosity is usually hundreds of times greater than the kinematic viscosity, so that the laminar shear can generally be neglected in comparison to the turbulent shear.

Notice that, in the laminar boundary layer, shear stress is dependent on a physical property of the fluid, namely, the coefficient of viscosity μ.

In the turbulent boundary layer, however, ϵ is not a physical property but is dependent on the flow itself and the nature of the turbulence in the flow. The inherent complexity of turbulent flow has so far prevented the derivation of an expression for ϵ and τ_t in terms of the mean properties of the flow; instead we shall have to resort to approximate methods for handling the turbulent boundary layer.

In an analysis of boundary flow over a flat plate or over any body, we are very much interested in the type of flow, laminar or turbulent, in the boundary layer along the body surface, since the flow characteristics and resultant shear forces acting on the plate are so different for the two types of flow. It is important at this point to establish a criterion for the occurrence of boundary layer transition. For pipe flow, a critical value of Reynolds number based on pipe diameter was used to determine the type of flow. For boundary flow over a body, a critical Reynolds number based on x, the distance along the surface on which the boundary layer grows, will be used as a criterion of transition (see Figure 8.6). The value of critical Reynolds number for transition depends on such factors as the rate of pressure change with x, the roughness of the body surface, any heat transfer to or from the flow, and the amount of turbulence initially present in the stream. In addition, it should be emphasized that there is no clear line of demarcation between laminar and turbulent boundary layer flows; rather, there exists a transition region, as shown in Figure 8.5, over which the disturbances present gradually build up to give a fully turbulent boundary layer flow. We can at best specify, therefore, a value of critical Reynolds number averaged over the transition region.

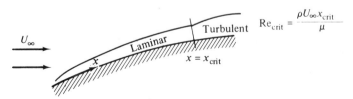

Figure 8.6

For flow over a flat plate, value of the critical Reynolds number for transition can range from 2×10^5 for a rough plate with a fairly high degree of turbulence in the approach flow, up to 3×10^6 for a very smooth plate with a low level of turbulence in the free stream flow. In the former case, the transition region might occur typically over the Reynolds number range 1.5×10^5 to 2.5×10^5. For purposes of calculation, we shall take the critical Reynolds number for flat plate boundary layer flow to be 5×10^5. For example, if a uniform flow of air at 30°C (86°F) and 1 atmosphere pressure, with a velocity of 30 m/s (98 fps), passes over a flat plate aligned to the flow, transition will occur at

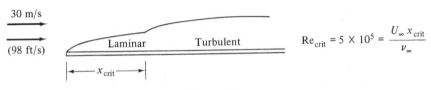

Figure 8.7

$$x = \frac{\text{Re}_{\text{crit}}\nu}{U_\infty}$$

$$= \frac{(5 \times 10^5)(1.60 \times 10^{-5} \text{ m}^2/\text{s})}{30 \text{ m/s}}$$

$$= 0.2667 \text{ m } (0.875 \text{ ft})$$

from the leading edge (see Figure 8.7).

Before ending this qualitative description of boundary layer flow, it is interesting to compare viscous flow in the boundary layer with viscous fully developed pipe flow of Chapter 6. The basic nature of the flow, that is, laminar or turbulent, is the same whether we have flow over a surface or flow though a pipe, so that, for example, in both cases we get fairly uniform velocity profiles for turbulent flow. However, for fully developed flow through a pipe, the viscous flow filled the entire pipe; no part of the flow was treated as nonviscous. After a certain inlet region, velocity profiles in the pipe did not change with distance, with such properties as friction factor independent of distance along the pipe. In contrast, with boundary layer flow, we have viscous flow and frictionless flow occurring in the same flow field. The boundary layer thickness continues to increase with distance x along the plate, so that the velocity profiles change with x. The implication is that the shear exerted by the fluid on the wall also varies with x. For this reason, it can be seen that boundary layer flow presents considerably more difficulties than pipe flow.

8.3 EQUATIONS OF MOTION
FOR THE LAMINAR BOUNDARY LAYER

In this section we shall first derive the continuity and momentum equations for the laminar boundary layer. The derivation of the continuity equation is exactly the same as that presented in Chapter 7; it will be repeated here for completeness. In the momentum equation, we shall now have to add the viscous fluid forces acting on the fluid inside the control volume.

In the following material, it will be assumed that the flow is steady and two-dimensional; that is, flow properties do not vary in the z direction. Furthermore, we shall assume incompressible flow ($\rho = $ constant), with no variation in physical properties such as μ throughout the flow field. The

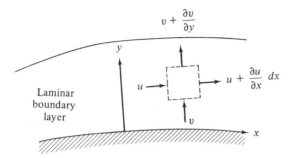

Figure 8.8

latter assumption requires that significant temperature variations not be present in the flow. Gravitational forces will be neglected. The radius of curvature of the body on which the boundary layer grows will be assumed large enough so that centrifugal forces need not be taken into account (for a flat plate, the radius of curvature is infinite). Finally, as was discussed in the previous section, the boundary layer will be assumed thin enough so that pressure does not vary in the direction normal to the wall surface.

Let x be the coordinate directed along the body surface, y the coordinate normal to the surface, with u the velocity component in the x direction and v the velocity component in the y direction (see Figure 8.8). Select a differential control volume as shown, with sides dx and dy. First let us consider a mass balance for the control volume. If the x component of velocity at the left-hand face is u, the mass flow crossing this face and entering the control volume is $\rho u\, dy$. The velocity components u and v vary throughout the flow field, so, from calculus, the x component of velocity at the right-hand face will be $u + (\partial u/\partial x)\, dx$. Thus the mass flow crossing the right-hand face will be $\rho[u + (\partial u/\partial x)\, dx]\, dy$. Similarly, if the velocity in the y direction at the bottom face is v, the mass flow entering the control volume across the bottom face is $\rho v\, dx$. The velocity in the y direction at the top face is $v + (\partial v/\partial y)\, dy$; therefore, the mass flow rate leaving the control volume across the top face is

$$\rho\left(v + \frac{\partial v}{\partial y}dy\right) dx$$

For steady flow, the rate at which mass enters the differential control volume is equal to the rate at which mass leaves the differential control volume. Summing, we obtain

$$\rho u\, dy + \rho v\, dx = \rho\left(u + \frac{\partial u}{\partial x}dx\right) dy + \rho\left(v + \frac{\partial v}{\partial y}dy\right) dx$$

Simplifying, we obtain the continuity equation in the form

$$\frac{\partial u}{\partial x} + \frac{\partial v}{\partial y} = 0 \tag{8.3}$$

The momentum equation for steady flow is

$$\Sigma \mathbf{F} = \iint \mathbf{V}(\rho \mathbf{V} \cdot \mathbf{dA})$$

or, in differential form,

$$\Sigma \, d\mathbf{F} = \mathbf{V}(\rho \mathbf{V} \cdot \mathbf{dA})$$

Momentum is a vector quantity. Therefore, for this two-dimensional flow, we should expect to have two momentum equations, one in the x direction and one in the y direction. However, as stated previously, the boundary layer is very thin; the velocity component in the y direction is very small in comparison with the velocity component in the x direction. In other words, the y component of momentum is an order of magnitude less than the x component of momentum. For this reason, we shall only consider the momentum equation in the x direction.

The differential forces acting on the control volume of Figure 8.8 in the x direction are shown below; neglecting gravity, there are only pressure forces and viscous forces. Note that the shear force acting on the bottom face, caused by fluid nearer to the wall and moving slower than that in the control volume, tends to slow down the fluid in the control volume and is in the negative x direction. The fluid above the top face is moving faster than that in the control volume; hence the shear force on the top surface is in the positive x direction. The expression for laminar shear stress is $\mu(\partial u/\partial y)$; allowance must be made for the variation of the shear throughout the flow field, so that the shear stress on the top face is

$$\mu \frac{\partial u}{\partial y} + \frac{\partial}{\partial y}\left(\mu \frac{\partial u}{\partial y}\right) dy$$

Summing the forces in the x direction shown in Figure 8.9, we have

$$\Sigma \, dF_x = p \, dy - \left(p + \frac{\partial p}{\partial x} \, dx\right) dy - \mu \frac{\partial u}{\partial y} \, dx + \mu\left(\frac{\partial u}{\partial y} + \frac{\partial}{\partial y} \frac{\partial u}{\partial y} \, dy\right) dx$$

$$= \left(-\frac{\partial p}{\partial x} + \mu \frac{\partial^2 u}{\partial y^2}\right) dx \, dy$$

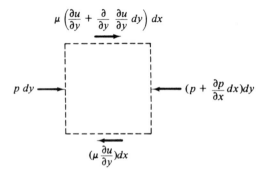

Figure 8.9

Next we shall consider the rate at which x momentum enters and leaves the control volume. Since the mass flow rate crossing the left face was $\rho u\, dy$, the rate at which x momentum crosses this face is $\rho u^2\, dy$. From calculus, the rate at which x momentum crosses the right face is

$$\rho\left(u^2 + \frac{\partial u^2}{\partial x}\, dx\right) dy$$

Also, the fluid flow crossing the bottom face, $\rho v\, dx$, has a velocity component in the x direction and hence brings into the control volume an x-momentum flux $\rho uv\, dx$, as shown in Figure 8.10.

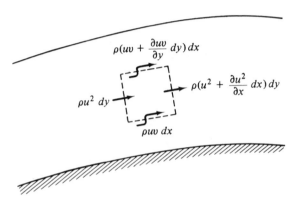

Figure 8.10

From calculus, the x-momentum flux leaving across the top face is

$$\rho\left(uv + \frac{\partial uv}{\partial y}\, dy\right) dx$$

The net rate of efflux of x momentum from the control volume is

$$\rho\left(uv + \frac{\partial uv}{\partial y}\, dy\right) dx + \rho\left(u^2 + \frac{\partial u^2}{\partial x}\, dx\right) dy - \rho u^2 dy - \rho uv\, dx$$

Canceling terms, we obtain

$$\frac{\partial uv}{\partial y}\, dy\, dx + \frac{\partial u^2}{\partial x}\, dx\, dy$$

But

$$\frac{\partial uv}{\partial y} = u\frac{\partial v}{\partial y} + v\frac{\partial u}{\partial y} \quad \text{and} \quad \frac{\partial u^2}{\partial x} = u\frac{\partial u}{\partial x} + u\frac{\partial u}{\partial x}$$

Also, from the continuity equation (8.3),

$$u\left(\frac{\partial u}{\partial x} + \frac{\partial v}{\partial y}\right) = 0$$

so that the net rate of efflux of x momentum takes the form

$$\rho\left(u\frac{\partial u}{\partial x} + v\frac{\partial u}{\partial y}\right) dx\,dy$$

Substituting into the momentum equation, we obtain

$$-\frac{\partial p}{\partial x} + \mu\frac{\partial^2 u}{\partial y^2} = \rho\left(u\frac{\partial u}{\partial x} + v\frac{\partial u}{\partial y}\right) \tag{8.4}$$

Equation (8.4) is the momentum equation for the laminar boundary layer. In order to obtain a solution for flow in a laminar boundary layer, Equation (8.4) must be solved in conjunction with the continuity equation (8.3).

The boundary conditions for the preceding equations are the following. At a fixed wall surface ($y = 0$), the velocity components u and v are both equal to zero. At the outer edge of the boundary layer ($y = \delta$), u approaches the free stream velocity U_∞, as shown in Figure 8.11. The free stream velocity U_∞ outside the boundary layer is given by the potential flow solution about the body in question. For flow over a flat plate, the free stream velocity U_∞ is constant along the plate; the Bernoulli equation for potential flow has the form

$$p_\infty + \tfrac{1}{2}\rho U_\infty^2 = \text{Constant}$$

so that there is no pressure variation outside the boundary layer. Since we have also assumed that there is no pressure variation in the boundary layer in a direction normal to the body surface, it follows that, for uniform flow over a flat plate aligned with the flow, there is no pressure variation throughout the entire flow field. The boundary layer equations (8.3) and (8.4), for the flat plate boundary layer, reduce to

$$\frac{\partial u}{\partial x} + \frac{\partial v}{\partial y} = 0 \tag{8.5}$$

$$u\frac{\partial u}{\partial x} + v\frac{\partial u}{\partial y} = \nu\frac{\partial^2 u}{\partial y^2}$$

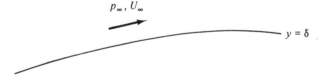

Figure 8.11

The foregoing system of complex partial differential equations must be solved simultaneously to obtain, for example, the velocity distribution in the boundary layer. A solution for the flat plate boundary layer was first worked out by Blasius in 1908. The details of the analysis used in the solution are beyond the scope of this text; however, we shall present the solution and discuss some of its ramifications.

Blasius was able to present a single curve of u/U_∞ versus the parameter

$$\frac{y}{x} \sqrt{\frac{\rho U_\infty x}{\mu}}$$

as shown in Figure 8.12. In other words, the velocity profiles are similar; at any distance x along the plate, the single curve shown can be used for determining u/U_∞.

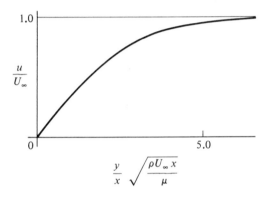

Figure 8.12

The velocity component u in the boundary layer approaches the free stream velocity U_∞ asymptotically; this makes the definition of a boundary layer thickness arbitrary, since u/U_∞ only approaches 1 as y approaches infinity. However, we can define a boundary layer thickness δ to be the distance y from the surface at which the value of u is 99 percent of the free stream value. From Figure 8.12, for $u/U_\infty = 0.99$, we obtain

$$\frac{y}{x} \sqrt{\frac{\rho U_\infty x}{\mu}} = 5.0$$

Therefore,

$$\delta = \frac{5.0x}{\sqrt{Re_x}} \tag{8.6}$$

where

$$Re_x = \frac{\rho U_\infty x}{\mu}$$

Attempts have been made to define a boundary layer thickness in a less arbitrary manner. For example, displacement thickness δ^* refers to the displacement of the external flow due to the presence of the boundary layer. To clarify this concept, consider again uniform flow with velocity U_∞ passing over a flat plate aligned to the flow (Figure 8.13). Due to the decrease in velocity in the boundary layer, the flow initially contained in the height H, namely, $U_\infty H$, will have to be displaced upward a distance δ^* to satisfy continuity. In other words,

$$U_\infty H = U_\infty \delta^* + U_\infty(H - \delta) + \int_0^\delta u\, dy$$

Simplifying,

$$U_\infty \delta^* = U_\infty \delta - \int_0^\delta u\, dy$$

or

$$\delta^* = \int_0^\delta \left(1 - \frac{u}{U_\infty}\right) dy$$

From $y = \delta$ to $y = \infty$, $u = U_\infty$, so we can express the previous result as

$$\delta^* = \int_0^\infty \left(1 - \frac{u}{U_\infty}\right) dy \tag{8.7}$$

Using the Blasius solution for u/U_∞, it can be shown that

$$\delta^* = \frac{1.73x}{\sqrt{\mathrm{Re}_x}} \tag{8.8}$$

for laminar incompressible flow over a flat plate aligned with the flow.

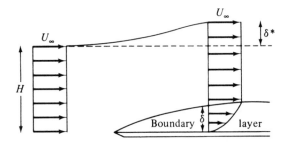

Figure 8.13

It is of great importance to be able to calculate the force exerted on a body by fluid passing over it. For example, the propulsion requirements for an airplane are determined by the drag force exerted by the air passing over its surfaces. For a laminar boundary layer, the fluid shear stress $\tau = \mu(\partial u/\partial y)$. At the surface of a flat plate,

$$\tau_0 = \mu\left(\frac{\partial u}{\partial y}\right)_{y=0}$$

From the Blasius solution given in Figure 8.12, the slope at the wall

$$\left(\frac{\partial u}{\partial y}\right)_{y=0}$$

is equal to

$$\left(\frac{\partial u}{\partial y}\right)_{y=0} = 0.332\frac{U_\infty}{x}\sqrt{Re_x}$$

so that

$$\tau_0 = 0.332\mu\frac{U_\infty}{x}\sqrt{Re_x}$$

This can be expressed in terms of a local skin friction coefficient C_{f_x}, defined by

$$C_{f_x} = \frac{\tau_0}{\frac{1}{2}\rho U_\infty^2}$$

so that, for laminar boundary layer flow over a flat plate,

$$C_{f_x} = \frac{0.664}{\sqrt{Re_x}} \tag{8.9}$$

The wall shear stress varies with x; thus in order to obtain the total force exerted by the fluid passing over the plate, we must integrate. The total drag force D on a plate of length L is equal to $\int_0^L \tau_0 b\ dx$, where b is the width of the plate (dimension into the page) and we have assumed the fluid to be passing over only one side of the plate (Figure 8.14).

U_∞

L **Figure 8.14**

Therefore,

$$D = \int_0^L 0.332\mu\frac{U_\infty}{x}\sqrt{\frac{\rho U_\infty x}{\mu}}\ b\ dx$$

$$= 0.332b\sqrt{\rho\mu}\ U_\infty^{3/2}\int_0^L \frac{dx}{\sqrt{x}}$$

$$= \frac{0.664b\rho U_\infty^2 L}{\sqrt{Re_L}}$$

where

$$Re_L = \frac{\rho U_\infty L}{\mu} \tag{8.10}$$

Define a dimensionless skin friction drag coefficient C_{D_f} to be

$$C_{D_f} = \frac{D}{\frac{1}{2}\rho U_\infty^2 bL}$$

$$D_f = \frac{0.664 b\rho U_\infty^2 L}{\frac{1}{2}\rho U_\infty^2 bL\sqrt{\mathrm{Re}_L}} \tag{8.11}$$

$$D_f = \frac{1.328}{\sqrt{\mathrm{Re}_L}}$$

cal skin friction coefficient, C_{D_f} is a coefficient g over a length L of surface.

ough air at atmospheric pressure with a velocity of 1.5 m/s. Assuming that the wings can be treated as flat plates with $L = 10$ cm and width equal to 25 cm, calculate the boundary layer thicknesses δ and δ^* at the trailing edge of the wing and the aerodynamic drag on the wing. Assume an air temperature of 10°C.

Solution   At 10°C and atmospheric pressure, the kinematic viscosity ν for air is 1.42×10^{-5} m²/s. Therefore, the Reynolds number at the trailing edge, $x = L$, is

$$\mathrm{Re} = \frac{U_\infty L}{\nu} = \frac{(1.5 \text{ m/s})(0.10 \text{ m})}{1.42 \times 10^{-5} \text{ m}^2/\text{s}} = 1.056 \times 10^4$$

This is less than the critical Reynolds number for transition to turbulent flow, 5×10^5, discussed in Section 8.1. Thus the boundary layer over the entire length of the wing is laminar, and the equations developed in this section can be used for analysis. The boundary layer thickness δ at which $u/U_\infty = 0.99$ is given by Equation (8.6):

$$\delta = \frac{5.0x}{\sqrt{\mathrm{Re}_x}} = \frac{5.0(0.10 \text{ m})}{\sqrt{1.056 \times 10^4}} = 0.4866 \text{ cm at the trailing edge}$$

From Equation (8.8),

$$\delta^* = \frac{1.73x}{\sqrt{\mathrm{Re}_x}} = \frac{1.73(0.10 \text{ m})}{\sqrt{1.056 \times 10^4}} = 0.1684 \text{ cm at the trailing edge}$$

From Equation (8.11),

$$C_{D_f} = \frac{1.328}{\sqrt{\mathrm{Re}_L}} = 0.01292$$

Therefore,

$$D = (\tfrac{1}{2}\rho U_\infty^2 bL)C_{D_f} \times 2$$

where the factor of 2 accounts for the air flowing over both sides of the wing. With $\rho = 1.247$ kg/m³,

$$D = [\tfrac{1}{2}(1.247 \text{ kg/m}^3)(1.5^2 \text{ m}^2/\text{s}^2)(0.10 \times 0.25 \text{ m}^2)][0.01292 \times 2]$$
$$= 9.063 \times 10^{-4} \text{ N}$$ ■

EXAMPLE 8.2

A model hydrofoil boat is moving in a "flying" condition (i.e., with its hull out of the water and solely supported by its foils) with a velocity of 5 fps. Treating the hydrofoils as flat plates with $L = 4$ in and width equal to 10 in, determine the boundary layer thicknesses δ and δ^* at the trailing edge of the foil and the drag on the foil for a water temperature of 50°F. Compare the results with those obtained in Example 8.1 for a similar foil moving in air.

Solution   At 50°F, the density of water is 62.41 lbm/ft^3 and the kinematic viscosity is 1.41×10^{-5} ft^2/s. The Reynolds number at the trailing edge, where $x = L$, is

$$\text{Re} = \frac{U_\infty L}{\nu} = \frac{(5 \text{ ft/s}) (\tfrac{4}{12} \text{ ft})}{1.41 \times 10^{-5} \text{ ft}^2/\text{s}} = 1.18 \times 10^5$$

which is less than the critical Reynolds number for transition, namely, 5×10^5, resulting in a laminar boundary layer over the entire hydrofoil. From Equation (8.6), we get the boundary layer thickness δ:

$$\text{Re} = \frac{U_\infty L}{\nu} = \frac{5 \text{ ft/s}(\tfrac{4}{12}) \text{ ft}}{1.41 \times 10^{-5} \text{ ft}^2/\text{s}} = 1.18 \times 10^5$$

$$\delta = \frac{5.0x}{\sqrt{\text{Re}_x}} = \frac{5(\tfrac{4}{12})}{\sqrt{1.18 \times 10^5}} = 0.00485 \text{ ft}$$

$$= \underline{0.058 \text{ in}}   \text{at the trailing edge}$$

From Equation (8.8),

$$\delta^* = \frac{1.73x}{\sqrt{\text{Re}_x}} = \frac{1.73 (\tfrac{4}{12})}{\sqrt{1.18 \times 10^5}} = 0.0168 \text{ ft}$$

$$= \underline{0.020 \text{ in}}   \text{at the trailing edge}$$

From Equation (8.11),

$$C_{D_f} = \frac{1.328}{\sqrt{\text{Re}_L}} = \frac{1.328}{\sqrt{1.18 \times 10^5}} = 0.00387$$

Therefore,

$$D = \left(\frac{\rho}{2} U_\infty^2 \, bL\right) C_{D_f} \times 2$$

$$= \frac{62.41}{2(32.17)} (5)^2 \left(\frac{10}{12} \times \frac{4}{12}\right)(0.00387)(2)$$

$$= 0.0521 \text{ lbf} = \underline{5.21 \times 10^{-2} \text{ lbf}}$$

In Example 8.1, we had $V = 1.5$ m/s $(= 4.92$ fps), $L = 10$ cm $(= 3.94$ in), and width $= 25$ cm $(= 9.84$ in), and

$$D_{air} = 9.063 \times 10^{-4} \text{ N} = (9.063 \times 10^{-4} \text{ N})(0.2248 \text{ lbf}/\text{N})$$
$$= 2.04 \times 10^{-4} \text{ lbf}$$

Thus a 255-fold increase in drag occurs in water, mainly due to the difference between the density of air and that of water (0.0794 lbm/ft^3 versus 62.41 lbm/ft^3), but mitigated by a decrease in viscosity and hence a decrease in drag coefficient C_{D_f} (0.01292 versus 0.00387). ∎

The Blasius solution presented in this section is valid only for laminar boundary layer flow over a flat plate. In many situations, however, we are also interested in the characteristics of the turbulent boundary layer. As discussed previously, we do not have available a discrete expression for τ for turbulent flow. Thus an exact solution to turbulent boundary layer flow is not possible. Instead we shall develop an approximate method for handling boundary layer flows, in which it is not necessary to have an expression for τ at each point in the boundary layer. This method, developed by von Kármán, is called the momentum integral method.

8.4 MOMENTUM INTEGRAL METHOD

According to the Kármán *momentum integral method*, we shall write the continuity and momentum equations for a control volume extending from the wall surface to the outer edge of the boundary layer, as shown in Figure 8.15. Whereas in Section 8.3 we wrote the equations of motion that would be applicable at each point inside the boundary layer, now we shall only be interested in satisfying the equations of motion for the fluid contained within a cross section from $y = 0$ to $y = \delta$.

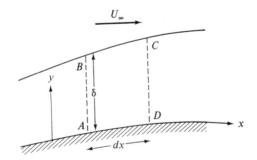

Figure 8.15

The velocity outside the boundary layer is U_∞, with U_∞ a function of x for flow over a curved body surface and U_∞ constant for flow over a flat plate. We shall now write the equations of motion for the control volume shown, assuming again that the flow is two-dimensional, incompressible, and steady.

The mass flow rate crossing the left-hand face AB in Figure 8.15 is $\rho \int_0^\delta u \, dy$. From calculus, the mass flow rate crossing the right face CD is

$$\rho \left[\int_0^\delta u \, dy + \frac{\partial}{\partial x} \left(\int_0^\delta u \, dy \right) dx \right]$$

The difference between these two is equal to

$$\rho \frac{\partial}{\partial x} \left(\int_0^\delta u \, dy \right) dx$$

and represents the mass flow rate that has entered the control volume across the top surface BC.

The momentum equation in the x direction is given by

$$\Sigma F_x = \int_{\substack{\text{control} \\ \text{surface}}} V_x (\rho \mathbf{V} \cdot \mathbf{dA})$$

The forces acting on the fluid in the control volume, pressure forces and viscous shear forces, are shown in Figure 8.16. There is no viscous shear at the outer edge of the boundary layer, for the velocity has been shown to approach asymptotically the free stream velocity U_∞. In other words, $\partial u / \partial y = 0$ at $y = \delta$.

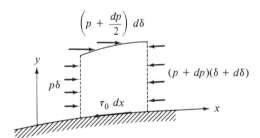

Figure 8.16

Summing forces, we obtain

$$\Sigma F_x = p\delta + \left(p + \frac{dp}{2} \right) d\delta - (p + dp)(\delta + d\delta) - \tau_0 \, dx$$

By dropping second-order terms and canceling, the above simplifies to

$$\Sigma F_x = -\delta \, dp - \tau_0 \, dx$$

The rate of momentum influx across AB is $\rho \int_0^\delta u^2 \, dy$. The rate of momentum efflux across CD is

$$\rho \left[\int_0^\delta u^2 \, dy + \frac{\partial}{\partial x} \left(\int_0^\delta u^2 \, dy \right) dx \right]$$

The flow into the control volume across BC is coming from a region of velocity U_∞ so that the rate of x-momentum influx across CD is

$$U_\infty \rho \frac{\partial}{\partial x}\left(\int_0^\delta u \, dy\right) dx$$

Substituting the foregoing expressions into the momentum equation, we obtain

$$-\delta \, dp - \tau_0 \, dx = -\rho \int_0^\delta u^2 \, dy + \rho \int_0^\delta u^2 \, dy + \rho \frac{\partial}{\partial x}\left(\int_0^\delta u^2 \, dy\right) dx$$

$$- U_\infty \rho \frac{\partial}{\partial x}\left(\int_0^\delta u \, dy\right) dx$$

Simplifying,

$$\rho \frac{d}{dx}\left(\int_0^\delta u^2 \, dy\right) - \rho U_\infty \frac{d}{dx}\left(\int_0^\delta u \, dy\right) = -\delta \frac{dp}{dx} - \tau_0 \qquad (8.12)$$

Equation (8.12) constitutes the momentum integral equation. It is valid for both laminar and turbulent boundary layer flows. Before proceeding, it is well to discuss the nature of the approximation that has been involved in the development of the equation. The continuity and momentum equations have been satisfied for the entire control volume ABCD of Figure 8.15. However, this does not mean that the equations of motion have been satisfied for each differential control volume that makes up ABCD (such as that shown in Figure 8.17). As a matter of fact, being an equation integrated over the entire boundary layer, Equation (8.12) does not yield any information about the details of the flow at a particular point in the boundary layer. For example, if we wish to proceed any further with Equation (8.12) and find, for example, τ_0 as a function of x, we must assume a variation of u with y. Naturally, the more accurate the assumption of velocity profile, the more accurate the determination of τ_0. Fortunately, it has been found that even a fairly rough approximation of velocity profile yields satisfactory results from the momentum integral method.

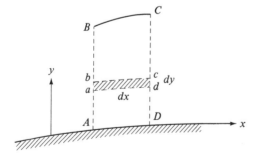

Figure 8.17

We shall now check the method by assuming a velocity profile for a laminar boundary layer on a flat plate and comparing with the exact Blasius solution. A simple velocity profile that satisfies the conditions $u = 0$ at $y = 0$ and $u = U_\infty$ at $y = \delta$ is the linear profile shown in Figure 8.18, given

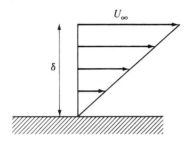

Figure 8.18

by $u/U_\infty = y/\delta$. For flow over a flat plate, $dp/dx = 0$, so that Equation (8.12) reduces to

$$\rho\frac{d}{dx}\int_0^\delta u^2\, dy - \rho U_\infty\frac{d}{dx}\left(\int_0^\delta u\, dy\right) = -\tau_0 \qquad (8.13)$$

For the assumed profile,

$$\tau_0 = \mu\left(\frac{\partial u}{\partial y}\right)_{y=0} = \mu\frac{U_\infty}{\delta}$$

Substituting,

$$\rho\frac{d}{dx}\int_0^\delta U_\infty^2\frac{y^2}{\delta^2}dy - \rho U_\infty\frac{d}{dx}\left(\int_0^\delta \frac{yU_\infty}{\delta}\, dy\right) = -\mu\frac{U_\infty}{\delta}$$

Combining,

$$\rho U_\infty^2\frac{d}{dx}\left(\frac{\delta}{3} - \frac{\delta}{2}\right) = -\mu\frac{U_\infty}{\delta}$$

or

$$\delta\, d\delta = \frac{6\mu\, dx}{\rho U_\infty}$$

Integrating the above with the boundary condition $\delta = 0$ at $x = 0$, we obtain

$$\delta = \sqrt{\frac{12\mu x}{\rho U_\infty}}$$

$$= \frac{3.46x}{\sqrt{\text{Re}_x}}$$

We can now derive expressions for δ^* and C_{f_x} as a function of x for the linear profile:

$$\delta^* = \int_0^\delta\left(1 - \frac{u}{U_\infty}\right) dy$$

$$= \int_0^\delta\left(1 - \frac{y}{\delta}\right) dy$$

$$= \frac{\delta}{2}$$

so that

$$\delta^* = \frac{1.73x}{\sqrt{\text{Re}_x}}$$

The skin friction coefficient [see Equation (8.9)] is given by

$$C_{f_x} = \frac{\tau_0}{\frac{1}{2}\rho U_\infty^2}$$

For the assumed profile,

$$\tau_0 = \frac{\mu U_\infty}{\delta}$$

so that

$$\begin{aligned}
C_{f_x} &= \frac{\mu U_\infty}{\frac{1}{2}\rho U_\infty^2 \delta} \\
&= \frac{\mu U_\infty}{\frac{1}{2}\rho U_\infty^2} \frac{\sqrt{\text{Re}_x}}{3.46x} \\
&= \frac{0.5780}{\sqrt{\text{Re}_x}}
\end{aligned}$$

A comparison of results for δ^* and C_{f_x} as computed from the momentum integral method using a rough velocity profile and as computed from the exact Blasius solution given by Equations (8.8) and (8.9) indicates the accuracy of the approximate method. A selection of velocity profile closer to the Blasius profile would give even more accurate results. In the next section we shall apply the same method to an analysis of the turbulent boundary layer on a flat plate.

EXAMPLE 8.3
Air at 30°C is blowing past a flat wall with a speed of 5 m/s. The plate length in the direction of air flow is one meter with a plate width of 0.5 m. Assume that the air velocity profile is given by the expression

$$\frac{u}{U_\infty} = 2\frac{y}{\delta} - 2\left(\frac{y}{\delta}\right)^3 + \left(\frac{y}{\delta}\right)^4$$

Compare the plate drag for this velocity distribution with that obtained using a linear velocity distribution.

Solution   For the given velocity profile, the wall shear τ_0 becomes

$$\tau_0 = \mu\left(\frac{\partial u}{\partial y}\right)_{y=0} = \mu\frac{2U_\infty}{\delta}$$

Substituting into Equation (8.13), we obtain

$$\rho \frac{d}{dx} \int_0^\delta U_\infty^2 \left[2\frac{y}{\delta} - 2\left(\frac{y}{\delta}\right)^3 + \left(\frac{y}{\delta}\right)^4 \right]^2 dy$$

$$- \rho U_\infty \frac{d}{dx} \int_0^\delta U_\infty \left[2\frac{y}{\delta} - 2\left(\frac{y}{\delta}\right)^3 + \left(\frac{y}{\delta}\right)^4 \right] dy = -\mu \frac{2U_\infty}{\delta}$$

Evaluating the two integrals, we get

$$\rho U_\infty^2 \frac{d}{dx}(0.5825\delta) - \rho U_\infty^2 \frac{d}{dx}(0.7000\delta) = -\mu \frac{2U_\infty}{\delta}$$

or

$$0.1175 \, \rho U_\infty^2 \delta \frac{d\delta}{dx} = 2\mu U_\infty$$

Integrating with $\delta = 0$ at $x = 0$, we obtain

$$\frac{\delta^2}{2} = \frac{2}{0.1175} \frac{\mu x}{\rho U_\infty}$$

and

$$\delta = 5.836 \frac{x}{\sqrt{Re_x}}$$

The expression for the displacement thickness δ^* is

$$\delta^* = \int_0^\delta \left(1 - \frac{u}{U_\infty} \right) dy = 0.3\delta$$

so that

$$\delta^* = 1.751 \frac{x}{\sqrt{Re_x}}$$

The local skin friction coefficient C_{f_x} is:

$$C_{f_x} = \frac{\tau_0}{\frac{1}{2}\rho U_\infty^2} = \frac{2\mu U_\infty}{\frac{1}{2}\rho U_\infty^2 \delta} = \frac{0.6854}{\sqrt{Re_x}}$$

The total skin friction drag coefficient C_{D_f} is obtained by integrating the expression for τ_0:

$$C_{D_f} = \frac{D}{\frac{1}{2}\rho U_\infty^2 bL} = \frac{\int_0^L \tau_0 b \, dx}{\frac{1}{2}\rho U_\infty^2 bL}$$

$$= \frac{\int_0^L C_{f_x} b \, dx}{bL}$$

$$= \frac{1.3708}{\sqrt{Re_L}}$$

For the given flow conditions,

$$\text{Re}_L = \frac{U_\infty L}{\nu} = \frac{(5 \text{ m/s})(1 \text{ m})}{1.60 \times 10^{-5} \text{ m}^2/\text{s}} = 3.125 \times 10^5$$

$$D = C_{D_f} \tfrac{1}{2}\rho U_\infty^2 bL \quad \text{where} \quad \rho = 1.164 \text{ kg/m}^3$$

$$= \underline{0.0171 \text{ N}}$$

For a linear velocity distribution,

$$C_{f_x} = \frac{0.577}{\sqrt{\text{Re}_x}}$$

and hence

$$C_{D_f} = \frac{1.156}{\sqrt{\text{Re}_L}} \quad \text{or} \quad D = \underline{0.0150 \text{ N}}$$

For the Blasius profile, $D = \underline{0.0173 \text{ N}}$.   ∎

8.5 TURBULENT BOUNDARY LAYER ON A FLAT PLATE

The success of the momentum integral method in predicting the characteristics of the laminar boundary layer leads us to attempt the same procedure for analysis of the turbulent boundary layer on a flat plate. Naturally the velocity profile selected must be different from that used for the laminar case. Since no exact analysis is available for the turbulent boundary layer, we shall pick a profile that is in reasonable agreement with experimental data. It has been pointed out that the effect of turbulence is to bring about a more uniform profile than would be obtained for the laminar boundary layer. It has been found that a profile based on

$$\frac{u}{U_\infty} = \left(\frac{y}{\delta}\right)^{1/7} \tag{8.14}$$

gives a very good correlation with experimental data over a wide range of turbulent Reynolds numbers. A plot of Equation (8.14) is given in Figure 8.19.

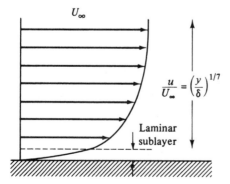

Figure 8.19

The one-seventh-power law profile is not valid in the immediate vicinity of the wall, since in this region there exists the laminar sublayer in which turbulence is damped out by the wall. It is therefore necessary to resort to experimental data to obtain an expression for τ_0. An equation that yields good agreement with data over the Reynolds number range $5 \times 10^5 < \text{Re}_x < 10^7$ is the Blasius resistance formula,

$$\frac{\tau_0}{\rho U_\infty^2} = 0.0225 \left(\frac{\nu}{U_\infty \delta}\right)^{1/4} \tag{8.15}$$

We can now substitute Equations (8.14) and (8.15) into the momentum integral equation (8.13) to obtain expressions for the rate of growth and the skin friction coefficient of the turbulent boundary layer on a flat plate:

$$\rho \frac{d}{dx}\left(\int_0^\delta u^2 \, dy\right) - \rho U_\infty \frac{d}{dx}\left(\int_0^\delta u \, dy\right) = -\tau_0$$

$$-\rho U_\infty^2 (0.0225)\left(\frac{\nu}{U_\infty \delta}\right)^{1/4} = \rho U_\infty^2 \frac{d}{dx}\int_0^\delta \left(\frac{y}{\delta}\right)^{2/7} dy - \rho U_\infty^2 \frac{d}{dx}\int_0^\delta \left(\frac{y}{\delta}\right)^{1/7} dy$$

$$(0.0225)\left(\frac{\nu}{U_\infty \delta}\right)^{1/4} = \frac{d}{dx}\left(\frac{7}{9}\delta - \frac{7}{8}\delta\right) = \frac{d}{dx}\frac{7}{72}\delta$$

$$\frac{7}{72}\frac{d\delta}{dx} = 0.0225\left(\frac{\nu}{U_\infty \delta}\right)^{1/4}$$

or

$$\delta^{1/4} \, d\delta = \frac{72}{7}(0.0225)\left(\frac{\nu}{U_\infty}\right)^{1/4} dx$$

Integrating the above, we obtain

$$\frac{4}{5}\delta^{5/4} = \frac{72}{7}(0.0225)\left(\frac{\nu}{U_\infty}\right)^{1/4} x + \text{constant}$$

If we assume the turbulent boundary layer starts at the leading edge of the plate, $\delta = 0$ at $x = 0$, and the integration constant is equal to zero. Actually, as explained in Section 8.1, up to a critical Reynolds number for transition of approximately 5×10^5, the boundary layer is laminar, so this assumption involves an approximation. A method for handling the actual case in which both types of boundary layer flows are present on a surface will be discussed later in this chapter.

With the constant equal to zero, we finally obtain

$$\delta = 0.37x\left(\frac{U_\infty x}{\nu}\right)^{-1/5} \tag{8.16a}$$

Since

$$\delta^* = \int_0^\delta \left(1 - \frac{u}{U_\infty}\right) dy \quad \text{and} \quad \frac{u}{U_\infty} = \left(\frac{y}{\delta}\right)^{1/7}$$

we can obtain a rate of growth of displacement thickness for the assumed turbulent profile. Thus

$$\delta^* = \int_0^\delta \left[1 - \left(\frac{y}{\delta} \right)^{1/7} \right] dy = \frac{\delta}{8}$$

and, for the turbulent boundary layer on a flat plate,

$$\delta^* = 0.046x(\text{Re}_x)^{-1/5} \tag{8.16b}$$

We are also interested in τ_0 as a function of x. Using the Blasius resistance formula [Equation (8.15)],

$$\tau_0 = \rho U_\infty^2 (0.0225) \left(\frac{\nu}{U_\infty \delta} \right)^{1/4}$$

where δ has been given by Equation (8.16a), we have

$$\tau_0 = \rho U_\infty^2 (0.0225) \left[\frac{\nu}{U_\infty} \frac{(U_\infty x/\nu)^{1/5}}{0.37x} \right]^{1/4}$$

$$= 0.029 \rho U_\infty^2 (\text{Re}_x)^{-1/5}$$

or

$$C_{f_x} = \frac{\tau_0}{\frac{1}{2}\rho U_\infty^2} = \frac{0.058}{\text{Re}_x^{1/5}} \tag{8.16c}$$

In order to obtain a drag coefficient for turbulent boundary layer flow over a plate of length L, the expression for τ_0 must be integrated:

$$D = \int_0^L \tau_0 \, dx$$

$$= 0.029 \rho U_\infty^2 \int_0^L \frac{b \, dx}{\text{Re}_x^{1/5}} \quad \text{(with } b = \text{plate width)}$$

$$= 0.036 \rho U_\infty^2 \frac{bL}{\text{Re}_L^{1/5}}$$

In terms of the dimensionless skin friction drag coefficient, C_{Df}, this becomes

$$C_{Df} = \frac{D}{\frac{1}{2}\rho U_\infty^2 bL} = \frac{0.072}{\text{Re}_L^{1/5}} \tag{8.17}$$

The expression for C_{Df} in Equation (8.17) has been derived using the approximate momentum integral method. It has been found that better agreement with experimental data is provided if the constant in Equation (8.17) is slightly altered, so that

$$C_{Df} = \frac{0.074}{\text{Re}_L^{1/5}} \tag{8.18}$$

The foregoing expression for skin friction drag coefficient is valid over the Reynolds number range $5 \times 10^5 < \text{Re}_L < 10^7$.

EXAMPLE 8.4

An airplane is flying through air at atmospheric pressure with a velocity of 150 fps. Assuming that the wings can be treated as flat plates, with the turbulent boundary layer growing from the leading edge of the wing, calculate the boundary layer thicknesses δ and δ^* at the trailing edge of the wing and the total skin friction drag on the wing. Assume the wing to be 5 ft long, with a width of 40 ft. The air temperature is 40°F.

Solution   At 40°F and atmospheric pressure, the kinematic viscosity for air is 1.47×10^{-4} ft^2/s. Therefore, the Reynolds number at the trailing edge is

$$\text{Re}_L = \frac{U_\infty L}{\nu} = \frac{(150 \text{ ft/s})(5 \text{ ft})}{1.47 \times 10^{-4} \text{ ft}^2/\text{s}} = 5.10 \times 10^6$$

Since this is greater than the critical Reynolds number for transition, it follows that the boundary layer is turbulent over the great majority of the plate. From Equation (8.16a),

$$\delta = 0.37x(\text{Re}_x)^{-1/5}$$

At $x = L$,

$$\delta = 0.37(5)(5.10 \times 10^6)^{-1/5}$$

or

$$\delta = 0.084 \text{ ft} = \underline{1.01 \text{ in}}$$

From Equation (8.16b),

$$\delta^* = \frac{\delta}{8} = 0.0105 \text{ ft} = \underline{0.126 \text{ in}}$$

Applying Equation (8.18),

$$C_{D_f} = \frac{0.074}{\text{Re}_L^{1/5}} = 0.00337$$

$$D = (\tfrac{1}{2}\rho U_\infty^2 bL)(C_{D_f})2$$

For air at 40°F and atmospheric pressure, $\rho = 0.0794$ lbm/ft^3. Therefore,

$$D = \frac{1}{2}\left(\frac{0.0794 \text{ lbm/ft}^3}{32.17 \text{ lbm/slug}}\right)(150^2 \text{ ft}^2/\text{s}^2)(5 \text{ ft} \times 40 \text{ ft})(0.00337)(2)$$

$$= \underline{37.43 \text{ lbf}} \qquad \blacksquare$$

In flow over a flat plate, even though Re_L is greater than the critical Reynolds number for transition, the boundary layer is laminar up to the point at which Re_x is equal to Re_{crit}. Therefore, by assuming that the turbulent boundary layer starts at $x = 0$, we are introducing an error into the expression for drag. An approximate method for eliminating this error was first introduced by Prandtl. According to this method, the total drag is first calculated from Equation (8.18), assuming the turbulent boundary layer starts at $x = 0$. We now subtract from this the turbulent drag from $x = 0$ to transition, then add the laminar drag [from Equation (8.11)] from $x = 0$ to transition:

$$\text{Drag} = \text{Turbulent drag from } x = 0 \text{ to } x = L$$
$$- \text{ Turbulent drag from } x = 0 \text{ to } x = x_{\text{crit}}$$
$$+ \text{ Laminar drag from } x = 0 \text{ to } x = x_{\text{crit}}$$

For example, if $\text{Re}_{\text{crit}} = 5 \times 10^5$, we obtain

$$C_D \frac{1}{2} \rho U_\infty^2 bL = \frac{0.074}{\text{Re}_L^{1/5}} \frac{1}{2} \rho U_\infty^2 bL - \frac{0.074}{(5 \times 10^5)^{1/5}} \frac{1}{2} \rho U_\infty^2 bL \frac{x_{\text{crit}}}{L}$$
$$+ \frac{1.328}{(5 \times 10^5)^{1/2}} \frac{1}{2} \rho U_\infty^2 bL \frac{x_{\text{crit}}}{L}$$

Since

$$\frac{x_{\text{crit}}}{L} = \frac{\text{Re}_{\text{crit}}}{\text{Re}_L}$$

the above simplifies to

$$C_D = C_{D_t} - \frac{1700}{\text{Re}_L} \tag{8.19}$$

where C_{D_t} is the turbulent drag coefficient from Equation (8.18).

The method used in deriving Equation (8.19) is approximate in that the characteristics of the turbulent boundary layer after the point of transition are assumed to be the same as if the turbulent boundary layer started at $x = 0$. Equation (8.19) is valid up to $\text{Re}_L = 1 \times 10^7$. An expression for C_{D_f} which seems to fit experimental data over a wider range of Reynolds numbers is the Prandtl-Schlichting expression:

$$C_{D_f} = \frac{0.455}{(\log_{10} \text{Re}_L)^{2.58}} - \frac{1700}{\text{Re}_L} \tag{8.20}$$

valid for turbulent Reynolds numbers up to 10^9. The second term on the right side of Equation (8.20) accounts for the laminar portion of the boundary layer up to the point of transition, and it is based on a critical Reynolds number of 5×10^5. From the results of Equations (8.11), (8.18), (8.19), and (8.20), it is possible to plot C_{D_f} versus Re_L for a flat plate at zero incidence (see Figure 8.20).

EXAMPLE 8.5

A barge is traveling down a river with a velocity of 1.0 m/s (Figure 8.21). Calculate the total drag force on the flat bottom surface of the barge, and the power required to push the bottom surface through the water at the given velocity. (This power is only a small part of the total power required to move the barge at a constant velocity; the drag force on the front surface of the barge, here a flat plate at right angles to the flow, is much greater than that on the bottom surface aligned with the flow. Drag on entire immersed

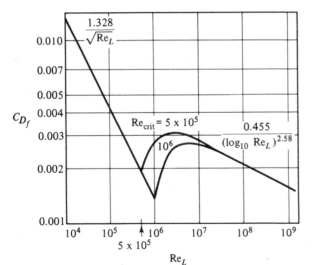

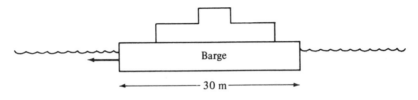

Figure 8.20

Figure 8.21

bodies will be considered in Chapter 9.) Take the bottom surface as rectangular, 10 m wide by 30 m long. Take the kinematic viscosity of water to be 1.00×10^{-6} m²/s, density 998 kg/m³.

Solution We shall first calculate Re_L in order to determine whether or not transition has occurred on the surface:

$$\mathrm{Re}_L = \frac{U_\infty L}{\nu} = \frac{(1.0 \text{ m/s})(30 \text{ m})}{10^{-6} \text{ m}^2/\text{s}} = 3 \times 10^7$$

Since this value is greater than 5×10^5, we shall use Equation (8.19) to determine C_{D_f}:

$$C_{D_f} = \frac{0.074}{\sqrt[5]{\mathrm{Re}_L}} - \frac{1700}{\mathrm{Re}_L}$$

$$= 0.002365 - 0.000057$$

$$= 0.002308$$

Therefore, total drag on the bottom surface is

$$D = C_{D_f}\tfrac{1}{2}\rho U_\infty^2(bL)$$

$$= (0.002308)\tfrac{1}{2}(998 \text{ kg/m}^3)(1.0^2 \text{ m}^2/\text{s}^2)(300 \text{ m}^2)$$

$$= \underline{345.5 \text{ N}}$$

The power required to push the bottom surface through the water with a steady velocity of 1.0 m/s is simply drag force × surface velocity, or

$$\text{Power} = (345.5 \text{ N})(1.0 \text{ m/s})$$
$$= \underline{345.5 \text{ W}}$$

From this example it can be seen that, for Re_L sufficiently large ($>10^7$), the correction for the laminar portion of the boundary layer becomes quite small. ∎

EXAMPLE 8.6

A 100-fps wind blows parallel to a billboard, as shown in Figure 8.22. If the air temperature is 40°F and the billboard is 15 ft × 30 ft, calculate the frictional force exerted by the wind on the billboard. Assume the wind and billboard are perfectly aligned.

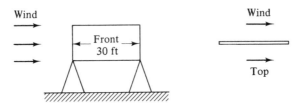

Figure 8.22

Solution At 40°F and 1 atmosphere pressure,

$$\nu = 1.47 \times 10^{-4} \text{ ft}^2/\text{s}$$

and

$$\rho = 0.0794 \text{ lbm/ft}^3 = 2.47 \times 10^{-3} \text{ slugs/ft}^3$$

Therefore,

$$\text{Re}_L = \frac{U_\infty L}{\nu} = \frac{(100 \text{ ft/s})(30 \text{ ft})}{1.47 \times 10^{-4} \text{ ft}^2/\text{s}} = 2.04 \times 10^7$$

From Equation (8.19),

$$C_{Df} = \frac{0.074}{\sqrt[5]{\text{Re}_L}} - \frac{1700}{\text{Re}_L}$$

$$= 0.002554 - 0.000083$$

$$= 0.002471$$

$$\text{Total frictional force} = C_{Df} \tfrac{1}{2}\rho U_\infty^2 bL \times 2$$

$$= (0.002471)(\tfrac{1}{2})(2.47 \times 10^{-3} \text{ slugs/ft}^3)$$
$$\times (100^2 \text{ ft}^2/\text{s}^2)(450 \text{ ft}^2)(2)$$

$$= \underline{27.5 \text{ lbf}}$$

From Figure 8.20 it can be seen that, at comparable Reynolds numbers, the skin friction drag coefficient for the laminar boundary layer is less than that for the turbulent boundary layer. Thus, in order to decrease skin friction, it is desirable to retain the laminar boundary layer over as great a portion of the surface as possible. By making the surface of the plate very smooth and by reducing any disturbances to a minimum, it is possible to increase the critical Reynolds number for transition above 5×10^5. For example, by raising Re_{crit} to 10^6, the distance from the leading edge to the point of transition is increased by a factor of 2 times the distance for Re_{crit} equal to 5×10^5.

∎

For a transition Reynolds number of 10^6, the expression for the drag coefficient comparable to Equation (8.20) becomes

$$C_{D_f} = \frac{0.455}{(\log_{10} Re_L)^{2.58}} - \frac{3300}{Re_L} \tag{8.21}$$

This result for a flat plate at zero incidence has been included in Figure 8.20.

PROBLEMS

8.1. Water at 28°C flows over a flat plate aligned with the flow. If the uniform velocity of the water outside the boundary layer is 10 m/s, determine the distance along the plate at which transition occurs. Assume a critical Reynolds number of 5×10^5.

8.2. Repeat Problem 8.1 for critical Reynolds numbers of 2.5×10^5 and 3×10^6.

8.3. At 68°F, glycerin has a specific gravity of 1.26 with a viscosity of 0.03 lbf s/ft². For flow of glycerin over a flat plate aligned with the flow, determine the distance along the plate at which transition occurs for a critical Reynolds number of 5×10^5. $U_\infty = 3$ fps.

8.4. Water flows slowly over a flat plate aligned with the flow. If the uniform water velocity outside the boundary layer is 10 cm/s and the water temperature is 5°C, find the shear stress at the surface of the plate at a point 50 cm from the leading edge. Calculate the local skin friction coefficient at the same point.

8.5. An airplane is in level flight at an altitude of 30,000 ft. Treating the wings as rectangular flat plates, determine the fraction of the boundary layer flow over the wing that is laminar. $Re_{crit} = 5 \times 10^5$, $U_\infty = 150$ fps, $L = 5$ ft.

8.6. Calculate the force required to push an oil dipstick into the oil reservoir shown in Figure P8.6 with a velocity of 1.0 cm/s. The dipstick has a rectangular cross section 1 cm × 0.25 cm. Assume the oil has specific gravity 0.86 with viscosity 7×10^{-3} Pa · s.

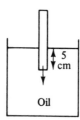

Figure P8.6

8.7. In Section 8.4 the boundary layer assumed a linear velocity profile for the laminar boundary layer for substitution into the Kármán momentum integral equation. As a better approximation, assume a velocity profile:

$$\frac{u}{U_\infty} = C_0 + C_1 \frac{y}{\delta} + C_2 \frac{y^2}{\delta^2} + C_3 \frac{y^3}{\delta^3}$$

Using appropriate boundary conditions to evaluate the constants C_0, C_1, C_2, and C_3, substitute into the Kármán momentum integral to get expressions for δ^* and C_{f_x} as functions of x. Compare with the Blasius solution.

8.8. Repeat Problem 8.7 for a sinusoidal profile

$$\frac{u}{U_\infty} = \sin\left(\frac{\pi}{2}\frac{y}{\delta}\right)$$

8.9. For Example 8.3, compute the portion of the drag due to the laminar part of the boundary layer, assuming transition at a critical Reynolds number of 5×10^5.

8.10. Consider an airplane in level flight at sea level and at high altitude. For the same forward velocity, would you expect the boundary layer to be thicker at sea level or at high altitude? Explain.

8.11. In the derivation of Equation (8.19), suppose the transition point to occur at Re $= 10^6$. Determine the correction factor that would have to be applied to account for the laminar portion of the boundary layer.

8.12. Repeat Problem 8.9 for a critical Reynolds number of 10^6. Use results of Problem 8.11.

8.13. A prototype of the model airplane described in Example 8.1 is 15 times as large as the model. Its operating speed is 80 m/s. Calculate the aerodynamic drag of the wing. Can the results of Example 8.1 be used to predict the drag of the prototype? Assume a critical Reynolds number of 5×10^5.

8.14. An airplane wing has a span of 60 ft and chord of 12 ft. Calculate the drag force on the wing when the plane is flying at 125 mph, with ambient conditions 13.5 psia and 20°F. Assume the wing to behave as a flat plate aligned with the flow. Calculate the horsepower required to push the wing through air at this velocity.

8.15. A ship is moving through water with a velocity of 10 m/s. Assuming that the surface of the ship in contact with the water can be treated as a flat plate, calculate the power required to maintain the above velocity. Take the surface to be 10 m wide by 25 m long, with the water temperature 20°C.

8.16. A 30-fps wind blows over a flat roof (Figure P8.16). If the air temperature is 32°F and the roof is 30 ft wide, calculate the wind drag force on the roof.

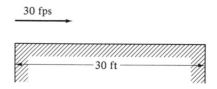

Figure P8.16

8.17. An oil with viscosity of 70 Pa · s and density 850 kg/m^3 flows over a flat plate with a free stream velocity of 10 cm/s. If the plate is 10 cm long and 5 cm wide,

(a) Calculate the maximum boundary layer thickness on the plate and the drag force exerted by the fluid on the plate.

(b) Repeat the above for water at 25°C.

(c) Repeat for air at 25°C, 1 atmosphere.

(d) Repeat for air at 25°C, 50 kPa. (Remember that the viscosity of a gas is relatively insensitive to pressure, but the density is directly proportional to pressure—i.e., the perfect gas law.)

(e) Repeat for hydrogen at 25°C, 1 atmosphere.

8.18. For a turbulent boundary layer, with Re > 10^7, assume the velocity profile can be approximated by

$$\frac{u}{U_\infty} = \left(\frac{y}{\delta}\right)^{1/8}$$

Using the expression for shear provided by Equation (8.15), determine δ^* and C_{f_x} as functions of x for this case.

8.19. Water flows over a slightly curved surface. At a point on the surface, it is found that an expression for free stream velocity is given by $U_\infty = 2.0 + 0.2x$. The boundary layer at the point in question is laminar, so approximately treat the velocity profile as linear, with $u/U_\infty = y/\delta$. Using the Kármán momentum integral approach, determine τ_0. Is this greater or less than τ_0 for a flat plate? Explain.

8.20. To get out of the wind on the ocean shore, it often helps to lie down. To show this, assume a boundary layer thickness of 4 ft with a wind velocity of 15 mph. Calculate the wind velocity at distances of 6 in and 12 in above the ground. Assume a turbulent velocity profile (e.g., the one-seventh-power law).

8.21. A streamlined passenger train is 150 m long. Assuming the flat plate boundary layer equations can be applied, determine the boundary layer thickness at the end of the train. The air temperature is 30°C, train velocity 60 m/s. Find the wall shear stress at $x = 150$ m and at $x = 75$ m.

8.22. A source is located a distance d from a flat wall. The free stream velocity along the wall can be determined as indicated in Problem 7.11. Using the Kármán momentum integral approach, determine the skin friction coefficient at the wall surface.

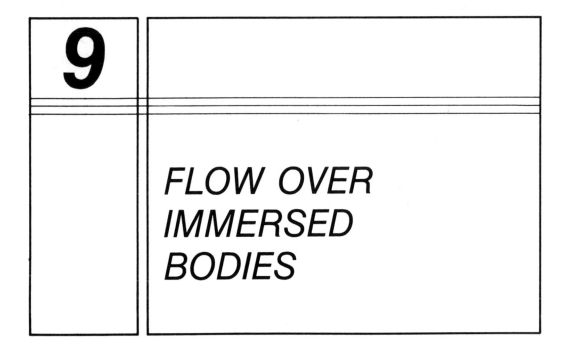

9

FLOW OVER IMMERSED BODIES

9.1 INTRODUCTION

In studying uniform flow past a body surface, the engineer is most often interested in the resultant fluid forces acting on the body. Such forces can be divided into two components, lift force acting upward and normal to the approach flow, and drag force acting in the same direction as the approach flow, as shown in Figure 9.1. In Chapter 7 frictionless, nonviscous flow about certain body shapes was discussed; it was found that the drag on a body in potential flow was zero. For an analysis of drag, therefore, it is necessary to examine real, viscous fluid effects, such as boundary layer growth and separation.

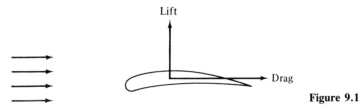

Lift

Drag

Figure 9.1

In Chapter 8 we restricted our study of viscous effects to boundary layer flow over a flat plate aligned with the flow direction. In this case, there were no pressure gradients in the flow field. For uniform flow past a finite body, such as an airfoil or cylinder, however, the boundary layer is subjected to both positive and negative pressure gradients. The effect of these pressure gradients on boundary layer behavior is extremely important in an analysis of the drag characteristics of a body shape.

In this chapter we shall first study the effect of pressure gradients on the boundary layer and shall demonstrate how separation of the flow from the body surface may result from a positive pressure gradient acting on the boundary layer. Then we shall show how the occurrence of flow separation on a body can alter the pressure distribution from that obtained in potential flow and lead to a pressure drag on the body (unlike potential flow). Finally, we shall discuss some viscous effects that alter the lift characteristics of an airfoil, discussed in Section 7.8.

It will be assumed in this chapter, as in Chapter 8, that the Reynolds numbers are large enough so that the flow can be divided into two parts: a viscous region consisting of a thin boundary layer, separation region, and wake, and a nonviscous region, where the effects of viscosity can be neglected, as shown in Figure 9.2. It is the purpose of this chapter to describe the flow of a real, viscous fluid over a body and to discuss the resultant fluid forces acting on the body.

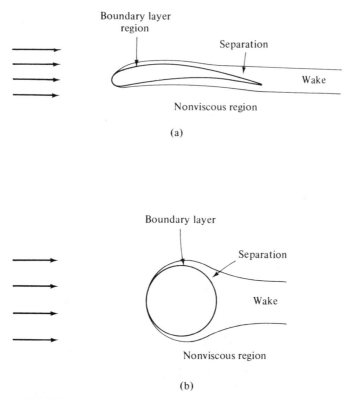

Figure 9.2. (a) Real fluid flow about an airfoil; (b) real fluid flow about a circular cylinder.

9.2 BOUNDARY LAYERS WITH PRESSURE GRADIENTS

The pressure distribution on the surface of a cylinder immersed in uniform steady flow was derived in Chapter 7 and is repeated in Figure 9.3. For a real fluid at high Reynolds numbers, a boundary layer grows along the body surface, starting at point A. This boundary layer is thin enough so that there is no pressure variation in the direction normal to the body surface. Along the surface, the pressure decreases from A to B, then starts to increase as the flow proceeds along the rear portion of the body surface (B to C). From A to B, the effect of the decreasing pressure is to accelerate the fluid particles in the boundary layer. Note that a fluid particle within the boundary layer is subjected to the same pressure variation as a fluid particle in the potential flow just outside the boundary layer, for there is no pressure variation in the direction normal to the surface. The acceleration of fluid particles in the boundary layer tends to oppose the action of viscous shear forces; hence a negative pressure gradient is called a *favorable pressure gradient*.

Starting at B, however, over the rear half of the cylinder, the pressure increases, thus slowing down the fluid particles in the boundary layer. The decelerating effect of the increasing pressure, felt through the boundary layer, has the same result on the fluid as viscous shear; hence a positive pressure gradient is called an *adverse pressure gradient*.

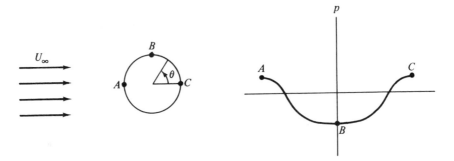

Figure 9.3

The fluid in the vicinity of the wall, however, has little kinetic energy to begin with. A decrease in kinetic energy due to an adverse pressure gradient, if great enough, may lead to a negative velocity, or flow reversal. The change in velocity profile in the boundary layer due to an adverse pressure gradient, such as exists on the rear half of the cylinder, is depicted in Figure 9.4. At the wall surface, due to viscosity, the flow velocity must always be zero. With the adverse pressure gradient acting to slow down the fluid in the boundary layer, particles of fluid near the wall are decelerated and, in profiles (d) and (e), actually reversed in direction. Profile (c), marking the onset of separation, has $\partial u/\partial y = 0$ at the wall surface. The region of flow reversal, outlined by the broken line of Figure 9.4, is called the *separation*

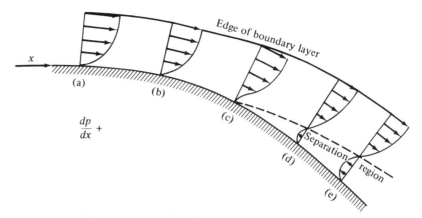

Figure 9.4.   Effect of adverse pressure gradient.

region, since the forward flow has separated from the boundary. The reverse motion of the fluid particles at the edge of the separation region gives rise to a vortex flow in the separated region. Downstream of the point of separation, there exists a region of eddying, irregular flow called the wake (Figure 9.5).

Figure 9.5

This region forms, essentially, a new boundary in the flow, different from that of the bounding surface. The wake can be clearly seen in the smoke tunnel photograph (Figure 9.6) of flow about a circular cylinder. Since the boundary between the separated region and wake is unknown, it is impossible to use potential flow theory to predict the flow properties downstream of the point of separation.

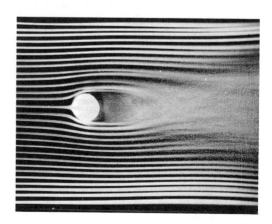

Figure 9.6.   Smoke tunnel photograph of flow about a circular cylinder. (*Courtesy* Benedum Instrument Co., Rockville, Maryland.)

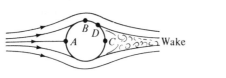 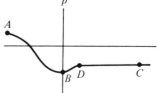

Figure 9.7

From experimental data, a typical pressure distribution on a circular cylinder immersed in a real, viscous flow is shown in Figure 9.7. It can be seen that the pressure in the separated region and that in the wake are reasonably uniform; this pressure is approximately equal to the pressure on the surface of the cylinder at the point of separation D. Due to boundary layer separation, then, the average pressure on the rear half of the cylinder is less than that on the front half. On the rear half of the cylinder, the pressure on the cylinder surface does not recover to the full stagnation pressure at C, as was the case for potential flow. Thus, by integrating the pressure forces over the cylinder surface for the real fluid case, we obtain a pressure drag on the cylinder. In order to determine this pressure drag, it is first necessary to find the location of the point of separation, point D, at which $(\partial u/\partial y)_{y=0} = 0$. Approximate methods have been worked out for predicting boundary layer behavior with varying external pressure; these methods, however, are complex and beyond the scope of this text. In general, in order to determine precisely the pressure drag on an arbitrary body, it is necessary to resort to experimental tests.

We can deduce some significant results from the discussion above. In order to keep pressure drag to a minimum, it is desirable to reduce the extent of the separation region; that is, to move the point of boundary layer separation as close as possible to the trailing edge of the body. Since separation has been found to result from adverse pressure gradients, it is desirable to reduce the magnitude of the adverse pressure gradient acting on a body surface. This technique is used in the streamlining of a body shape to reduce pressure drag. For example, consider the symmetrical airfoil shown in Figure 9.8.

Streamlined body shape

Figure 9.8

The potential flow pressure distribution is given in Figure 9.9. Again stagnation points are located at points A and C, with minimum pressure at B, the point of maximum thickness. The body is reduced in thickness very gradually,

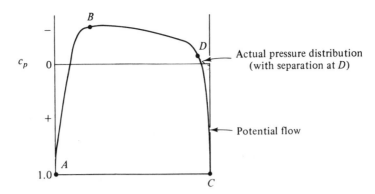

Figure 9.9

from B to the trailing edge, so that no large adverse pressure gradients are incurred. In this case, the point of separation can be delayed almost until the trailing edge itself is reached and the resultant pressure drag reduced to a minimum.

An important factor in the determination of the point of boundary layer separation on a circular cylinder or other body is whether the boundary layer is laminar or turbulent. This can be seen by an examination of typical velocity profiles for the two types of boundary layers, shown in Figure 9.10. If we consider a point A, near the wall, in the profiles, it is to be noted that there is considerably more kinetic energy at this point in the turbulent profile. Due to the turbulent mixing action, energy is more evenly distributed over each flow cross section. If an adverse pressure gradient is applied to the two profiles, slowing down the particles in the boundary layer, it is clear that flow reversal and separation will occur first for the laminar boundary layer. In other words, a turbulent boundary layer is better able to resist the decelerative effects of the adverse pressure gradient.

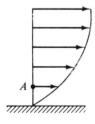

Laminar profile

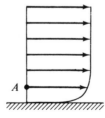

Turbulent profile   **Figure 9.10**

The resultant effect on pressure drag for flow about a cylinder is shown in Figure 9.11. With a laminar boundary layer, Case 1, separation occurs approximately at the midpoint of the cylinder, with the pressure on the rear half of the cylinder approximately equal to the minimum pressure on the cylinder surface. In Case 2, it is assumed that transition from a laminar to

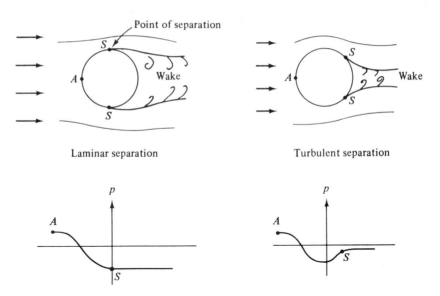

Case 1   Case 2

Point of separation

Laminar separation   Turbulent separation

Figure 9.11

a turbulent boundary layer has occurred somewhere on the front portion of the cylinder. The resultant turbulent boundary layer is better able to resist separation, so that there is a pressure recovery on the rear half of the cylinder up to point S. Clearly, the pressure drag on the cylinder of Case 2 is much less than that of Case 1. It follows that by roughening the cylinder of Case 1 and causing transition to a turbulent boundary layer on the front half of the cylinder, the drag can be reduced.

Before proceeding, let us review briefly the information we have presented on pressure drag, or form drag as it is sometimes called. First, pressure drag is associated with the growth and separation of the boundary layer; for nonviscous flow, pressure drag would be zero. Second, pressure drag can be reduced by shaping or streamlining the body so as to place the point of boundary layer separation in the vicinity of the trailing edge of the body. Third, Reynolds number of the flow, which determines whether the boundary layer at separation is laminar or turbulent, is an important parameter in describing pressure drag. Finally, due to the complexity of the nonlinear partial differential equations of boundary layer flow, it is necessary to resort to experimental data for a determination of the drag characteristics of a particular body shape.

9.3 TOTAL DRAG OF VARIOUS SHAPED BODIES

In Section 9.2 we discussed pressure drag; in Chapter 8 we studied skin friction drag. The total drag on a body is due to the sum of the two types of drag. In many cases, one or the other of the two is predominant. For example, for uniform flow over a flat plate aligned with the flow direction

288

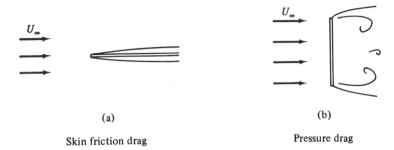

(a)   (b)

Skin friction drag   Pressure drag

Figure 9.12.   (a) Flat plate aligned with flow; (b) flat plate normal to flow.

(Figure 9.12a) with no pressure gradients, the entire drag is due to skin friction. For a bluff body such as a circular cylinder, over 90 percent of the drag is pressure drag, with only a small fraction due to skin friction. For a flat plate normal to the flow direction (Figure 9.12b), the entire drag is pressure drag.

It is convenient to express the drag of a bluff body in terms of a nondimensional parameter, C_D, called *drag coefficient:*

$$C_D = \frac{D}{\frac{1}{2}\rho U_\infty^2 A} \tag{9.1}$$

with A the projected frontal area of the bluff body normal to the flow direction.* For a circular cylinder of length L and diameter D, as shown in Figure 9.13, the projected area A would be DL.

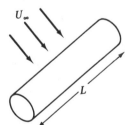

Figure 9.13.   Uniform flow over circular cylinder with axis normal to flow.

The drag coefficient C_D for a smooth circular cylinder in uniform flow with L/D very large is shown in Figure 9.14. It can be seen that, over the Reynolds number (Re $= \rho U_\infty D/\mu$) range 1000 to 3×10^5, the value of C_D is fairly constant and has a value between 1 and 1.2. For this range of Reynolds numbers, laminar boundary layer separation is occurring on the cylinder. At a Reynolds number of about 4×10^5, however, transition to a turbulent boundary layer occurs before separation, with the separation point moved back on the rear portion of the cylinder. A dramatic drop of drag

---

* Remember that C_D for a flat plate aligned with the flow was based on surface area.

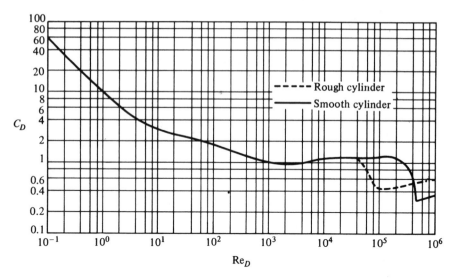

Figure 9.14. Drag coefficient of circular cylinders. (Adapted from Schlichting, *Boundary Layer Theory*, 6th ed., McGraw-Hill, Inc., New York, 1968, and reprinted by permission of the publisher.)

coefficient to $C_D = 0.3$ follows, as was explained in Section 9.2. The effect of roughening the cylinder is shown by the dotted curve of Figure 9.14.

In general, for bluff bodies at sufficiently high Reynolds numbers, the drag coefficient C_D becomes constant, independent of Reynolds number.

A curve of drag coefficient versus Reynolds number based on diameter for a sphere is given in Figure 9.15. Whereas the overall variation on C_D can be seen to be similar, the drag on three-dimensional bodies such as the sphere tends to be less than that of two-dimensional shapes. This is because flow can go around the sides of a finite three-dimensional shape and reduce the extent of the separation region; with a two-dimensional shape, which is essentially infinitely long, flow cannot go around the ends. For example, if a finite cylinder of L/D equal to 5 is placed normal to the flow, the drag coefficient is only 0.8 over the Reynolds number range 10^3 to 10^5.

For Reynolds numbers less than 1000, the viscous forces become comparable to the inertial forces throughout the flow field; the flow can no longer be divided simply into viscous and nonviscous regions. An analytical solution is available for flow about a sphere at Reynolds numbers less than 1 in which the inertial forces can be neglected in comparison to the viscous forces. This analysis, initially derived by Stokes, showed that for Re < 1,

$$C_D = \frac{24}{\text{Re}} \quad \text{with} \quad \text{Re} = \frac{U_\infty D}{\nu} \tag{9.2}$$

Except for this relatively narrow range of Reynolds numbers, however, curves such as Figures 9.14 and 9.15 must be constructed from experimental data.

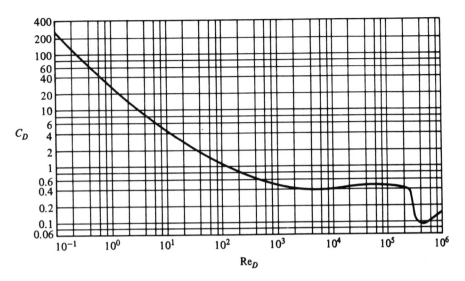

Figure 9.15. Drag coefficient of spheres. (Adapted from Schlichting, *Boundary Layer Theory*, 6th ed., McGraw-Hill, Inc., New York, 1968, and reprinted by permission of the publisher.)

EXAMPLE 9.1

Air blows over a cylindrical smokestack with a velocity of 10 m/s (see Figure 9.16). Determine the aerodynamic force on the stack per meter of height, assuming stack height to be much greater than stack diameter, so that the data of Fig. 9.14 are applicable. Stack diameter is 1.25 m; take ν for air = 1.4×10^{-5} m^2/s, ρ for air = 1.25 kg/m^3.

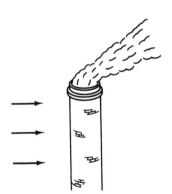

Figure 9.16

Solution   For the conditions given, the Reynolds number is

$$\text{Re} = \frac{U_\infty D}{\nu} = \frac{(10 \text{ m/s})(1.25 \text{ m})}{1.4 \times 10^{-5} \text{ m}^2/\text{s}} = 8.929 \times 10^5$$

From Figure 9.14,

$$C_D = 0.6 \qquad \text{for a rough cylinder}$$

$$
\begin{aligned}
\text{Aerodynamic force} &= \text{Drag force} \\
&= C_D \tfrac{1}{2}\rho U_\infty^2 (D \times 1) \\
&= (0.6)\tfrac{1}{2}(1.25 \text{ kg/m}^3)(10^2 \text{ m}^2/\text{s}^2)(1.25 \text{ m}) \\
&= \underline{46.88 \text{ N/m}}
\end{aligned}
$$

At the other extreme from a bluff body is a highly streamlined body shape. For a thin airfoil aligned with the flow direction, the pressure drag can be of the same order of magnitude or even less than the skin friction drag. In this case, it sometimes becomes desirable to maintain a laminar flow over as much of the foil surface as possible, so as to reduce skin friction drag. The skin friction drag of an airfoil can be approximated by the flat plate formulas of Chapter 8. It can be seen that in order to reduce pressure drag, we would prefer a turbulent boundary layer with a delay in separation; to reduce skin friction drag, we prefer a laminar boundary layer.

Smoke tunnel photographs of flow over an airfoil shape aligned with the flow and of flow over a body with smoothly curved front, but with the rear half removed, are shown in Figure 9.17a and b. For the airfoil shape, there is only a very small wake region; most of the drag will be skin friction drag. For the shape without a streamlined trailing half, there is a large wake region clearly visible; most of the drag for this shape is pressure drag.

Typical drag coefficients for several bodies are given in Table 9.1.

The following examples illustrate some of the principles discussed in this section.

EXAMPLE 9.2

A parachutist weighs 175 lb and has a projected frontal area of 2 ft^2 in free fall. His drag coefficient based on frontal area is found to be 1.0. If the air temperature is 70°F, determine his terminal velocity.

Solution At terminal velocity, neglecting the buoyant force due to the displaced air, the gravity force due to the parachutist's mass is balanced by his drag:

$$Mg = C_D \tfrac{1}{2}\rho V_{\text{term}}^2 A$$

The density of air at normal atmospheric pressure and 70°F is 0.0749 lbm/ft^3. Therefore,

$$175 \text{ lbf} = (1.0)\tfrac{1}{2}\frac{(0.0749 \text{ lbm/ft}^3)}{(32.2 \text{ lbm/slug})}V_{\text{term}}^2 (2.0 \text{ ft}^2)$$

$$V_{\text{term}} = \underline{274.3 \text{ ft/s}}$$

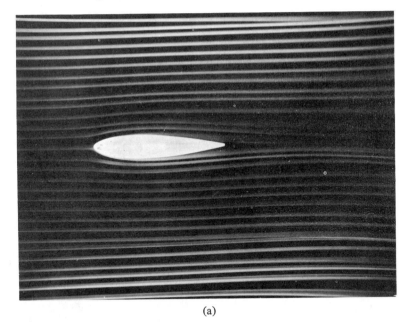

(a)

(b)

Figure 9.17. (a) Flow over streamlined shape; (b) flow over same shape with rear half removed. (*Courtesy* Benedum Instrument Co., Rockville, Maryland.)

TABLE 9.1　Drag Coefficients for Several Shapes

2-Dimensional Shapes*		$C_D^\dagger$
Semitubular	$\longrightarrow$ (	1.2
Semitubular	$\longrightarrow$)	2.3
Square Cylinder	$\longrightarrow$ □	2.0
Flat Plate	$\longrightarrow$ ▯	2.0

3-Dimensional Shapes**		$C_D^\dagger$
Hollow Hemisphere	$\longrightarrow$ (	0.35
Hollow Hemisphere	$\longrightarrow$)	1.35
Circular Disk (diameter D)	$\longrightarrow$ |↕D	1.11
Rectangular Plate (b × h)	$\longrightarrow$ |↕h	
$\dfrac{b}{h} = 1$		1.16
$\dfrac{b}{h} = 4$		1.17
$\dfrac{b}{h} = 8$		1.23
$\dfrac{b}{h} = 25$		1.57

† Based on dimension normal to flow. $Re > 10^4$ with Re also based on dimension normal to flow.

* Source: Lindsey, W. F., *Drag of Cylinders of Simple Shapes,* NACA TR 619, 1938

** Source: Baumeister, T., Avallone, E. A., Baumeister, T., *Marks' Standard Handbook for Mechanical Engineers,* Eighth Edition, McGraw-Hill Book Company, New York, 1978, p. 11–68.

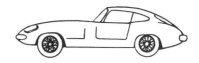

1920 car   Streamlined car

Figure 9.18

EXAMPLE 9.3

A large percentage of the aerodynamic drag on an automobile is due to pressure drag. Whereas a convertible open car, vintage 1920, with boxlike structure may have $C_D = 0.9$, a modern streamlined car with rounded front and tapered rear will have $C_D = 0.40$ (Figure 9.18). For a car traveling at 90 km/h, calculate the power required to overcome aerodynamic drag for the two cases. Assume a frontal area of 2.0 m² for both cars, with air temperature 10°C.

Solution From Equation (9.1), Drag $= C_D \frac{1}{2} U_\infty^2 A$; Power required = Drag × Velocity. For air at atmospheric pressure and 10°C, we find $\rho = 1.247$ kg/m³. Therefore, for $C_D = 0.9$,

$$\text{Power} = (0.9)\left(\frac{1}{2}\right)(1.247 \text{ kg/m}^3)\left(\frac{90 \times 1000}{3600}\right)^3 (\text{m}^3/\text{s}^3)(2.0 \text{ m}^2)$$

$$= \underline{17.54 \text{ kW}}$$

For the streamlined car,

$$\text{Power} = \underline{7.80 \text{ kW}} \qquad \blacksquare$$

The advantage of streamlining at high velocities can be appreciated from the above example. A smoke photograph of flow over a late-model automobile is shown in Figure 9.19. Notice the irregular flow wake region at the rear end of the auto.

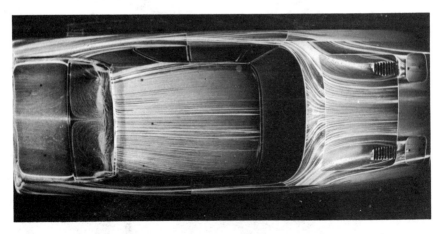

Figure 9.19. Flow over an automobile. (*Courtesy* Ford Motor Co.)

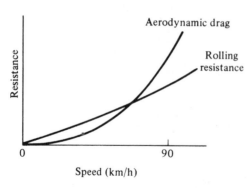

Figure 9.20

The power required to maintain a car at constant velocity on a level road surface is the sum of the power required to overcome aerodynamic drag and the power required to overcome rolling resistance between the tires and the road. At low speed, rolling resistance predominates; at very high speeds, aerodynamic drag exceeds rolling resistance (Figure 9.20).

9.4 AIRFOILS AND LIFT

A typical pressure distribution for a symmetric airfoil was given in Figure 9.9. As the angle of attack of the foil is increased from zero, the fluid moving over the top surface must accelerate more rapidly, yielding a more negative pressure coefficient (see Figure 9.21). The fluid traveling over the lower surface undergoes a much more gradual acceleration. The resultant difference in pressure between upper and lower surfaces yields a positive lift force on the foil, generally expressed in terms of a lift coefficient C_L with

$$C_L = \frac{L}{\frac{1}{2}\rho V^2 A}$$

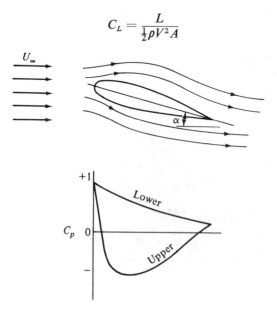

Figure 9.21

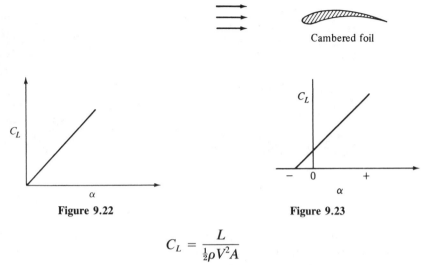

Cambered foil

Figure 9.22   **Figure 9.23**

$$C_L = \frac{L}{\frac{1}{2}\rho V^2 A}$$

For small angles of attack, the lift coefficient varies linearly with angle of attack, with $C_L = 0$ for $\alpha = 0$ for a symmetric airfoil (Figure 9.22). If the foil is not symmetric but is provided with curvature, or camber, as shown in Figure 9.23, lift can be produced at zero angle of attack.

A similar curve of C_L versus α was presented on the basis of potential flow theory in Section 7.8.

As the angle of attack of the foil is increased, however, the flow accelerates more and more over the upper surface, with the point of minimum pressure on the upper surface moving toward the leading edge of the foil. This process increases the region of adverse pressure gradient on the upper surface and moves the point of separation forward toward the leading edge (see Figure 9.24). Eventually the point of separation moves up almost to the leading

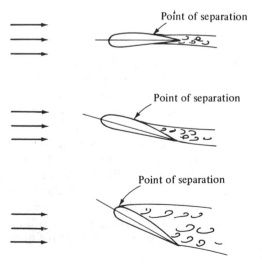

Point of separation

Point of separation

Point of separation

Figure 9.24

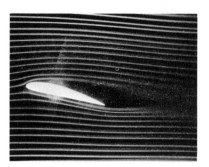

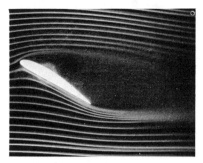

Figure 9.25. Flow over an airfoil at low angle of attack. (*Courtesy* Benedum Instrument Co., Rockville, Maryland.)

Figure 9.26. Flow over an airfoil at high angle of attack. (*Courtesy* Benedum Instrument Co., Rockville, Maryland.)

edge, leaving the entire upper surface in a separation region. Note also the smoke tunnel photographs, of Figures 9.17(a), 9.25, and 9.26. Under these conditions, the pressure in the separation region on the upper surface of the foil is approximately the same as that in the undisturbed flow ahead of the foil, p_∞. With no underpressure on the top surface, there is a drop in lift coefficient, as shown in Figure 9.27, obviously detrimental to the operation of an airfoil. This condition is called *stall*.

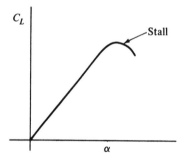

Figure 9.27

The design and operation of an airfoil or any lifting device must involve not only a consideration of lift but also the resultant drag on the foil. We have discussed pressure drag and skin friction drag. From our previous discussion, it can be seen that as the angle of attack of the foil increases and the separation point moves forward to the leading edge, the pressure drag must also increase. For a finite, three-dimensional airfoil, there is a third type of drag, called *induced drag*, which is incurred by the circulatory flow set up about the airfoil in the generation of lift (see Section 7.8). The induced drag increases with circulation and lift. With induced drag and pressure drag increasing with angle of attack, there is a resultant increase of C_D with α (C_D is based on wetted area A, expressed as span times chord of the foil). Curves of C_D and C_L versus α for the NACA 23012 airfoil, used

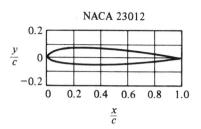

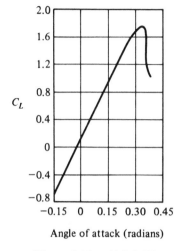

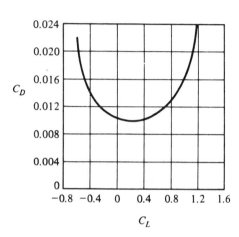

Figure 9.28. Airfoil lift and drag characteristics of NACA 23012. (Data from NACA Report 524 at Reynolds number of 6×10^6.)

on the Douglas DC-6 airplane, are shown in Figure 9.28. From such curves, the angle of attack for maximum L/D can be found, a prime operating point for an airfoil.

EXAMPLE 9.4

The cruising speed of a jet aircraft is 350 mph at an altitude of 30,000 ft ($\rho = 0.000706$ slugs/ft^3). Using the data of Figure 9.28, determine the wing lift and drag under cruise conditions at an angle of attack of 8°. Assume the wing has an aspect ratio (ratio of span to average chord) of 10, with a total span b of 115 ft. Approximate the wing as a rectangle of area bc, with C_L and C_D invariant across the entire span.

Solution From Figure 9.28, $C_L = 1.0$ and $C_D = 0.017$. The wing area $A = b^2/\mathcal{R}$, with $\mathcal{R}$ the aspect ratio; $A = 115^2/10 = 1323$ ft^2.

$$\text{Lift} = C_L \tfrac{1}{2} \rho V^2 A$$

$$= (1.0)\tfrac{1}{2}(0.000706 \text{ slugs/ft}^3)[(350 \text{ mi/h})(5280 \text{ ft/mi})(1 \text{ h}/3600 \text{ s})]^2(1323 \text{ ft}^2)$$

$$= \underline{12.3 \times 10^4 \text{ lbf}}$$

$$\text{Drag} = C_D \tfrac{1}{2} \rho V^2 A$$

$$= \underline{2090 \text{ lbf}} \qquad \blacksquare$$

PROBLEMS

9.1. Determine the velocity at which a small metal sphere, 5 mm in diameter, will settle in SAE 10 oil at 20°C. Repeat for a sphere of diameter 5 cm. Metal density = 8700 kg/m³.

9.2. SAE 10 oil at 40°F flows at 1.2 ft/s over a circular tube of 1.0 in diameter (see Figure P9.2). Determine the force of the oil on the tube in lbf/ft.

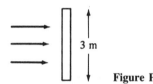

Oil flow

Figure P9.2

9.3. Calculate the wind force on a rectangular signboard 3 m by 15 m if the wind speed is 50 km/h. Take air at standard conditions and assume the flow is perpendicular to the board, as shown in Figure P9.3.

3 m

Figure P9.3

9.4. Suppose the wind of Example 8.6, Figure 8.22, were blowing normal to the billboard. Find the force exerted by the wind on the billboard (Figure P9.4). Compare with the result of Example 8.6.

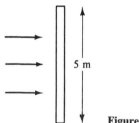

5 m

Figure P9.4

9.5. Air flows over a circular smokestack with a velocity of 50 mph, as shown in Figure 9.16. Determine the aerodynamic force on the stack per foot of height. Stack diameter is 5 feet, air temperature is 20°F.

9.6. Consider a square smokestack having the same cross-sectional area as that of Problem 9.5. Under the same conditions as Problem 9.5, determine the aerodynamic force on the stack. (See Figure P9.6.)

Figure P9.6

9.7. Paint spray droplets leave a high-pressure spray gun with a velocity of 50 m/s. The spray gun is located 1.0 m from the surface to be painted. Determine the velocity of the droplets just prior to impingement on the surface. Assume a mean droplet diameter of 50 microns, with air temperature 30°C. (1 micron = 10^{-6} m, ρ = 1314 kg/m³.)

9.8. Compare the drag and power requirements to pull the following two-dimensional bodies through the air: a circular cylinder, a flat plate normal to the direction of motion, and a square cylinder aligned with the flow. For each body, take the maximum dimension normal to the flow as 2.0 ft, with the velocity of the body 2.5 ft/s, air temperature 0°F.

9.9. Calculate the drag coefficient for a circular cylinder as shown in Figure P9.9 in uniform flow based on the following: Up to point of separation, the potential flow pressure distribution is valid; the pressure in the separation region is uniform and equal to the local pressure on the cylinder surface at the point of separation; separation occurs at θ_s = 90°.

Figure P9.9

9.10. Repeat the previous problem for θ_s = 120° and θ_s = 135°. Compare your results with those of Figure 9.14.

9.11. A baseball 2.95 inches in diameter with a mass of 0.34 lbm leaves the pitcher's hand with a velocity of 90 mph. Determine the velocity of the ball when it crosses home plate, 60 ft 6 in away. Air temperature is 90°F.

9.12. An airplane with total mass of 2500 kg is in level flight at a velocity of 60 m/s. If the wing area is 25m², with air density 1.2 kg/m³, determine the wing angle of attack, assuming the airfoil of Figure 9.28.

9.13. Determine the wing drag of Problem 9.12.

9.14. An automobile radio aerial is 3 ft long and consists of three sections, each 1 ft long and having diameters $\frac{1}{4}$ in, $\frac{3}{8}$ in, and $\frac{1}{2}$ in, respectively. Calculate the movement about the base of the aerial for wind speeds of 30 mph and 55 mph.

9.15. For the NACA 23012 airfoil of Figure 9.28, find the angle of attack for maximum C_L/C_D.

9.16. A submarine has the shape of an 8:1 elliptic cylinder. Calculate the power required to maintain an underwater velocity of 20 ft/s. Assume a water temperature of 40°F, with a frontal area of 50 ft², C_D = 0.15.

9.17. Determine the terminal velocity of a person falling with parachute, assuming a parachute diameter of 4 m. Use the properties of air at 5000 m; mass of person and chute is 100 kg (see Figure P9.17).

Figure P9.17

9.18. A helium-filled balloon is to be used to measure wind velocity, the angle of the string indicative of wind velocity (see Figure P9.18). Plot V as a function of θ over the range $V = 0$ to $V = 60$ mph. The balloon has a diameter of 3 ft; the material of which the balloon is made has a mass of 0.3 lbm.

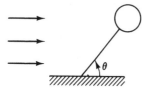

Figure P9.18

9.19. Determine the terminal ascent velocity of the balloon of Problem 9.18; assume an air temperature of 50°F.

9.20. A tractor-trailer truck is moving down a highway with a velocity of 23 m/s directly into a 15-m/s wind. The drag coefficient of the track is 0.80, based on an area equal to the product of the distance from the roof of the trailer to the ground (4 m) and the trailer width (2.2 m). If the local temperature is −10°C, determine the aerodynamic drag force on the truck. How much power is required to overcome aerodynamic drag and maintain a constant truck velocity of 23 m/s into the wind?

9.21. A $\frac{3}{8}$-in-diameter cable is strung between two poles 100 ft apart. An 80-mph wind is blowing normal to the cable. Determine the air force on the entire length of cable. (Air temperature is 20°F.)

9.22. A sphere of diameter 2.0 cm and specific gravity 2.5 is dropped into a tank of water at 20°C. Determine the terminal velocity of the sphere and estimate the distance from the tank surface at which terminal velocity will be reached.

9.23. It is estimated that, at 55 mph, 70% of the total power for an automobile is required to overcome aerodynamic drag, 30% to overcome rolling resistance. For a streamline car, with $C_D = 0.40$ and frontal area of 18.0 ft², determine the engine horsepower required for a steady velocity of 55 mph.

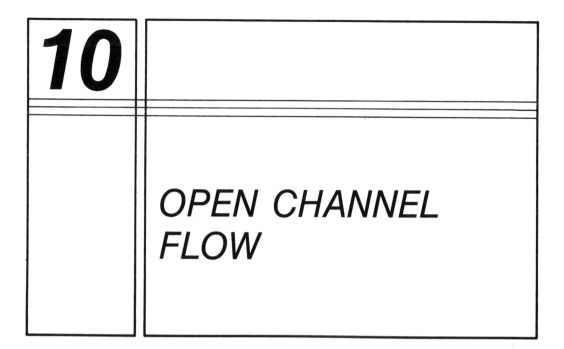

10

OPEN CHANNEL FLOW

10.1 INTRODUCTION

Open channel flow is the term applied to fluid motion in which a liquid in a conduit has part of its boundary exposed to atmospheric pressure. The conduit may be completely open to the atmosphere, as in natural river beds or artificial canals and channels, or it may be closed, as in drainage and sewerage pipes that are not running full. Various examples of open channels are shown in Figure 10.1.

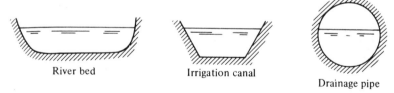

River bed   Irrigation canal

Drainage pipe

Figure 10.1. Examples of open channel cross sections.

Open channel flow is distinguished, on one hand, from pipe flow (discussed in Chapter 6), in which the fluid is completely surrounded by a solid boundary; and, on the other hand, from free jet flow (discussed in Chapter 4), in which the fluid is completely surrounded by the atmosphere. The presence of the solid boundary in pipe flow can sustain a pressure variation along the surface of the pipe, while the presence of a free surface in open channel flow ensures constancy of pressure along the boundary exposed to the atmosphere. In

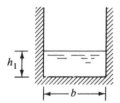
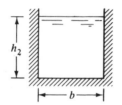

Figure 10.2. Flow through rectangular cross section.

the case of flow with friction through a constant-area pipe, a pressure difference has to exist to sustain the flow. In the case of flow with friction in an open channel of constant cross section, the effect of gravity—that is, changes in elevation in the channel—sustains the flow through the channel. As contrasted with pipe flow, the cross-sectional area of the flow can change with different flow condition, subject to the confines of the channel. For example, for flow through a rectangular cross section, a change in the depth of the liquid results in a different rectangular cross section (Figure 10.2).

Just as in pipe flow, the flow may be laminar or turbulent in character, with turbulent flow constituting the most frequent type when the liquid is water. The analysis of open channel flows is important in situations relating to the determination and control of river flow and in establishing design requirements for irrigation canals.

In this chapter we shall treat the various aspects of steady open channel flow, using the techniques and fundamental equations of fluid flow obtained in preceding chapters. In particular, we shall discuss flows with varying cross-sectional areas (due to changes in water depths or width of channel), using the one-dimensional frictionless approximation found useful in analyzing flow through nozzles as presented in Chapter 4. The flow of a thin film down an inclined plane will be presented as an example of laminar open channel flow; and, finally, several examples of one-dimensional turbulent flow (uniform and nonuniform) will be given.

Before proceeding, a few definitions associated with open channel flow will be given. When the cross section of the flow does not vary along the direction of the flow, the flow is called *uniform*. Otherwise it is called *nonuniform* flow. Thus in uniform open channel flow, the liquid depth is constant and the surface of the liquid is parallel to the channel bottom (Figure 10.3). When the depth of the liquid varies, we have nonuniform open channel flow (also called *varied* flow). This type of flow occurs when the shape of the channel cross section changes or when the shape of the channel bottom changes. If the depth increases in the downstream direction, the velocity of the flow slows down and we have retarded flow (Figure 10.4a); while for depth decreasing in the downstream direction, we obtain velocity increases and have accelerated flow (Figure 10.4b).

Uniform flow is achieved in any channel that is sufficiently long and that has a constant channel slope and cross section. An example of how uniform flow is established will now be given. An irrigation canal has a

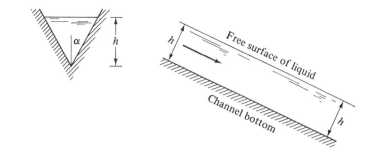

Figure 10.3. Uniform flow in a triangular channel.

(a)   (b)

Figure 10.4. Nonuniform flow in an open channel: (a) retarded flow; (b) accelerated flow.

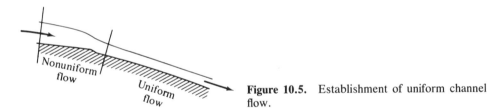

Figure 10.5. Establishment of uniform channel flow.

change in slope, as shown in Figure 10.5. Due to the change in the slope of the channel flow, the flow experiences an acceleration because the component of the gravity force along the flow is greater than the shear force along the channel wall retarding the flow. As the flow accelerates, the shear force will increase as a result of the increased velocity until a value of the velocity is reached when the shear force equals the gravity component. At this condition of equilibrium, the velocity and liquid depth remain constant and the flow becomes uniform open channel flow.

Nonuniform flow caused by a change in the slope of the channel floor is illustrated in Figure 10.5. When a barrier, such as a spillway, is placed in the path of uniform flow along a sloping channel, we obtain another example of gradually varied flow. The resulting profile of the water surface behind the barrier is called the *backwater curve* and is shown in Figure 10.6.

An example of rapidly varied flow occurs when water flows down a spillway from a reservoir, as shown in Figure 10.7.

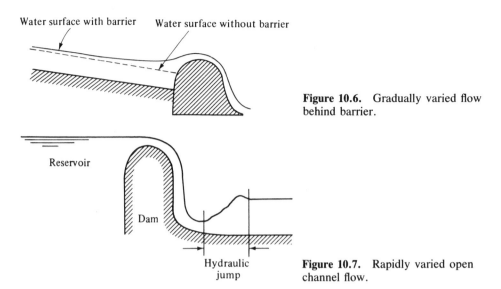

Water surface with barrier   Water surface without barrier

Figure 10.6. Gradually varied flow behind barrier.

Reservoir

Dam

Hydraulic jump

Figure 10.7. Rapidly varied open channel flow.

The establishment of uniform flow in the open channel (Figure 10.5) corresponds to the establishment of fully developed flow inside a pipe as described in Section 6.1. The occurrence of a hydraulic jump (Figure 10.7) has its equivalence in pipe flow at a sudden enlargement in pipe diameter as described in Section 6.4.

10.2 FRICTIONLESS OPEN CHANNEL FLOW

When changes in cross-sectional area take place in open channel flow, the influence of the friction forces can be neglected if the effects of the cross-sectional changes predominate. An example of this situation occurs in the flow through short flumes, which are channels for conveying water for irrigation or power. In this section we will utilize one-dimensional frictionless flow to treat several such examples. As pointed out in Section 4.1, the one-dimensional approximation becomes more exact the more gradual the area changes. Since the flow is assumed to be frictionless, the Bernoulli equation is applicable here and will be used in conjunction with the continuity equation to treat several examples of open channel flow with variable cross-sectional area. The changes in cross section may be due to changes in liquid depth, channel width, or cross-sectional shape.

10.2(a) Flow in a Venturi Flume

First consider the flow in a flume in which the width of the channel is deliberately changed for the purpose of measuring the flow rate through the channel. A typical flume is shown in Figure 10.8. The flume has vertical walls, thereby making the cross section rectangular. The channel floor is taken to be horizontal. In an actual flow case with friction the channel floor would have to possess a small slope to overcome the effects of friction along the channel walls. Bernoulli's equation can be taken along any streamline of the flow, for example, the bottom of the channel; however, it is simplest to take it along the streamline constituting the free surface. Bernoulli's equation (4.21) along the free surface where p is constant becomes

$$\frac{V_1^2}{2} + gh_1 = \frac{V_2^2}{2} + gh_2$$

From the continuity equation (4.6), we obtain the volume flow rate Q:

$$Q = V_1 b_1 h_1 = V_2 b_2 h_2$$

Hence V_1 can be expressed in terms of V_2:

$$V_1 = V_2 \frac{b_2 h_2}{b_1 h_1}$$

and can be substituted into the Bernoulli equation to give

$$\frac{V_2^2}{2} \frac{b_2^2 h_2^2}{b_1^2 h_1^2} + gh_1 = \frac{V_2^2}{2} + gh_2$$

Thus

$$h_2 - h_1 = \frac{V_2^2}{2g}\left(\frac{b_2^2 h_2^2}{b_1^2 h_1^2} - 1\right)$$

$$= \frac{Q^2}{2g}\left(\frac{1}{b_1^2 h_1^2} - \frac{1}{b_2^2 h_2^2}\right)$$

Finally, we obtain the expression for volume flow rate:

$$\frac{Q^2}{2g} = \frac{h_1 - h_2}{1/b_2^2 h_2^2 - 1/b_1^2 h_1^2} \tag{10.1}$$

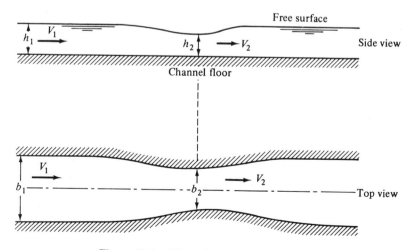

Figure 10.8.   Flow through Venturi flume.

Thus the volume flow rate in an open channel can be obtained by inserting a restriction into the flow and by measuring the liquid depth at the approach and at the minimum width section.

10.2(b) Flow over a Channel Rise

As the next example, consider the case of frictionless flow along a horizontal rectangular channel floor with a small rise in the floor, as indicated in Figure 10.9. Channel width is constant.

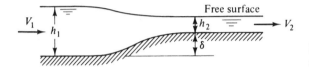

Figure 10.9. Open channel flow past rise.

Again taking Bernoulli's equation along the free surface, we obtain

$$\frac{V_1^2}{2} + gh_1 = \frac{V_2^2}{2} + g(h_2 + \delta)$$

where δ is the elevation of the rise above the channel floor. The velocities can be eliminated from the preceding equation by using the continuity equation

$$Q = bhV$$

where b is the width of the rectangular channel. Thus we obtain

$$\frac{Q^2}{2b^2h_1^2} + gh_1 = \frac{Q^2}{2b^2h_2^2} + g(h_2 + \delta)$$

or

$$\frac{Q^2}{2gb^2h_1^2} + h_1 = \frac{Q^2}{2gb^2h_2^2} + (h_2 + \delta) \tag{10.2}$$

The sum of the two terms

$$h + \frac{Q^2}{2gb^2h^2} = H \tag{10.3}$$

is called *specific head* and assigned the symbol H. Plots of the depth of liquid (h) against the specific head (H) for given flow rates are shown in Figure 10.10. It can be seen that, for a given specific head and flow rate, two different depths h are possible (e.g., points x and y of flow Q_r). Also note there there is a minimum value of specific head at a given flow. The critical depth h_c for minimum specific head can be found by differentiating Equation (10.3) with respect to h and setting the result equal to zero:

$$1 - \frac{2Q^2}{2gb^2h_c^3} = 0$$

Thus the critical depth is

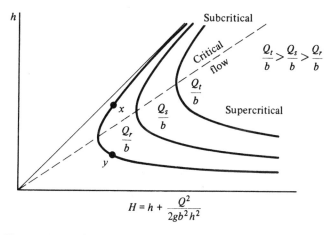

$$H = h + \frac{Q^2}{2gb^2h^2}$$

Figure 10.10. Specific head diagram for flow over channel rise.

$$h_c = \sqrt[3]{\frac{Q^2}{gb^2}}$$

The corresponding value for H is

$$H_c = h_c + \frac{h_c^3}{2h_c^2} = \frac{3}{2}h_c$$

For depths less than h_c, the given flow cannot occur.

A simple transformation in terms of velocity can be effected using the continuity equation, namely, $V = Q/bh$. Thus

$$h_c^3 = \frac{b^2h_c^2V^2}{gb^2}$$

or

$$\frac{V^2}{gh_c} = 1$$

The quantity V^2/gh_c will be recognized as the Froude number, described in Chapter 5. Thus we see that the critical depth of flow in an open channel occurs when the Froude number (V^2/gh) equals unity. When the Froude number of the flow is less than unity, we have subcritical flow, while for Froude numbers greater than 1, we obtain supercritical flow. This can also be stated in terms of volume flow rate as follows:

$$\frac{Q^2}{b^2g} < h^3 \qquad \text{Subcritical flow}$$

$$\frac{Q^2}{b^2g} = h^3 \qquad \text{Critical flow}$$

$$\frac{Q^2}{b^2g} > h^3 \qquad \text{Supercritical flow}$$

Thus for depths exceeding the critical depth (given by $h_c = \sqrt[3]{Q^2/b^2g}$) at a specified flow rate, the open channel flow is called *subcritical open channel flow.* For depths less than the critical depth, the flow is called *supercritical open channel flow.* In Figure 10.10, point X is in the subcritical flow region, point Y is in the supercritical flow region. Since, for a given flow rate, velocities are higher in the supercritical region, this is called *rapid flow*; flow in the subcritical region is called *tranquil flow.*

To present the variation of specific head H in compact, nondimensional form, divide Equation (10.3) by the critical depth h_c to obtain

$$\frac{H}{h_c} = \frac{h}{h_c} + \frac{Q^2}{2gb^2h^2h_c} = \frac{h}{h_c} + \frac{Q^2}{2gb^2(h/h_c)^2h_c^3}$$

With $h_c^3 = \dfrac{Q^2}{gb^2}$, we get

$$\frac{H}{h_c} = \frac{h}{h_c} + \frac{1}{2}\left(\frac{h}{h_c}\right)^{-2} \tag{10.3a}$$

Values of nondimensionalized specific head as obtained from Equation (10.3a) have been tabulated as a function of nondimensionalized channel depth together with the Froude number of the flow in Table 1 of Appendix D. In the subcritical region ($h/h_c > 1$), the value of specific head approaches the line $h/h_c = H/h_c$ asymptotically at large values of h/h_c. At small values of h/h_c in the supercritical region ($h/h_c < 1$), the relation between h/h_c and H/h_c reduces to

$$\frac{H}{h_c} = \frac{1}{2(h/h_c)^2}$$

The expression for the Froude number reduces to

$$\text{Fr} = \frac{V^2}{gh} = \frac{Q^2}{gb^2h^3} = \left(\frac{h_c}{h}\right)^3 = \left(\frac{h}{h_c}\right)^{-3}$$

Let us now return to flow past the rise in the channel bottom of Figure 10.9. If the approach flow is rapid, with $h_1 < h_c$, then as the channel bottom rises, decreasing the specific head (see Equation 10.2), the depth of flow will increase, as shown in Figure 10.11. However, if the approach flow is tranquil,

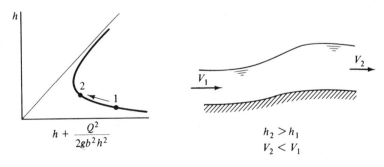

Figure 10.11. Supercritical flow over channel rise.

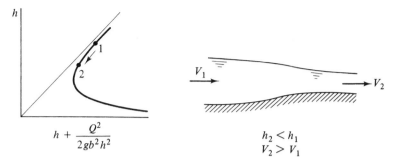

Figure 10.12.   Subcritical flow over channel rise.

with $h_1 > h_c$, then as the channel bottom rises, h will decrease, as shown in Figure 10.12.

EXAMPLE 10.1

Water flows through a rectangular open channel, 1 meter wide, at a rate of 5 m³/s. A submerged obstruction is placed across the channel floor. Determine the increase or decrease in water level due to the obstruction for two cases:

$$(1)\ h_1 = 2\ \text{m}, \qquad \delta = 0.2\ \text{m}$$

$$(2)\ h_1 = 1\ \text{m}, \qquad \delta = 0.2\ \text{m}$$

Assume frictionless flow.

Solution   First plot the specific head diagram. For $Q = 5.0$ m³/s, $b = 1$ m, we obtain

$$H = \frac{Q^2}{2gb^2h^2} + h = \frac{1.2742}{h^2} + h$$

This is shown in Figure 10.13. Note that

$$h_c = \sqrt[3]{\frac{Q^2}{gb^2}} \qquad \text{or} \qquad h_c = \sqrt[3]{\frac{(5\ \text{m}^3/\text{s})^2}{(9.81\ \text{m/s}^2)(1\ \text{m}^2)}} = 1.366\ \text{m}$$

It follows that, for Case (1), we have subcritical or tranquil flow at 1; for Case (2), we have supercritical or rapid flow.

Case (1): For $h_1 = 2$ m, $H_1 = 2.3186$ m. From Equation (10.2), with $\delta = 0.2$ m, $H_2 = 2.1186$ m. The value of h_2 can be found from Figure 10.13 or by solving $2.1186 = 1.2742/h_2^2 + h_2$. It follows that $h_2 = 1.651$ m.

Case (2): For $h_1 = 1$ m, $H_1 = 2.2742$ m. With $\delta = 0.2$ m, $H_2 = 2.0742$ m. From Figure 10.13 or solving $2.0742 = 1.2742/h_2^2 + h_2$, we find $h_2 = 1.225$ m. (See Figure 10.14.)

This example can also be solved by using the Specific Head Tables of Appendix D. For Case (1), we have $H_2/h_c = 2.1186/1.366 = 1.5510$. Interpolating between H/h_c values of 1.5472 and 1.5700 gives $h_2/h_c = 1.2083$

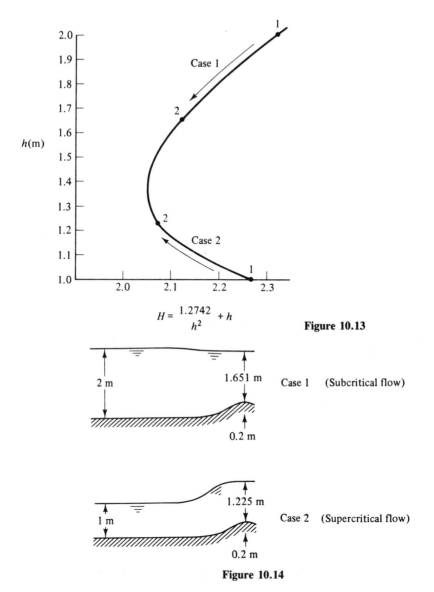

$$H = \frac{1.2742}{h^2} + h$$

Figure 10.13

Case 1   (Subcritical flow)

Case 2   (Supercritical flow)

Figure 10.14

and $h_2 = 1.650$ m. For Case (2) we have $H_2/h_c = 2.0742/1.366 = 1.5184$. Interpolating we get $h_2/h_c = 0.8978$ and $h_2 = 1.226$ m. ∎

10.2(c) Specific Head and Critical Depth

In Section 10.2(b) we discussed specific head and critical depth for a rectangular channel cross section. In general, the expression for the specific head is given by

$$H = h + \frac{Q^2}{2gA^2} \tag{10.4}$$

For a rectangular channel, the cross-sectional area is $A = bh$; hence the second term becomes $Q/2gb^2h^2$ [see Equation (10.3)]. For more general cross sections, for example, a circular cross section, we can determine the critical depth by setting the derivative of Equation (10.4) with respect to the depth h equal to zero:

$$\frac{dH}{dh} = 1 - \frac{Q^2}{gA^3}\frac{dA}{dh} = 0$$

The derivative dA/dh (i.e., the rate of change of cross-sectional area with respect to depth h) can be obtained by actually differentiating the analytical expression for the cross section or, in a simpler fashion, by considering the infinitesimal strip of width dh as shown in Figure 10.15, where it is seen that $dA = c\,dh$, where c is the width of the cross section at the water surface. Thus the critical depth h_c and the minimum specific head H_c occur when the relation

$$\frac{Q^2 c_c}{gA_c^3} = 1$$

is satisfied. For a given cross section, both c and A will be a function of the water depth h:

$$H_c = h_c + \frac{Q^2}{2gA_c^2} = h_c + \frac{1}{2}\frac{A_c}{c_c}$$

or

$$\frac{H_c}{h_c} = 1 + \frac{1}{2}\frac{A_c}{c_c h_c}$$

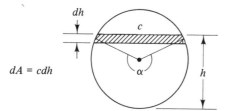

$dA = c\,dh$

Figure 10.15

For a rectangular cross section, we have shown in Section 10.2b that $A_c = bh_c$, $c_c = b$, and $h_c^3 = Q^2/gb^2$. Hence, $H_c/h_c = \frac{3}{2}$, as was shown in Section 10.2b.

Geometric relations for circular segments can be found in any engineering or physics handbook. For the cross-sectional area of a circular segment, we have

$$A = R^2\left[-\sqrt{\frac{h}{R}\left(2 - \frac{h}{R}\right)}\left(1 - \frac{h}{R}\right) + \cos^{-1}\left(1 - \frac{h}{R}\right)\right]$$

For the width of cross section at the water surface, we have

$$c = 2R \sqrt{\frac{h}{R}\left(2 - \frac{h}{R}\right)}$$

For critical flow in a circular cross section, we have

$$\frac{Q^2 c_c}{g A_c^3} = 1$$

or

$$\frac{A^3}{R^6}\left(\frac{R}{c}\right) = \frac{Q}{g R^5}$$

Next, by plotting $(A^3/R^6)(R/c)$ versus h/R on log-log paper, we can show that the relationship between the two quantities is

$$\frac{h_c}{R} = 0.842 \left(\frac{A^3/R^6}{c/R}\right)^{0.26} = 0.842 \left(\frac{Q}{g R^5}\right)^{0.26}$$

Hence, the critical depth h_c for a circular cross section is given by

$$h_c = 0.842 R \left(\frac{Q^2}{g R^5}\right)^{0.26} \tag{10.5}$$

EXAMPLE 10.2

Determine the critical depth in a 10-ft-diameter circular conduit for a volume flow rate of 200 cfs. Compare the value of critical depth with that in a rectangular channel 10 ft wide and with that in a triangular channel with included angle (2α) equal to $90°$.

Solution The critical depth of a circular conduit is

$$h_c = 0.842 R \left(\frac{Q^2}{g R^5}\right)^{0.26}$$

$$= 0.842(5 \text{ ft}) \left[\frac{200^2 \text{ ft}^6/\text{s}^2}{(32.17 \text{ ft/s}^2)(5^5 \text{ ft}^5)}\right]^{0.26}$$

$$= \underline{3.31 \text{ ft}}$$

For a 10-ft rectangular channel, the critical depth is

$$h_c = \sqrt[3]{\frac{Q^2}{g b^2}}$$

$$= \sqrt[3]{\frac{200^2 \text{ ft}^6/\text{s}^2}{(32.17 \text{ ft/s}^2)(10^2 \text{ ft}^2)}}$$

$$= \underline{2.32 \text{ ft}}$$

For a triangular channel (see Figure 10.3), the cross-sectional area A equals

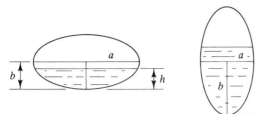

Figure 10.16. Elliptic cross sections.

$h^2 \tan \alpha$. In this case, we can obtain the critical depth easily by differentiating A with respect to h:

$$\frac{dA}{dh} = 2h \tan \alpha$$

Hence

$$\frac{Q^2}{gA^3} \frac{dA}{dh} = \frac{Q^2 2h_c \tan \alpha}{gh_c^6 \tan^3 \alpha} = 1$$

or

$$h_c^5 = \frac{2Q^2}{g \tan^2 \alpha} = \frac{2(200^2 \text{ ft}^6/\text{s}^2)}{(32.17 \text{ ft/s}^2)(1)^2}$$

$$h_c = \underline{4.78 \text{ ft}}$$

Alternatively, using the relation $Q^2 c_c / gA_c^3 = 1$ and the expression for c ($c = 2h \tan \alpha$) we get

$$\frac{Q^2 2h_c \tan \alpha}{gh_c^6 \tan^3 \alpha} = 1$$

as before.   ∎

For elliptic cross sections (Figure 10.16), the corresponding geometric relations can be shown to be

$$A = ab \left[-\sqrt{\frac{h}{b} \left(2 - \frac{h}{b} \right)} + \cos^{-1} \left(1 - \frac{h}{b} \right) \right]$$

and

$$c = 2a \sqrt{\frac{h}{b} \left(2 - \frac{h}{b} \right)}$$

Thus, for critical flow in an elliptic cross section, we obtain

$$\frac{A^3}{a^3 b^3} \left(\frac{a}{c} \right) = \frac{Q^2}{ga^2 b^3} = \frac{Q^2}{gb^5} \left(\frac{b}{a} \right)^2$$

and hence

$$h_c = 0.842b \left(\frac{b}{a} \right)^{0.52} \left(\frac{Q^2}{gb^5} \right)^{0.26} \tag{10.6}$$

In Table 3 of Appendix D, the geometric parameters for circular and elliptic cross sections are presented. Note that for circular sections, $a = b = R$. Elliptic integrals of the second kind were used to calculate the wetted perimeter P (i.e., the perimeter in contact with the solid boundary of the channel) of the elliptic sections.

10.2(d) Flow Through a Sluice Gate

As another example, consider frictionless flow generated when a gate that retains water in a pond is partially raised (Figure 10.17). In this case, Bernoulli's equation, after combining with the continuity equation, gives

$$h_0 = \frac{Q^2}{2b^2gh^2} + h \tag{10.7}$$

since the conditions designated with subscript 0 are taken at a large distance from the sluice gate, where the velocity of flow can be assumed to be negligible. Rewriting the preceding equation, we obtain

$$2gh^2h_0 - 2gh^3 = \left(\frac{Q}{b}\right)^2$$

or

$$2gh^2(h_0 - h) = \left(\frac{Q}{b}\right)^2$$

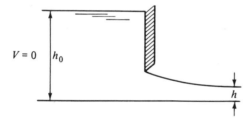

$V = 0$ h_0

h

Figure 10.17.   Flow through sluice gate.

Since we are interested in the variation of flow rate with gate position, let us plot downstream water depth as a function of volume flow rate. First, nondimensionalizing with respect to the reservoir depth h_0, we obtain

$$2g\left(\frac{h}{h_0}\right)^2\left(1 - \frac{h}{h_0}\right) = \left(\frac{Q}{b}\right)^2\frac{1}{h_0^3}$$

This equation is plotted in Figure 10.18. The maximum discharge through the gate will occur when

$$\frac{d(Q^2/b^2h_0^3)}{d(h/h_0)} = 0$$

Thus

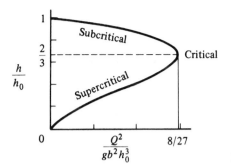

Figure 10.18. Flow rate diagram for sluice gate.

$$2g\left[2\frac{h}{h_0}\left(1 - \frac{h}{h_0}\right) - \left(\frac{h}{h_0}\right)^2\right] = 0$$

whence

$$2\frac{h}{h_0} - 3\left(\frac{h}{h_0}\right)^2 = 0$$

and

$$\frac{h}{h_0} = \frac{2}{3}$$

The nondimensional discharge is therefore given by:

$$\left(\frac{Q}{b}\right)^2_{max}\frac{1}{gh_0^3} = 2\left(\frac{2}{3}\right)^2\left(1 - \frac{2}{3}\right) = \frac{8}{27}$$

or

$$\left(\frac{Q}{b}\right)^2_{max} = \frac{8}{27}gh_0^3 \tag{10.8}$$

Next, let us evaluate the Froude number for the condition of maximum flow rate:

$$Fr = \frac{V^2}{gh} = \frac{Q^2}{gb^2h^3} = \frac{8}{27}h_0^3\left(\frac{3}{2}\right)^3\frac{1}{h_0^3} = 1$$

This indicates that the Froude number of the downstream flow is unity when maximum discharge through the sluice gate takes place. As indicated in Figure 10.18, subcritical and supercritical flows occur at discharge less than critical.

    Summarizing, it has been shown that by raising the gate from the closed position, the flow discharge is gradually increased until a maximum discharge is obtained when the depth downstream is two-thirds of the reservoir depth. Raising the sluice gate beyond the position for maximum discharge reduces the flow rate.

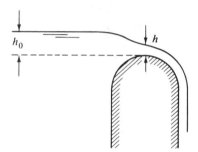

Figure 10.19.   Flow over crest of dam.

The result obtained for the flow through a sluice gate is also applicable to flow over the crest of a dam (Figure 10.19), if we let h_0 in Equation (10.7) be the water level in the reservoir measured from the crest of the dam and h be the depth of water above the crest. Equation (10.8) gives the maximum volume flow rate over the dam.

10.3 LAMINAR OPEN CHANNEL FLOW

In the preceding section the equations of motion of open channel flow were applied to several problems involving frictionless flow. In this section we shall consider open channel flow of a viscous liquid in which the presence of frictional forces is significant and must be taken into account in treating the problem. Such flow occurs in the thin sheet of rain runoff on a sloping roof. Let us therefore examine the case of steady laminar flow down a wide inclined plane. In this case, the flow will be two-dimensional; that is, no vertical sidewalls will be present.

As indicated in Figure 10.20, the x coordinate is taken along the inclined plane with the y coordinate normal to the plane. Let us restrict our consideration to uniform flow; hence there will be no velocity in the y direction and no velocity changes in the direction of flow (x direction); the momentum equation (4.7) yields $\Sigma F_x = 0$. Next, let us evaluate the forces acting on an infinitesimal volume of the flow (Figure 10.21); τ is the shear stress acting on the control surface, p is the pressure, and $g\,dy\,dx$ is the gravity force. The velocity gradient is taken as positive in the positive y direction; the shear stress τ acting on the lower surface of the element will be in the negative x direction, since the velocity of fluid adjacent to it will be less. On the other hand, the shear stress $(\tau + d\tau)$ acting on the upper surface of the element will be in the positive x direction, for the velocity of the fluid above it will be greater. The force balance in the x direction yields

$$\Sigma F_x = -\tau\,dx \times 1 + (\tau + d\tau)dx \times 1 + \rho g \sin \alpha\,dy\,dx \times 1$$

$$+ p\,dy \times 1 - \left(p + \frac{\partial p}{\partial x}dx\right)dy \times 1 = 0$$

and hence

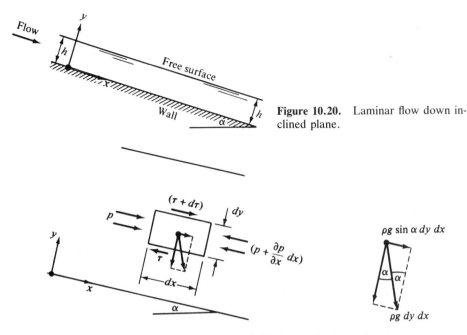

Figure 10.20. Laminar flow down inclined plane.

Figure 10.21. Forces acting on fluid element in the x direction.

$$d\tau\ dx - \frac{\partial p}{\partial x}dx\ dy + \rho g \sin \alpha\ dy\ dx = 0$$

Next, let us consider the momentum equation in the y direction. The forces acting on a fluid element in the y direction are depicted in Figure 10.22. Since the flow velocity in the y direction is zero, the only forces acting are the pressure forces. The force balance in the y direction yields

$$\Sigma F_y = p\ dx - \left(p + \frac{\partial p}{\partial y}\ dy \right) dx - \rho g \cos \alpha\ dy\ dx = 0$$

Thus

$$\frac{\partial p}{\partial y} = -\rho g \cos \alpha$$

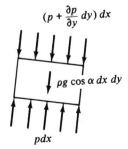

Figure 10.22. Forces acting on fluid element in the y direction.

In addition to satisfying the two differential equations, the flow must also satisfy certain conditions on the boundaries:

1. The pressure along the free surface remains constant and equal to atmospheric pressure; that is,

$$p = p_a = \text{constant} \quad \text{at} \quad y = h$$

2. The shear force exerted by the air in the atmosphere on the liquid is negligibly small, so that we may write

$$\tau = 0 \quad \text{at} \quad y = h$$

3. The velocity of fluid relative to the fixed wall must be zero; that is,

$$u = 0 \quad \text{at} \quad y = 0$$

First integrate the y-momentum equation with respect to y to obtain

$$p = -\rho g y \cos \alpha + f(x) + C \quad \text{where } C \text{ is a constant}$$

Using boundary condition 1, which expresses the constancy of the pressure along the entire free surface, that is, at $y = h$, we obtain

$$p_a = -\rho g h \cos \alpha + f(x) + C$$

whence

$$f(x) = 0 \quad \text{and} \quad C = p_a + \rho g h \cos \alpha$$

Hence from the y-momentum equation we obtain

$$p - p_a = \rho g (h - y) \cos \alpha$$

which is the hydrostatic equation (with $z = y \cos \alpha$) expressing the change in static pressure p due to change in elevation.

From the preceding hydrostatic equation we have $\partial p / \partial x = 0$. Hence the momentum equation in the x direction becomes, upon division by the volume of the element $(dx \, dy)$,

$$\frac{d\tau}{dy} + \rho g \sin \alpha = 0$$

Integration with respect to y gives

$$\tau = -\rho g y \sin \alpha + C_1 \quad \text{with } C_1 \text{ a constant.}$$

From boundary condition 2, $\tau = 0$ at $y = h$; hence

$$C_1 = \rho g h \sin \alpha \quad \text{and} \quad \tau = \rho g (h - y) \sin \alpha$$

Since the flow is in the x direction only, we have for the shear stress

$$\tau = \mu \frac{du}{dy}$$

Thus

$$\mu \frac{du}{dy} = \rho g (h - y) \sin \alpha$$

and

$$u = -\frac{\rho g \sin \alpha}{\mu}\frac{(h - y)^2}{2} + C_2 \quad \text{with} \quad C_2 \text{ a constant}$$

From boundary condition 3, $u = 0$ at $y = 0$; therefore,

$$C_2 = \frac{\rho g \sin \alpha}{\mu}\frac{h^2}{2}$$

and

$$u = \frac{\rho g}{2\mu}(2h - y)y \sin \alpha \tag{10.9}$$

The resulting velocity distribution is shown in Figure 10.23. The volume flow per unit width (q) is

$$q = \int u \, dA = \int_0^h u \, dy \quad \text{for unit channel width}$$

$$q = \frac{\rho g}{3\mu}h^3 \sin \alpha = \frac{g}{3\nu}h^3 \sin \alpha$$

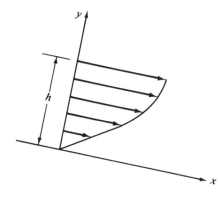

Figure 10.23.   Velocity distribution in laminar flow down inclined plane.

Thus the volume flow in laminar motion is proportional to the slope of the channel floor and acceleration due to gravity but inversely proportional to the kinematic viscosity coefficient ν.

10.4 TURBULENT OPEN CHANNEL FLOW

We treated an example of laminar open channel flow in the preceding section. Just as in the flow through pipes, there exists in open channel flows a critical Reynolds number that predicts transition between laminar and turbulent flow. The flow is laminar below the critical Reynolds number and turbulent above the critical value. It has been found that, for open channel flow, the flow is definitely turbulent if the Reynolds number of the flow (based on the hydraulic radius of the channel section) exceeds 3000. To achieve this Reynolds number in a wide channel 1 m (about 3 ft) deep, for a water temperature of

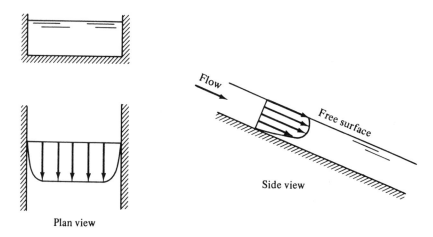

Figure 10.24. Velocity distribution in rectangular open channel.

20°C (68°F), the average flow velocity will only be about 1.4 mm/s (0.06 in/s). Since the water velocity of open channels is usually much greater than this, the Reynolds number of the flow is so high that not only is the flow turbulent but also the frictional characteristics of the flow will be relatively independent of the Reynolds number and will be a function of the roughness of the channel only. A typical velocity distribution in a rectangular channel is shown in Figure 10.24, where it is seen that the maximum velocity does not occur at the free surface (due to the shear effects of the atmospheric air) but some distance below the free surface, usually from about 5 to 25 percent of the liquid depth. As in flow through pipes, fully developed turbulent open channel flow can be approximated by one-dimensional flow.

10.4(a) Uniform Flow

Consider a liquid flowing down a sloping channel of constant cross section at steady-state conditions. The Reynolds number of the flow is sufficiently high so that the flow is fully turbulent. Hence we assume one-dimensional flow; that is, the velocity will be constant over a cross section. For steady uniform flow, the conditions at the various sections also remain the same. Figure 10.25 shows a channel with constant cross section and inclination. Taking the x coordinate in the direction of the flow, the x-momentum equation (4.7) then states $\Sigma F_x = 0$; that is, the sum of the external forces acting on the infinitesimal control volume is zero; τ_w is the wall shear stress due to friction acting on the channel bottom, dA_s is the differential surface area of the channel over which the shear stress acts, and A is the cross-sectional area of the channel. The frictional force due to the atmospheric air at the free surface is neglected. Next, we will show that the pressure gradient in

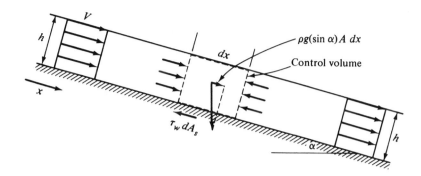

Figure 10.25.   One-dimensional uniform flow.

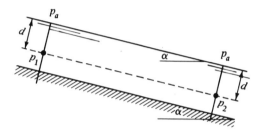

Figure 10.26

the direction of flow is zero for uniform open channel flow. Since the flow is uniform, the free surface will be parallel to the channel bottom and thus its inclination will also be equal to α, as shown in Figure 10.26. Since the pressure distribution at every cross section is the same in uniform flow, the pressure at a given depth d below the free surface must be the same for every cross section. Therefore, $p_1 = p_2$, and the pressure gradient in the direction of flow is zero everywhere in the channel. The pressure variation at a given cross section is that due to the hydrostatic head and thus will vary linearly with distance from the free surface as expressed by the hydrostatic equation. Summing the external forces (shear, pressure, and gravity forces) acting on the control volume, we obtain

$$\Sigma F_x = -\tau_w \, dA_s + \rho g(\sin \alpha) A \, dx + pA - pA = 0$$

or

$$\tau_w \, dA_s = \rho g(\sin \alpha) A \, dx$$

Thus, in steady uniform one-dimensional open channel flow, the component of the gravity force driving the flow and acting in the direction of the flow is balanced by the frictional force acting in the direction opposite to the flow. Replacing dA_s by the product of the wetted perimeter P and element width dx, we obtain

$$\tau_w P \, dx = \rho g(\sin \alpha) A \, dx$$

or

$$\tau_w = \rho g(\sin \alpha)\frac{A}{P} \tag{10.10}$$

Here P is the wetted perimeter of the channel cross section defined as the perimeter of fluid in contact with the solid boundary of the channel (Figure 10.27). For a rectangular channel P is $2h + b$, where h is the depth of the channel and b is its width. Next, express the shear stress τ_w in terms of the friction factor f, as was done in pipe flow in Equation (6.3):

$$\tau_w = \frac{f}{4}\frac{\rho}{2}V^2$$

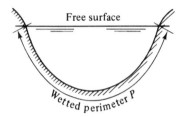

Figure 10.27

Substituting into Equation (10.10), we obtain

$$\frac{f}{4}\frac{\rho}{2}V^2 = \rho g(\sin \alpha)\frac{A}{P}$$

or

$$V = \sqrt{\frac{8g}{f}} \sqrt{\frac{A}{P}} \sqrt{\sin \alpha} \tag{10.11}$$

Equation (10.11) is the *Chezy formula*, in which the $\sqrt{8g/f}$ was originally thought to be constant. However, it has been found by experimental measurement that $\sqrt{8g/f}$ (called the *Chezy constant*) is actually a function of the roughness of the channel (similar to the roughness coefficient ϵ/D for pipe flow) and the hydraulic radius ($= 2A/P$). One of the most frequently used relationships for $\sqrt{8g/f}$ is due to Manning. In SI units, it is given by

$$\sqrt{\frac{8g}{f}} = \frac{0.820}{n}\left(\frac{A}{P}\right)^{1/6} \tag{10.12a}$$

In English units, it is given by

$$\sqrt{\frac{8g}{f}} = \frac{1.486}{n}\left(\frac{A}{P}\right)^{1/6} \tag{10.12b}$$

where n is the roughness coefficient. Typical values of the Manning roughness coefficient for various channel materials are given in Table 10.1.

The hydraulic radius R_h of open channel flow is defined as the ratio of 2 times the cross-sectional area A and wetted perimeter P, as was done for

TABLE 10.1 Average Value of Manning Roughness Coefficient (n)

Nature of surface	$n(m^{1/6})$	$n(ft^{1/6})$
Brick	0.013	0.016
Cast iron	0.010	0.015
Concrete, finished	0.010	0.012
Concrete, unfinished	0.011	0.014
Earth, good condition	0.021	0.025
Earth, with stones or weeds	0.029	0.035
Gravel	0.023	0.028
Rubble	0.021	0.025
Sewer pipe, vitrified	0.011	0.013
Steel, riveted	0.012	0.015
Wood, planed	0.010	0.012
Wood, unplaned	0.011	0.013

closed conduit (pipe) flow in Equation (6.1).* For a circular duct flowing half-full of liquid, this definition results in the statement that the hydraulic radius R_h equals the physical radius of the pipe or the depth of liquid:

$$R_h = 2\frac{A}{P} = 2\frac{(\pi/2)R^2}{\pi R} = R$$

Substituting the Manning relation [Equation (10.12a)] into Equation (10.11), we obtain the *Manning formula* in SI units:

$$V = \frac{0.820}{n}\left(\frac{A}{P}\right)^{2/3}\sqrt{\sin\alpha} \qquad (10.13a)$$

Here the average flow velocity V in the channel is in m/s, A the cross-sectional area is in m², P the wetted perimeter is in m, n the roughness coefficient is in $m^{1/6}$, and $\sin\alpha$ is nondimensional. Note for small inclination α, $\sin\alpha$ can be taken equal to the slope ($\tan\alpha$). The discharge Q (in m³/s) through the open channel is obtained by multiplying Equation (10.13a) by the cross-sectional area A (m²).

In the English system of units, Manning's formula is given by

$$V = \frac{1.486}{n}\left(\frac{A}{P}\right)^{2/3}\sqrt{\sin\alpha} \qquad (10.13b)$$

The average flow velocity is in fps, A is in ft², P is in ft, and the roughness coefficient n is in $ft^{1/6}$. The discharge Q (in ft³/s) is obtained by multiplying Equation (10.13b) by the cross-sectional area A (in ft²).

---

* In some texts, the ratio A/P is called the *hydraulic radius.* The definition given here is preferred, since it becomes identical to the radius for a circular cross section.

EXAMPLE 10.3

A semicircular channel made of unplaned wood has a radius of 5 ft. For a channel slope of 0.002, compute the rate of discharge for uniform flow using the Manning formula.

Solution   From Table 10.1, the average Manning roughness coefficient for unplaned wood is 0.013. The cross-sectional area A is $(\pi/2)(5^2) = 39.3$ ft^2, and the wetted perimeter P is $\pi R = \pi(5) = 15.7$ ft. Hence the average channel velocity is, from Equation (10.13b),

$$V = \frac{1.486}{0.013} \left(\frac{39.3}{15.7}\right)^{2/3} \sqrt{0.002}$$

$$= 9.42 \text{ fps}$$

Thus the discharge through the channel is

$$Q = VA = (9.42 \text{ ft/s})(39.3 \text{ ft}^2) = \underline{370.2 \text{ ft}^3/\text{s}}   \blacksquare$$

When a channel is running full, the friction factor obtained from Manning's roughness coefficient should equal that obtained for fully developed turbulent flow in a pipe (Figure 6.18b). An examination of this figure shows that the friction factor f can be approximated by the relation $f = 0.185(\epsilon/D_h)^{1/3}$ for ϵ/D_h ranging from 0.001 to 0.05. From Equation (10.12) we get with $D_h = 4A/P$, $f = 0.185(10^9 n^6/D_h)^{1/3}$. Thus, the roughness height ϵ and Manning's roughness coefficient n are related by $\epsilon = 10^9 n^6$, with ϵ in m and n in m$^{1/6}$ in SI units, and ϵ in ft and n in ft$^{1/6}$ in English units.

Hydraulically optimum cross sections.   Based on Manning's formula [Equation (10.13)] for the velocity in an open channel, the discharge of the channel becomes

$$Q = \frac{0.820 \, A^{5/3}}{n \, P^{2/3}} \sqrt{\sin \alpha}   \tag{10.14a}$$

where Q is in m^3/s, A is in m^2, P is in m, and n is obtained from Table 10.1. In English units, the discharge of the channel becomes

$$Q = \frac{1.486 \, A^{5/3}}{n \, P^{2/3}} \sqrt{\sin \alpha}   \tag{10.14b}$$

where Q is in ft^3/s, A in ft^2, P is in ft, and n is given in Table 10.1.

In design problems relating to open channels, it is usually required to find the channel of minimum cross-sectional area for a specified rate of discharge, or to find the channel that allows maximum discharge for a specified channel cross-sectional area. From Equation (10.13) it is seen that in either case the wetted perimeter P must be a minimum. The cross section thus obtained is called a *hydraulically optimum section*. Furthermore, since the optimum section has the least wetted perimeter, it will require the least lining material and thus will probably be the most economical section.

It is known from geometry that the circle has the smallest perimeter for a given area. Therefore an open channel of semicircular cross section (when its free surface line equals the diameter of the circle) should be chosen as the optimum cross-sectional shape. However, for certain materials used to construct the channel, the use of a semicircular cross section may be impractical. For example, channels excavated from the soil have trapezoidal cross sections with sides sloped so as to avoid cave-ins. Semicircular channels are usually manufactured of steel or sometimes are made of wooden planks. Generally, wooden channels have rectangular cross sections for ease of construction. Oval (elliptic) cross sections are used for sewers having large variations in the rate of discharge.

Consequently, it is of practical interest to determine the hydraulically optimum sections for certain cross-sectional shapes. As an example, consider a rectangular cross section of width b and depth h. For this cross section, the wetted perimeter P is

$$P = 2h + b$$

Since the area of the section is $A = hb$, $b = A/h$. Thus we can express the perimeter P in terms of the depth h:

$$P = \frac{A}{h} + 2h$$

The variation of the wetted perimeter P with depth h, for a cross-sectional area of 2, is plotted in Figure 10.28, where it is seen that the minimum perimeter occurs when the depth is unity, corresponding to a channel width of 2. The occurrence of such a minimum can also be shown analytically.

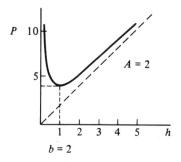

Figure 10.28.   Variation of wetted perimeter P with depth h for a given area ($A = 2$).

Differentiate the expression for P with respect to h and set the derivative equal to zero to obtain

$$\frac{dP}{dh} = -\frac{A}{h^2} + 2 = 0$$

Hence

$$h^2 = \frac{A}{2} = \frac{hb}{2}$$

and

$$h = \frac{b}{2}$$

Thus for a given cross-sectional area, the optimum proportion for a rectangular channel occurs when the depth of flow is one-half the width of the channel (Figure 10.29).

As another example of a hydraulically optimum cross section, consider a trapezoidal cross section, as shown in Figure 10.30. For a trapezoidal cross section, the wetted perimeter P is

$$P = b + \frac{2h}{\sin \alpha}$$

while the cross-sectional area of the trapezoid is

$$A = bh + h^2 \operatorname{ctn} \alpha$$

Hence

$$b = \frac{A - h^2 \operatorname{ctn} \alpha}{h}$$

Thus we can express P in terms of α and h:

$$P = \frac{A - h^2 \operatorname{ctn} \alpha}{h} + \frac{2h}{\sin \alpha} = \frac{A}{h} - h \operatorname{ctn} \alpha + \frac{2h}{\sin \alpha}$$

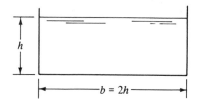

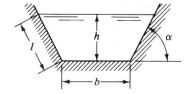

Figure 10.29. Optimum rectangular cross section.

Figure 10.30. Trapezoidal cross section.

Here the perimeter is a function of the depth h and the angle α. Figure 10.31a shows the variation of P with h for several angles holding the area A constant at 2 m². For each angle α, a minimum value of P is obtained. In Figure 10.31b, these minima have been plotted as a function of α. It is seen that the smallest minimum is obtained for α equal to $\pi/3$ rad. This can also be shown analytically as follows: Differentiating P first with respect to h and setting equal to zero, we obtain

$$\frac{dP}{dh} = -\frac{A}{h^2} + \frac{2}{\sin \alpha} - \operatorname{ctn} \alpha = 0$$

and

$$h^2 = A \frac{\sin \alpha}{2 - \cos \alpha}$$

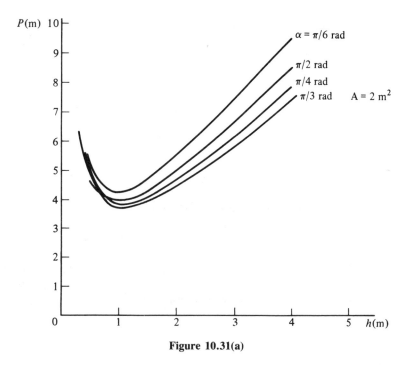

Figure 10.31(a)

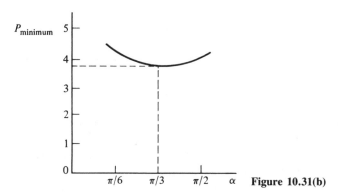

Figure 10.31(b)

Then differentiating h with respect to α and setting equal to zero results in

$$2h\frac{dh}{d\alpha} = A\left[\frac{\cos \alpha}{2 - \cos \alpha} - \frac{\sin^2 \alpha}{(2 - \cos \alpha)^2}\right] = A\frac{2\cos \alpha - \cos^2 \alpha - \sin^2 \alpha}{(2 - \cos \alpha)^2} = 0$$

Hence

$$2\cos \alpha - 1 = 0$$
$$\cos \alpha = \tfrac{1}{2}$$
$$\alpha = \frac{\pi}{3}$$

Substituting this result into the expression for h^2, we obtain

$$h^2 = A\frac{\frac{1}{2}\sqrt{3}}{2 - \frac{1}{2}} = \frac{A}{\sqrt{3}}$$

Therefore,

$$A = \sqrt{3}h^2 = bh + \frac{1}{\sqrt{3}}h^2$$

and

$$b = \sqrt{3}h - \frac{1}{\sqrt{3}}h = \frac{2}{3}\sqrt{3}h$$

Next, determine the length of the sloping sides (l) of the channel

$$l = \frac{h}{\sin \alpha} = \frac{2}{\sqrt{3}}h = \frac{2}{\sqrt{3}}\frac{3}{2\sqrt{3}}b = b$$

Thus the sloping sides of the channel have the same length as the width of the horizontal bottom. For a given cross-sectional area, therefore, the optimum proportion for a trapezoidal section occurs when the cross section is a semihexagon (Figure 10.32).

Figure 10.32.   Optimum trapezoidal cross section.

10.4(b) Nonuniform Flow

In the preceding section it was shown that for steady uniform flow—that is, flow in which the liquid surface remains parallel to the channel bed—the calculations of the flow quantities are relatively simple and the flow discharge, for example, can be predicted by use of the Manning formula. In the more general case of steady flow, the depth does not remain constant but varies along the channel and thus the liquid surface is not parallel to the channel bed. This effect becomes particularly important if the channel is long.

    Let us now develop the basic equation for steady nonuniform flow in which the change in depth is gradual (gradually varied flow). In gradually varied flow, the depth, the cross-sectional area, the bottom slope, and the roughness of the channel change slowly along the channel. We therefore assume that the frictional losses at a given section of the nonuniform flow are given by the Manning formula for uniform flow for the same depth and flow rate. As pointed out previously, for turbulent flow in a channel, it is reasonable to assume one-dimensional flow. The momentum equation (4.9)

$$\Sigma F_x = \int V_x(\rho V_n \, dA)$$

therefore becomes for one-dimensional flow

$$\Sigma F_x = \rho Q(V_2 - V_1)$$

Take the control volume for nonuniform flow as shown in Figure 10.33. As was done for uniform open channel flow, the x coordinate is taken along the channel bottom. Here V is the average velocity at the cross section, p is the average pressure, τ_w is the channel wall shear stress, and α is the angle of inclination of the channel bed. Substituting expressions for the forces acting on the control volume, the momentum equation becomes

$$-\tau_w \, dA_s + \rho g(\sin \alpha)A \, dx + pA - (p + dp)(A + dA) = \rho Q(V + dV - V)$$

where dA_s is the differential surface area taken along the wetted perimeter and A is the cross-sectional area of the channel. Neglecting $dp \, dA$ as of second order relative to $p \, dA$ or $A \, dp$, we obtain

$$-\tau_w P \, dx + \rho g(\sin \alpha)A \, dx - d(Ap) = \rho AV \quad dV$$

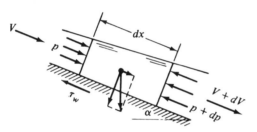

Figure 10.33

Next, replace $V \, dV$, using the continuity equation,

$$AV = (A + dA)(V + dV)$$
$$= AV + dA \cdot V + dV \cdot A + \underset{\text{second order}}{dA \cdot dV}$$

so that

$$V \, dA + A \, dV = 0$$

or

$$dV = -\frac{V \, dA}{A} - \tau_w P \, dx + \rho g(\sin \alpha)A \, dx - d(Ap) = -\rho V^2 \, dA \quad (10.15)$$

Then restrict the consideration to rectangular channels, where $A = hb$, and evaluate the average pressure over each cross section of the control volume. From the hydrostatic equation, the average pressure for a rectangular cross section (Figure 10.34) will be one-half the maximum, since pressure increases linearly with submergence. Thus

$$p_{av} = \tfrac{1}{2}\rho g h \cos \alpha$$

and

$$d(Ap) = d(hb\tfrac{1}{2}\rho g h \cos \alpha) = \rho g b h \cos \alpha \, dh$$

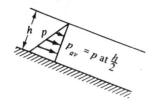

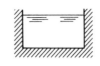

Figure 10.34. Average hydrostatic pressure for rectangular section.

Substituting into Equation (10.15) and dividing by A, we obtain

$$\left(\rho g \sin \alpha - \tau_w \frac{P}{A}\right) dx - \rho g (\cos \alpha) \, dh = -\rho V^2 \frac{dh}{h}$$

Rearranging further, we have

$$\left(\cos \alpha - \frac{V^2}{gh}\right) dh = \left[\sin \alpha - \frac{\tau_w(P/A)}{\rho g}\right] dx$$

Letting

$$\tau_w = \frac{f}{4}\rho\frac{V^2}{2}$$

and using Manning's formulation for $\sqrt{8g/f}$, τ_w becomes, in SI units,

$$\tau_w = \rho V^2 g\left(\frac{n}{0.820}\right)^2\left(\frac{P}{A}\right)^{1/3}$$

The slope of the channel is small for gradually varied flow, and therefore we can replace $\cos \alpha$ by 1. We then obtain the differential equation for gradually varied nonuniform flow in SI units:

$$\left(1 - \frac{V^2}{gh}\right) dh = \left[\sin \alpha - \frac{n^2}{0.6724}\frac{V^2}{(A/P)^{4/3}}\right] dx \qquad (10.16)$$

or

$$\frac{dh}{dx} = \frac{\sin \alpha - (n^2/0.6724)V^2/(A/P)^{4/3}}{1 - (V^2/gh)} \qquad (10.17a)$$

In English units, the differential equation for gradually varied flow becomes

$$\frac{dh}{dx} = \frac{\sin \alpha - (n^2/2.21)V^2/(A/P)^{4/3}}{1 - (V^2/gh)} \qquad (10.17b)$$

Because of its nonlinearity, Equation (10.17) cannot be readily integrated analytically, and numerical means or stepwise integration is used to obtain the variation of depth along the channel. However, certain general characteristics of the behavior of the free surface can be obtained from Equation (10.17) without actual integration. It is seen from Equation (10.17a) that the rate of change of depth (dh/dx) will be equal to zero (uniform flow), for $V^2/gh \neq 1$, when

$$\sin \alpha = \frac{n^2}{0.6724}\frac{V^2}{(A/P)^{4/3}} = \frac{n^2}{0.6724}Q^2\frac{P^{4/3}}{A^{10/3}}$$

This equation is identical to Equation (10.14a), derived in Section 10.4a for uniform flow. In English units, we have

$$\sin \alpha = \frac{n^2}{2.21} Q^2 \frac{P^{4/3}}{A^{10/3}}$$

Thus for a given channel cross section, rate of discharge, channel roughness, and channel slope, there corresponds only one depth at which the flow is uniform. Next consider subcritical (i.e., $V^2/gh < 1$) nonuniform flow. When the slope of the channel floor is greater than that for uniform flow at the same depth, the rate of change of depth is positive and the slope of the liquid surface increases in the downstream direction. On the other hand, if the slope of the channel floor is less than that for uniform flow, the rate of change of depth will be negative, and the free surface slope, and hence the liquid depth, decreases in the downstream direction. For supercritical flow (i.e., $V^2/gh > 1$), the change in the surface slopes will be opposite to that in subcritical flow.

Let us now illustrate the application of the differential equation for gradually varied flow to the problem of determining the depth variation for open channel flow in which a barrier such as a dam has been placed. The resulting curve of the free surface is called the backwater curve. The actual behavior of the flow behind a barrier will depend on the value of the velocity of the flow approaching the barrier. If the approach velocity V is less than $\sqrt{gh}$, then the backwater curve mentioned above will be obtained as shown in Figure 10.35. If the approach velocity V is greater than $\sqrt{gh}$, Equation (10.17) indicates that the depth increases in an upstream direction, as indicated in Figure 10.36. This is a physically inconsistent situation, for the backwater curve does not coalesce into the free surface at large distances from the barrier. Under these conditions, the flow adjustment between upstream and

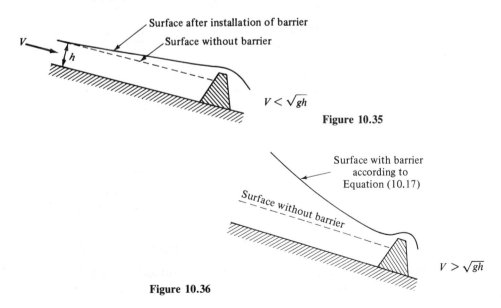

Figure 10.35

Figure 10.36

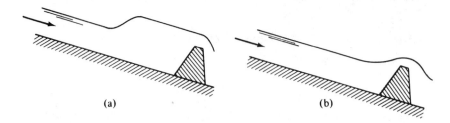

(a)   (b)

Figure 10.37. (a) Hydraulic jump ahead of barrier; (b) swell at barrier.

barrier flows must occur in the vicinity of the barrier; that is, a rapidly varied flow takes place, as indicated in Figure 10.37a or b, depending on the height of the barrier.

EXAMPLE 10.4

The flow rate and water depth just upstream of the dam shown in Figure 10.38 are 10 m³/s and 3 m, respectively. The channel cross section is rectangular with a width of 5 m. The channel has a slope of 0.001 and a roughness coefficient n of 0.020. Determine the variation of water depth along the channel.

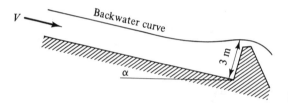

Figure 10.38

Solution At a large distance upstream from the dam, the velocity of the flow will be that corresponding to uniform flow. From Equation (10.13a) we can write

$$V = \frac{0.820}{n} \left(\frac{A}{P}\right)^{2/3} \sqrt{\sin \alpha} = \frac{Q}{bh} = \frac{10 \text{ m}^3/\text{s}}{5h \text{ m}^2}$$

$$= \frac{0.820 \text{ m}^{1/2}/\text{s}}{0.02 \text{ m}^{1/6}} \left[\left(\frac{5h}{5 + 2h}\right)^{2/3} \text{m}^{2/3}\right] \sqrt{0.001} = \frac{2}{h} \text{m/s}$$

or

$$\frac{5 + 2h}{5h} = (0.64821h)^{3/2}$$

Solving, far upstream, $h = 1.577$ m, and the water depth in nonuniform flow will vary from 1.577 m to 3 m. The actual depth variation is obtained from Equation (10.16), which is written here as a difference equation:

$$\left(1 - \frac{V_{av}^2}{gh_{av}}\right)\Delta h = \left(\sin \alpha - \frac{n^2}{0.6724} \frac{V_{av}^2}{(A/P)_{av}^{4/3}}\right)\Delta x$$

For evaluation of the preceding equation, divide the depth range 1.577 m to 3 m into increments as indicated in Table 10.2. The value of the quantities

TABLE 10.2

h (m)	A (m²)	P (m)	$\dfrac{A}{P}$ (m)	V (m/s)	$\dfrac{V^2}{g}$ (m)	$\left(\dfrac{A}{P}\right)_{av}$	$\left(\dfrac{V^2}{g}\right)_{av}$	h_{av}	$\left[1 - \dfrac{(V^2)_{av}}{g h_{av}}\right]\Delta h$
3	15	11	1.3636	0.6667	0.0453				
						1.3422	0.0487	2.9	0.1966
2.8	14	10.6	1.3208	0.7143	0.0520				
						1.2977	0.0562	2.7	0.1958
2.6	13	10.2	1.2745	0.7692	0.0603				
						1.2495	0.0656	2.5	0.1948
2.4	12	9.8	1.2245	0.8333	0.0708				
						1.1974	0.0775	2.3	0.1933
2.2	11	9.4	1.1702	0.9091	0.0842				
						1.1407	0.0931	2.1	0.1911
2.0	10	9.0	1.1111	1	0.1019				
						1.0788	0.1139	1.9	0.1880
1.8	9	8.6	1.0465	1.1111	0.1258				
						1.0111	0.1426	1.7	0.1832
1.6	8	8.2	0.9756	1.2500	0.1593				
						0.9713	0.1616	1.589	0.0207
1.577	7.885	8.154	0.9670	1.2682	0.1639				

$(V^2)_{av}$	$\dfrac{n^2}{0.6724}\dfrac{(V^2)_{av}}{\left(\dfrac{A}{P}\right)_{av}^{4/3}}$	Δx (m)
0.4774	0.0001918	243.3
0.5510	0.0002316	254.8
0.6431	0.0002843	272.2
0.7605	0.0003558	300.1
0.9133	0.0004558	351.2
1.1173	0.0006007	470.8
1.3985	0.0008198	1016.6
1.5854	0.0009805	1061.5
		$L = 3970.5$ m

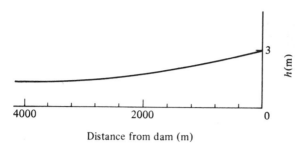

Figure 10.39. Backwater curve behind dam.

with subscript av are taken as the average over the interval. The velocity of the flow at each assumed depth is obtained from the continuity equation

$$V = \frac{Q}{bh} = \frac{2}{h} \text{ m/s}$$

The other quantities are evaluated as indicated in the table, resulting in an incremental distance Δx along the channel over which the selected change in depth occurs. Summing all Δx's, we obtain the total length of the backwater equal to 3970.5 m. A plot of the backwater curve behind the dam is given in Figure 10.39. ∎

10.4(c) Losses at Channel Entrances

When a liquid flows from a reservoir into an open channel, entrance losses will occur. These entrance losses are similar to the losses incurred in flow from a reservoir into a pipe flowing full as discussed in Section 6.4, where the loss factor varied from 0.5 for a square-edged inlet to 0.05 for a well-rounded inlet. In a similar fashion, the loss at the entrance to a channel from a reservoir or pond varies with the type of transition section used. A well-designed transition section has a loss of about 5 percent of the velocity head, whereas the loss can be as much as 25 percent of the velocity head for a poorly designed transition.

Using the modified Bernoulli equation (6.7) evaluated at the free surface, we obtain with constant pressure at the surface

$$gh_0 = gh_1 + \frac{V_1^2}{2} + K\frac{V^2}{2}$$

or

$$h_0 = h_1 + (1 + K)\frac{V_1^2}{2g} \tag{10.18}$$

where K is the loss factor ranging from 0.05 to 0.25, h_0 the elevation of the liquid surface in the reservoir above the channel floor, and h_1 the depth of liquid at the channel entrance (Figure 10.40).

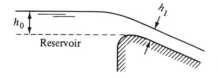

Figure 10.40. Transition from reservoir into open channel.

EXAMPLE 10.5

A rectangular flume 10 ft wide is connected to a reservoir as shown in Figure 10.41. The flume is made of finished concrete ($n = 0.012$), has a slope of 0.001, and is 500 ft long. The water surface in the reservoir is 5 ft above the upper flume bed. The water depth at the downstream end of the flume is 3 ft. Assuming a flume entrance loss of 25 percent of the velocity head, determine the discharge and depth variation in the flume.

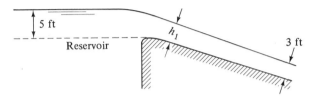

Figure 10.41. Nonuniform flow from a reservoir.

Solution   With an entrance loss of $0.25(V_1^2/2g)$, Equation (10.18) becomes

$$h_0 = h_1 + 1.25\frac{V_1^2}{2g} \quad \text{or} \quad V_1 = 7.17\sqrt{5 - h_1} \text{ ft/s}$$

As a first estimate, take h_1 equal to 4 ft and divide the total change in depth along the flume into two increments of 0.5 ft each. With $h_1 = 4$ ft we obtain $V_1 = 7.18$ ft/s, resulting in the discharge of 287 ft³/s, since $Q = V_1 h_1 b$. The velocity of flow at the other depths is obtained from the relation

$$V = \frac{Q}{bh} = \frac{287 \text{ ft}^3/\text{s}}{(10 \text{ ft})h}$$

The basic equation used to obtain the total length of the flume is the difference equation based on Equation (10.17b) used in Example 10.4. The procedure utilized in this example is to assume a depth h_1 and to compare the total length of flume thus obtained with the specified length. As is seen from Table 10.3, two tries were required to obtain a length of about 500 ft.

The discharge through the flume will be equal to 7.43 ft/s × 3.93 ft × 10 ft = 292 ft³/s with a depth variation as given in Table 10.3. In this example the depth decreases in the downstream direction, for the liquid accelerates as it moves downstream. For a more refined depth curve, smaller increments than 0.5 ft would have to be used. Good approximation for the discharge can be obtained by using a single increment in h.   ■

10.5 HYDRAULIC JUMP

In contrast to the gradually varied flow just discussed, the properties of a rapidly varied flow change at a rapid rate with respect to the distance along the channel. Examples of such rapidly varied flows are the hydraulic jump, the hydraulic drop that occurs at a sudden drop in the bottom of a channel,

TABLE 10.3 First Try

h (ft)	V (fps)	A (ft²)	P (ft)	$\frac{A}{P}$ (ft)	V^2	V^2_{av}	$\left(\frac{A}{P}\right)_{av}$	$\left(\frac{A}{P}\right)^{4/3}_{av}$
4	7.18	40	18	2.22	51.5			
3.5	8.20	35	17	2.06	67.2	59.4	2.14	2.75
3	9.60	30	16	1.88	92.0	79.6	1.96	2.47

h	$\frac{n^2}{2.21}\frac{V^2_{av}}{(A/P)^{4/3}_{av}}$	$\sin\alpha - \frac{n^2}{2.21}\frac{V^2_{av}}{(A/P)^{3/4}_{av}}$	h_{av}	$\left(\frac{V}{gh}\right)_{av}$	$1-\left(\frac{V}{gh}\right)_{av}$	Δh	Δx
4							
3.5	1.41×10^{-3}	-0.41×10^{-3}	3.75	0.493	0.507	0.5	618
3	2.10×10^{-3}	-1.10×10^{-3}	3.25	0.760	0.240	0.5	$\dfrac{109}{727}$

$$L = 727 \text{ ft}$$

Second Try

h (ft)	V (fps)	A (ft²)	P (ft)	$\frac{A}{P}$ (ft)	V^2	V^2_{av}	$\left(\frac{A}{P}\right)_{av}$	$\left(\frac{A}{P}\right)^{4/3}_{av}$
3.93	7.43	39.3	17.86	2.20	55.3			
3.5	8.35	35	17	2.06	69.7	62.5	2.13	2.74
3	9.74	30	16	1.88	94.8	82.3	1.97	2.47

h	$\frac{n^2}{2.21}\frac{V^2_{av}}{(A/P)^{4/3}_{av}}$	$\sin\alpha - \frac{n^2}{2.21}\frac{V^2_{av}}{(A/P)^{4/3}_{av}}$	h_{av}	$\left(\frac{V}{gh}\right)_{av}$	$1-\left(\frac{V}{gh}\right)_{av}$	Δh	Δx
3.93							
3.5	1.49×10^{-3}	-0.49×10^{-3}	3.72	0.522	0.478	0.43	420
3	2.20×10^{-3}	-1.20×10^{-3}	3.25	0.788	0.212	0.5	$\dfrac{88.5}{508.5}$

$$L = 508.5 \text{ ft}$$

and the free surface flow around obstructions like bridge piers. Here we shall treat the hydraulic jump as an example of rapidly varied flows. In a hydraulic jump there occurs a sudden change in liquid depth from less-than-critical to greater-than-critical depth. The velocity of the flow changes from supercritical to subcritical as a result of the jump. This transition takes place over a relatively short distance, usually less than 5 times the depth of flow after the jump, over which the height of the liquid increases rapidly, incurring a considerable loss of energy. An example of a hydraulic jump can be observed when a jet of water from a faucet strikes the horizontal surface of the kitchen sink. The water flows radially outward and a circular jump occurs.

We shall restrict the derivation of the basic equations of the hydraulic jump to rectangular horizontal channels. First, we shall determine the down stream depth of the jump by using the momentum and continuity equations for one-dimensional flow. Then the energy loss due to the jump will be evaluated, using the energy equation.

Consider the control volume as shown in Figure 10.42, extending over the channel width b. Here it can be assumed that the friction forces at the channel walls are negligible due to the short length of channel over which the jump takes place. Sides of the control volume are taken far enough upstream and downstream of the jump that uniform velocity profiles can be taken at 1 and 2. Furthermore, since we have taken the channel bed to be horizontal, the component of the gravity forces in the flow direction is zero.

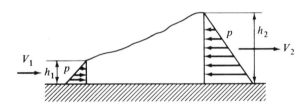

Figure 10.42. Forces acting on control volume in hydraulic jump.

The momentum equation for one-dimensional flow,

$$\Sigma F_x = \rho Q(V_2 - V_1)$$

becomes, using the average hydrostatic pressures acting on the cross sections as the only forces,

$$\tfrac{1}{2}\rho g h_1 h_1 b - \tfrac{1}{2}\rho g h_2 h_2 b = \rho Q(V_2 - V_1)$$

From the continuity equation, we have

$$Q = V_1 h_1 b = V_2 h_2 b$$

Substituting the expressions for $V_1 = Q/bh_1$ and $V_2 = Q/bh_2$ into the momentum equation results in

$$\frac{h_2^2 - h_1^2}{2} = \frac{Q^2}{b^2 g}\left(\frac{1}{h_1} - \frac{1}{h_2}\right) = \frac{Q^2}{b^2 g}\frac{h_2 - h_1}{h_1 h_2}$$

Divide by $h_2 - h_1$, multiply by 2, and rearrange terms to obtain the quadratic equation for h_2:

$$h_2^2 + h_1h_2 - \frac{2Q^2}{b^2gh_1} = 0$$

Finding the root of this equation gives the relation between the two depths of the jump and the channel discharge:

$$h_2 = -\frac{h_1}{2} + \sqrt{\frac{2Q^2}{b^2gh_1} + \frac{h_1^2}{4}} \qquad (10.19)$$

where the negative sign in front of the radical has been deleted, for a negative h_2 has no physical meaning. Equation (10.19) is plotted in Figure 10.43. The value of the downstream velocity V_2 can now be evaluated from

$$V_2 = \frac{Q}{bh_2}$$

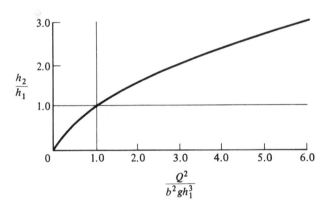

Figure 10.43

From Equation (10.19), it can be seen that

$$\text{If} \quad \frac{Q^2}{b^2gh_1^3} = 1 \quad \text{then} \quad \frac{h_2}{h_1} = 1$$

Note that

$$\frac{Q^2}{b^2gh_1^3} = \frac{V^2}{gh_1} = \text{Fr}_1 \quad \text{and} \quad \text{Fr}_2 = \frac{V_2^2}{gh_2} = \frac{Q^2}{b^2gh_2^3} = \frac{\text{Fr}_1}{(h_2/h_1)^3}$$

so that if the Froude number of the approach flow is 1, the jump is infinitesimally weak. Also, if $\text{Fr}_1 > 1$, corresponding to supercritical approach flow, it can be seen from Figure 10.43 that $h_2/h_1 > 1$ and hence $\text{Fr}_2 < 1$, or the flow changes from supercritical to subcritical across the jump. If $\text{Fr}_1 < 1$, then $\text{Fr}_2 > 1$ and the flow changes from subcritical to supercritical across the jump.

Now we can write Bernoulli's equation for a streamline along the free surface. In this case, we must take into account friction losses Δh_f caused by shear within the fluid as it goes through the jump:

$$h_1 + \frac{V_1^2}{2g} = h_2 + \frac{V_2^2}{2g} + \Delta h_f \tag{10.20}$$

It can be shown that, using Equations (10.19) and (10.20), for $\text{Fr}_1 > 1$ and transition from supercritical to subcritical flow, the term Δh_f is positive. For $\text{Fr}_1 < 1$ and transition from subcritical to supercritical flow, Δh_f is negative. We recognize that Δh_f must be a positive term (due to friction) so that Fr_1 must be greater than 1. It follows that, for a hydraulic jump, the approach flow must be supercritical, with flow after the jump subcritical. This is shown in Figure 10.44 on a specific head diagram.

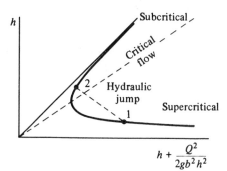

$$h + \frac{Q^2}{2gb^2h^2}$$

Figure 10.44.   Specific head diagram for hydraulic jump.

By using the continuity equation and the relationship between h_1 and h_2 given in Equation (10.19), we obtain, after appropriate simplification, an expression for Δh_f at the jump in terms of the liquid depths h_1 and h_2:

$$\Delta h_f = \frac{(h_2 - h_1)^3}{4h_1h_2} \tag{10.21}$$

This is usually expressed as an equivalent loss of energy per unit mass, Δe_f, where

$$\Delta e_f = \frac{g(h_2 - h_1)^3}{4h_1h_2} \tag{10.22}$$

For a given discharge Q, the rate of energy loss can be computed as $\rho Q\,\Delta e_f$.

Equations (10.21) and (10.22) are tabulated in Table 2 of Appendix D as h_2/h_1, Fr_2, and $\Delta e_f/h_1g$ versus Fr_1.

EXAMPLE 10.6

A hydraulic jump occurs in a rectangular channel having a discharge of 3 m³/s per meter width. The approach depth is 0.75 m. Calculate the depth in the channel after the jump, as well as the energy dissipated per meter of channel width.

Solution The channel depth after the jump is obtained from Equation (10.19):

$$h_2 = -\frac{0.75}{2} + \sqrt{\frac{2(3)^2 \text{ m}^6/\text{s}^2}{(1 \text{ m}^2)(9.81 \text{ m/s}^2)(0.75 \text{ m})} + \frac{0.75^2}{4}} \text{ m}^2$$

$$= -0.375 + 1.6084$$

$$= 1.2334 \text{ m}$$

The energy dissipated per kilogram mass is obtained from Equation (10.22):

$$\Delta e_f = \frac{g(h_2 - h_1)^3}{4h_1 h_2}$$

$$= \frac{(9.81 \text{ m/s}^2)[(1.2334 - 0.75)^3 \text{m}^3]}{[4(0.75) \text{ m}](1.2334 \text{ m})}$$

$$= 0.2995 \text{ J/kg}$$

For a discharge of 3 m³/s per meter width, the rate of energy loss is

$$(1000 \text{ kg/m}^3)(3 \text{ m}^3/\text{s})(0.2995 \text{ J/kg}) = \underline{898.4 \text{ W}}$$

The velocity V_1 is

$$V_1 = \frac{Q}{bh_1} = \frac{3}{0.75} = 4.0 \text{ m/s}$$

while

$$V_2 = \frac{3}{1.2334} = 2.4323 \text{ m/s}$$

The approach flow is supercritical since $V_1^2/gh_1 = 2.175 > 1$, while the flow downstream of the jump is subcritical since $V_2^2/gh_2 = 0.489 < 1$.

The solution can also be obtained by use of the Hydraulic Jump Tables of Appendix D. For $\text{Fr}_1 = V_1^2/gh_1 = 2.175$, obtain by interpolation that $h_2/h_1 = 1.6438$, hence $h_2 = \underline{1.233 \text{ m}}$, and $\text{Fr}_2 = 0.489$. Also, $\Delta e_f/h_1 g = 0.040762$ and there is an energy loss of 0.2999 J/kg, resulting in a rate of energy loss of $\underline{899.7 \text{ W.}}$ ■

PROBLEMS

10.1. A rectangular Venturi flume is inserted into a channel having a width of 1 m. The flow is horizontal and may be assumed to be frictionless. The width at the throat (minimum width section) is 0.5 m. The depth upstream of the flume is 0.7 m, while the depth at the throat is 0.5 m. Compute the volume flow rate through the flume.

10.2. A rectangular channel is 3 m wide and carries a volume flow rate of 6 m³/s. Plot depth h versus specific head H for this open channel flow. Determine the critical depth h_c. Repeat for flows of 5 m³/s and 7 m³/s.

10.3. In Problem 10.2, calculate the rise in the channel floor that will give critical flow for a channel flow of 6 m^3/s and an approach water depth of 0.95 m. Repeat for an approach water depth of 0.53 m. In each case, determine whether the approach flow is subcritical or supercritical.

10.4. Assuming that the downstream water depth h can be taken as approximately 0.60 of the actual sluice gate opening h_s, sketch a plot of flow through a sluice gate versus sluice gate opening. Take the upstream water depth h_0 to be 25 ft; gate is 10 ft wide. (See Figure P10.4.)

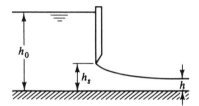

Figure P10.4

10.5. The water level in a reservoir is 1 m above the crest of a dam. What is the maximum volume flow rate per unit width passing over the dam?

10.6. A film of viscous liquid flows down a vertical wall. The velocity at the free surface of the liquid is 1 in/s. The film is 0.02 in thick. Plot the velocity distribution in the liquid film.

10.7. A thin layer of viscous liquid is transported up a moving inclined wall as shown in Figure P10.7. Find the expression for the velocity distribution in the liquid layer. What is the velocity of the wall for the special case when the velocity of the fluid at the free surface is zero?

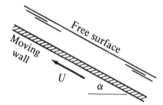

Figure P10.7

10.8. Obtain an expression for the drag exerted in uniform two-dimensional open channel flow on a channel floor of length L for (a) laminar flow and (b) turbulent flow. Assume channel width b to be much greater than liquid depth h.

10.9. Evaluate the hydraulic radius for the following channel cross sections:
   (a) Optimum rectangular section
   (b) Optimum trapezoidal section
   (c) Triangular section with included angle of $\pi/2$ rad
   (d) Very wide rectangular section
   (e) Very narrow rectangular section
   Plot R_h/h as a function of h/b for rectangular channels.

10.10. A rectangular flume made of timber has a slope of 0.001. Find the volume flow rate through the flume for (a) a width of 5 ft and a water depth of 2.5

ft, and (b) a width of 2.5 ft and a water depth of 5 ft. Also, determine which of the two cross sections requires less timber.

10.11. Uniform open channel flow takes place in a channel made of concrete (n = 0.010) and having a rectangular section with width of 3 m and depth of 2 m. The slope of the bed is 0.05. Compute the volume flow rate through the channel.

10.12. Water flows through the triangular canal shown in Figure P10.12. The canal is made of brick and has a slope of 0.10. Find the volume flow rate for a water depth 50 cm.

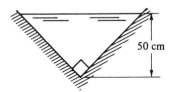

Figure P10.12

10.13. For Problem 10.12, compute the friction factor f from Equation (10.12a). Compare with the value obtained from the Moody diagram of Figure 6.20.

10.14. A circular finished concrete conduit (n = 0.012 ft$^{1/6}$) has a radius of 5 ft and a slope of 0.0015. Calculate the average conduit velocity and discharge rates for various water depths in the conduit. What value of water depth results in the highest velocity and what value in the highest discharge rate?

10.15. A rectangular channel lined with brick is 3 m wide; the channel slope is 0.05. Compute the flow depth for uniform open channel flow of 100 m^3/s.

10.16. The trapezoidal canal shown in Figure P10.16 carries a volume flow rate of 200 cfs. The slope of the canal is 0.0005 and the water depth is to be 4 ft or less. Determine the bottom dimension b of the canal.

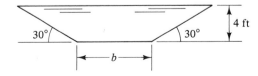

Figure P10.16

10.17. Design an optimum open channel with a trapezoidal cross section. The channel is to be made of brick and is to carry 300 m^3/s. The bottom slope is 0.0008.

10.18. A circular conduit of brick flowing half full carries 250 cfs at an average velocity of 12 fps. Determine the slope of the conduit.

10.19. A concrete rectangular channel carries water at a volume rate of 50 m^3/s. Its width is 3 m. For a channel roughness coefficient n of 0.014, compute:
(a) The channel slope necessary to maintain a uniform depth 1 m
(b) The channel slope necessary to maintain a uniform depth of 2 m
(c) The critical channel depth and velocity

10.20. Compute the volume flow rate through the river channel and floodway shown in Figure P10.20. Assume steady uniform flow with n = 0.020 for the river channel and n = 0.030 for the floodway. The slope of channel and floodway is 0.0005. (*Hint:* Determine the flow rate of channel and floodway separately and add to obtain total flow.)

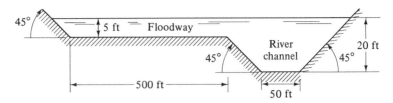

Figure P10.20

10.21. Derive the equation for the volume rate of flow through a triangular channel having an included angle of $\pi/2$. How does the flow rate vary with change in depth?

10.22. Determine the hydraulically optimum triangular cross section shown in Figure P10.22.

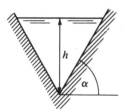

Figure P10.22

10.23. An open channel has a triangular cross section as shown in Figure P10.23. What is the depth h for maximum discharge for a given roughness n and channel bed slope?

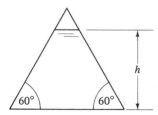

Figure P10.23

10.24. What would be the discharge through the channel of Example 10.2 if the slope were changed to 0.50 m/km?

10.25. The flow in the channel of Example 10.2 is decreased to 2.0 m³/s, all other conditions remaining the same. Determine the resulting water surface.

10.26. For a channel with triangular cross section (see Figure P10.22), determine an expression for the critical depth and hence show that the expression for the specific head becomes

$$\frac{H}{h_c} = \frac{h}{h_c} + \frac{1}{4}\left(\frac{h_c}{h}\right)^4$$

10.27. Nonuniform open channel flow takes place in a channel having an upward slope as shown in Figure P10.27. Calculate the depth h_2 if the channel is very wide and the slope is -0.003.

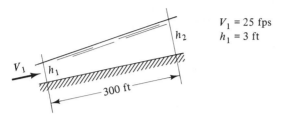

$V_1 = 25$ fps
$h_1 = 3$ ft

Figure P10.27

10.28. Calculate the specific head for a given flow of 150 cfs in a channel having a circular cross section (radius = 5 ft) at depths of 1, 2.5, 5, and 7.5 ft.

10.29. Derive an equation similar to Equations (10.17a) and (10.17b) for nonuniform flow through a triangular cross section.

10.30. Far upstream of the dam shown in Figure P10.30, the uniform flow velocity is 1.0 m/s, with a depth of 1.5 m. Just upstream of the dam, the water depth is 2.5 m. The channel cross section is rectangular, channel slope is 0.0012, gravel-lined. Determine the variation of water depth along the channel. Take $b = 10$ m.

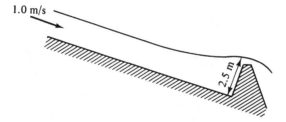

Figure P10.30

10.31. A rectangular channel 8 ft wide carries a volume flow of 200 cfs. Calculate the critical depth of the channel and the critical velocity.

10.32. A hydraulic jump occurs in uniform open channel flow. The rectangular channel carries 10 m³/s with a slope of 0.075. Determine the depth before the jump if the depth after the jump is 1.5 m. The channel width is 3 m.

10.33. A trapezoidal canal with 60° side slopes and a bottom width of 7.5 ft carries a volume flow of 500 cfs. Calculate the critical depth of the canal and the critical velocity.

10.34. The water depth upstream of a hydraulic jump is measured at 1 m, while the depth in the rectangular channel after the jump is 2 m. Calculate the upstream and downstream velocities, and the energy dissipated per meter width.

10.35. The depth of water in a rectangular channel is 1 ft. The channel is 5 ft wide and carries a volume flow of 100 cfs. What will be the water depth after a hydraulic jump? Calculate the loss of energy due to the jump.

10.36. A triangular canal having the cross section shown in Figure P10.36 experiences a hydraulic jump. The water flow rate in the canal is 0.3 m³/s. The depth ahead of the jump is 0.25 m. Determine the depth after jump and the rate of energy loss due to the hydraulic jump.

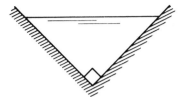

Figure P10.36

10.37. Show that, for $Fr_1 < 1$, the term Δh_f of Equation (10.20) is negative.

11

COMPRESSIBLE FLOW

11.1 INTRODUCTION

In this chapter we shall study compressible flows, in which appreciable density changes occur as the result of the flow. In our study of the flow of liquids, we were justified in assuming incompressible flow, for a liquid can sustain large external pressures with only a slight change in density. For example, the imposition of a pressure of 200 atmospheres on liquid water raises the density by 1 percent. However, with a gas, significant density changes can occur as the result of an externally applied pressure. This fact can be seen by looking at the equation of state for a gas; for example, for a perfect gas, $p = \rho RT$, indicating the dependence of density on pressure. For certain cases of low-velocity gas flows, the pressure changes are small enough so that the flow can be approximated as incompressible. However, in the more general case, density variation must be taken into account; the effect of compressibility may completely alter the nature of the flow.

Naturally, by the introduction of density as a flow variable, we increase the complexity of the flow equations. An equation of state for the fluid, expressing p as a function of ρ and T, must be introduced. Furthermore, this equation of state brings in temperature as another variable. For compressible flows, then, we shall have four flow variables—velocity, pressure, temperature, and density—and we shall have to deal with four equations—continuity, momentum, energy, and an equation of state.

348

11.2 WAVE PROPAGATION THROUGH COMPRESSIBLE MEDIA

The difference between compressible and incompressible flows can be seen by the following example. A fluid is contained in a long cylinder to the right of a piston (Figure 11.1). The piston is suddenly given an increment of velocity dV toward the right. If the fluid were to be incompressible, the entire body of fluid must move instantaneously to the right with velocity dV, since no piling up of the fluid and resultant density changes can occur. In other words, the motion of the piston is felt instantaneously throughout the incompressible fluid.

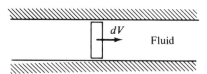

Figure 11.1

In the real case, however, a fluid is compressible. At the instant the piston is given the velocity dV, the fluid piles up next to the piston face (since the piston is impervious to the fluid), causing a slight density increase in the fluid near the piston. Farther down the tube, however, the fluid is at rest and does not immediately move with the piston. As time passes, the layer of compressed fluid at the piston face compresses, in turn, the layer next to it and so on; thus the disturbance caused by the motion of the piston is propagated through the fluid at finite velocity. In a study of compressible flow, it is important to develop an expression for the velocity of propagation of an infinitesimal disturbance; this velocity is called the sonic velocity. Returning to the piston-cylinder arrangement just described, let the sonic wave velocity be a, as shown in Figure 11.2. Use a coordinate system moving with the wave, as shown in Figure 11.3; in other words, take all velocities with respect to the constant-velocity wave.

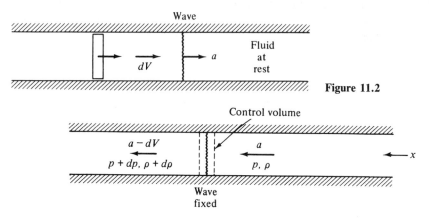

Figure 11.3. Coordinate system fixed to wave.

If we let p and ρ be the pressure and density in the undisturbed fluid ahead of the wave, then the pressure and density behind the wave and next to the piston face are, respectively, $p + dp$ and $\rho + d\rho$. Select a control volume as shown in Figure 11.3; the wave can be assumed thin enough so that there is no change in the flow cross-sectional area. The continuity equation for this one-dimensional steady flow yields

$$\rho a = (\rho + d\rho)(a - dV)$$

Eliminating second-order terms, we find that $a\, d\rho = \rho\, dV$. The momentum equation for steady flow in the x direction [Equation (4.9)] is

$$\Sigma\, F_x = \dot{m}(V_2 - V_1)$$

which reduces to

$$\Sigma\, F_x = -\rho A a\, dV$$

Since the only forces acting on the control volume in the x direction are pressure forces, the above can be written as

$$pA - (p + dp)A = -\rho A a\, dV$$

Canceling and simplifying, we obtain

$$dp - \rho a\, dV = 0$$

Combining with the continuity equation,

$$dp - a^2\, d\rho = 0 \qquad \text{or} \qquad \frac{dp}{d\rho} = a^2 \qquad (11.1)$$

In order to evaluate this expression for sonic velocity, the manner in which pressure varies with density must be determined.

Uniform flow exists upstream and downstream of the wave, and so, with no temperature gradients at the control volume boundaries (temperature gradients are inside the control volume), there is no heat transfer into or out of the control volume; the sound wave is adiabatic. In addition, with only infinitesimal changes taking place across the wave, the departure of the fluid from thermodynamic equilibrium is negligible; this satisfies the definition of a reversible process. The definition of entropy is

$$ds = \left(\frac{dq}{T}\right)_{\text{reversible}}$$

Therefore a process that is adiabatic and reversible involves no change in entropy; it is isentropic. Equation (11.1) can be written as

$$a = \sqrt{\left(\frac{\partial p}{\partial \rho}\right)_s} \qquad (11.2)$$

It is to be remembered from thermodynamics that for a perfect gas with constant specific heats undergoing an isentropic process,

$$\frac{p}{\rho^\gamma} = \text{constant}$$

where

$$\gamma = \frac{c_p}{c_v}$$

Evaluating the partial derivative in Equation (11.2), we obtain

$$\left(\frac{\partial p}{\partial \rho}\right)_s = \frac{\gamma p}{\rho} = \gamma RT$$

Therefore, for a perfect gas, the sound velocity can be written as

$$a = \sqrt{\gamma RT} \tag{11.3}$$

For air at 0°C (273 K),

$$a = \sqrt{(1.4)(0.2870 \text{ kJ/kg} \cdot \text{K})(273 \text{ K})}$$
$$= \sqrt{(1.4)(287.0 \text{ N} \cdot \text{m/kg} \cdot \text{K})(273 \text{ K})}$$
$$= 331.2 \text{ m/s}$$

In the English system, at 32°F (492°R),

$$a = \sqrt{(1.4)(53.35 \text{ ft-lbf/lbm°R})(32.2 \text{ lbm/slug})(1.0 \text{ slug-ft/s}^2\text{-lbf})(492°R)}$$
$$= 1087.8 \text{ ft/s}$$

A solid is much less compressible than a gas. For copper, the bulk modulus β_s defined as

$$\beta_s = \rho\left(\frac{\partial p}{\partial \rho}\right)_s$$

is equal to 123 GPa (123,000 MPa) with the density of copper 8880 kg/m³. The velocity of sound in copper is

$$a = \sqrt{\frac{\beta_s}{\rho}} = \sqrt{\frac{123 \times 10^9 \text{ kg} \cdot \text{m/s}^2 \cdot \text{m}^2}{8880 \text{ kg/m}^3}}$$
$$= 3722 \text{ m/s}$$

For an incompressible fluid, the effect of a small disturbance is propagated instantaneously throughout the entire body of fluid. In other words, the velocity of sound in an incompressible fluid is infinite. [This can be seen from Equation (11.1) with $d\rho = 0$.] It follows that the more compressible the substance, the lower the sonic velocity, as was demonstrated by our sample calculation of the sound velocity in copper and air.

The fact that small disturbances are propagated at the velocity of sound indicates the importance of sonic velocity in the motion of bodies through air. An object traveling through air acts somewhat like the piston of Figure 11.1, sending out pressure waves as it compresses the air in front of it. If the body itself is traveling at subsonic velocity, these pressure waves move ahead of the body, in a sense signaling the air that the body is present. In this way, the air ahead of the body is able to sense its presence and adjust to the body before it gets there. The resultant flow over the body is smooth,

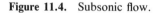

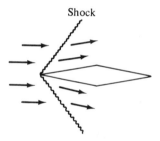

Shock

Figure 11.4.   Subsonic flow.   **Figure 11.5.**   Supersonic flow.

with gradually changing streamlines (e.g., Figure 11.4). Note that the fluid in front of the body has been displaced and has started to adjust to the presence of the body before reaching it. A body traveling at supersonic velocity is moving faster than the pressure or signal waves. In this case, the air ahead of the object cannot sense the presence of the object; rather, the flow must adjust very suddenly to the presence of the body. The flow still has to turn, due to the body, but the turning in a supersonic flow is brought about by a shock wave, a very abrupt change in flow properties, shown in Figure 11.5 (the shock is only a small fraction of an inch in thickness).

Figure 11.6 depicts supersonic flow over a sphere. Observe the curved shock caused by the presence of the sphere in the supersonic stream. The

Figure 11.6.   Supersonic flow of air at 1220 m/s over a 1.3-cm-diameter sphere. (*Courtesy* AVCO Corp.)

flow region ahead of the shock is completely unaffected by the presence of the ball. Note also the wake region downstream of the ball. When an airplane is traveling supersonically in the vicinity of the ground, the sudden pressure change caused by the shock manifests itself very dramatically in the sonic boom, heard and felt by an observer on the ground (see Figure 11.7).

Thus, in compressible flow, we have two different regimes, characterized by different flow phenomena. A criterion of the two regimes is *Mach number M*, the ratio of flow velocity to sound velocity:

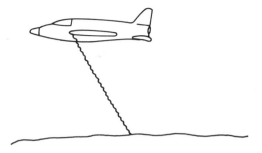

Figure 11.7. Sonic boom.

$$M = \frac{V}{a} \tag{11.4}$$

For $M < 1$, we have subsonic flow, characterized by smooth, gradual adjustments of the flow to imposed conditions; for $M > 1$, we have supersonic flow, characterized by shocks and sudden, rapid adjustments of the flow.

11.3 ISENTROPIC FLOW THROUGH A CHANNEL OF VARYING AREA

In this section we shall treat compressible flow inside a channel of varying area, as might occur in a rocket nozzle, jet engine intake, or turbine blade passage. To simplify the equations of fluid motion, one-dimensional flow will be assumed in the channel, with properties varying only in the direction of flow. From the discussion in Section 4.1, the student should recognize the nature of this approximation; only gradual area changes can be allowed, with solutions yielding mean flow properties at a given cross section. In addition, we shall take the flow to be isentropic, that is, adiabatic and reversible, the latter assumption indicating the flow to be frictionless. It is realized that no internal flow is truly frictionless; however, in many cases friction has a much smaller effect on flow properties than, for example, area change. Furthermore, we shall deal only with gas flows, so that gravitational forces can be neglected. In this section, then, we shall primarily be concerned with demonstrating the effect on a compressible flow of a change in channel cross-sectional area.

For steady one-dimensional flow through a varying-area channel, select the control volume shown in Figure 11.8. The continuity equation for steady flow [Equation (4.3)] yields

$$\int\limits_{\substack{\text{control} \\ \text{surface}}} \rho V_n \, dA = 0$$

or

$$-\rho A V + (\rho + d\rho)(A + dA)(V + dV) = 0$$

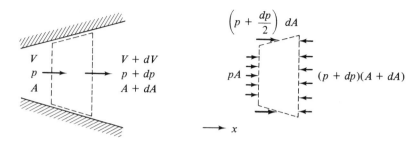

Figure 11.8

Combining and dropping second-order terms, we obtain

$$V\rho \, dA + VA \, d\rho + \rho A \, dV = 0$$

or

$$\frac{d\rho}{\rho} + \frac{dA}{A} + \frac{dV}{V} = 0 \qquad (11.5)$$

The momentum equation for steady one-dimensional flow in the x direction is given by Equation (4.9):

$$\Sigma F_x = \int\limits_{\substack{\text{control} \\ \text{surface}}} V_x \rho V_n \, dA$$

For our case, with friction and gravity neglected, pressure forces are the only external forces acting on the control surface. As shown in Figure 11.8, let $p + dp/2$, the average of the pressures acting on opposite sides, be the pressure acting on the slanted side walls of the control surface. Summing forces, we obtain

$$\Sigma F_x = pA + \left(p + \frac{dp}{2} \right) dA - (p + dp)(A + dA)$$

Dropping second-order terms and simplifying,

$$\Sigma F_x = -A \, dp$$

The momentum flux for this steady flow, with $\rho A V$ constant, is given by

$$\int\limits_{\substack{\text{control} \\ \text{surface}}} V_x \rho V_n \, dA = \rho A V \, dV$$

Equating the summation of forces to momentum flux, we have

$$dp + \rho V \, dV = 0 \qquad (11.6)$$

Combining the continuity equation (11.5) and the momentum equation (11.6), we obtain

$$dp + \rho V^2 \left(-\frac{d\rho}{\rho} - \frac{dA}{A} \right) = 0$$

In Section 11.2 we found that $a^2 = (\partial p / \partial \rho)_s$. Therefore, for this isentropic flow,

$$dp + \rho V^2 \left(-\frac{dp}{\rho a^2} - \frac{dA}{A} \right) = 0$$

or, since the Mach number $M = V/a$,

$$dp(1 - M^2) = \rho V^2 \frac{dA}{A} \tag{11.7}$$

Equation (11.7) shows the effect of Mach number on compressible flow through a channel of varying area. For subsonic low ($M < 1$), $1 - M^2$ is positive, so we find from Equation (11.7) that an increase in area brings about an increase in pressure and, from Equation (11.6), a decrease in velocity. Also, for subsonic flow, a decrease in area brings about a decrease in pressure and an increase in velocity (see Figure 11.9).

For supersonic flow, however, the opposite variation occurs. With $1 - M^2$ negative, an increase in area brings about an increase in velocity, with a decrease in area causing a decrease in velocity.

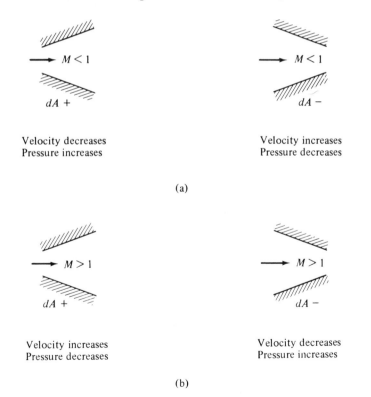

$M < 1$

$dA +$

Velocity decreases
Pressure increases

$M < 1$

$dA -$

Velocity increases
Pressure decreases

(a)

$M > 1$

$dA +$

Velocity increases
Pressure decreases

$M > 1$

$dA -$

Velocity decreases
Pressure increases

(b)

Figure 11.9.   Effect of Mach number on flow through channel of varying area: (a) subsonic flow; (b) supersonic flow.

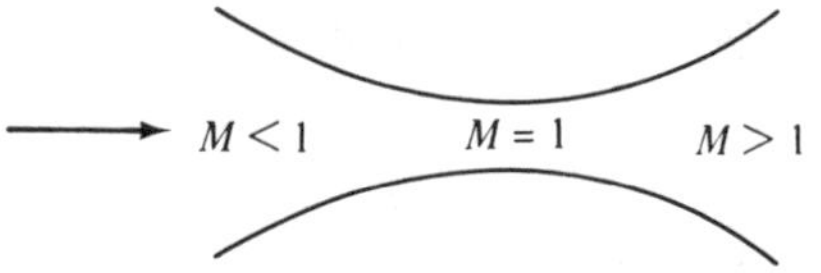

Figure 11.10.   Converging-diverging nozzle.

It follows, then, that a subsonic flow cannot be accelerated to velocities greater than the velocity of sound in a converging nozzle, irrespective of the pressure difference imposed on the nozzle. Acceleration to supersonic velocities can be achieved only in a converging-diverging nozzle (Figure 11.10).

We wish next to develop the necessary equations to determine quantitatively the variation of flow properties with area for isentropic flow. Two properties that will prove extremely useful for treating a compressible flow are stagnation pressure and stagnation temperature. Stagnation or total temperature T_t is defined as the temperature attained by bringing a flow *adiabatically* to rest at a point in a steady flow process. This is not to be confused with static temperature T, the temperature of a flowing stream as measured by an observer or instrument moving at the local stream velocity. In the absence of potential energy changes, heat transfer or work, the energy equation (4.17) becomes

$$h + \frac{V^2}{2} = \text{constant}$$

At the stagnation state, V is equal to zero. Moreover, assuming we are dealing with a perfect gas with constant specific heats (c_p, c_v), so that $\Delta h = c_p \, \Delta T$, we obtain

$$c_p T_t = c_p T + \frac{V^2}{2}$$

or

$$T_t = T + \frac{V^2}{2c_p} = T\left(1 + \frac{V^2}{2c_p T}\right)$$

But, since $c_p = R\gamma/(\gamma - 1)$ and $a^2 = \gamma RT$, there results

$$T_t = T\left(1 + \frac{\gamma - 1}{2}M^2\right) \tag{11.8}$$

Equation (11.8) is plotted in Appendix B as T/T_t versus M, for $\gamma = 1.3$, 1.4, and $\frac{5}{3}$. To demonstrate the use of stagnation temperature, consider an adiabatic flow in a channel (Figure 11.11). Let us write the energy equation for the control volume contained between cross sections 1 and 2:

$$q = h_2 + \frac{V_2^2}{2} - h_1 - \frac{V_1^2}{2}$$

$$= c_p T_2 + \frac{V_2^2}{2} - c_p T_1 - \frac{V_1^2}{2}$$

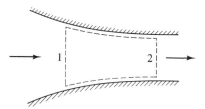

Figure 11.11

For adiabatic flow, $q = 0$, we have the result $T_{t_1} = T_{t_2}$; in other words, stagnation temperature is constant for adiabatic flow and can be used as a reference property. For example, suppose that a flow of air ($\gamma = 1.4$) is expanded adiabatically from a large reservoir at 500 K through a nozzle, as shown in Figure 11.12. For adiabatic flow, the stagnation temperature at any cross section of the nozzle is equal to 500 K. As the flow expands, the static temperature decreases according to Equation (11.8). For example, if the flow is expanded to $M = 0.8$, the static temperature at this point is 0.8865(500 K) $= 443$ K (Appendix B, Table 1).

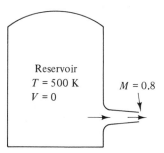

Figure 11.12

Stagnation or total pressure p_t of a flowing stream is defined as the pressure attained at a point by bringing the flow *isentropically* to rest in a steady flow process. For a perfect gas with constant specific heats undergoing an isentropic process, it is to be remembered from thermodynamics that p/ρ^γ = constant; since $p = \rho RT$,

$$\frac{p_2}{p_1} = \left(\frac{T_2}{T_1}\right)^{\gamma/(\gamma-1)}$$

With state 2 the stagnation state, we have

$$\frac{p_t}{p} = \left(\frac{T_t}{T}\right)^{\gamma/(\gamma-1)}$$

or

$$\frac{p_t}{p} = \left(1 + \frac{\gamma - 1}{2}M^2\right)^{\gamma/(\gamma-1)} \tag{11.9}$$

Equation (11.9) is tabulated in Appendix B as p/p_t, versus M for $\gamma = 1.3$, 1.4, and $\frac{5}{3}$. For isentropic flow, the stagnation pressure remains constant.

If the flow through the nozzle of Figure 11.12 were isentropic, then the stagnation pressure at any cross section of the flow would be the same as the pressure in the reservoir; the static pressure would decrease according to Equation (11.9). For example, if the pressure in the reservoir were 500 kPa and the flow expanded isentropically to $M = 0.8$, the static pressure at that point would be $0.6560(500) = 328$ kPa (see Appendix B, Table 1).

Note that for M^2 small, Equation (11.9) can be expanded in a series, yielding $p_t = p[1 + (\gamma/2)M^2 + \cdots]$, where $(\gamma p/2)M^2 = \frac{1}{2}\rho V^2$. In other words, as $M \to 0$, $p_t \to p + \frac{1}{2}\rho V^2$, the incompressible expression derived in Chapter 4; for $M^2 \to 0$, the effects of compressibility become negligible.

The constancy of stagnation pressure and stagnation temperature for isentropic flow is useful in developing an expression for mass flow through a channel in terms of Mach number. From the continuity equation for steady flow, $\dot{m} = \rho A V$. For a perfect gas with constant specific heats, this reduces to

$$\dot{m} = \frac{p}{RT}AM\sqrt{\gamma RT}$$

Substituting expressions for p_t and T_t, we obtain

$$\dot{m} = \frac{p_t}{\sqrt{RT_t}}A\sqrt{\gamma}\,M\left(1 + \frac{\gamma - 1}{2}M^2\right)^{(1/2)-(\gamma/\gamma - 1)}$$

Since $\dot{m}$, p_t, and T_t are constants, we now have a relationship between cross-sectional area and Mach number for steady isentropic flow of a perfect gas. Selecting the area at which $M = 1$ as a reference area A^*, we have

$$\frac{A}{A^*} = \frac{1}{M}\left(\frac{\dfrac{\gamma + 1}{2}}{1 + \dfrac{\gamma - 1}{2}M^2}\right)^{(1/2)-(\gamma/\gamma - 1)} \tag{11.10}$$

Again, this result is tabulated in Appendix B for $\gamma = 1.3$, 1.4, and $\frac{5}{3}$.

EXAMPLE 11.1
An airflow ($\gamma = 1.4$) is expanded isentropically in a nozzle from $M_1 = 0.3$, $A_1 = 1000$ cm^2, to a Mach number M_2 of 3.0. Determine (1) the minimum nozzle area, (2) A_2, (3) p_2/p_1, and (4) T_2/T_1 (see Figure 11.13).

Figure 11.13

Solution

1. Since flow in the nozzle goes from subsonic to supersonic speeds, the flow must pass through a minimum area A^* at which $M = 1$. From Appendix B, at $M_1 = 0.3$, $A_1/A^* = 2.0351$ so that the minimum area $= 1/2.0351 = \underline{491 \text{ cm}^2}$.

2. At $M_2 = 3.0$, $A_2/A^* = 4.235$ and so $A_2 = \underline{2079 \text{ cm}^2}$.

3. For this isentropic flow, T_t and p_t are constants.

$$\frac{p_2}{p_1} = \frac{p_2/p_{t2}}{p_1/p_{t1}} = \frac{0.0272}{0.9395} = \underline{0.0290} \qquad \left(\text{see Appendix B, Table 1 for } \frac{p}{p_t} \right)$$

4. $$\frac{T_2}{T_1} = \frac{T_2/T_{t2}}{T_1/T_{t1}} = \frac{0.3571}{0.9823} = \underline{0.3635} \qquad \left(\text{see Appendix B, Table 1 for } \frac{T}{T_t} \right)$$

■

Notice from the above that two solutions are possible, for each A/A^*, one subsonic and the other supersonic. This follows from our previous discussion in that a converging-diverging nozzle is required to accelerate a stagnant flow to supersonic velocities.

11.4 COMPRESSIBLE NOZZLE FLOW, CHOKING

Let us first consider flow through a simple converging nozzle supplied from a large reservoir, in which conditions are maintained constant and velocity is negligible. It is desired to investigate the flow through the nozzle as a function of back pressure p_b, the pressure outside the nozzle. Certainly, if the back pressure is equal to the reservoir pressure p^r, there is no flow (curve a of Figure 11.14). As p_b is reduced below p_r, flow is initiated through the nozzle, the flow increasing as p_b is reduced (curves b and c of Figure 11.14). For this subsonic flow, pressure in the nozzle decreases with x. If the nozzle back pressure is allowed to decrease further, we eventually reach the point at which sonic velocity is achieved at the exit plane of the nozzle (curve d). We know that velocities greater than the velocity of sound cannot be achieved in the converging nozzle. Let us discuss in physical terms what happens when p_b is reduced below the value necessary to achieve sonic velocity.

We have shown that the presence of a small disturbance in compressible flow is propagated by a wave traveling at the velocity of sound relative to the flow into which it moves. A change in back pressure can be considered such a small disturbance; the wave must reach the reservoir in order for there to be an adjustment of flow with change in back pressure. When the flow velocity reaches sonic velocity at the exit plane, however, with the wave itself traveling at sonic velocity with respect to the flow, the absolute wave velocity is zero, as shown in Figure 11.15. With the wave thus unable

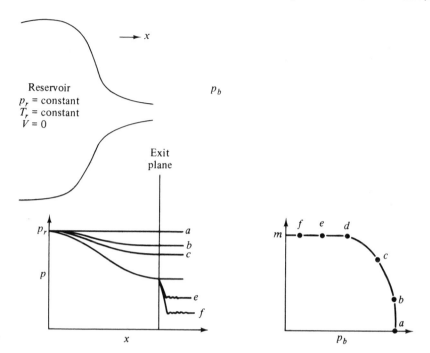

Figure 11.14.   Flow through converging nozzle.

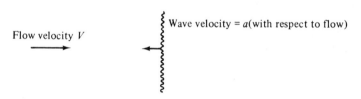

Figure 11.15.   Absolute wave velocity (from right to left) = $a - V$. If $V = a$, absolute wave velocity $= 0$.

to move past the exit plane, the reservoir cannot sense any further decrease in back pressure. For all back pressures below that of curve d in Figure 11.14, the reservoir continues to send out the same flow as that for curve d, with the pressure distribution in the nozzle unchanged (see curves e and f of Figure 11.14). For all back pressures below d, the nozzle is choked, with the pressure ratio p_b/p_r corresponding to d called the *critical pressure ratio*. To compute the critical pressure ratio, we can use Equation (11.9) with $M = 1$:

$$\left(\frac{p_b}{p_r}\right)_{\text{crit}} = \left[\frac{1}{1 + (\gamma - 1)/2}\right]^{\gamma/(\gamma - 1)}$$

or

$$\left(\frac{p_b}{p_r}\right)_{\text{crit}} = \left(\frac{2}{\gamma + 1}\right)^{\gamma/(\gamma - 1)} \tag{11.11}$$

For example,

$$\text{for } \gamma = 1.4, \quad \left(\frac{p_b}{p_r}\right)_{\text{crit}} = 0.5283$$

$$\text{for } \gamma = 1.3, \quad \left(\frac{p_b}{p_r}\right)_{\text{crit}} = 0.5457$$

$$\text{for } \gamma = \frac{5}{3}, \quad \left(\frac{p_b}{p_r}\right)_{\text{crit}} = 0.4871$$

It follows from our discussion that for all back pressures below that of curve d, the pressure at the nozzle exit plane remains the same as that of d. In other words, the flow in the nozzle cannot adjust to the back pressure; the exit plane pressure is greater than the back pressure. The flow pressure eventually does decrease to the back pressure outside the nozzle by means of expansion waves. The analysis of expansion waves, two- or three-dimensional in nature, is beyond the scope of the present study. However, the student should recognize that, over a range of back pressures, there will be a difference between nozzle exit plane pressure and back pressure.

EXAMPLE 11.2
A converging nozzle of exit area 50 cm^2 is supplied from a reservoir in which the presure is 200 kPa, temperature 300 K (see Figure 11.16). Calculate the mass flow rate of air through the nozzle for back pressures of 0, 100, and 150 kPa.

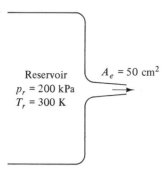

Reservoir
$p_r = 200$ kPa
$T_r = 300$ K

$A_e = 50$ cm^2

Figure 11.16

Solution   The nozzle is choked for all back pressures below 0.5283(200) or 105.7 kPa. With a choked nozzle, the velocity at the exit plane is equal to the velocity of sound, or $M_e = 1$. For a choked nozzle, $p_e = 0.5283(200)$ = 105.7 kPa. From Appendix B, for $\gamma = 1.4$, $T_e/T_t = 0.8333$ at $M_e = 1$, so that $T_e = 0.8333(300) = 250$K. Over the range $0 \le p_b \le 105.7$ kPa, we have

$$\dot{m} = \frac{p_e}{RT_e} A_e V_e = \frac{p_e}{RT_e} A_e \sqrt{\gamma R T_e}$$

$$= \frac{(105.7 \text{ kN/m}^2)(50 \times 10^{-4} \text{ m}^2)\sqrt{1.4(287 \text{ N} \cdot \text{m/kg} \cdot \text{K})250 \text{ K}}}{(0.2870 \text{ kN} \cdot \text{m/kg} \cdot \text{K})250 \text{ K}}$$

$$= 2.335 \text{ kg/s}$$

With a back pressure of 150 kPa, the nozzle is not choked and flow at the exit plane is subsonic; therefore, the exit plane pressure is equal to the back pressure. From Appendix B, $\gamma = 1.4$, at $p_e/p_t = 150/200 = 0.75$, we obtain

$$M_e = 0.654 \quad \text{and} \quad \frac{T_e}{T_t} = 0.9212 \quad \text{or} \quad T_e = 276.4 \text{ K}$$

with

$$V_e = M_e \sqrt{\gamma R T} = 0.654\sqrt{1.4(287 \text{ N} \cdot \text{m/kg} \cdot \text{K})276.4 \text{ K}}$$

With $p_b = 150$ kPa,

$$\dot{m} = \left(\frac{150 \times 50 \times 10^{-4}}{0.2870 \times 250}\right)[0.654\sqrt{1.4(287)276.4}]$$

$$= \underline{2.278 \text{ kg/s}} \qquad \blacksquare$$

EXAMPLE 11.3

A rocket nozzle has a converging exhaust nozzle of exit area 20 in². (See Figure 11.17.) If $p_c = 100$ psia and $T_c = 2500°$R, find the rocket thrust for $p_b = 20$ psia and $p_b = 0$ psia. Assume the rocket exhaust gases to have $\gamma = 1.4$ with a mean molecular weight of 20. Take the flow in the nozzle to be isentropic.

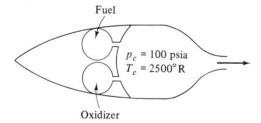

Fuel

$p_c = 100$ psia
$T_c = 2500°$R

Oxidizer   **Figure 11.17**

Solution   Select a control volume as shown in Figure 11.18. The only forces acting on the fluid inside the control volume are pressure forces and the thrust force (see Figure 11.19). From the momentum equation

$$\Sigma F_x = \int_{\substack{\text{control} \\ \text{surface}}} V_x(\rho V_n dA)$$

$$\Sigma F_x = T - (p_e - p_b)A_e$$

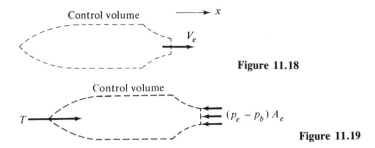

Figure 11.18

Figure 11.19

For both values of p_b,

$$\frac{p_b}{p_c} < \left(\frac{p_b}{p_c}\right)_{crit}$$

so the nozzle is choked and

$$M_e = 1 \quad \text{or} \quad V_e = \sqrt{\gamma R T_e}$$

$$R = \frac{\overline{R}}{\text{Molecular mass}} = \frac{1545.33 \text{ ft-lbf}/\text{lbm-mol}°R}{20 \text{ lbm}/\text{lbm-mol}}$$

$$= 77.27 \text{ ft-lbf}/\text{lbm}°R$$

$$T_e = \frac{T_e}{T_t} T_t$$

$$T_e = 0.833(2500°R)$$

$$= 2083°R$$

and

$$V_e = \sqrt{(1.4)(77.27 \text{ ft-lbf}/\text{lbm}°R)(32.2 \text{ lbm}/\text{slug})(2083°R)}$$

$$= 2694 \text{ ft/s} \quad \text{for} \quad p_b = 0 \text{ and } 20 \text{ psi}$$

$$p_e = 0.5283(100 \text{ psi}) = 52.83 \text{ psi for } M_e = 1$$

$$\dot{m} = \rho_e A_e V_e$$

$$= \frac{p_e}{R T_e} A_e V_e$$

$$= \frac{(52.83 \text{ lbf}/\text{in}^2)(20 \text{ in}^2)(2694 \text{ ft/s})}{(77.27 \text{ ft-lbf}/\text{lbm}°R)(2083°R)}$$

$$= 17.69 \text{ lbm/s} \quad \text{for} \quad p_b = 0 \quad \text{and} \quad 20 \text{ psi}$$

$$T = \dot{m} V_e + (p_e - p_b) A_e$$

$$\dot{m} V_e = (17.69 \text{ lbm/s})(2694 \text{ ft/s})\left(\frac{1}{32.2} \text{ slugs/lbm}\right)$$

$$= 1480 \text{ lbf}$$

For $p_b = 0$ psi:

$$T = 1480 \text{ lbf} + [(52.83 - 0) \text{ lbf}/\text{in}^2](20 \text{ in}^2)$$
$$= 1480 \text{ lbf} + 1057 \text{ lbf}$$
$$= \underline{2537 \text{ lbf}}$$

For $p_b = 20$ psi:

$$T = 1480 \text{ lbf} + [(52.83 - 20)\text{lbf}/\text{in}^2](20 \text{ in}^2)$$
$$= 1480 \text{ lbf} + 657 \text{ lbf}$$
$$= \underline{2137 \text{ lbf}}$$

The thrust calculated represents the force of the rocket on the fluid; the force of fluid on rocket is equal in magnitude and opposite in direction.   ∎

Let us now replace the converging nozzle of Figure 11.14 with a converging-diverging nozzle and investigate the effect on the resultant flow of decreasing the back pressure (Figure 11.20). Again, conditions in the reservoir will be maintained constant. With the back pressure equal to the reservoir pressure, there will be no flow, and we obtain curve a of Figure 11.20. As the back pressure is decreased slightly below the reservoir pressure, flow is induced through the nozzle, with subsonic flow in both converging and diverging sections. According to Figure 11.9, pressure decreases in the converging section up to the throat, then increases in the diverging section, as shown by curve b of Figure 11.20. Further decreases in back pressure induce more and more flow through the nozzle; eventually sonic flow is achieved at the throat, with subsonic flow again in the diverging section (see curve c). Additional decreases of back pressure cannot be sensed by the fluid in the reservoir; the wave cannot move past the throat. Therefore, for all back pressures below that of curve c, the mass flow rate through the nozzle remains equal to that of c; the nozzle is choked. Equations developed in the previous sections have shown that two isentropic solutions exist for a given area ratio A/A^*, one subsonic and the other supersonic. Therefore, if the back pressure is lowered to that of curve d, supersonic flow will exist at the nozzle exit plane. For back pressures below that of curve d and between those of c and d, one-dimensional isentropic solutions are not possible. Some of these nonisentropic solutions will be discussed in the next section, which deals with shock waves.

EXAMPLE 11.4

A converging-diverging nozzle is supplied from a large, constant-pressure air reservoir, as shown in Figure 11.20. The exit area of the nozzle is 100 cm², throat area 50 cm². The reservoir pressure and temperature are, respectively, 400 kPa and 100°C. (1) Find the maximum back pressure at which the nozzle is choked. (2) Find the mass flow rate for back pressures of 0, 200, and 300 kPa. (3) Determine the back pressure and exit Mach number for perfectly

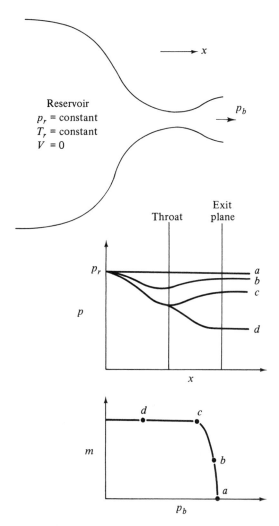

Figure 11.20. Flow through converging-diverging nozzle.

expanded supersonic flow in the nozzle, corresponding to curve d of Figure 11.20.

Solution

1. The maximum back pressure to choke the nozzle is that of curve c, for which $M = 1$ at the nozzle throat followed by subsonic flow in the diverging section. From Appendix B, with $A/A^* = 2.0$, we find a subsonic solution $M = 0.31$ and $p/p_t = 0.936$. Therefore, this nozzle is choked for all back pressures less than $0.936(400) = \underline{374.4 \text{ kPa}}$.

2. The mass flow rate through the nozzle will be the same for back pressures of 0, 200, and 300 kPa. Evaluating the mass flow at the throat for a throat Mach number of 1, we find

$$\dot{m} = \rho A V$$

$$= \frac{p}{RT} A \sqrt{\gamma RT} \qquad \text{where } p = 0.528p_r \quad \text{and} \quad T = 0.833T_r$$

$$= \frac{(0.528 \times 400 \text{ kPa})(50 \times 10^{-4} \text{ m}^2)}{(0.2870 \text{ kN} \cdot \text{m/kg} \cdot \text{K})(0.833 \times 373\text{K})} \sqrt{(1.4)(287.0)(0.833 \times 373 \text{ m/s}}$$

$$= 4.184 \text{ kg/s}$$

3. From Appendix B, at $A/A^* = 2.0$, we find the supersonic solution $M = 2.20$ and $p/p_t = 0.0935$. The back pressure for perfectly expanded supersonic exit flow is $0.0935(400) = \underline{37.4 \text{ kPa}}$. ■

EXAMPLE 11.5

A converging-diverging nozzle is designed to produce an exit Mach number of 1.8 with perfectly expanded isentropic flow. Under a certain off-design operating condition, the nozzle is supplied from a large air reservoir at 200 psia and 150 psia (see Figure 11.21). Is the nozzle choked? Repeat for back pressures of 180 psia, 100 psia, and 50 psia. For an inlet stagnation temperature of 500°R, determine the flow rates for each of the specified back pressures.

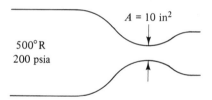

Figure 11.21

Solution   For $M = 1.8$, from Appendix B, $A/A^* = 1.439$. At this area ratio, for subsonic flow, $p/p_t = 0.868$. Thus, the nozzle is choked for all back pressures below $200(0.868$ psia$)$, or 174 psia; in other words, the nozzle is choked for back pressures of 100 psia and 50 psia.

$$\dot{m} = \rho A V$$

$$= \frac{p}{RT} A (M \sqrt{\gamma RT})$$

At the nozzle throat, $M = 1$ and $p = 0.528(200) = 105.6$ psia,

$$T = \left(\frac{T}{T_t}\right) T_t = 0.8333(500) = 416.7°\text{R}$$

Therefore,

$$\dot{m} = \frac{(105.6 \text{ lbf/in}^2)(10 \text{ in}^2)(1)}{(53.3 \text{ ft-lbf/lbm°R})(416.7°\text{R})}$$
$$\times \sqrt{1.4(53.3 \text{ ft-lbf/lbm°R})(32.2 \text{ lbm/slug})(416.7°\text{R})}$$
$$= \underline{47.6 \text{ lbm/s}} \qquad \text{for} \qquad \text{back pressures of 100 psia}$$
$$\text{and 50 psia}$$

For a back pressure of 180 psia, there is subsonic flow at nozzle throat and exit. To find the exit Mach number, from Appendix B, at $p/p_t = 180/200 = 0.9$, $M = 0.391$ and $T/T_t = 0.9703$. The exit area $= 10(1.439) = 14.39$ in^2.

$$\dot{m} = \frac{p}{RT}AM\sqrt{\gamma RT}$$

where we will evaluate $\dot{m}$ at the nozzle exit:

$$\dot{m} = \frac{(180 \text{ lbf}/\text{in}^2)(14.39 \text{ in}^2)(0.391)}{(53.3 \text{ ft-lbf}/\text{lbm}^\circ R)(0.9703 \times 500^\circ R)}$$
$$\times \sqrt{1.4(53.3 \text{ ft-lbf}/\text{lbm}^\circ R)(32.2 \text{ lbm}/\text{slug})(0.9703 \times 500^\circ R)}$$

$$= \underline{42.3 \text{ lbm/s}} \text{ for a back pressure of 180 psia.} \qquad ■$$

11.5 NORMAL SHOCK WAVES

In Section 11.1 the shock was introduced as a very sudden change of fluid properties occurring in a supersonic flow. In this section we shall present the equations of motion for a normal shock wave, a wave normal to the flow direction, and derive expressions for the changes of properties that occur across such a wave.

Consider a steady one-dimensional flow in which a normal shock occurs, as shown in Figure 11.22. The shock itself is very thin, only a small fraction of an inch in thickness; the sudden, finite changes taking place inside the wave involve departures from thermodynamic equilibrium. The finite changes of, for example, temperature and velocity lead to large gradients and therefore heat conduction and viscous dissipation inside the wave, which render the shock process internally irreversible. Furthermore, we shall select our control volume to include the wave (Figure 11.23) with property gradients confined to the interior of the control volume. With no temperature gradients at the control volume boundaries, the shock process can be taken as adiabatic.

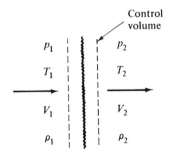

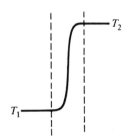

Figure 11.22. Normal shock wave.

Figure 11.23. Temperature gradient in shock.

From the second law of thermodynamics, entropy ($ds \overset{\text{irrev}}{\underset{\text{rev}}{\geq}} dq/T$)
must increase for an adiabatic and irreversible process; in other words,
$s_2 > s_1$. With this in mind, we can now proceed with the equations of motion
for a fixed shock.

The continuity equation for this steady flow is $\rho A_1 V_1 = \rho_2 A_2 V_2$ where,
for the very thin shock, $A_1 = A_2$. Therefore:

$$\rho_1 V_1 = \rho_2 V_2 \tag{11.12}$$

or, for a perfect gas with constant specific heats, this becomes

$$\frac{p_1}{RT_1} M_1 \sqrt{\gamma R T_1} = \frac{p_2}{RT_2} M_2 \sqrt{\gamma R T_2}$$

The only external forces acting on the fluid in the control volume are pressure
forces, so the momentum equation yields

$$\Sigma F_x = \int_{\substack{\text{control} \\ \text{surface}}} V_x (\rho V_n \, dA)$$

$$p_1 A_1 - p_2 A_2 = \rho_2 A_2 V_2^2 - \rho_1 A_1 V_1^2$$

where $A_1 = A_2$. Canceling, we find

$$p_1 + \rho_1 V_1^2 = p_2 + \rho_2 V_2^2 \tag{11.13}$$

or

$$p_1 \left(1 + \frac{V_1^2}{RT_1} \right) = p_2 \left(1 + \frac{V_2^2}{RT_2} \right)$$

For a perfect gas with constant specific heats, we can write $a^2 = \gamma RT$ to
obtain

$$p_1 (1 + \gamma M_1^2) = p_2 (1 + \gamma M_2^2) \tag{11.14}$$

For adiabatic flow, the stagnation temperature is constant, $T_{t1} = T_{t2}$.
For a perfect gas with constant specific heats, we find

$$T_1 \left(1 + \frac{\gamma - 1}{2} M_1^2 \right) = T_2 \left(1 + \frac{\gamma - 1}{2} M_2^2 \right) \tag{11.15}$$

Combining Equations (11.13), (11.14), and (11.15), we have an equation for
M_2 in terms of M_1:

$$\frac{M_1}{1 + \gamma M_1^2} \sqrt{1 + \frac{\gamma - 1}{2} M_1^2} = \frac{M_2}{1 + \gamma M_2^2} \sqrt{1 + \frac{\gamma - 1}{2} M_2^2} \tag{11.16}$$

A trivial solution of Equation (11.16) is $M_2 = M_1$, corresponding to isentropic
constant-area flow. We are interested here in the irreversible solution, with
changes taking place across the wave. Solving for M_2, we find

$$M_2^2 = \frac{M_1^2 + \dfrac{2}{\gamma - 1}}{\dfrac{2\gamma}{\gamma - 1}M_1^2 - 1} \tag{11.17}$$

This result is plotted in Figure 11.24 for $\gamma = 1.4$.

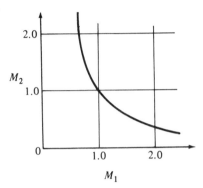

Figure 11.24

It can be seen that for $M_1 > 1$, M_2 will be <1 and, from Equation (11.14), we have a compression shock ($p_2 > p_1$). For $M_1 < 1$, M_2 will be >1 and we have an expansion shock. It is important now to determine the variation of entropy for these two cases. From thermodynamics, recall that

$$T\, ds = dh - \frac{dp}{p}$$

For a perfect gas with constant specific heats, this becomes

$$T\, ds = c_p\, dT - RT\frac{dp}{p}$$

or

$$ds = c_p\frac{dT}{T} - R\frac{dp}{p}$$

Integrating,

$$s_2 - s_1 = c_p \ln\frac{T_2}{T_1} - R \ln\frac{p_2}{p_1}$$

With M_2 already determined as a function of M_1 [Equation (11.17)] and, from Equations (11.14) and (11.15), p_2/p_1 and T_2/T_1 found as functions of M_1 and M_2, we can express p_2/p_1 and T_2/T_1, and hence $s_2 - s_1$, as a function of M_1 alone. The result is shown in Figure 11.25.

It can be seen that for $M_1 > 1$, $s_2 - s_1 > 0$, whereas for $M_1 < 1$, $s_2 - s_1 < 0$. In other words, the only solutions to the normal shock equations which do not violate the second law of thermodynamics are those for which

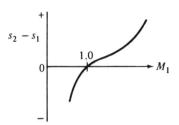

Figure 11.25

$M_1 > 1$. An expansion shock is impossible; for a shock to occur, there must be supersonic flow.

From the equations derived, we can determine

$$M_2, \frac{p_2}{p_1}, \frac{T_2}{T_1}, \frac{\rho_2}{\rho_1} = \frac{V_1}{V_2}, \quad \text{and} \quad \frac{p_{t2}}{p_{t1}}$$

as a function of M_1 alone, for a given γ. These results are tabulated in Appendix C, for $\gamma = 1.3$, 1.4, and $\frac{5}{3}$. The following examples will illustrate the use of these tables.

EXAMPLE 11.6

A normal shock occurs at the inlet or diffuser of a jet engine (Figure 11.26) with the engine traveling at a Mach number of 1.8 at an altitude where the ambient pressure is 30 kPa. Determine the Mach number, static pressure, and stagnation pressure of the flow leaving the diffuser. The area ratio of the diffuser is 3 to 1. Assume isentropic flow in the diffuser downstream of the shock.

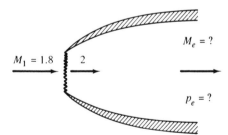

$M_1 = 1.8$

$M_e = ?$

$p_e = ?$

Figure 11.26

Solution First we shall find the flow properties just downstream of the shock, using Appendix C, with $\gamma = 1.4$. For $M_1 = 1.8$: $M_2 = 0.617$,

$$\frac{p_2}{p_1} = 3.613, \qquad \frac{p_{t2}}{p_{t1}} = 0.813$$

From Appendix B, at $M_1 = 1.8$, $p_1/p_{t1} = 0.174$. Therefore $p_{t1} = 172.4$ kPa. We find that $p_{t2} = 0.813(172.4) = 140.2$ kPa. Since the flow downstream of the shock is isentropic,

$$p_{te} = p_{t2} = \underline{140.2 \text{ kPa}}$$

For the isentropic flow from 2 to e, with A^*_{2e} the value of A^* for this flow, we can write

$$\frac{A_e}{A^*_{2e}} = \frac{A_e}{A_2}\frac{A_2}{A^*_{2e}}$$

where A/A^* at $M = 0.617$ is 1.169 (Appendix B). Therefore,

$$\frac{A_e}{A^*_{2e}} = (3)(1.169) = 3.507$$

and $M_e = \underline{0.17}$. For $M_e = 0.17$, $p_e/p_{te} = 0.980$; so

$$p_e = 0.98(140.2) = \underline{137.4 \text{ kPa}} \qquad \blacksquare$$

EXAMPLE 11.7
A normal shock moves through still air (14.7 psi, 60°F) with a steady velocity of 4000 fps (Figure 11.27a). Calculate the velocity of the air behind the wave and the static pressure and temperature in this air stream.

Solution   In order to treat the shock with the steady-state equations developed in this section, we shall take all velocities with respect to an observer "sitting" on the wave. In other words, impose a velocity of 4000 fps from right to left on the wave and air flow shown. (See Figure 11.27b.)

$$M_1 = \frac{4000 \text{ ft/s}}{\sqrt{1.4(53.3 \text{ ft-lbf}/\text{lbm}°R)(32.2 \text{ lbm/slug})(520°R)}}$$
$$= 3.58$$

Figure 11.27

From Appendix C,

$$\frac{p_2}{p_1} = 14.79, \quad \frac{T_2}{T_1} = 3.426, \quad \frac{\rho_2}{\rho_1} = \frac{V_1}{V_2} = 4.316$$

Therefore the pressure behind the wave is $14.79(14.7) = \underline{218 \text{ psi}}$, temperature behind the wave is $3.426(520) = \underline{1780°R}$, and $4000 - V = 4000/4.316 = 925$ fps, so that $V = 4000 - 925 = \underline{3075 \text{ fps}}$. $\qquad \blacksquare$

We can now return to the converging-diverging nozzle of Figure 11.20 and fill in some of the gaps. For a back pressure slightly below that of curve c, a normal shock appears just downstream of the nozzle throat, as shown

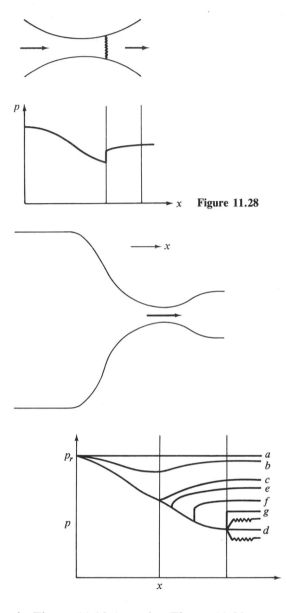

Figure 11.28

Figure 11.29. Converging-diverging nozzle with shocks.

in Figure 11.28 (see also Figure 11.29, curve *e*). A further decrease in back pressure causes the shock to move downstream (curve *f*) until eventually, for a low enough back pressure, it reaches the nozzle exit plane (curve *g*). Note that with a normal shock in the nozzle and subsonic flow at the exit plane, the pressure at the exit plane is equal to the back pressure. If the nozzle back pressure is decreased slightly below that of *g*, the shock moves outside the nozzle, in the form of oblique shock waves inclined to the flow direction (see Figure 11.30), with the exit plane pressure less than the back pressure. With oblique shocks at the exit, the nozzle is said to be *overexpanded;* in other words, the nozzle has expanded the flow to a pressure less than the back pressure. Oblique shocks will appear at the exit for back pressures between those of curves *g* and *d*.

For back pressures below that of *d*, the back pressure is less than the exit pressure and expansion waves appear at the nozzle exit; in this condition, the nozzle is said to be *underexpanded* (Figure 11.31). Again, an analysis of expansion waves and oblique shock waves, two- and three-dimensional in nature, is beyond the scope of this text. For better coverage, the student should refer to more advanced texts in this area.*

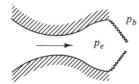

Figure 11.30.   Oblique shocks at nozzle exit for overexpanded nozzle.

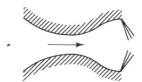

Figure 11.31.   Expansion waves at nozzle exit for underexpanded nozzle.

EXAMPLE 11.8

A converging-diverging nozzle is supplied from a large constant-pressure air reservoir, as shown in Figure 11.20 and described in Example 11.4. The exit area of the nozzle is 200 cm^2, throat area 100 cm^2. The reservoir pressure and temperature are, respectively, 300 kPa and 300 K. Determine the range of back pressures over which a normal shock will appear in the nozzle.

Solution   For a shock just downstream of the throat, the back pressure must be slightly less than that of curve *c*, Figure 11.29. For curve *c*, with $A/A^* = 2.0$, the subsonic solution from Appendix B, $\gamma = 1.4$, yields $M = 0.31$, $p/p_t = 0.936$. Therefore, $p_b = 280.8$ kPa.

For a shock at the nozzle exit plane, with isentropic flow in the nozzle up to the exit plane, the Mach number M_1 just before the shock can be found from Appendix B, from the supersonic solution with $A_1/A^* = 2.0$. (See Figure 11.32.) We find that $M_1 = 2.20$ and $p_1/p_r = 0.0935$, or $p_1 = 28.05$ kPa. From the Normal Shock Tables, Appendix C, for $M_1 = 2.20$, $p_2/p_1 =$

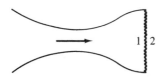

Figure 11.32

* See, for example, J. E. A. John, *Gas Dynamics, 2nd Edition* (Boston: Allyn and Bacon, 1984); J. D. Anderson, *Modern Compressible Flow* (New York: McGraw-Hill Book Company, 1982).

5.48, so that $p_2 = 5.48(28.05) = 153.7$ kPa. Therefore, a normal shock will appear in the nozzle for 153.7 kPa $< p_b <$ 280.8 kPa. Oblique shocks will appear at the exit for 28.05 kPa $< p_b <$ 153.7 kPa and expansion waves for $p_b <$ 28.05 kPa. ■

EXAMPLE 11.9
Let the rocket nozzle of Example 11.3 have a converging-diverging exhaust nozzle of exit area 50 in² and throat area 20 in². With $p_c = 100$ psia and $T_c = 2500°$R, find the rocket thrust for $p_b = 20$ psia and $p_b = 0$ psia. (See Figure 11.33.) Assume the rocket exhaust gases to have $\gamma = 1.4$ with a mean molecular weight of 20.

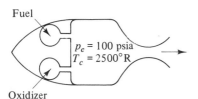

Fuel

$p_c = 100$ psia
$T_c = 2500°$R

Oxidizer   **Figure 11.33**

Solution   The nozzle is choked for both back pressures, so that $M_{throat} = 1$. Using the throat conditions,

$$\dot{m} = \frac{p}{RT}A\sqrt{\gamma RT} \qquad \text{with} \qquad T = 0.8333(2500) = 2083°\text{R}$$

$$p = 0.5283(100 \text{ lbf}/\text{in}^2) = 52.83 \text{ lbf}/\text{in}^2$$

and

$$R = \frac{1545.33 \text{ ft-lbf}/\text{lbm-mol}°\text{R}}{20 \text{ lbm}/\text{lbm-mol}} = 77.27 \text{ ft-lbf}/\text{lbm}°\text{R}$$

Solving, we find

$$\dot{m} = \frac{(52.83 \text{ lbf}/\text{in}^2)(20 \text{ in}^2)}{(77.27 \text{ ft-lbf}/\text{lbm}°\text{R})(2083°\text{R})}$$
$$\times \sqrt{1.4(77.27 \text{ ft-lbf}/\text{lbm}°\text{R})(32.2 \text{ lbm}/\text{slug})(2083°\text{R})}$$

$$= 17.69 \text{ lbm}/\text{s} \qquad \text{for} \qquad p_b = 0 \qquad \text{and} \qquad p_b = 20 \text{ psia}$$

For $A/A^* = 50/20 = 2.5$, the supersonic solution to the isentropic flow equations yields $M = 2.44$ and $p/p_t = 0.0643$ (see Appendix B). Therefore the back pressure corresponding to curve d of Figure 11.29 is 6.43 psia. In order to determine the flow pattern at the nozzle exit, we must next find the back pressure corresponding to curve g of Figure 11.29. For $M_1 = 2.44$, we find from the Normal Shock Tables that $p_2/p_1 = 6.78$, so that $p_2 = 6.78(6.43) = 43.5$ psia. Therefore for a back pressure of 0 psia, expansion waves appear at the nozzle exit; for a back pressure of 20 psia, oblique shock waves appear at the nozzle exit. For both back pressures, flow adjustment occurs outside the nozzle; flow inside the nozzle is isentropic with supersonic flow at the nozzle exit plane. For both cases, $M_e = 2.44$ and $p_e = 6.43$ psia. From the momentum equation,

$$T = \dot{m}V_e + (p_e - p_b)A_e$$

where $V_e = M_e \sqrt{\gamma R T_e}$.

For $M_e = 2.44$, $T_e/T_c = 0.457$, so that

$$T_e = 0.457(2500) = 1143°R$$

and

$$V_c = 2.44\sqrt{1.4(77.27 \text{ ft-lbf}/\text{lbm°R})(32.2 \text{ lbm/slug})(1143°R)} = 4869 \text{ ft/s}$$

For $p_b = 20$ psia,

$$T = \frac{(17.69 \text{ lbm/s})(4869 \text{ ft/s})}{(32.2 \text{ lbm/slug})} + [(6.43 - 20)\text{lbf}/\text{in}^2](50 \text{ in}^2)$$

$$= 2675 - 679$$

$$= \underline{1996 \text{ lbf}}$$

For $p_b = 0$ psia,

$$T = 2675 + [(6.43 - 0) \text{ lbf}/\text{in}^2](50 \text{ in}^2)$$

$$= \underline{2997 \text{ lbf}}$$ ■

PROBLEMS

11.1. Determine the velocity of sound in helium at one atmospheric pressure and 25°C. Repeat for nitrogen.

11.2. An aircraft is to fly at Mach 10 at an altitude of 100,000 ft. Determine the speed of the aircraft in mph.

11.3. Calculate the velocity of sound in superheated steam at 20 psia, 1000°F; assume the steam to behave as a perfect gas with $\gamma = 1.3$, mean molecular mass 18.

11.4. A projectile is moving at Mach 4.5 through helium. Determine the projectile speed in m/s for a helium temperature of 100 K.

11.5. An airflow at Mach 0.25 passes through a circular channel of inner diameter 10 cm. The static pressure of the flow is 100 kPa, the static temperature is 30°C. Calculate the mass flow through the channel.

11.6. An airflow at Mach 0.22 passes through a channel at cross sectional area 10 in². The stagnation temperature is 500°R. Calculate the mass flow rate through the channel.

11.7. The cross-sectional area of the flow in Problem 11.6 is reduced to 6 in². Assuming isentropic flow, calculate the Mach number at the reduced area. What percentage reduction in area would be required to reach Mach 1 in the channel (starting from 10 in²)?

11.8. A converging nozzle of exit area 12 cm² is supplied from a helium reservoir in which the pressure is 500 kPa, temperature 300 K. Calculate the mass flow rate of helium for back pressures of 0, 100, 300, and 400 kPa. Let $\gamma = \frac{5}{3}$.

11.9. Steam enters a converging passage with a velocity of 50 m/s, static pressure 3 MPa, static temperature 1500 K. In the passage, the cross-sectional flow area is reduced from 150 cm^2 to 40 cm^2. Determine the mass flow rate of the steam, the velocity at the passage exit, and the static pressure at the passage exit. Assume steady isentropic flow with the steam behaving as a perfect gas ($\gamma = 1.3$, $\overline{M} = 18$).

11.10. A converging nozzle of exit area 4 in^2 is supplied from a nitrogen reservoir in which the pressure is 60 psia, temperature 600°R. Calculate the mass flow rate of the nitrogen for back pressures of 0, 20, and 40 psia. If air were the working fluid, would the mass flow be greater or less?

11.11. A converging nozzle has an exit area of 100 cm^2. Air stored in a reservoir is to be discharged through the nozzle to an ambient pressure of 101 kPa. Calculate the flow rate through the nozzle for reservoir pressures of 200, 400, 600, 800, and 1000 kPa. Assume isentropic flow in the nozzle, with reservoir air temperature maintained constant at 300°C.

11.12. Air is stored in a high-pressure tank at 3 MPa and 25°C. A leak develops in an outlet valve on the tank, causing the pressure in the tank to decrease with time. If the volume of the tank is 5 m^3 and the leak can be represented by a nozzle of throat area 15 mm^2, estimate the time required for the pressure in the tank to drop to 500 kPa. Assume tank air temperature remains constant.

11.13. Consider isentropic flow of air in the varying-area channel shown in Figure P11.13. Determine the Mach number at A_2 and A_3. If the static pressure at 1 is 20 psia, determine the stagnation and static pressures at sections 2 and 3.

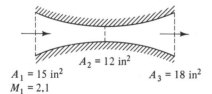

$A_1 = 15$ in^2

$M_1 = 2.1$

$A_2 = 12$ in^2

$A_3 = 18$ in^2

Figure P11.13

11.14. A wind tunnel is to be designed to provide isentropic test section airflow at Mach 3, static pressure 1 psia, and static temperature 460°R (see Figure P11.14). Determine the area ratio of the nozzle required and the reservoir conditions that must be maintained.

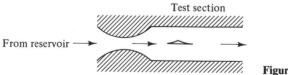

Test section

From reservoir →

Figure P11.14

11.15. A hypersonic tunnel is to be designed to provide Mach 8 flow at 3.0 kPa, 80 K. With helium as the working fluid ($\gamma = \frac{5}{3}$), determine the nozzle area ratio required and the reservoir pressure and temperature. What is the Reynolds number of the tunnel, assuming a characteristic length of 25 cm? $\mu = 6.8 \times 10^6$ Pa·s.

11.16. A converging-diverging nozzle has a ratio of exit area of 1.6. The nozzle is supplied from an air reservoir in which the pressure is 500 kPa, temperature 100°C. Determine the range of back pressures over which the nozzle is choked. Determine the maximum velocity possible at the nozzle exit plane. Calculate the flow rate through the nozzle for a back pressure of 50 kPa and for a back pressure of 5.0 kPa.

11.17. A converging-diverging nozzle has a ratio of exit area to throat area of 1.2. The nozzle is supplied from an air reservoir in which the pressure is 30 psia, temperature 100°F. Determine the range of back pressure over which a shock appears in the nozzle, the range of back pressure over which the nozzle is choked, and the mass flow rate for back pressures of 20 psia and 25 psia.

11.18. A converging-diverging nozzle is to be designed to operate supersonically at an altitude at which the ambient pressure is 5 psia. If the stagnation pressure of the flow at the nozzle inlet is 100 psia, calculate the nozzle area ratio required. Determine the nozzle throat area for a mass flow of 10 lbm/s and stagnation temperature 1000°R. Assume the working fluid to have $\gamma = 1.4$ with a molecular weight of 16.

11.19. For the nozzle of Problem 11.18, determine the mass flow rate and exit plane pressure when operating at sea level (ambient pressure 14.7 psi).

11.20. A converging-diverging nozzle is to be designed to operate supersonically at an altitude of 10,000 m. If the stagnation pressure of the flow at the nozzle inlet is 600 kPa, calculate the nozzle area ratio required. Determine the nozzle throat area for a mass flow of 3 kg/s and stagnation temperature 600 K. Assume the working fluid to have $\gamma = 1.3$ with a molecular mass of 16.

11.21. For the nozzle of Problem 11.20, determine the mass flow rate and exit plane pressure when operating at sea level (ambient pressure 101 kPa).

11.22. A normal shock moves into still air (60°F, 14.7 psia) with a velocity of 3000 ft/s. Calculate the velocity of the air behind the wave and the static and stagnation pressure and temperature in the airflow behind the wave. Explain why there is a difference in stagnation temperature between the air at rest in front of the wave and the air moving behind the wave.

11.23. A normal shock moves into still air (20°C, 101 kPa) with a velocity of 1000 m/s. Calculate the velocity of the air behind the wave and the static and stagnation pressure and temperature in the airflow behind the wave. Repeat for carbon dioxide as the working fluid (take $\gamma = 1.30$).

11.24. Consider airflow in the varying-area channel of Figure P11.24. Determine the Mach number, static pressure, and stagnation pressure at section 3. Assume isentropic flow except for shocks.

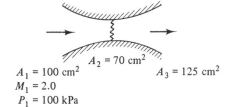

$A_1 = 100$ cm^2
$M_1 = 2.0$
$P_1 = 100$ kPa

$A_2 = 70$ cm^2 $A_3 = 125$ cm^2

Figure P11.24

11.25. A converging-diverging nozzle has a ratio of exit area to throat area of 1.6. The nozzle is supplied from an air reservoir in which the pressure is 80 psia, temperature 100°F. Determine the range of back pressures over which a normal shock will appear in the nozzle.

11.26. Which of the two cases shown in Figure P11.26 will involve the larger loss of stagnation pressure? Air is the working fluid.

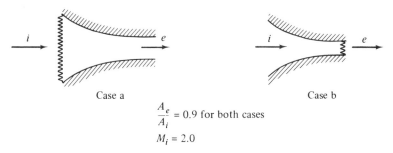

Case a   Case b

$$\frac{A_e}{A_i} = 0.9 \text{ for both cases}$$

$$M_i = 2.0$$

Figure P11.26

11.27. Determine the value of back pressure necessary for the shock to position itself as shown in Figure P11.27.

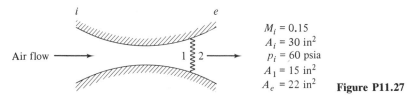

$M_i = 0.15$
$A_i = 30 \text{ in}^2$
$p_i = 60 \text{ psia}$
$A_1 = 15 \text{ in}^2$
$A_e = 22 \text{ in}^2$ **Figure P11.27**

11.28. A converging-diverging nozzle of area ratio 4 to 1 exhausts to standard atmospheric pressure (101 kPa). The nozzle is supplied from a large air reservoir in which pressure can be maintained at a constant value. Determine the range of reservoir pressures over which the nozzle will be choked, with Mach number 1 at the throat. Determine the range of reservoir pressures over which there will be supersonic flow at the nozzle exit plane.

11.29. For Problem 11.28, determine the range of reservoir pressures over which a normal shock will appear in the nozzle.

11.30. Helium is expanded in a converging-diverging nozzle of area ratio 1.48 to 1. The nozzle inlet velocity is negligible, inlet pressure and temperature are 100 psia and 300°F. Determine the nozzle exit velocity for back pressure of 90 psia, 60 psia, 20 psia, and 0 psia.

11.31. Oxygen is expanded in a converging-diverging nozzle of area ratio 3 to 1. The nozzle inlet velocity is negligible, inlet pressure and temperature are 750 kPa and 600 K. Determine the nozzle exit velocity for back pressures of 500 kPa, 250 kPa, and 0 kPa.

11.32. Conditions at the inlet of a jet engine nozzle are velocity 30 m/s, static pressure 250 kPa, static temperature 700 K, and area 1.0 m². If the nozzle is to be designed to operate in the perfectly expanded, supersonic mode at an altitude of 5000 m, calculate the nozzle throat and exit area required. Determine the design thrust developed by the nozzle. Assume the working fluid to have the same properties as air.

11.33. Calculate the thrust developed by the nozzle of Problem 11.32 when operating at sea level, assuming conditions at the nozzle inlet remain the same.

11.34. A rocket is to be designed to provide 100,000 pounds of thrust in space. A propellant combination is to be used that is capable of providing a chamber temperature of 3500°R, with pumps available for maintaining a chamber pressure of 200 psia. Because of space limitations, the nozzle is restricted to an area ratio of 2 to 1. Determine the nozzle throat and exit areas. Assume the exhaust gases behave as a perfect gas with constant specific heats, $\gamma = 1.3$, and mean molecular mass 18. Also calculate the thrust of the rocket at takeoff from the earth's surface.

11.35. For the rocket of Problem 11.34, sketch a plot of thrust versus altitude from sea level to 100,000 ft.

11.36. Determine the loss of stagnation pressure for the supersonic inlet shown in Figure P11.36 ($\gamma = 1.4$).

$M_1 = 2.0$
$p_1 = 10$ kPa

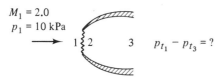

$p_{t_1} - p_{t_3} = ?$

Figure P11.36

11.37. Air at 0°F is moving through a 3-in-diameter tube with a velocity of 100 ft/s. A valve at the end of the tube is suddenly closed, causing a normal shock to move back into the airflow. Determine the velocity of this shock wave and the pressure ratio across the wave.

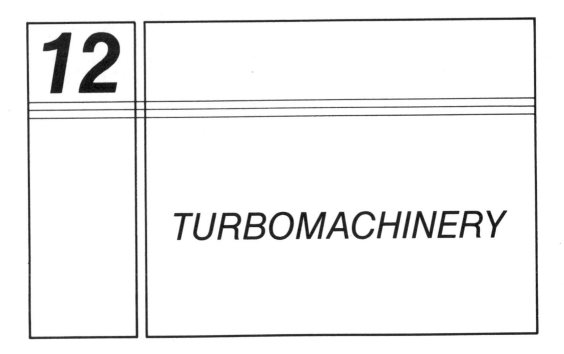

TURBOMACHINERY

12.1 INTRODUCTION

The application of fluid flow to the transfer of energy between a fluid and the rotating parts of turbomachines is the topic under discussion here. Examples of turbomachines abound in all phases of our technological society: the compressor and turbine in the jet engines that are used to power airplanes, the steam turbine in electric power stations, the rotary pumps in water supply systems, the fans in hot-air furnaces and air conditioning units, and the windmills that are still utilized as auxiliary power units on farms.

Although the design of turbomachines involves many aspects of engineering knowledge, the importance of fluid mechanics in turbomachinery lies in its ability to predict general performance characteristics, and to establish the range of operating conditions for different types of turbomachines.

Turbomachines are called *pumps* when they absorb power; they are called *turbines* when they produce power. Thus, in pumps, work is done on the fluid; whereas in turbines, the fluid does work on the machine. The work is done on or by the rotating member of the machine (called rotor, runner, or impeller) which carries vanes or blades. The rotor is attached to a shaft that transmits the mechanical energy to or from the machine. The rotor is usually housed in a casing, thereby allowing the pressure of the working fluid to vary within the machine and the flow to be completely separated from the environment in which the turbomachine is placed. On the other hand, there are installations in which the rotor is completely exposed to the fluid environment in which it operates. Examples of such turbomachines are circulating fans and windmills.

The path of the fluid particle through the rotor may be mainly in the axial direction (i.e., along the shaft axis of the machine), or may be mainly in a radial direction (i.e., in a plane perpendicular to the shaft axis), or may be a combination of both directions. The type of flow and the shapes of the vanes of the rotor depend largely on the conditions of application, the mass flow, pressure rise, and so on, desired for a pump, or the horsepower, space limitations, variation in operating conditions, and similar factors for a turbine.

Thus turbomachines are categorized into machines that have shaft work done on them by the fluid, called turbines, and machines that do work on the fluid, such as pumps, compressors, fans, and blowers. The distinction in the latter category is based on the type of fluid being pumped and the magnitude of pressure rise in the fluid as it passes through the machine. Specifically, the machines are commonly called *pumps* when the fluid is a liquid. A *fan* is a machine that imparts a sufficient pressure rise to the gas to cause flow of the gas. Examples are circulating and ventilation fans. Machines imparting a higher pressure rise and substantial velocities to the fluid are called *blowers*. Finally, *compressors* are machines whose main function is to cause a pressure rise in the fluid with little increase in velocity.

These two categories of turbomachines may be further subdivided into *axial-flow machines*, in which the flow is along the axis and the radius of the flow remains constant, and *radial-flow machines*, in which the fluid passes through the rotating portion of the machine mainly in a radial direction. Further classification of turbomachines can be made on the basis of the working fluid utilized—for example, gas and steam turbines, hydraulic (water) pumps and turbines, air compressors, and so forth.

Examples of radial flow machines are illustrated in Figures 12.1 and 12.2. Figure 12.1 shows a centrifugal pump, while Figure 12.2 shows a radial fan commonly utilized in hot-air heating systems in the home. In both machines, the flow enters the machine in an axial direction but leaves in an outward

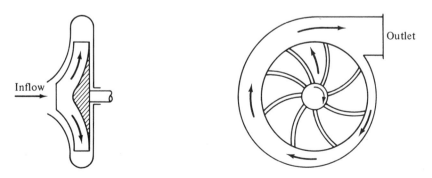

Figure 12.1. Centrifugal pump.

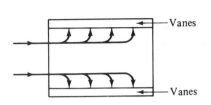

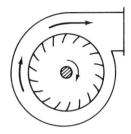

Figure 12.2. Radial-flow ventilation fan.

direction (perpendicular to the axis of rotation of the machine). The change in the flow direction occurs in the first example in the moving impeller (rotor with vanes); in the second case it occurs prior to entering the vanes. In this case, the flow also enters the vanes in a radial direction.

An example of a two-stage axial-flow compressor is shown in Figure 12.3. Each stage consists of a stationary and a moving row of vanes. Such a compressor is typical of air compressors used in airplane jet engines. Another example of an axial-flow machine is the so-called propeller turbine shown in Figure 12.4. Although the fluid enters in a radial direction, the inflow to and the outflow from the vanes of the turbine are in an axial direction, the radial direction of the flow having been changed in the fixed housing of the turbine.

Figure 12.4. Propeller turbine.

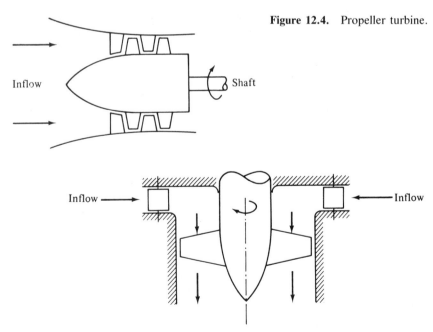

Figure 12.3. Axial-flow compressor.

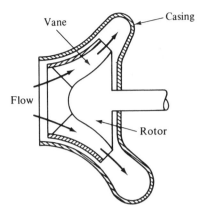

Vane

Casing

Flow

Rotor

Figure 12.5. Mixed-flow pump.

In *mixed-flow machines*, the fluid velocity leaving the rotor has both axial and radial components (Figure 12.5).

12.2 BASIC EQUATIONS OF TURBOMACHINERY

One of the basic equations describing the fluid dynamics of turbomachines is the angular momentum Equation (4.10) derived in Chapter 4, Section 4.2b. This equation states that the sum of all externally applied torques acting on the fluid in a control volume equals the rate of change of angular momentum:

$$T_0 = \frac{\partial}{\partial t} \int_{\substack{\text{control} \\ \text{volume}}} rV_t \ dM + \int_{\substack{\text{control} \\ \text{surface}}} rV_t(\rho V_n \, dA)$$

This equation becomes for steady flow

$$T_0 = \int_{\substack{\text{control} \\ \text{surface}}} rV_t(\rho V_n \, dA)$$

The applied torques T_0 acting on the fluid may be due to pressure forces, viscous forces, gravity forces, magnetic forces, shaft torque, and so forth. As in the case of continuity and linear momentum equations, the flow of angular momentum into the control volume is taken as negative and the efflux of angular momentum is taken as positive. Thus an increase in the angular momentum of the fluid in the control volume corresponds to a net positive external torque in the direction of rotation (as occurs in pumps and compressors). A decrease in angular momentum will be obtained when the net external torque acting on the control volume is in the direction opposite to that of rotation (as occurs in turbines).

The total external torque T_0 exerted on the control volume of a turbomachine will, in general, consist of the torque exerted by the shaft of the turbomachine and of the torque due to pressure and shear forces exerted at the entry and exit sections of the control volume. The control surface can usually be taken such that the torque due to shear will be small. Furthermore, because of circumferential symmetry, there will be no net torque exerted on the control volume by the pressure forces. Thus T_0 will normally represent the shaft torque (T_s) of the turbomachine and is given for machines operating at constant speed (steady flow) by

$$T_s = \int\limits_{\substack{\text{control} \\ \text{surface}}} rV_t(\rho V_n \, dA) \tag{12.1}$$

Let us recall the significance of velocities V_t and V_n appearing in Equation (12.1): V_n is the component of the velocity of the fluid in a direction normal to the surface of the control volume, while V_t is the component of the velocity of the fluid in a tangential direction, that is, located in a plane perpendicular to the axis of rotation of the machine and in a direction perpendicular to a radius drawn from the axis of rotation. As is customary, the velocity V_t is taken as positive when in the direction of rotation. Contributions to the integral of Equation (12.1) taken over the control surface are due only at those openings in the control surface that have tangential components of fluid velocity. A typical control volume for turbomachines is shown in Figure 12.6.

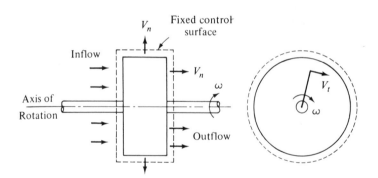

Figure 12.6. Control volume for turbomachine.

Assuming average velocities of fluid flow at inlet and outlet of the control surface and designating the inlet with subscript 1 and the outlet with subscript 2, we obtain from the continuity equation (4.3) for steady incompressible flow

$$\int\limits_{\substack{\text{control} \\ \text{surface}}} \rho V_n \, dA = -\rho V_{n1}A_1 + \rho V_{n2}A_2 = 0$$

Hence

$$\rho V_{n1}A_1 = \rho V_{n2}A_2 = \rho Q$$

Substituting into Equation (12.1), we have for the shaft torque

$$T_s = \underbrace{-\rho Q R_1 V_{t1}}_{\text{inflow}} + \underbrace{\rho Q R_2 V_{t2}}_{\text{outflow}} = \rho Q (R_2 V_{t2} - R_1 V_{t1}) \qquad (12.2)$$

where R_1 and R_2 are actual or average radial distances of the inlet and outlet areas and V_{t1} and V_{t2} are actual or average tangential velocities. Equation (12.2) is the basic equation describing the behavior of the fluid as it passes through a turbomachine and is known as *Euler's turbine equation*. The shaft torque T_s is taken as positive when it is exerted by the shaft on the fluid (pumps) and as negative when exerted by the fluid on the shaft (turbines).

As an example of the application of Equation (12.2), consider the case of the radial ventilation fan shown in Figure 12.7. Choose as control surface the annulus described by the two coaxial circular cylinders shown. Here V_n is in the radial direction and hence is equal to V_r (i.e., the component of the absolute velocity of the fluid in the radial direction). The shaft torque T_s is given by Equation (12.2) with $Q = 2\pi R_1 V_{r1} b = 2\pi R_2 V_{r2} b$, where b is the width of the fan.

In addition to obtaining an expression for the shaft torque of turbomachines as given in Equation (12.2), it is also of interest to derive an expression for the rate of energy transfer (i.e., work per unit time) and for the specific energy rate (i.e., work per unit mass of fluid).

It is known from dynamics that for a rotor turning at constant speed, the work done by a torque equals the product of torque and angle over which work is done. Hence the rate of work (dW_s/dt) due to a torque equals the product of torque (T_s) and angular velocity (ω):

$$-\frac{dW_s}{dt} = T_s \omega \qquad (12.3)$$

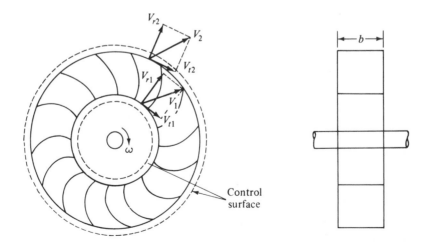

Figure 12.7.   Control volume for radial turbomachine.

Substituting Equation (12.2) into Equation (12.3), we obtain

$$-\frac{dW_s}{dt} = \rho Q \omega (R_2 V_{t2} - R_1 V_{t1})$$

It will be noted that the product ωR represents the tangential velocity of the rotor at a radius R. Designating this velocity as U_t, we have

$$-\frac{dW_s}{dt} = \rho Q (U_{t2} V_{t2} - U_{t1} V_{t1}) \tag{12.4}$$

The work done per unit mass of fluid is obtained by dividing Equation (12.4) by the mass rate of flow (ρQ):

$$\frac{-dW_s/dt}{\rho Q} = \frac{T_s \omega}{\rho Q} = (U_{t2} V_{t2} - U_{t1} V_{t1}) \tag{12.5}$$

The preceding equation simply states that the work per unit mass of fluid is proportional to the difference of the products of two velocities, the tangential velocity of the rotor and the tangential component of the absolute fluid velocity.

Thus far we have utilized the angular momentum equation to obtain a relation between shaft torque and change in angular momentum in the working fluid of a turbomachine. In addition to the momentum equation, the flow of the fluid through the rotor must also obey the energy equation (4.19). For incompressible one-dimensional flow without addition of heat, the energy equation becomes [see Equation (4.19) and Example 4.9]:

$$-\frac{dW}{dt} = \rho Q \left[\left(\frac{p_2}{\rho} + \frac{V_2^2}{2} + g z_2 \right) - \left(\frac{p_1}{\rho} + \frac{V_1^2}{2} + g z_1 \right) \right]$$

As defined in Section 4.3, W consists of all work done by the fluid, such as shaft work and viscous shear work. W is taken as numerically positive if the work is done by the fluid in the control volume. Let the work (W) consist of shaft work (W_s) and frictional work (W_f); hence the energy equation can be written as

$$-\frac{dW_s}{dt} - \frac{dW_f}{dt} = \rho Q g \left[\left(\frac{p_2}{\rho g} + \frac{V_2^2}{2g} + z_2 \right) - \left(\frac{p_1}{\rho g} + \frac{V_1^2}{2g} + z_1 \right) \right] \tag{12.6}$$

where $[p/\rho g + (V^2/2g) + z]$ is called *total head H*. Defining the change in total head ($H_2 - H_1$) occurring across the inlet and outlet of the rotor as ΔH, we obtain for the energy equation

$$-\frac{dW_s}{dt} - \frac{dW_f}{dt} = \rho Q g \, \Delta H \tag{12.7}$$

where ΔH is in units of meters or feet. When work is done by the fluid (as in turbines), W_s will be numerically positive; when work is done on the fluid (as in pumps), W_s will be negative. W_f is always numerically positive, for it is the shear work done by the fluid.

If we assume the frictional work to be negligible, then we can combine the momentum equation (12.5) with the energy equation (12.7) to obtain

$$U_{t2}V_{t2} - U_{t1}V_{t1} = g\,\Delta H_i \qquad (12.8)$$

where ΔH_i represents the ideal change in total head across the turbomachine. The actual change in total head, in the case of pumps, will be smaller than the ideal one because of fluid dynamic losses; that is,

$$\Delta H_{\text{pump}} = \Delta H_i - h_{\text{losses}}$$

Or, defining a "hydraulic" pump efficiency, we have

$$e_{\text{pump}} = \frac{\Delta H_{\text{pump}}}{\Delta H_i}$$

For turbines, the actual change in total head must be greater than the ideal one to account for internal fluid dynamic losses in the turbine; that is,

$$\Delta H_{\text{turbine}} = \Delta H_i + h_{\text{losses}}$$

Or, defining a "hydraulic" turbine efficiency, we have

$$e_{\text{turbine}} = \frac{\Delta H_i}{\Delta H_{\text{turbine}}}$$

The "hydraulic" efficiency is further reduced by leakage of fluid around the rotor without passing through the rotor vanes (termed volumetric losses) and by mechanical friction in the bearings and glands of the machine (termed mechanical losses).

In general, the absolute fluid velocity V occurring in the energy equation will be the vector sum of the components V_t, V_r, and V_a, where V is the absolute velocity of the fluid entering or leaving the impeller, V_t is the tangential component of the absolute velocity of the fluid, V_r is the radial component of the absolute velocity of the fluid, and V_a is the axial component of the absolute velocity of the fluid.

In this chapter we shall treat only those cases in which either V_r or V_a is zero. For example, at the entrance of a stage of an axial machine (consisting of a fixed row of vanes and a moving row of vanes) as shown in Figure 12.8,

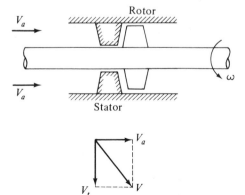

Figure 12.8.   Stage of axial-flow machine.

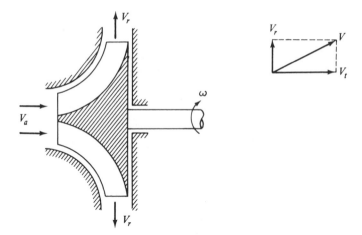

Figure 12.9.   Radial-flow machine showing rotor.

the absolute velocity V has two components, V_a and V_t. At the exit of the radial machine shown in Figure 12.9, the absolute velocity V has two components, V_r and V_t.

The absolute velocity of the fluid in an impeller can also be resolved into two alternate components, one component being the tangential velocity of the blade of the rotor (U_t), the other component the velocity of the fluid relative to the rotor blade (V_{rel}). Thus the velocity V at the exit of the radial fan shown in Figure 12.9 can be represented in two ways (Figure 12.10) in terms of parallelograms. Sometimes an alternative triangular representation is also utilized (Figure 12.11).

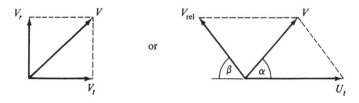

Figure 12.10.   Velocity vector parallelogram.

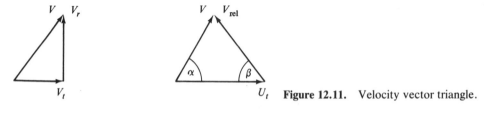

Figure 12.11.   Velocity vector triangle.

In addition to the energy equation written in terms of absolute velocities, it is of interest to write Bernoulli's equation in rotating coordinates; that is, we wish to use relative velocities in lieu of absolute velocities in the equation. This form, in conjunction with the continuity equation $\rho Q = \rho V_{rel} A$, is needed to determine the shape of rotor vane passages. In the absence of elevation changes, Equation (12.8) can be written as

$$U_{t2}V_{t2} - U_{t1}V_{t1} = \frac{p_2 - p_1}{\rho} + \frac{V_2^2 - V_1^2}{2} \tag{12.9}$$

Using the law of cosines, we have, from the velocity vector diagram of Figure 12.12,

$$V^2 = V_{rel}^2 + U_t^2 - 2U_t V_{rel} \cos \beta$$

It can also be seen from the velocity diagram that V_t (the tangential component of the absolute velocity) is

$$V_t = V \cos \alpha = U_t - V_{rel} \cos \beta$$

Hence, substituting $V_{rel} \cos \beta = U_t - V_t$ in the cosine law equation, we obtain

$$V^2 = V_{rel}^2 + U_t^2 + 2U_t(V_t - U_t)$$

Forming the difference in absolute kinetic energy

$$\frac{V_2^2 - V_1^2}{2g} = \frac{1}{2g}[V_{rel2}^2 - V_{rel1}^2 + U_{t1}^2 - U_{t2}^2 + 2(U_{t2}V_{t2} - U_{t1}V_{t1})]$$

and substituting into Equation (12.9), we have

$$\frac{(p_2 - p_1)}{\rho g} + \frac{V_{rel2}^2 - V_{rel1}^2}{2g} - \frac{U_{t2}^2 - U_{t1}^2}{2g} = 0$$

$$\frac{p}{\rho g} + \frac{1}{2g}(V_{rel2}^2 - U_t^2) = \text{Constant} \tag{12.10}$$

The preceding expression is Bernoulli's equation for rotating coordinates. Note the occurrence of the rotational velocity U_t in the equation. A special case occurs when stations 1 and 2 are at the same radius; then $U_{t1} = U_{t2}$ and we have

$$\frac{p}{\rho g} + \frac{1}{2g}V_{rel}^2 = \text{Constant} \tag{12.11}$$

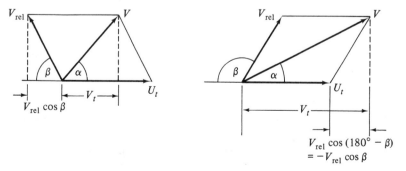

Figure 12.12.   Velocity diagram indicating flow angles.

Comparison of Equation (12.11) with Equation (4.21) shows that in this special case the Bernoulli equation for rotating coordinates is identical to that for fixed coordinates.

12.3 AXIAL-FLOW MACHINES

In axial-flow machines, the flow enters and leaves the machine mainly in an axial direction. In the rotating row of vanes, a change in the tangential fluid velocity causes the transfer of energy between rotor and fluid. The stationary row of blades (stator), if present, merely causes a change in the velocity of the fluid (in magnitude and direction) and no transfer of energy; that is, no work is done on or by the blades, for they are fixed. Multistaging is used in axial-flow machines to achieve greater power in the case of turbines or higher pressure ratios in the case of compressors. A photograph of a rotor of an axial-flow compressor for a jet engine is shown in Figure 12.13. A cutaway view of an axial-flow steam turbine for a nuclear power plant is shown in Figure 12.14.

Since the radius of axial-flow machines remains essentially unchanged along the direction of the flow, such machines have virtually no radial component of velocity. An exception occurs in the case of compressible flow in multistage machines, where the blade lengths vary along the axis to counteract the density changes that occur as the fluid is compressed (as in a compressor) or expands (as in a turbine). However, since the change of density in each stage is usually small, the fluid can be assumed to be incompressible in each stage. However, different densities are taken for successive stages. For axial-flow machines, Equation (12.2) can therefore be written as

$$T_s = \rho Q R (V_{t2} - V_{t1}) \tag{12.12}$$

provided that the length of the blades is small in comparison to the radius of the machine.

A stage of an axial-flow machine is composed of a rotating set of blades (*rotor*) and a stationary set of blades (*stator*). Often a fixed set of blades (*guide vanes*) is utilized ahead of the rotor to guide the fluid to the rotor blades at a suitable angle. Changes in the pressure in the rotor are accompanied by changes in relative velocity, a measure of the so-called degree of reaction. If there is no change in pressure across the rotor, we have zero degree of reaction. The moving blades (*vanes*) of turbomachines perform in a twofold manner: they change the direction of the flow and they can effect a change in the relative fluid velocity. Thus the moving blades act as moving nozzles as well as deflectors of the fluid stream.

As a simple example of the action of the rotor blades, consider the case of a single-stage turbine. As shown in Figure 12.15, the turbine stage consists of a stationary nozzle or set of stationary nozzles whose function is to convert the energy of the high-pressure working medium into a stream of high-velocity

Figure 12.13. Rotor of an axial-flow compressor for jet engine. (*Courtesy* General Electric Co.)

Figure 12.14. Steam turbine generator for nuclear power plant (1800 rpm, 600,000 kW). (*Courtesy* General Electric Co.)

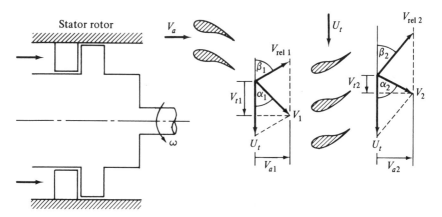

Figure 12.15. Axial-flow turbine stage showing velocity diagram for rotor.

fluid and to direct the stream of the fluid against the row of moving vanes. The flow leaves the fixed nozzles with an absolute velocity V_1, making an angle α_1 with the direction of the tangential velocity of the blade U_t. The velocity of the fluid relative to the moving blade is V_{rel}, making an angle β. The magnitude of the relative velocity increases as it passes through the blades. Since the cross-sectional area of the turbine does not change along the turbine axis, we obtain from the continuity equation for incompressible flow

$$V_{a1} = V_{a2} = V_a$$

From the velocity diagram of Figure 12.15, it can be seen that

$$V_{t1} = V_a \operatorname{ctn} \alpha_1$$

$$V_{t2} = V_a \operatorname{ctn} \alpha_2$$

and therefore Equation (12.12) can be expressed as

$$T_s = \rho Q R V_a(\operatorname{ctn} \alpha_2 - \operatorname{ctn} \alpha_1) \tag{12.13}$$

Since by convention the shaft torque T_s is taken as negative for a turbine, we obtain $\operatorname{ctn} \alpha_2 < \operatorname{ctn} \alpha_1$ and hence angle α_2 is greater than α_1. For a pump, T_s is positive; hence $\operatorname{ctn} \alpha_2 > \operatorname{ctn} \alpha_1$ and the angle α_2 is smaller than α_1. For zero shaft torque, angle α_2 equals α_1. As illustrated in Figure 12.16, we therefore obtain that the absolute velocity V_2 is less than V_1 for turbines, while V_2 exceeds V_1 for pumps.

In lieu of the expression for shaft torque as given in Equation (12.13), it is more desirable to give an expression for the shaft torque in terms of flow angles that do not vary when the flow velocities are changed. Such

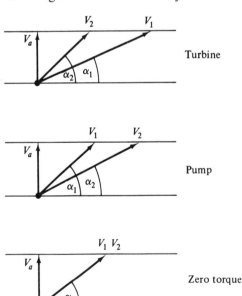

Turbine

Pump

Zero torque

Figure 12.16. Absolute velocities in turbomachines.

angles are α_1 and β_2. The absolute inlet flow angle α_1 is determined by the setting of the stationary nozzles or blades, whereas the relative flow angle β_2 at exit is determined by the direction of the trailing edges of the rotor blades. From the velocity diagram of Figure 12.15, it can be seen that

$$V_{rel} \cos \beta = V_a \, \text{ctn} \, \beta$$

and

$$V_t = V_a \, \text{ctn} \, \alpha$$

Hence

$$V_{t2} - V_{t1} = U_t - V_{rel2} \cos \beta_2 - V_a \, \text{ctn} \, \alpha_1$$
$$= U_t - V_a \, \text{ctn} \, \beta_2 - V_a \, \text{ctn} \, \alpha_1$$

By substitution into Equation (12.12), the shaft torque becomes

$$T_s = \rho Q R[U_t - V_a(\text{ctn} \, \alpha_1 + \text{ctn} \, \beta_2)] \tag{12.14}$$

To obtain a similar expression for the change in total head ΔH, we need to relate it to the shaft torque T_s. From Equations (12.3) and (12.7), we have

$$T_s \omega = \rho Q g \, \Delta H$$

or

$$g \, \Delta H = \frac{T_s \omega}{\rho Q} \tag{12.15}$$

Combining it with Equation (12.14), we obtain for the change in total head

$$g \, \Delta H = U_t^2 - V_a U_t(\text{ctn} \, \alpha_1 + \text{ctn} \, \beta_2)$$

If we are interested in determining the performance of axial-flow turbomachines at constant rotor speed (U_t), we rewrite the foregoing equation

$$\frac{g \, \Delta H}{U_t^2} = 1 - \frac{V_a}{U_t}(\text{ctn} \, \alpha_1 + \text{ctn} \, \beta_2) \tag{12.16}$$

Thus the ideal performance characteristics of axial-flow machines can be obtained by plotting the quantity $g \, \Delta H/U_t^2$ against V_a/U_t for a given set of values of angles α_1 and β_2. As can be seen from Figure 12.17, a straight line having three distinct slopes (negative, horizontal, and positive) is obtained.

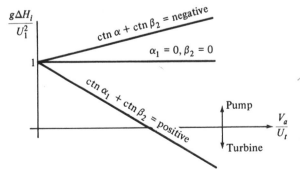

Figure 12.17. Ideal performance characteristics of axial-flow machines.

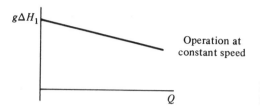

constant speed

Figure 12.18. Ideal performance characteristics of axial compressor.

The slope depends on the value of the quantity ($\operatorname{ctn} \alpha_1 + \operatorname{ctn} \beta_2$). As indicated
in Figure 12.17, the curves lying above the horizontal axis apply to pumps,
since pumps have positive values of ΔH. Turbines have negative values of
ΔH, and hence curves below the horizontal axis apply to turbine operation.
Most pumps have flow angles α_1 and β_2 such that ΔH decreases with increasing
velocity ratio V_a/U_t. Furthermore, pumps are usually connected to synchronous
(constant-speed) motors; thus U_t remains constant with change in flow rate.
Since $Q = V_a A$, we obtain from Equation (12.16) for a given axial compressor
the typical ideal performance curve shown in Figure 12.18.

In the case of axial-flow turbines, the flow rate available often remains
fixed and hence we are interested in determining the torque and change in
total head as a function of rotor speed. Thus we write Equation (12.14) as

$$T_s = \rho Q R V_a \left[\frac{U_t}{V_a} - (\operatorname{ctn} \alpha_1 + \operatorname{ctn} \beta_2) \right]$$

or

$$\frac{T_s}{\rho Q R V_a} = \frac{U_t}{V_a} - (\operatorname{ctn} \alpha_1 + \operatorname{ctn} \beta_2) \qquad (12.17)$$

Again, we can consider the angles α_1 and β_1 to remain constant at varying
rotor speeds, and we can plot the performance curve as given in Figure 12.19.
The expression ($\operatorname{ctn} \alpha_1 + \operatorname{ctn} \beta_2$) in Equation (12.17) becomes an intercept
instead of a slope as in Equation (12.16). The ideal ΔH is obtained by
rewriting Equation (12.16) in terms of $g \, \Delta H/V_a^2$:

$$\frac{g \, \Delta H_i}{V_a^2} = \left(\frac{U_t}{V_a} \right)^2 - \frac{U_t}{V_a}(\operatorname{ctn} \alpha_1 + \operatorname{ctn} \beta_2) = \frac{U_t}{V_a} \frac{T_s}{\rho Q R V_a}$$

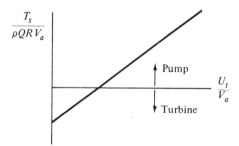

Figure 12.19. Ideal torque of axial
turbomachine.

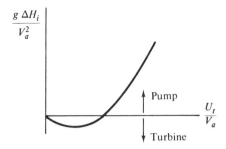

Figure 12.20.   Ideal change in total head of axial turbomachine.

The quantity $g\,\Delta H_i/V_a^2$ has been plotted against velocity ratio U_t/V_a in Figure 12.20. The performance variation of a given turbomachine at constant flow rate with rotor speed is shown in Figure 12.21. It is seen that a minimum ΔH_i value is obtained in the turbine regime.

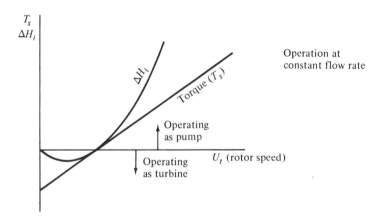

Figure 12.21.   Variation of ideal performance with rotor speed.

A special case of axial-flow turbine is the so-called *impulse turbine*. This type of turbine has no change in pressure across the rotor and thus has zero reaction. If we neglect frictional effects in the moving blades, it follows from Equation (12.10) that there is no change in the relative velocity across the rotor. Thus the rotor blades will be shaped such that $V_{\text{rel}_1} = V_{\text{rel}_2}$. From the velocity diagram of Figure 12.22, it is seen that

$$V_{\text{rel}_1} \sin \beta_1 = V_a$$

and

$$V_{\text{rel}_2} \sin \beta_2 = V_a$$

Thus $\sin \beta_1 = \sin \beta_2$. Two solutions are possible: $\beta_1 = \beta_2$ or $\beta_1 = 180° - \beta_2$. The first solution corresponds to the case where the blades are flat and parallel to the relative velocity V_{rel_1}; hence no force would be exerted on the

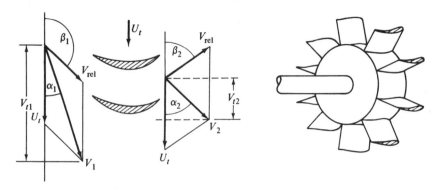

Figure 12.22. Impulse turbine rotor with velocity diagram.

blades. Therefore, we must choose the second solution. Now, from trigonometry, we have $\cos \beta_1 = -\cos (180° - \beta_1)$; hence $\cos \beta_1 = -\cos \beta_2$. Thus we can express the change in tangential velocities as

$$V_{t2} - V_{t1} = U_t - V_{rel} \cos \beta_2 - (U_t - V_{rel} \cos \beta_1)$$
$$= U_t - V_{rel} \cos \beta_2 - (U_t + V_{rel} \cos \beta_2)$$
$$= -2V_{rel} \cos \beta_2$$

Hence from Equation (12.12) the shaft torque becomes

$$T_s = -2\rho QRV_{rel} \cos \beta_2$$

or, since $V_{rel} \cos \beta = V_a \operatorname{ctn} \beta$,

$$T_s = -2\rho QRV_a \operatorname{ctn} \beta_2 \qquad (12.18)$$

Equation (12.18) indicates that the torque of an impulse turbine is proportional to the product of axial velocity and cotangent of the blade angle. The negative torque indicates that the fluid does work on the shaft (a turbine). In actual impulse turbines, the magnitude of the angle α_1 (the angle between absolute entrance velocity and the direction of motion of the blades) is about 20°, while β_2 (the angle between relative exit velocity and local tangential velocity of the rotor) usually has a value of about 30°.

The ideal torque characteristics of an axial impulse turbine with $\beta_2 = 30°$ are shown in Figure 12.23.

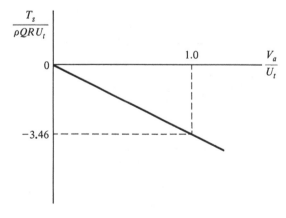

Figure 12.23. Torque characteristics of axial-flow impulse turbine.

396

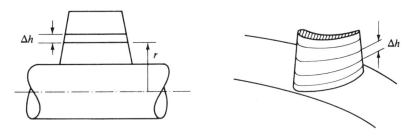

Figure 12.24.   Rotor blade.

When the blades of the rotor are not short relative to the radius of the machine, we need to account for the variation in radial distance when integrating Equation (12.1) to obtain the shaft torque. Thus, instead of using Equation (12.12) directly, we divide the blade into elements as shown in Figure 12.24 and add the contribution (dT_s) of each annulus of thickness Δh [computed by using Equation (12.12)] to obtain total shaft torque, that is, $T_s = \Sigma\, dT_s$. Furthermore, at different radial distances from the axis of rotation, the tangential velocity of the blades varies according to $U_t = \omega r$. Hence the relative blade angle β_1 must be varied along the blade to account for the changes in U_t. Otherwise the direction of the relative fluid velocity will not be in the direction of the blades at entrance, thereby resulting in flow losses due to separation.

EXAMPLE 12.1
An axial-flow impulse turbine rotates at a speed of 60 rad/s with a mass flow rate of 10 kg/s. The rotor hub radius is 0.1 m, while that of the tip is 0.5 m. The axial-flow velocity V_a is 10 m/s. Absolute inflow velocities to the rotor are constant along the radial distance. The rotor exit blade angle β_2 is 0.5 rad at a radial distance of 0.3 m. Find the ideal torque and power developed by the turbine using the blade element procedure and taking Δh equal to 0.05 m.

Solution   Using Equation (12.18), the contribution of each blade element to the total torque is

$$\Delta T_s = \frac{2\rho Q r V_a\, \text{ctn}\, \beta_2\, \Delta h}{R_{\text{tip}} - R_h}$$

$$= -\frac{2(10)r(10)(\text{ctn}\,\beta_2)0.05}{0.4}$$

$$= -25r\, \text{ctn}\, \beta_2$$

Since $U_t = \omega r$, the relative velocity V_{rel}, and hence blade angles β_1 and β_2, will vary along the radial distance. From Figure 12.22 we get

$$V_t - U_t = -V_a\, \text{ctn}\, \beta_1$$

$$\tan \beta_1 = \frac{V_a}{\omega r - V_{t_1}}$$

At $r = 0.3$ m, $\beta_2 = 0.5$ rad; hence $\beta_1 = \pi - 0.5 = 2.6416$ rad. Thus, tan $2.6416 = 10/[60(0.3) - V_{t_1}]$ and therefore $V_{t_1} = 36.3$ m/s. The results of calculations for each blade element are given in tabular form:

r (m)	r_{av} (m)	β_1 (rad)	β_2 (rad)	$25\, r_{av}\, ctn\, \beta_2$
.10–.15	.125	2.8074	.3342	9.000
.15–.20	.175	2.7540	.3698	10.717
.20–.25	.225	2.7283	.4133	12.825
.25–.30	.275	2.6739	.4677	13.613
.30–.35	.325	2.6047	.5369	13.650
.35–.40	.375	2.5145	.6271	12.938
.40–.45	.425	2.3946	.7470	11.475
.45–.50	.475	2.2332	.9084	9.263
				93.481 N·m

Thus, the ideal torque is $T_s = -93.48$ N · m, while the power of the turbine is $(93.48$ N · m$)(60$ rad/s$) = $ 5.61 kW. ∎

Other examples of axial-flow machines are the propeller turbine and the axial-flow pump shown in Figures 12.25 and 12.26. In the case of the axial-flow (propeller) hydraulic turbine, the inflow to the turbine is in a plane perpendicular to the shaft and is guided by vanes as indicated in Figure 12.25. The flow is turned into an axial direction by the stationary ducting. To determine the inflow to the rotating blades, use Equation (4.10) for steady flow without external torque:

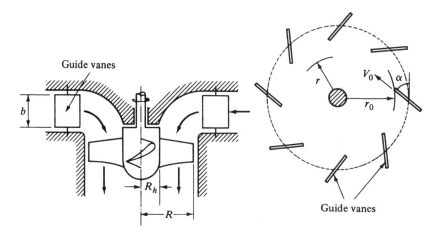

Figure 12.25. Axial-flow hydraulic turbine.

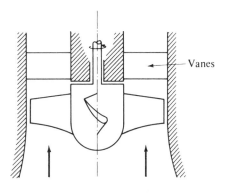

Vanes

Figure 12.26.   Axial-flow pump.

$$0 = \int\limits_{\substack{\text{control}\\\text{surface}}} rV_t(\rho V_n \, dA)$$

$$= -r_0 V_{t0}\rho Q + rV_t\rho Q$$

Hence

$$V_t r = V_{t0} r_0$$

Thus the distribution of the tangential inflow velocity component over the blade of the turbine is given by

$$V_t = \frac{V_0 r_0 \cos \alpha}{r} \tag{12.19}$$

This flow will be recognized as free vortex flow, discussed in Section 7.4. The flow rate through the turbine is

$$Q = 2\pi r_0 b V_r = 2\pi r_0 b V_0 \sin \alpha$$

while the axial velocity at inlet to the rotor blades is given by

$$V_a = \frac{Q}{\pi(R^2 - R_h^2)} \tag{12.20}$$

where R_h and R are the hub and tip radius of the rotor blade, respectively.

EXAMPLE 12.2

The guide vanes of an axial-flow propeller turbine are set at an angle of 30° with respect to the radial direction. The inner radius of the guide vanes is 5 ft; the vanes have a height of 1.5 ft. The fluid velocity at the vanes is 10 fps. The turbine blades have a tip radius of 2.5 ft and a hub radius of 0.5 ft. The rotor speed is 300 rpm. Determine the blade angles at the leading edge of the propeller blades.

Solution   The tangential velocity component at the tip of the blades is, from Equation (12.19),

$$V_t = \frac{(10 \text{ ft/s})(5 \text{ ft}) \cos (90° - 30°)}{2.5 \text{ ft}} = 10 \text{ ft/s}$$

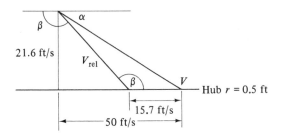

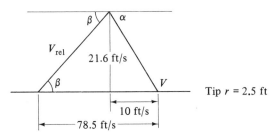

Figure 12.27

while at the hub

$$V_t = \frac{(10 \text{ ft/s})(5 \text{ ft}) \cos 60°}{0.5 \text{ ft}} = 50 \text{ ft/s}$$

The axial velocity is, from Equation (12.20),

$$V_a = \frac{2\pi r_0 b V_0 \sin \alpha}{\pi (R^2 - R_h^2)} = \frac{2(5 \text{ ft})(1.5 \text{ ft})(10 \text{ ft/s}) \sin 60°}{6.25 \text{ ft}^2 - 0.25 \text{ ft}^2}$$

$$= 21.6 \text{ ft/s}$$

The tangential velocity of the blades at the tip is

$$U_t = 2\pi(\tfrac{300}{60})(2.5 \text{ ft}) = 78.5 \text{ ft/s}$$

while at the hub

$$U_t = 2\pi(\tfrac{300}{60})(0.5 \text{ ft}) = 15.7 \text{ ft/s}$$

The velocity diagrams can now be drawn (Figure 12.27).

At the tip: $\tan \beta = \dfrac{V_a}{U_t - V_t} = \dfrac{21.6}{78.5 - 10} = 0.316$

$$\beta = \underline{17.5°}$$

At the hub: $\tan (180° - \beta) = \dfrac{21.6}{50 - 15.7} = 0.630$

$$180° - \beta = 32.2°$$

$$\beta = \underline{147.7°}$$

■

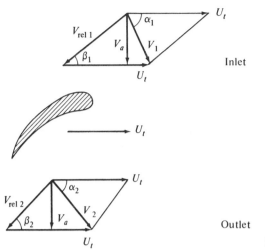

Inlet

Outlet

Figure 12.28

Up to now we have utilized one-dimensional flow theory to determine the performance of turbomachines. This assumption is reasonable when the number of blades of the machine is large. However, when the number of blades of the turbomachine is small, the distance between blades is sufficiently large so that each blade acts like an airfoil instead of each pair of blades acting as a moving nozzle. To examine the behavior of the flow past a moving airfoil, let the flow approach the moving airfoil with absolute velocity V_1 and leave with absolute velocity V_2 (Figure 12.28). We can construct the two velocity diagrams shown, one at the inflow to the foil, the other outflow from the airfoil. To apply airfoil theory, we use an average relative velocity $(V_{\text{rel}})_{\text{av}}$ such that

$$(V_{\text{rel}})_{\text{av}} = \tfrac{1}{2}(V_{\text{rel}_1} + V_{\text{rel}_2})$$

The flow past the airfoil therefore has velocity $(V_{\text{rel}})_{\text{av}}$ and an angle of incidence of γ (see Figure 12.29). According to airfoil theory, discussed in Section 9.4, the foil will experience a lift force L perpendicular to the direction of $(V_{\text{rel}})_{\text{av}}$. However, in order to evaluate the forces on the rotor, we need to

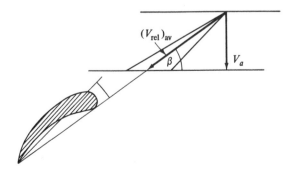

Figure 12.29

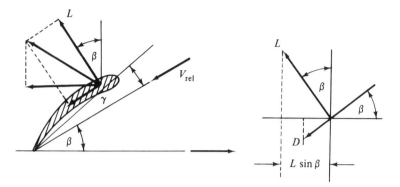

Figure 12.30

resolve the lift L and drag D into components parallel to the plane of rotation and in the axial direction. The component of the total force along the plane of rotation (F_t) is (see Figure 12.30)

$$F_t = L \sin \beta + D \cos \beta$$

while the component in the axial direction (F_a) is

$$F_a = L \cos \beta - D \sin \beta$$

The characteristics of airfoils are usually specified in terms of lift and drag coefficients, namely,

$$L = C_L \tfrac{1}{2} \rho V_{\text{rel}}^2 A \tag{12.21}$$

and

$$D = C_D \tfrac{1}{2} \rho V_{\text{rel}}^2 A \tag{12.22}$$

with A the planform area of the airfoil, usually written as chord times span. Next, divide the blade into elements of width Δh and chord length c (Figure 12.31). This step is necessary because the magnitude of relative flow velocity will vary with radial distance from the axis. Hence we write

$$F_t = (C_L \sin \beta + C_D \cos \beta) \tfrac{1}{2} \rho V_{\text{rel}}^2 c \, \Delta h$$

and

$$F_a = (C_L \cos \beta - C_D \sin \beta) \tfrac{1}{2} \rho V_{\text{rel}}^2 c \, \Delta h$$

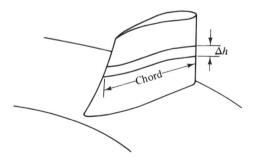

Figure 12.31. Rotor blade indicating element Δh.

Next we need to relate the forces exerted on the foil to the characteristic quantities of the flow, such as static pressure and velocity. Using the momentum equation in the tangential direction, we obtain

$$nF_t = \dot{m}(V_{t2} - V_{t1})$$

where $\dot{m}$ is the mass flow rate in the annulus of thickness Δh, r is the radial distance of Δh from the centerline, and n is the number of blades in the rotor; $\dot{m}$ equals $\rho V_a 2\pi r\, \Delta h$. Hence the rate of change of momentum in the tangential direction is

$$\rho V_a(V_{t2} - V_{t1})2\pi r\, \Delta h$$

Thus

$$n(C_L \sin \beta + C_D \cos \beta)\tfrac{1}{2}\rho V_{\text{rel}}^2 c\, \Delta h = \rho V_a(V_{t2} - V_{t1})2\pi r\, \Delta h$$

Rewriting, we have, with $V_{\text{rel}} = V_a/(\sin \beta)$,

$$(V_{t2} - V_{t1}) = \frac{1}{2}V_{\text{rel}}\frac{cn}{2\pi r}(C_L + C_D \operatorname{ctn} \beta) \tag{12.23}$$

Equation (12.23) gives the relationship between the tangential fluid velocities of the rotor and the foil characteristics of the rotor blades. Since for efficient airfoil sections the lift is much greater than the drag of the section and β is usually not greater than $\pi/4$, Equation (12.23) can be simplified to

$$(V_{t2} - V_{t1}) = \frac{1}{2}V_{\text{rel}}C_L c\frac{n}{2\pi r} \tag{12.24}$$

To obtain a relationship in the axial direction, neglect changes in momentum in the axial direction to write

$$nF_a = (p_2 - p_1)2\pi r\, \Delta h$$

Thus we have

$$n(C_L \cos \beta - C_D \sin \beta)\frac{\rho}{2}V_{\text{rel}}^2 c\, \Delta h = (p_2 - p_1)2\pi r\, \Delta h$$

and the relation between the pressure change across the rotor and the foil characteristics of the rotor blades is

$$p_2 - p_1 = \frac{\rho}{2}V_{\text{rel}}^2 \frac{cn}{2\pi r}(C_L \cos \beta - C_D \sin \beta) \tag{12.25}$$

EXAMPLE 12.3

A ducted axial-flow fan consisting of a rotor only is to be designed to deliver 8.4 m³/s of air with a static pressure rise of 0.35 kPa across the fan blades shaped like airfoil sections. The rotor speed is 100 rad/s with a tip radius of 0.5 m and a hub radius of 0.30 m. The air density is 1.2 kg/m³. Assuming a nine-bladed fan, calculate the chord length of the blades and their angle of inclination with respect to the direction of rotation of the blades.

Solution   First compute the tangential blade velocity (U_t):

At the tip: $U_t = \omega R = (100 \text{ rad}/\text{s})(0.5 \text{ m}) = 50 \text{ m}/\text{s}$

At the hub: $U_t = \omega R = (100 \text{ rad}/\text{s})(0.30 \text{ m}) = 30 \text{ m}/\text{s}$

Next, compute the tangential fluid velocity (V_t). Since the fan consists of a rotor only, the inflow to the rotor is in the axial direction and hence $V_{t1} = 0$. Assume initially that $\Delta H = \Delta p/\rho g$; that is, assume the change in kinetic energy head is much smaller than the change in static pressure head, and substitute into Equation (12.8) to obtain at the tip:

$$V_{t2} = \frac{g \, \Delta H}{U_{t2}} = \frac{9.81 \text{ m/s}^2}{50 \text{ m/s}} (29.7 \text{ m}) = 5.827 \text{ m/s}$$

At the hub:

$$V_{t2} = \frac{(9.81 \text{ m/s}^2)(29.7 \text{ m})}{30 \text{ m/s})} = 9.712 \text{ m/s}$$

where a pressure change of 0.35 kPa is equivalent to a change in head of

$$\frac{\Delta p}{\rho g} = \frac{350 \text{ N/m}^2}{(1.2 \text{ kg/m}^3)(9.81 \text{ m/s}^2)} = 29.73 \text{ m}$$

The axial-flow velocity is given by

$$V_a = \frac{Q}{\pi(R_{tip}^2 - R_{hub}^2)} = \frac{8.4 \text{ m}^3/\text{s}}{\pi(0.25 \text{ m}^2 - 0.09 \text{ m}^2)} = 16.71 \text{ m/s}$$

To check on the assumption of $\Delta H \approx \Delta p$, let us compute the change in kinetic energy based on the V_t components just obtained. Specifically, from the definition of total head without changes in elevation, we have, from Equation (12.9),

$$\Delta H = \frac{p_2 - p_1}{\rho g} + \frac{V_2^2 - V_1^2}{2g}$$

In this case, where $V_1 = V_a$, the second term of the above equation becomes $V_{t2}^2/2g$. Therefore,

At the tip: $\dfrac{V_{t2}^2}{2g} = \dfrac{5.827^2 \text{ m}^2/\text{s}^2}{2(9.81) \text{ m/s}^2} = 1.7306 \text{ m}$

At the hub: $\dfrac{V_{t2}^2}{2g} = \dfrac{9.712^2 \text{ m}^2/\text{s}^2}{2(9.81) \text{ m/s}^2} = 4.8075 \text{ m}$

Thus ΔH becomes

At the tip: 29.73 m + 1.7306 m = 31.46 m

At the hub: 29.73 m + 4.8075 m = 34.54 m

We can then recompute V_{t2}:

At the tip: $V_{t2} = \dfrac{g \, \Delta H}{U_{t2}} = \dfrac{(9.81 \text{ m/s}^2)(31.46 \text{ m})}{50 \text{ m/s}} = 6.172 \text{ m/s}$

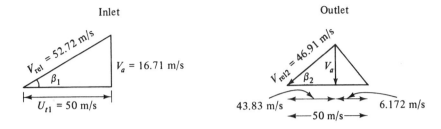

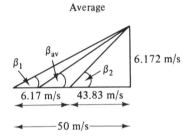

Figure 12.32

At the hub: $V_{t2} = \dfrac{g \, \Delta H}{U_{t2}} = \dfrac{(9.81 \text{ m/s}^2)(34.54 \text{ m})}{30 \text{ m/s})} = 11.29 \text{ m/s}$

Next let us compute the relative velocities (V_{rel}) from the velocity diagrams. At the tip (Figure 12.32):

$$V_{\text{rel}_1} = \sqrt{V_a^2 + U_{t1}^2} = \sqrt{16.71^2 + 50^2} = 52.72 \text{ m/s}$$

$$V_{\text{rel}_2} = \sqrt{V_a^2 + (U_{t2} - V_{t2})^2} = \sqrt{16.71^2 + (50 - 6.172)^2} = 46.91 \text{ m/s}$$

Combining the two triangles, we obtain the average relative velocity ($V_{\text{rel}})_{\text{av}}$ and average angle β_{av}:

$$(V_{\text{rel}})_{\text{av}} = \tfrac{1}{2}(52.72 + 46.91) = 49.82 \text{ m/s}$$

and

$$\tan \beta_{\text{av}} = \frac{16.71 \text{ m/s}}{\tfrac{1}{2}(50 + 43.83) \text{ m/s}} = 0.3562$$

$$\beta_{\text{av}} = 0.3422 \text{ rad}$$

At the hub (Figure 12.33):

$$V_{\text{rel}_1} = \sqrt{16.71^2 + 30^2} = 34.34 \text{ m/s}$$

$$V_{\text{rel}_2} = \sqrt{16.71^2 + (30 - 11.29)^2} = 25.09 \text{ m/s}$$

$$(V_{\text{rel}})_{\text{av}} = \frac{34.34 + 25.09}{2} = 29.72 \text{ m/s}$$

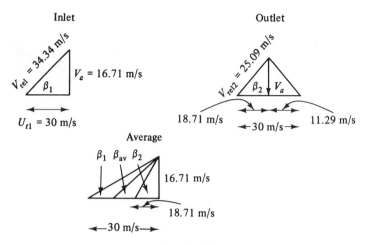

Figure 12.33

and

$$\tan \beta_{av} = \frac{16.71 \text{ m/s}}{\frac{1}{2}(30 + 18.71)\text{m/s}} = 0.6861$$

$$\beta_{av} = 0.6013 \text{ rad}$$

Having obtained the values of the relative velocities V_{rel}, we can now utilize Equations (12.24),

$$V_{t2} = \frac{1}{2} V_{rel}\frac{n}{2\pi r}(C_L c)$$

and (12.8), $\Delta H = V_t U_t/g$, to obtain $C_L c$. Combining these two equations,

$$\Delta H = \frac{1}{2g} V_{rel}U_t\frac{n}{2\pi r}(C_L c)$$

or

$$C_L c = \frac{2g \, \Delta H}{V_{rel}U_t}\frac{2\pi r}{n}$$

Hence:

At the tip: $$C_L c = \frac{2(9.81 \text{ m/s}^2)(31.46 \text{ m})}{(49.82 \text{ m/s})(50 \text{ m/s})}\frac{2\pi(0.5 \text{ m})}{9}$$

$$= 0.08650 \text{ m}$$

At the hub: $$C_L c = \frac{2(9.81 \text{ m/s}^2)(34.54 \text{ m})}{(29.72 \text{ m/s})(30 \text{ m/s})}\frac{2\pi(0.3 \text{ m})}{9}$$

$$= 0.1592 \text{ m}$$

Choosing an airfoil with characteristics such that $C_L = 0.5$ at $\gamma = 0.035$ rad, we obtain $c = 0.173$ m at the tip and $c = 0.3184$ m at the hub. The angle of the fan blades with the direction of U_t is $\gamma + \beta = 0.035$ rad $+ 0.3422$ rad $= 0.3772$ rad at the tip, and 0.035 rad $+ 0.6013$ rad $= 0.6363$ rad at the hub.

Using the foregoing procedure, the chord length of foil section and the blade angles $(\gamma + \beta)$ can be computed for a number of radial locations from hub to tip.   ∎

The preceding discussion applies to the situation when the distance between the blades is sufficiently large; specifically, when the ratio

$$\frac{2\pi r/n}{c}$$

(the ratio of circumferential distance between adjoining blades and chord length) is greater than 1. When this ratio is less than 1, the blades no longer can be treated as isolated airfoils. The effect of the proximity of neighboring airfoils (called *cascade effect*) should then be taken into account. This effect results in reduced lift coefficients as compared to isolated airfoils.

12.4 RADIAL-FLOW MACHINES

In radial-flow turbomachines, the fluid flows mainly in a direction perpendicular to the axis of rotation. In the case of radial-flow compressors, the flow is outward, that is, toward the larger radius. Radial-flow machines are used mainly for pumps, although use is made of this type in gas turbines. In that case, the flow is in an inward direction, that is, toward the smaller radius.

In the centrifugal pump shown in Figure 12.34a and b, the fluid enters the rotor axially in the central part of the rotor, is forced radially outward by the rotating vanes, and is discharged at its periphery without axial velocity component into a casing. In the rotating vanes, the fluid pressure, as well as the absolute fluid velocity, is increased. In the casing, most of the absolute kinetic energy is then converted into pressure. A cutaway view of a multistage centrifugal compressor is shown in Figure 12.35.

As an example of a radial-flow machine, let us analyze a centrifugal compressor having blades of constant width. As in the analysis of an axial-flow machine, let us assume frictionless flow and radial symmetry. Let subscript 2 indicate conditions at the outer radius and subscript 1 conditions at the inner radius. Flow takes place in a radially outward direction. The shaft torque supplied to the fluid by the rotor is given by Equation (12.2) as

$$T_s = \rho Q(\underset{\text{outlet}}{R_2 V_{t2}} - \underset{\text{inlet}}{R_1 V_{t1}})$$

Again, if we assume thin, closely spaced vanes, the flow angles will be equal to the blade angles. From the continuity equation, we have

$$V_{r1}A_1 = V_{r2}A_2$$

where V_r is the radial component of the absolute velocity and A is the area

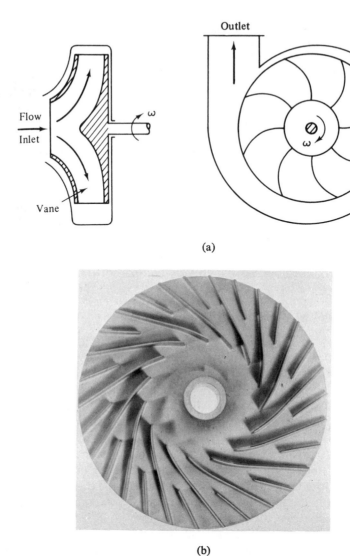

(a)

(b)

Figure 12.34. (a) Centrifugal pump. (b) Impeller for Freon compressor used on Boeing 707 (5.15 in diameter, 23,500 rpm). (*Courtesy* Air Research Manufacturing Co.)

around the periphery of the rotor open to the flow. For thin vanes, we can write

$$Q = V_{r1}2\pi R_1 b = V_{r2}2\pi R_2 b$$

or

$$\frac{V_{r1}R_1}{V_{r2}R_2} = 1$$

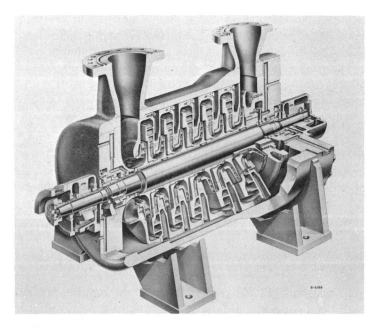

Figure 12.35.   Centrifugal compressor. (*Courtesy* Ingersoll-Rand.)

From the velocity diagram shown in Figure 12.36, we have for radial machines

$$V_t = V_r \text{ ctn } \alpha$$

Thus from Equation (12.2) we obtain

$$T_s = \rho Q(R_2 V_{r2} \text{ ctn } \alpha_2 - R_1 V_{r1} \text{ ctn } \alpha_1)$$

Substituting $V_{r1}R_1 = V_{r2}R_2$ obtained above, we simplify the expression to

$$T_s = \rho Q R_2 V_{r2}(\text{ctn } \alpha_2 - \text{ctn } \alpha_1) \tag{12.26}$$

This expression for shaft torque will be recognized as almost being identical to Equation (12.13) for the axial-flow case. The difference between the two

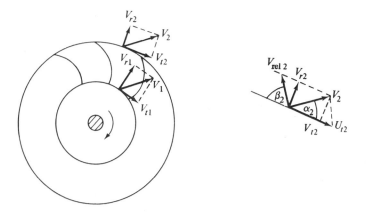

Figure 12.36.   Velocity diagram for centrifugal pump.

equations is in the use of the radial velocity component V_{r2} instead of the axial velocity component V_a. The velocity components V_r and V_a, respectively, are associated with the flow through the turbomachines.

The vanes of radial-flow machines are usually designed such that α_1 equals 90°; that is, the absolute velocity into the rotor is in a radial direction and has no tangential component (i.e., $V_1 = V_{r1}$). Hence the expression for the torque becomes

$$T_s = \rho Q R_2 V_{r2} \operatorname{ctn} \alpha_2 \qquad (12.27)$$

This equation clearly indicates that for nonzero shaft torque the angle α_2 cannot be equal to 90°. Since, from the velocity diagram, the tangential component of the absolute fluid exit velocity V_{t2} can be expressed as

$$V_{t2} = U_{t2} - V_{r2} \operatorname{ctn} \beta_2$$

then Equation (12.26) can also be expressed in terms of exit blade angle β_2:

$$T_s = \rho Q R_2 U_{t2}\left(1 - \frac{V_{r2}}{U_{t2}} \operatorname{ctn} \beta_2\right) \qquad (12.28)$$

Using Equations (12.15) and (12.28), we obtain an expression for the change in total head for a radial-flow machine with purely radial inflow:

$$g\,\Delta H = U_{t2}^2\left(1 - \frac{V_{r2}}{U_{t2}} \operatorname{ctn} \beta_2\right) \qquad (12.29)$$

Three types of blades are used for radial-flow machines, as shown in Figure 12.37: radial blades ($\beta_2 = 90°$), backward-curved blades ($\beta_2 < 90°$), and forward-curved blades ($\beta_2 > 90°$). The first two types are utilized in pumps, whereas forward-curved blades are used in radial turbines. In the last case, the flow of fluid is in the inward direction (i.e., toward the shaft).

In order to establish the typical performance curve of an outward-flowing radial machine, consider a centrifugal pump operating in connection with a constant-speed motor. The flow exit angle β_2 can generally be assumed to remain constant and independent of the flow velocities. Thus for a constant pump speed of a given size the change in total head can be expressed, using Equation (12.29), as

$$g\,\Delta H = U_t^2 - \frac{U_t \operatorname{ctn} \beta_2}{A_2}Q$$

where A_2 is the discharge area of the rotor. This equation is plotted in Figure 12.38 for three values of blade angle β_2. For a pump, β_2 is usually less than 90° and the bottom curve would describe the performance of the pump.

EXAMPLE 12.4

A centrifugal water pump rotates at 900 rpm. The inlet radius of the impeller is 3 in, while its outlet radius is 6 in. The width of the impeller is 2 in. The blade angles are $\beta_1 = 25°$ and $\beta_2 = 15°$. For smooth radial inflow and

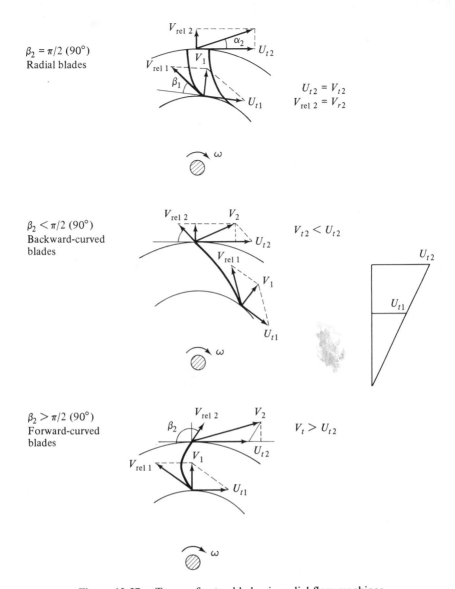

$\beta_2 = \pi/2\ (90°)$
Radial blades

$U_{t2} = V_{t2}$
$V_{rel\ 2} = V_{r2}$

$\beta_2 < \pi/2\ (90°)$
Backward-curved blades

$V_{t2} < U_{t2}$

$\beta_2 > \pi/2\ (90°)$
Forward-curved blades

$V_t > U_{t2}$

Figure 12.37. Types of rotor blades in radial-flow machines.

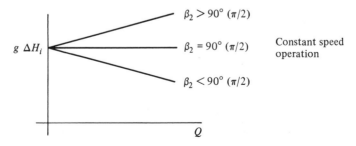

$g\ \Delta H_i$

$\beta_2 > 90°\ (\pi/2)$

$\beta_2 = 90°\ (\pi/2)$

$\beta_2 < 90°\ (\pi/2)$

Constant speed operation

Q

Figure 12.38. Ideal performance characteristics of radial-outflow machine.

411

neglecting losses, determine the pump discharge, ideal change in total head ΔH, pump power, and pressure rise in the impeller.

Solution   As in Figure 12.36, choose the inlet as section 1 and the outlet as section 2. The inflow velocity diagram will be as follows (Figure 12.39).

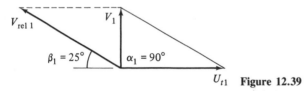

Figure 12.39

The tangential velocity of the impeller at inlet is

$$U_{t1} = 2\pi(\tfrac{900}{60})R_1 = 94.2(\tfrac{3}{12}) = 23.5 \text{ ft/s}$$

From the velocity diagram, we have

$$\frac{V_1}{U_{t1}} = \tan \beta_1 \quad \text{or} \quad V_1 = (23.5 \text{ ft/s}) \tan 25° = 10.96 \text{ ft/s}$$

The flow rate through the pump is $Q = V_{r1}2\pi R_1 b$. Here $V_1 = V_{r1}$; therefore,

$$Q = (10.96 \text{ ft/s})(2\pi)(\tfrac{3}{12} \text{ ft})(\tfrac{2}{12} \text{ ft})$$
$$= 2.87 \text{ ft}^3/\text{s}$$

The ideal change in total head is given by a single term of Equation (12.8):

$$\Delta H_i = \frac{U_{t2}V_{t2}}{g}$$

since $V_{t1} = 0$ due to radial inflow. With $V_{t2} = U_{t2} - V_{r2} \operatorname{ctn} \beta_2$, we obtain from Equation (12.29)

$$\Delta H_i = \frac{U_{t2}^2}{g}\left(1 - \frac{V_{r2}}{U_{t2}} \operatorname{ctn} \beta_2\right)$$

The radial component of the absolute exit velocity (V_{r2}) is obtained from the continuity relation; that is,

$$V_{r2} = \frac{Q}{2\pi R_2 b} = \frac{2.87 \text{ ft}^3/\text{s}}{2\pi(\tfrac{6}{12} \text{ ft})(\tfrac{2}{12} \text{ ft})} = 5.48 \text{ ft/s}$$

and

$$U_{t2} = \omega R_2 = 94.2(\tfrac{6}{12}) = 47 \text{ ft/s}$$

Hence

$$\Delta H_i = \frac{U_{t2}^2}{g}\left(1 - \frac{V_{r2}}{U_{t2}} \operatorname{ctn} \beta_2\right)$$

$$= \frac{47^2 \text{ ft}^2/\text{s}^2}{32.17 \text{ ft/s}^2}\left(1 - \frac{5.48}{47} \operatorname{ctn} 15°\right)$$

$$= \underline{38.8 \text{ ft}}$$

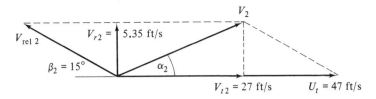

Figure 12.40

Next, to determine the absolute velocity V_2, we need V_{t2}:

$$V_{t2} = U_{t2}\left(1 - \frac{V_{r2}}{U_{t2}} \text{ctn } \beta_2\right)$$

$$= 47\left(1 - \frac{5.48}{47} \text{ctn } 15°\right)$$

$$= 26.5 \text{ ft/s}$$

Thus

$$V_2 = \sqrt{V_{r2}^2 + V_{t2}^2} = \sqrt{5.48^2 + 26.5^2} = 27.1 \text{ ft/s}$$

and

$$\text{ctn } \alpha_2 = \frac{V_{t2}}{V_{r2}} = \frac{26.5}{5.48} = 4.84$$

or $\alpha_2 = 11.7°$ (Figure 12.40). To determine pressure rise in the impeller, use Equation (12.9):

$$U_{t2}V_{t2} = g \, \Delta H = \frac{1}{\rho}(p_2 - p_1) + \frac{V_2^2 - V_1^2}{2}$$

Hence the pressure rise in the impeller

$$p_2 - p_1 = \rho g \, \Delta H - \frac{\rho}{2}(V_2^2 - V_1^2)$$

$$= \frac{62.4 \text{ lbm/ft}^3}{32.17 \text{ lbm/slug}} (32.17 \text{ ft/s}^2)(38.8 \text{ ft})$$

$$- \frac{1}{2} \frac{62.4 \text{ lbm/ft}^3}{32.17 \text{ lbm/slug}} (27.1^2 \text{ ft}^2/\text{s}^2 - 10.96^2 \text{ ft}^2/\text{s}^2)$$

$$= 2421 \text{ psf} - 596 \text{ psf}$$

$$= 1825 \text{ psf} = \underline{12.7 \text{ psi}}$$

The ideal pump power is obtained from Equation (12.7) with frictional work rate set equal to zero:

$$-\frac{dW_s}{dt} = \rho Q g \, \Delta H = \frac{62.4 \text{ lbm/ft}^3}{32.17 \text{ lbm/slug}} (2.87 \text{ ft}^3/\text{s}) (32.17 \text{ ft/s}^2)(38.8 \text{ ft})$$

$$= 6949 \text{ ft-lbf/s} = \frac{6949 \text{ ft-lbf/s}}{(550 \text{ ft-lbf/s})/\text{hp}}$$

$$\frac{dW_s}{dt} = -12.6 \text{ hp}$$

where the negative sign indicates that work is being done on the fluid (pump). ∎

As indicated earlier, the rotor of a centrifugal pump increases the pressure and the absolute fluid velocity of the fluid. The kinetic energy leaving the rotor is further converted into pressure in the pump casing. Let us use the basic momentum equation to examine the behavior of the fluid in the casing of the radial pump. Assuming frictionless flow, Equation (12.1) becomes, in the absence of an externally applied torque,

$$0 = \int_{\substack{\text{control} \\ \text{surface}}} rV_t(\rho V_n \, dA)$$

Here V_n equals V_r and inflow takes place at the surface area 2, while outflow occurs at the undesignated area. Since we have radial flow symmetry, we can write

$$\int_{\substack{\text{control} \\ \text{surface}}} rV_t(\rho V_n \, dA) = \underset{\text{inflow}}{-r_2\rho QV_{t2}} + \underset{\text{outflow}}{r\rho QV_t} = 0$$

or

$$V_t = \frac{V_{t2}r_2}{r} = \frac{c}{r}$$

The preceding equation will again be recognized as representing free vortex flow (discussed in Section 7.4). Hence the tangential component of the flow in the pump casing is free vortex flow. In addition, there is a radial component V_r, which satisfies the following relation derived from the continuity equation. The continuity equation is

$$2\pi r V_r b = 2\pi r_2 V_{r2} b$$

or

$$V_r = \frac{r_2 V_{r2}}{r} = \frac{c'}{r}$$

From the velocity diagram of Figure 12.41, we obtain

$$\tan \alpha = \frac{V_r}{V_t} = \frac{c'/r}{c/r} = \text{Constant}$$

This relation states that the angle of inclination of the flow in the casing remains constant at all radii. The resulting flow pattern in the pump casing is shown in Figure 12.42.

The resulting change of pressure in the casing can be obtained by use of the Bernoulli equation (4.21), which becomes upon neglecting changes in elevation

$$\frac{p - p_2}{\rho} = \frac{V_2^2 - V^2}{2}$$

Substituting into the Bernoulli equation the expression for V obtained above,

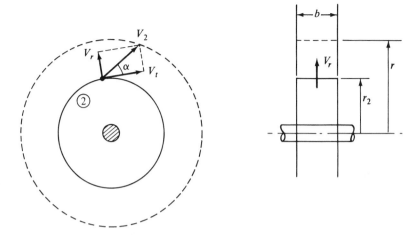

Figure 12.41.   Schematic of radial pump casing.

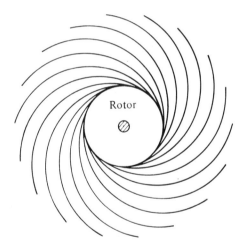

Figure 12.42.   Flow pattern in pump casing.

$$V^2 = V_r^2 + V_t^2 = \frac{V_{r2}^2 r_2^2}{r^2} + \frac{V_{t2}^2 r_2^2}{r^2} = \frac{r_2^2}{r^2} V_2^2$$

there results

$$\frac{p - p_2}{\rho} = \frac{V_2^2}{2}\left(1 - \frac{r_2^2}{r^2}\right) \tag{12.30}$$

Since r_2 is smaller than r, an increase in pressure takes place in the pump casing, converting a large portion of the kinetic energy at section 2 into pressure.

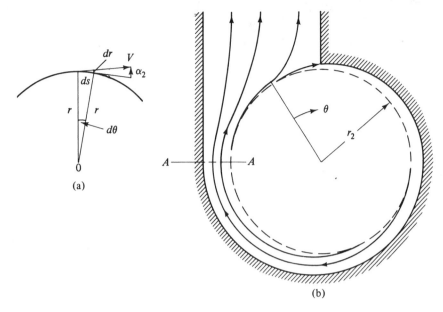

Figure 12.43

EXAMPLE 12.5

It is required to design a housing for an air blower of a hot-air furnace (see Figure 12.2). The blower rotor has an outer diameter D_2 of 0.6 m. The exit velocity V_2 is 10 m/s with an exit angle α_2 of 0.05 rad. In the design, allow a minimum clearance of 10 mm between rotor and housing.

Solution To obtain the shape of the blower housing, use the streamlines shown in Figure 12.42. Along each streamline tan α is constant, where α is the angle of inclination of the flow leaving the blower rotor. The differential equation of the streamlines is (see Figure 12.43a):

$$\tan \alpha_2 = \frac{dr}{ds} = \frac{dr}{r \, d\theta}$$

or

$$\frac{dr}{r} = \tan \alpha_2 \cdot d\theta$$

Since α_2 remains constant at different radial distances r, we can integrate the equation to obtain

$$\ln r = \tan \alpha_2 \cdot \theta + C$$

Taking $\theta = 0$ at $r = r_2$, the constant C becomes $\ln r_2$ or $\ln r/r_2 = \tan \alpha_2 \cdot \theta$. Here the rotor radius r_2 is 0.3 m and the minimum clearance is 10 mm; therefore, the minimum radius of the housing is $r = 0.310$ m. For this radial distance, $\theta = \ln (0.310/0.3)/\tan 0.05 = 0.655$ rad. Therefore the streamline which represents the outline of the housing has a radial distance of 0.310 m at $\theta = 0.655$ rad. To provide an outlet for the flow from the rotor, the streamline shape of the housing is terminated after an included angle of $\frac{3}{2}\pi$ rad ($\theta = 0.655$ rad $+ \frac{3}{2}\pi$ rad $= 5.367$ rad) and the housing outlet is taken

416

to be a straight wall tangent to the streamline (Figure 12.43b). At the junction point, the radial distance of the housing is

$$\ln\left(\frac{r}{0.3}\right) = (5.367)\tan 0.05$$

or

$$\frac{r}{0.3} = 1.308 \quad \text{and} \quad r = 0.3924 \text{ m}$$

Radial distances of the housing calculated for other θ values are shown in the table:

θ (rad)	r/r_2	r (mm)	θ (rad)	r/r_2	r (mm)
0	1.000	300.0	π	1.170	351.1
$\pi/6$	1.027	308.0	$7\pi/6$	1.201	360.4
0.655	1.033	310.0	$4\pi/3$	1.233	370.0
$\pi/3$	1.054	316.1	$3\pi/2$	1.266	379.8
$\pi/2$	1.082	324.5	$5\pi/3$	1.300	389.9
$2\pi/3$	1.110	333.1	5.367	1.308	392.4
$5\pi/6$	1.140	342.0			

Other streamlines in the gap between housing and rotor can be drawn (Figure 12.43b), since all streamlines are identical in shape, as seen in Figure 12.42.

Values of velocity and pressure within the space between rotor and housing can be evaluated from the relations $V^2 = r_2^2 V_2^2/r^2 = 0.3^2 10^2/r^2$ and $p - p_2 = \rho V_2^2(1 - r_2^2/r^2)/2 = 1.2(100)(1 - 0.3^2/r^2)/2$, where the density of the air was taken as 1.2 kg/m³. Velocity and pressure distribution at Section A-A (outer radius: 0.3924 m; inner radius: 0.3 m) were evaluated and are shown in Figure 12.44.

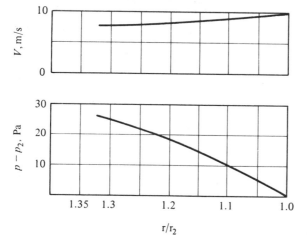

Figure 12.44

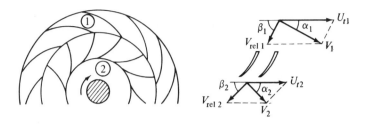

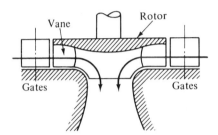

Figure 12.45.   Francis reaction turbine.

In some centrifugal pumps, a set of stationary vanes is added, which performs the function of reducing the kinetic energy into pressure. The stationary vanes can effect this reduction in a shorter radial distance than the vaneless pump casing discussed above.

Another example of a radial-flow machine is the Francis hydraulic turbine, in which the flow is inward and essentially in a radial direction. However, in the modern version of the Francis turbine, the runner has been modified to result in mixed flow. Figure 12.45 shows the original version of the Francis reaction turbine. At its nominal operating condition, the stationary guide vanes (gates) deflect the incoming flow into the rotor parallel to the rotating vanes. The gates are also used to control the flow of water to the turbine. However, the condition of parallelism will not be satisfied at off-design operation. At the exit from the rotor, optimum flow will exist when the absolute velocity V_2 is in a radial direction (i.e., when $\alpha_2 = \pi/2$, or 90°). Furthermore, V_2 should be kept small, for its kinetic energy is lost. This is done by making the angle β_2 as small as possible, with values as low as 0.35 rad (20°) being used.

The kinetic energy of the absolute velocity leaving the rotor can be converted into pressure energy by using a diverging tube called the draft tube (Figure 12.46). The addition of a draft tube allows the pressure at the rotor exit to be reduced below atmospheric pressure, thus increasing the total head across the turbine rotor.

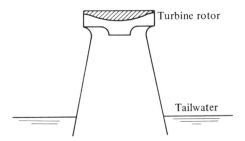

Turbine rotor

Tailwater

Figure 12.46.   Draft tube of hydraulic turbine.

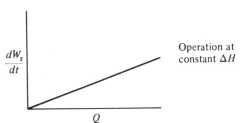

$\dfrac{dW_s}{dt}$

Operation at constant ΔH

Q

Figure 12.47.   Ideal shaft power for hydraulic turbine.

In the case of hydraulic turbines, the head supplied to the turbine is fixed by nature and usually remains constant. Hence we can assume the change in total head to be constant. From Equation (12.15), that is, $T_s \omega = \rho Q g\, \Delta H$, we therefore obtain that the ideal shaft power output of the turbine is a linear function of the flow rate through the turbine, as shown in Figure 12.47.

EXAMPLE 12.6

A Francis hydraulic reaction turbine discharges 130 ft^3/s under a total head of 250 ft. The diameter of the turbine rotor is 4 ft and its rotational speed is 360 rpm. The absolute water velocity leaving the stationary inlet gates makes an angle (α_1) of 20° with the tangential velocity. The area perpendicular to this absolute flow velocity (A_1) is 1.50 ft^2.

Assuming that the absolute velocity leaving the rotor is in a radial direction, determine the torque and power applied to the turbine shaft. What is the hydraulic efficiency of the turbine?

Solution   The schematic arrangement of the turbine systems is shown in Figure 12.48. For radial outflow from the rotor (i.e., $\alpha_2 = \pi/2$), we have from Equation (12.26)

$$T_s = -\rho Q R_1 V_{r1} \operatorname{ctn} \alpha_1$$

or since

$$V_{t1} = V_{r1} \operatorname{ctn} \alpha_1 = V_1 \cos \alpha_1$$

we obtain

$$T_s = -\rho Q R_1 V_1 \cos \alpha_1$$

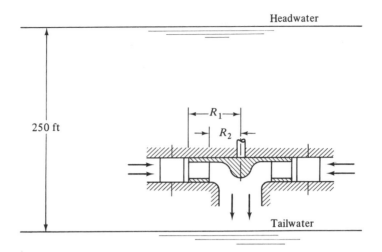

Figure 12.48

Now

$$V_1 = \frac{Q}{A_1} = \frac{130 \text{ ft}^3/\text{s}}{1.5 \text{ ft}^2} = 86.67 \text{ ft/s}$$

Note that A_1 is the area perpendicular to the flow velocity leaving the stationary gates and not the circumferential area at a radius of 2 ft.

$$T_s = -\frac{62.4 \text{ lbm/ft}^3}{32.17 \text{ lbm/slug}}(130 \text{ ft}^3/\text{s})(2 \text{ ft})(86.67 \text{ ft/s})(\cos 20°)$$

$$= -41{,}073 \text{ ft-lbf}$$

Since T_s is positive when a torque is applied to the fluid by the shaft, the negative sign indicates that the torque is applied to the shaft (turbine).

From Equation (12.3), we have

$$-\frac{dW_s}{dt} = T_s\omega$$

Thus,

$$-\frac{dW_s}{dt} = -(41{,}073 \text{ ft-lbf})\left(\frac{2\pi}{60}\frac{m}{s}\right)(360 \text{ m}^{-1}) = -1{,}548{,}400 \text{ ft-lbf/s}$$

$$= -\frac{1{,}548{,}400 \text{ ft-lbf/s}}{550 \text{ ft-lbf/s} \cdot \text{hp}} = \underline{-2815 \text{ hp}}$$

or

$$\frac{dW_s}{dt} = 2815 \text{ hp}$$

where dW_s/dt is the power applied by the fluid to the turbine shaft.

From the definition of the hydraulic efficiency of a turbine, we have

$$e_h = \frac{\Delta H}{\text{Available head}}$$

To determine ΔH, we use Equation (12.15):

$$\frac{T_s \omega}{\rho Q} = g \, \Delta H$$

Therefore,

$$\Delta H = -\frac{(1{,}548{,}400 \text{ ft-lbf}/\text{s})(32.17 \text{ lbm}/\text{slug})}{(62.4 \text{ lbm}/\text{ft}^3)(130 \text{ ft}^3/\text{s})(32.17 \text{ ft}/\text{s}^2)}$$

$$= -190.9 \text{ ft}$$

Here ΔH is negative, for this turbomachine is a turbine. Thus

$$e_h = \tfrac{190.9}{250} = \underline{0.764}$$ ∎

12.5 DIMENSIONAL ANALYSIS AND PERFORMANCE OF TURBOMACHINES

In the preceding sections we have discussed and analyzed the ideal performance of turbomachines based largely on one-dimensional frictionless flow assumptions. The actual performance of a turbomachine will differ from the ideal situation because of the occurrence of losses due to fluid dynamic drag, leakage, and mechanical friction. For example, the actual performance curve of a centrifugal pump running at constant speed is given in Figure 12.49, where the head developed by the pump is shown as a function of delivered volume flow rate. This performance curve differs from the ideal curve shown in Figure 12.50 by the effects of nonuniform velocity distribution, the frictional losses in the stationary and moving vanes of the pump, and the separation losses that occur at off-design conditions. As shown in Figure 12.50, at off-design conditions the flow through the pump, and hence the radial velocity V_r, differs from that at design, and hence the angle of relative rotor inflow velocity will be different from the blade angle β_1. The ensuing flow separation causes a loss that is proportional to the square of the difference in flow rates,

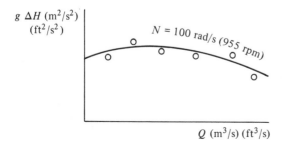

Figure 12.49. Performance characteristics of centrifugal pump.

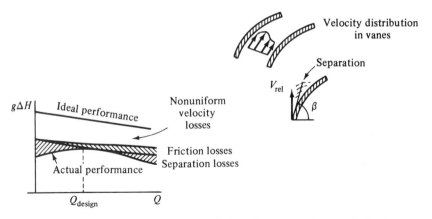

Figure 12.50. Performance characteristics of centrifugal pump indicating various losses.

$(Q - Q_{\text{design}})^2$. The frictional losses will be proportional to the square of the flow rate Q^2, while the flow nonuniformity losses are proportional to ΔH. As a result of these losses, the actual performance of the pump is reduced from the ideal by the amount indicated in Figure 12.50.

In order to avoid testing each pump for performance characteristics, use can be made of similitude considerations to establish such characteristics for classes of turbomachines having geometric similarity. To this end, the techniques of dimensional analysis described in Section 5.2 are utilized. As a first step in obtaining the applicable nondimensional parameters, the physical variables considered significant in this case must be listed. Specifically, we expect that the change in total head* ($g \, \Delta H$) should be a function of the size of the machine—say, its diameter D—the rotational speed of the rotor N, the flow rate through the machine Q, and the kinematic viscosity of the working fluid ν. This statement takes the form

$$g \, \Delta H = f(D, N, Q, \nu)$$

Carrying out the procedure discussed in Section 5.2, we obtain the functional relation between nondimensional parameters, namely,

$$\frac{g \, \Delta H}{N^2 D^2} = f\left(\frac{Q}{ND^3}, \frac{ND^2}{\nu}\right)$$

The nondimensional group Q/ND^3 is called the *flow coefficient* and represents the dependence of the flow rate through the machine on rotational speed and size. Since velocity is proportional to Q/D^2 and ND is proportional to the linear rotor speed U_t, the flow coefficient can be thought of as representing the ratio of fluid velocity to rotor velocity (i.e., V/U_t). The group $g \, \Delta H/N^2 D^2$

* Here we use $g \, \Delta H$ to assure representation as specific energy (energy per unit mass) rather than a linear dimension if we used ΔH.

is called the *head coefficient* and represents the dependence of the change in total head on the square of the rotor speed or on the change of VU_t [as expressed in Equation (12.8)], since U_t has just been shown to be proportional to the fluid velocity V. The third group, ND^2/ν, is the rotational Reynolds number, for ND is proportional to U_t and thus can be expressed as U_tD/ν, a Reynolds number. For sufficiently large Reynolds numbers, the effect of Reynolds number on the performance of the machine is less important than the other parameters. Consequently, we can write with sufficient accuracy

$$\frac{g\,\Delta H}{N^2D^2} = f\!\left(\frac{Q}{ND^3}\right)$$

From experimental tests on geometrically similar turbomachines over a range of speeds and sizes, we obtain the single curve shown in Figure 12.51. In the case of pumps, we are usually interested in the performance of the pump at different conditions of total head change. Specifically, since the driving motors of pumps are generally synchronous (constant-speed), we want to know the change in flow rate incurred by a change in total head against which the pump works. For example, in Example 4.9 we might ask what the volume flow rate would be for the installed pump if it were required to pump the water to a building at a higher elevation than that given in the example. Hence we can replot Figure 12.51 to obtain the curve ΔH versus Q for a given speed N as shown, for example, in Figure 12.49.

Similarly, we can perform a dimensional analysis for the overall efficiency (e) of a turbomachine under the assumption that the efficiency is a function of size (D), rotative speed (N), and flow rate (Q). In this case, we obtain the functional relation

$$e = f\!\left(\frac{Q}{ND^3}\right)$$

Results of experimental tests on a family of geometrically similar turbomachines are shown in Figure 12.52. This figure indicates that the overall efficiency

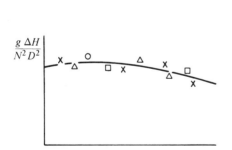

Figure 12.51. Head coefficients of turbomachines.

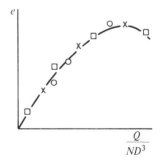

Figure 12.52. Overall efficiency of turbomachines.

(*e*) of the machines is zero at zero flow rate and increases to a maximum value corresponding to near design flow rates.

In selecting the type of turbomachine to be used for given conditions of operation, it has been found useful to utilize a nondimensional parameter not containing the size of the machine. Such a parameter can be obtained through the elimination of the diameter *D* from the flow and head coefficients by forming the following ratio:

$$\frac{\sqrt{Q/ND^3}}{\sqrt[3/4]{g\,\Delta H/N^2D^2}} = \frac{N\sqrt{Q}}{(g\,\Delta H)^{3/4}} = N_s$$

This dimensionless parameter is called *specific speed* and is designated N_s. When overall efficiencies of turbomachines are plotted as a function of N_s, the maximum efficiency will occur at different N_s for different types of machines. As is seen from Figure 12.53, axial-flow machines have large values of specific speed, whereas radial-flow machines exhibit low specific speeds. Mixed-flow machines have intermediate values of specific speed.

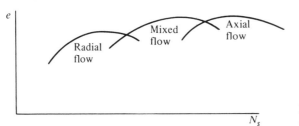

Figure 12.53. Overall efficiency of turbomachines.

Axial-flow machines are adaptable to large flow rates, for the inlet is unrestricted and larger areas can admit large quantities of fluid without excessively high velocities. However, each row of blades can transfer only a limited quantity of energy with reasonable efficiency. Thus we have large values of *Q* accompanied by small ΔH for axial flow machines, resulting in large specific speeds.

On the other hand, centrifugal machines have small inlet sizes, for the inlet radius is small relative to the overall radius of the machine. Hence the flow rate must be kept small to avoid high inlet velocities. Since the exit flow occurs at a larger radius than at inlet, a large change in total head can be developed. Thus, for radial-flow machines, we have low values of *Q* accompanied by large values of ΔH, resulting in low specific speeds.

Mixed-flow machines operate at medium levels of *Q* and ΔH, resulting in intermediate values of specific speeds.

In order to achieve reasonable efficiency for a given type of turbomachine, it should be operated at its applicable range of specific speed.

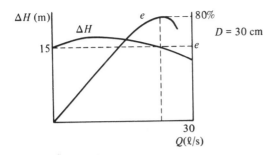

Figure 12.54. Measured characteristics of water pump.

EXAMPLE 12.7

A centrifugal water pump has the characteristic curves shown in Figure 12.54. Its impeller diameter is 30 cm and it rotates at a speed of 200 rad/s.

Determine the characteristic curves for a geometrically similar centrifugal pump having an impeller diameter of 15 cm and operating at 100 and 400 rad/s, respectively.

Solution   Use the procedure indicated in Section 5.3. To achieve dynamic similarity, the head and flow coefficients of the two pumps must be equal at corresponding conditions. Thus we have

$$\frac{g\,\Delta H_1}{N_1^2 D_1^2} = \frac{g\,\Delta H_2}{N_2^2 D_2^2}$$

and

$$\frac{Q_1}{N_1 D_1^3} = \frac{Q_2}{N_2 D_2^3}$$

or

$$\Delta H_2 = \Delta H_1 \left(\frac{N_2}{N_1}\right)^2 \left(\frac{D_2}{D_1}\right)^2$$

and

$$Q_2 = Q_1 \left(\frac{N_2}{N_1}\right)\left(\frac{D_2}{D_1}\right)^3$$

Now

$$\frac{N_2}{N_1} = \frac{100}{200} = \frac{1}{2}$$

$$\frac{D_2}{D_1} = \frac{15}{30} = \frac{1}{2}$$

Hence

$$\Delta H_2 = \Delta H_1 \tfrac{1}{4}\tfrac{1}{4} = \tfrac{1}{16}\,\Delta H_1$$

$$Q_2 = Q_1 \tfrac{1}{2}\tfrac{1}{8} = \tfrac{1}{16}\,Q_1$$

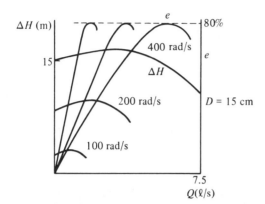

Figure 12.55. Computed characteristics of water pump.

Next, consider the second condition:

$$\frac{N_2}{N_1} = \frac{400}{200} = 2$$

$$\frac{D_2}{D_1} = \frac{15}{30} = \frac{1}{2}$$

Hence

$$\Delta H_2 = \Delta H_1 \, 4(\tfrac{1}{4}) = \Delta H_1$$

$$Q_2 = Q_1 \, 2(\tfrac{1}{8}) = \tfrac{1}{4} Q_1$$

The overall efficiencies will be equal at the same value of flow coefficient. Thus

$$e_1 = e_2$$

and

$$\frac{Q_1}{N_1 D_1^3} = \frac{Q_2}{N_2 D_2^3}$$

or

$$e_1 = e_2$$

at

$$Q_2 = Q_1 \left(\frac{N_2}{N_1}\right)\left(\frac{D_2}{D_1}\right)^3$$

Hence we can construct the new performance characteristics shown in Figure 12.55. ∎

PROBLEMS

12.1. An incompressible fluid flows through a stationary row of short blades of an axial turbomachine, as indicated in Figure P12.1. The absolute flow velocities

V_0 and V_1 make angles α_0 and α_1 with the plane of the stator. Pressures at inlet and outlet of the stator are p_0 and p_1, respectively. Find the torque and axial force required to hold the stator fixed.

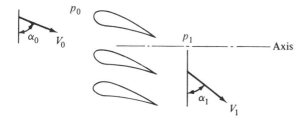

Figure P12.1

12.2. The average absolute velocity leaving the stator of an axial-flow hydraulic turbine, taken at a radius of 2 m, is 6.5 m/s. The angle α_1 between this velocity and the direction of tangential velocity of the rotor (U_t) is 1.1 rad. The rotor, rotating at 12 rad/s, discharges 20 m^3/s of water in an axial direction. Determine the torque exerted on the rotor and the average axial velocity.

12.3. A hydraulic propeller turbine is to produce 10,000 horsepower at a flow rate of 750 cfs. The rotor operates at a speed of 240 rpm with flow leaving in the axial direction. The inner radius of the control gates (guide vanes) is 5 ft with a gate length of 2 ft. Find the angle that the control gate makes with the radial direction.

12.4. Consider an impulse turbine stage as shown in Figure P12.4. Axial velocity is 100 m/s, $U_t = 175$ m/s. For a mass flow of 20 kg/s, find the power output. If the turbine is used in a jet engine with an inlet stagnation temperature of 1200 K, find the outlet stagnation temperature and stagnation pressure ratio for the stage. Assume isentropic flow with $\gamma = 1.35$, $R = 0.287$ kJ/kg $\cdot$ K.

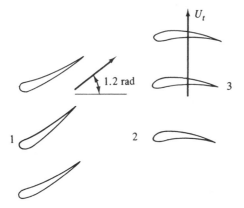

Figure P12.4

12.5. Shown in Figure P12.5 is an axial compressor rotor blade. For $C_L = 1.2$ and C_D negligible, find the force components in the axial and tangential directions acting on the blade. For a mean radius of one foot, calculate the torque exerted by the fluid on the blade. Assume an axial velocity of 400 fps, with chord 1.0 in, blade length 2 in, and average air density of 0.1 lbm/ft^3.

$\beta = 10°$

$\beta = 60°$

Figure P12.5

12.6. An axial-flow pump has a rotor with blade angles $\beta_1 = 0.45$ rad and $\beta_2 = 0.75$ rad. A downstream stator is used to remove the rotational component in the flow. The flow enters the rotor axially with an absolute velocity of 4 m/s. Determine the ideal change in total head and the inlet angle of the stator row.

12.7. The rotor blades of an axial compressor operate at a lift coefficient of 0.5. The resulting velocity diagram is shown in Figure P12.7. Compute the ideal rise in total head across the rotor.

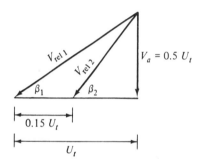

$V_{rel\,1}$

$V_{rel\,2}$

$V_a = 0.5\,U_t$

β_1

β_2

$0.15\,U_t$

U_t

Figure P12.7

12.8. An axial turbine impulse stage is designed to produce an axial exit velocity. For $U_t = 1225$ fps, sketch the velocity diagram and determine the work output of the stage. Assume a constant axial velocity of 500 fps, with an inlet stagnation temperature of 1220°R, and isentropic flow. Find the stage stagnation pressure ratio. (Take $c_p = 0.24$ Btu/lbm°R and $R = 53.3$ ft-lbf/lbm°R.)

12.9. An axial-flow fan is placed inside a ventilator shaft. The duct is made of galvanized iron and has a diameter of 0.7 m and length of 150 m. The fan rotates at 120 rad/s and moves 3 m³/s of air with density 1.2 kg/m³. Determine the pressure drop in the duct and the ideal power required to drive the fan, and design the fan blades using the blade element method. Use an airfoil section that has a lift coefficient of 0.4 at zero angle of attack and a lift curve slope of 3.4 per rad; take the hub radius to be 0.20 m.

12.10. A single-stage gas turbine consists of a set of stator blades (nozzles) and a set of rotor blades. The inlet conditions to the nozzles are stagnation pressure of 45 psia, stagnation temperature of 1200°F with a mass flow rate of 25 lbm/s. The rotor blade speed is 500 fps, while the axial velocity at rotor exit is 350 fps. Determine the exit angle of the nozzles, the blade angles of the rotor, and the stagnation pressure at the rotor exit (take $\gamma = 1.25$). Take $\alpha_1 = 15°$ and the gas constant $R = 35.3$ ft-lbf/lbm°R; the absolute rotor exit velocity is in the axial direction.

12.11. The radial-flow blower in a hot-air furnace rotates at 80 rad/s and moves 0.20m³/s of air with density 1.2 kg/m³. The rotor has a width of 0.3 m, an outer diameter of 0.3 m, and an inner diameter of 25 cm. The blades are designed for radial inflow ($\alpha_1 = \pi/2$). Compute the blade angle α_2, the ideal head produced, and the ideal power required. Blade angle $\beta_2 = 0.2$ rad.

12.12. Incompressible fluid flows through the stationary guide vanes of a radial-flow turbomachine, as shown in Figure P12.12. The fluid enters with velocity V_1 and leaves with velocity V_2. Derive an expression for the torque exerted on the guide vanes by the flow. What is the static pressure change ($p_2 - p_1$) across the vanes?

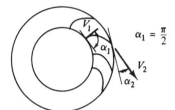

Figure P12.12

12.13. A centrifugal compressor with straight radial blades operates at 10,000 rpm. Assuming axial flow at inlet and a tip diameter of 32 in, find the work input to the rotor in Btu/lbm. If the inlet stagnation temperature is 65°F, find the stage pressure ratio for isentropic flow (take $\gamma = 1.4$, with $c_p = 0.24$ Btu/lbm°R, $R_1 = 8$ in).

12.14. A centrifugal pump delivers 15 liters/s of kerosene while operating at 120 rad/s. The rotor is 2 cm wide at exit, with backward-curved blades having $\beta_2 = 0.60$ rad. The radial velocity at exit is 2.5 m/s. Is this pump capable of producing a change in head of 15 m when operating at 150 rad/s?

12.15. A centrifugal water pump shown in Figure P12.15 is to be designed with the following characteristics:

Design speed: 1200 rpm

Design flow rate: 200 gpm

Backward-curved blades with $\beta_2 = 65°$

Axial inlet velocity less than 8 fps

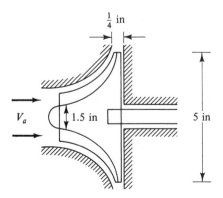

Figure P12.15

Determine the ideal change in total head, the absolute and relative discharge velocities, and ideal horsepower requirements at design conditions. Find the pump characteristic curve (ΔH_i versus Q) for constant-speed operation at 1200 rpm. Determine the ideal change in total head, flow rate, and ideal horsepower requirements for a speed of 600 rpm. What are the specific speeds at these two speed conditions?

12.16. Tests were conducted on a pump to establish its performance characteristics. However, the speed of the driving motor used in the tests decreased with increase in pump load. Show how you can compute from the tests the performance characteristics of the pump when driven by a synchronous motor.

12.17. Determine the ideal performance characteristics (ΔH_i versus Q) for a centrifugal pump having the following dimensions: rotor inlet radius 7.5 cm, rotor outlet radius 12.5 cm, inlet width of rotor 3.75 cm, outlet width 2.5 cm. The pump has radial inflow and operates at 150 rad/s, with $\beta_2 = 0.40$ rad. If $\beta_1 = 0.25$ rad, what is the design flow rate of the pump?

12.18. The speed of a hydraulic turbine is 240 rpm. The head at the site is 250 ft and the flow rate through the turbine is 500 cfs. Determine its specific speed for a hydraulic turbine efficiency of 85 percent.

12.19. In the preceding problem, what is the tangential component of the water leaving the stator at a radius of 2.0 ft? Determine the shaft torque and power. Assume the absolute velocity at rotor exit has no tangential component.

12.20. The hydraulic turbine of Example 12.2 is to be designed so that the flow leaves the blades without tangential component (i.e., $\alpha_2 = \pi/2$). What will be the power and torque developed by the turbine? Determine the variation of α_1, β_1, and β_2 with radial distance. Sketch blade shapes at hub, mean radius, and tip.

12.21. In designing radial-flow machines it is necessary, in addition to establishing the inlet and exit blade angles and flow velocities, to determine the blade shape between inlet and outlet. Determine the shape of blades in a radial-flow machine in which the relative blade angle β is taken as constant between inlet and outlet (i.e., $\beta_1 = \beta = \beta_2$). Take the ratio of outer to inner radius (R_2/R_1) as 1.1.

12.22. What will be the effect of doubling the minimum clearance between rotor and housing on velocity and pressure distribution at Section *A-A* in Example 12.5? Discuss the effect of clearance on the flow at the exit of the straight-walled housing.

12.23. What will be the increase in static pressure rise by adding a stator after the rotor of a horizontal axial-flow blower (see Figure 12.22)? The fluid enters and leaves the blower in an axial direction. The cross-sectional area of the blower remains constant. Take $V_{t2} = 0.5\, U_t$, neglect frictional losses, and assume constant fluid density.

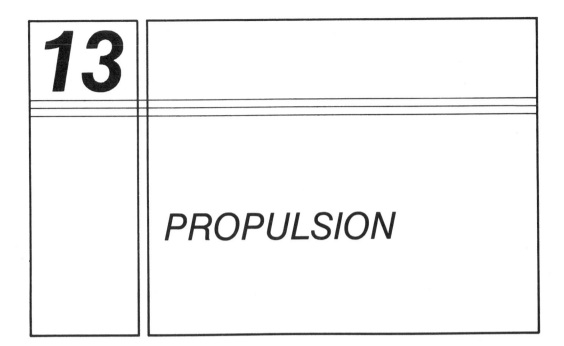

PROPULSION

13.1 INTRODUCTION

In Chapter 12 we discussed the behavior of fluids in machines in which an exchange of work occurs between the fluid and rotating parts of the machine. The purpose of the exchange was to get shaft work out of the fluid (turbines) or to increase the energy of the fluid (pumps). In order to describe the behavior of the fluid and the performance of the machines, we utilized the angular momentum equation as well as the energy equation. In this chapter we shall discuss devices whose purpose is to generate a force (thrust) by changing the linear momentum of the working fluid. The technique may involve the generation of a high-velocity fluid stream from material carried along with the device or may involve the alteration of momentum of the surrounding fluid. The propulsive force generated can be utilized to overcome the drag of a body moving at constant speed through an ambient medium, or it may be used to accelerate (or decelerate) a body, or to overcome the effect of gravity.

The supply of energy to the fluid can be accomplished by mechanical or chemical means, or by means of energy stored at high pressure. In every case, however, the propulsive forces are generated by a change of linear momentum of the working fluid. In the case of propellers, mechanical energy is supplied to a shaft of a bladed rotor, which increases the velocity of the ambient fluid passing through the propeller. The windmill is the opposite case, in that the velocity of the ambient air is reduced on passing through the blades, enabling the extraction of power. In jet propulsion, chemical energy is supplied to the device in order to obtain a high-velocity jet of fluid. Specifically, the propellant required to create the high-velocity stream in rockets is carried along as part of the rocket system. In air-breathing jet engines, only the fuel is carried along, with the surrounding air providing the working fluid.

The basic equation describing the performance of these propulsive devices is the linear momentum equation developed in Chapter 4. For the case in which the control volume experiences linear acceleration in the x direction, the momentum equation is given by Equation (4.25):

$$\Sigma F_x = M\frac{dV_{x\text{C.V.}}}{dt} + \frac{\partial}{\partial t}(MV_{x\text{rc}})_{\substack{\text{control} \\ \text{volume}}} + \int\limits_{\substack{\text{control} \\ \text{surface}}} V_{x\text{rc}}\rho V_{n\text{rc}}\,dA$$

Here $V_{x\text{C.V.}}$ is the velocity of the moving control volume relative to a fixed inertial system (i.e., an absolute velocity), while the other velocities are taken relative to the moving control volume. V_n is the component of the relative velocity of the fluid in a direction normal to the surface of the control volume. ΣF_x represents the sum of all externally applied forces acting on the control volume. The first term on the right-hand side represents the inertia force of the entire mass (M) within the control volume due to its acceleration, $dV_{x\text{C.V.}}/dt$. The second term refers to the rate of storage of linear momentum inside the control volume with velocities expressed relative to the control volume. Finally, the last term of Equation (4.25) denotes the net rate of efflux of linear momentum from the control volume.

In most applications the flow through the propulsive system is steady and the rate of storage of linear momentum within the control volume is zero. Equation (4.25) then reduces to

$$\Sigma F_x = M\frac{dV_{x\text{C.V.}}}{dt} + \int\limits_{\substack{\text{control} \\ \text{surface}}} V_x(\rho V_n\,dA) \tag{13.1}$$

When the control volume is moving with constant velocity so that $dV_{x\text{C.V.}}/dt$ equals zero, Equation (13.1) further simplifies to

$$\Sigma F_x = \int\limits_{\substack{\text{control} \\ \text{surface}}} V_x(\rho V_n\,dA) \tag{13.2}$$

13.2 JET PROPULSION

The most widely used jet propulsion devices are the turbojet engine and the rocket engine. In turbojets, the surrounding fluid is utilized as propellant, whereas rockets carry their own propulsive fluid. Specifically, in a turbojet engine (Figure 13.1), ambient air is taken in through the inlet and is compressed in the axial-flow compressor. Heat is then added by combustion in the

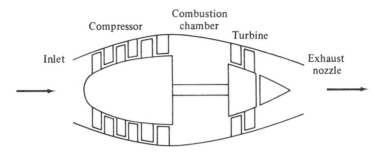

Figure 13.1. Turbojet engine.

combustion chamber. The products of combustion are expanded first in the axial-flow turbine, which drives the compressor, and then in the exhaust nozzle. In a rocket engine, no fluid is ingested from the surrounding medium; the rocket carries its own propulsive fluid. The propellant gas is generated at high pressure in the combustion chamber of the rocket engine. The source of the gas can be a solid fuel or a liquid fuel, or two liquids. Instead of using a combustion process, the propellant may also be heated by an external source such as a nuclear reactor. The high-pressure gas is then expanded through a converging-diverging nozzle, where it attains supersonic speeds at exit. An example of a liquid-propellant rocket system is shown in Figure 13.2.

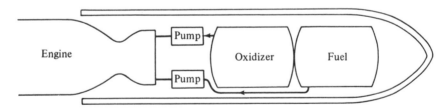

Figure 13.2. Liquid-propellant rocket.

For low thrust levels, as required in the control of the attitude of a spacecraft, compressed gas stored in containers is often utilized to generate small jets without addition of heat.

In order to analyze the performance of a jet propulsion system, consider the schematic shown in Figure 13.3. As indicated, the device is moving with steady velocity $V_{\text{C.V.}}$ in the positive x direction. One-dimensional flow will be assumed. Air is taken in at the inlet with uniform velocity V_i. In most applications the inlet velocity of the air will be equal to the forward speed $V_{\text{C.V.}}$. Exceptions arise when the device is placed in the flow field of another body. Then the magnitude of inlet velocity must be computed based on inviscid flow techniques discussed in Chapter 7. The air is compressed within the device; its temperature is raised by the combustion process and is expanded

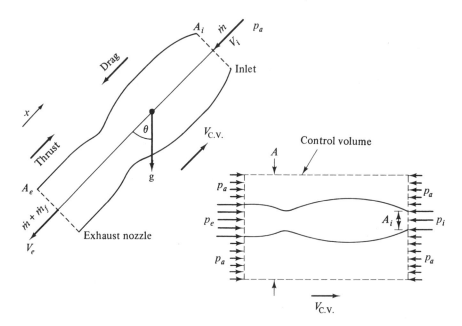

Figure 13.3. Schematic of jet propulsion system.

in the exhaust nozzle. The gas leaves the nozzle with uniform exit velocity V_e. In Figure 13.3, $V_{\text{C.V.}}$ is the absolute velocity of the propulsion system moving in a positive x direction, while V_i and V_e are inlet and exhaust velocities, respectively, taken relative to the moving system. The x axis is inclined at an angle of θ with respect to the vertical direction of gravity. From Figure 13.3 we also see that the pressure force is

$$p_e A_e + p_a(A - A_e) - p_a(A - A_i) - p_i A_i = (p_e - p_a)A_e - (p_i - p_a)A_i$$

so that ΣF_x can be expressed as

$$\Sigma F_x = \underbrace{-D}_{\substack{\text{viscous} \\ \text{force}}} \underbrace{- Mg \cos \theta}_{\substack{\text{body} \\ \text{force}}} + \underbrace{(p_e - p_a)A_e - (p_i - p_a)A_i}_{\text{pressure force}}$$

Substituting into Equation (13.1), we obtain

$$-D - Mg \cos \theta + (p_e - p_a)A_e - (p_i - p_a)A_i$$
$$= M \frac{dV_{\text{C.V.}}}{dt} - \dot{m}(V_e - V_i) - \dot{m}_f V_e$$

where $\dot{m}_f$ = mass flow rate of fuel, $\dot{m}$ = mass flow rate of air. Here the inflow was taken as positive and outflow as negative, for the velocities V_i and V_e are in the negative x direction. Rearranging the equation, we get

$$-D - Mg \cos \theta - M \frac{dV_{\text{C.V.}}}{dt} + \dot{m}(V_e - V_i)$$
$$+ \dot{m}_f V_e + (p_e - p_a)A_e - (p_i - p_a)A_i = 0$$

The first three terms are forces that have to be counteracted by the thrust force exerted by the propulsive device; that is,

$$-D - Mg \cos \theta - M\frac{dV_{\text{C.V.}}}{dt} + \text{Thrust} = 0$$

Hence the expression for the propulsive thrust becomes

$$\text{Thrust} = \dot{m}(V_e - V_i) + \dot{m}_f V_e + (p_e - p_a)A_e - (p_i - p_a)A_i \quad (13.3)$$

The thrust is exerted in a direction opposite to the exhaust velocity V_e. For example, in level flight ($\theta = 0$), if the thrust is utilized just to overcome the drag of a body, we obtain motion of the body at constant velocity. When the thrust exceeds the drag, the body will be accelerated.

13.2(a) Turbojets

Figure 13.4 shows a cutaway view of a typical *turbojet* engine, the Pratt and Whitney JT4 engine. Figure 13.5 shows the JT4 engine installed on a Douglas DC-8 jet transport. Air is taken in through the air inlet or diffuser and compressed in two axial-flow compressors, each having six stages. The pressure increase across the front (low-speed) compressor is fourfold, and across the rear (high-speed) compressor it is 3.2-fold. The airflow rate is 115 kg/s. The combustion chamber is located between the compressors and the turbines and consists of eight cylindrical cans arranged circumferentially around the shafts. The fuel consists of a mixture of kerosene and gasoline and is burnt in each combustion cylinder. The fuel flow rate is 1.85 kg/s. The hot gases of combustion are expanded in the two single-stage, axial-flow turbines, each driving one of the compressors. The turbines are connected to the compressors by two hollow, concentric drive shafts. The gases are further expanded to high velocity in the exhaust nozzle. The takeoff thrust

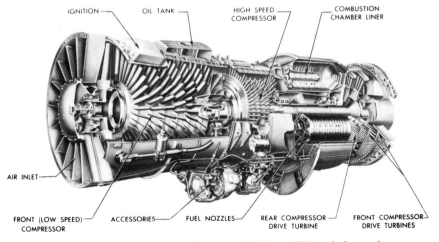

Figure 13.4. Cutaway view of Pratt and Whitney JT4 turbojet engine. (*Courtesy* United Aircraft Corp.)

Figure 13.5. Douglas DC-8 jet transport. (*Courtesy* McDonnell-Douglas Corp.)

of this engine is 77.8 kN. JT4 engines have been used in the Boeing 707 and Douglas DC-8 jet aircraft.

An example of a turbofan jet engine, Pratt & Whitney PW 4000, is shown in Figure 13.6. The *turbofan* engine is a basic jet engine to which a fan has been added. The fan pulls in air from outside the engine, increasing the velocity of that air and ejecting it through the fan exhaust. This process increases the total thrust and efficiency of the engine. The overall compressor pressure ratio is 32:1, thrust is 52,000–60,000 pounds, weight is 9200 pounds, and dimensions are 97 in diameter × 133 in long. For this turbofan, the

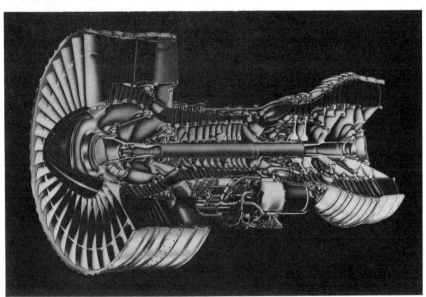

Figure 13.6. The Pratt & Whitney PW4000 turbofan engine. (*Courtesy* United Technologies, Pratt & Whitney.)

Figure 13.7.   Pratt & Whitney 2037 turbofan engine installed on Boeing 757 jet aircraft. (*Courtesy* United Technologies, Pratt & Whitney.)

Figure 13.8.   Pratt & Whitney 2037 turbofan engine installed on Boeing 757 jet aircraft. (*Courtesy* United Technologies, Pratt & Whitney.)

bypass ratio is 5:1. Figures 13.7 and 13.8 show the Pratt & Whitney 2037 turbofan engine installed on Boeing 757 jet aircraft.

Next, let us examine the performance of a turbojet in constant-velocity level flight. For subsonic flight speeds, the pressure at the inlet can be taken as the ambient pressure p_a with inlet velocity V_i equal to forward speed $V_{\text{C.V.}}$. For supersonic flights, consideration must be given to shock formation in the diffuser. Restricting our consideration to subsonic flight, we have, from Equation (13.3),

$$\text{Thrust} = \dot{m}(V_e - V_{\text{C.V.}}) + \dot{m}_f V_e + (p_e - p_a)A_e \qquad (13.4)$$

When the gas in the exit nozzle is expanded to ambient pressure, the pressure term becomes zero. Otherwise the contribution of the pressure thrust can be in the negative or positive x direction, depending on occurrence of over-expansion or underexpansion in the nozzle.

Usually the fuel mass flow rate is small in comparison to the airflow rate $\dot{m}$, so that we can write for a perfectly expanded nozzle ($p_e = p_a$):

$$\text{Thrust} = \dot{m}(V_e - V_{\text{C.V.}}) \qquad (13.5)$$

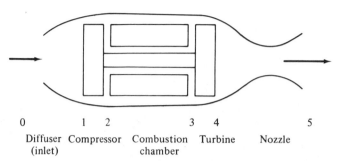

0	1	2		3	4		5
Diffuser	Compressor		Combustion		Turbine	Nozzle	
(inlet)			chamber				

Figure 13.9.

EXAMPLE 13.1

Consider the performance of an ideal turbojet engine shown in Figure 13.9, with a steady forward velocity of 125 m/s at 3 km altitude ($p = 84$ kPa, $T = 278$ K). Assume that the flow in the diffuser, compressor, turbine, and nozzle is isentropic, that the diffuser slows the inlet air down to negligible velocity with respect to the engine, that the nozzle is perfectly expanded, and that there is no pressure drop in the combustion chamber. Take the maximum allowable inlet turbine temperature that can be tolerated without damage to the blades to be 1200 K with a compressor pressure ratio of 8 to 1. The heating value of the fuel is 42,000 kJ/kg of fuel. (1) Determine the thrust specific fuel consumption (thrust per unit mass flow of fuel). (2) Find the thrust for an airflow of 25 kg/s.

Assume that the fuel flow rate can be neglected in comparison to the airflow rate; also assume that air can be treated as a perfect gas with constant specific heats ($c_p = 1.0$ kJ/kg · K, $\gamma = 1.4$, $R = 0.287$ kJ/kg · K).

Solution   With subsonic flow at the inlet, the inlet velocity $V_0 = 125$ m/s, $T_0 = 278$ K, and $p_0 = 84$ kPa. Assuming one-dimensional flow in the engine, the energy equation for adiabatic flow between 0 and 1 takes the form

$$\frac{V_0^2}{2} + c_p T_0 = \frac{V_1^2}{2} + c_p T_1$$

For V_1 negligible, we have

$$T_1 = T_0 + \frac{V_0^2}{2c_p} = 278 \text{ K} + \frac{125^2 \text{ m}^2/\text{s}^2}{2(1.0 \text{ kJ/kg} \cdot \text{K})(1000 \text{ J/kJ})}$$

$$= 278 + 7.81 = 285.81 \text{ K}$$

Since the flow between 0 and 1 has been assumed isentropic,

$$\frac{p_1}{p_0} = \left(\frac{T_1}{T_0}\right)^{\gamma/(\gamma - 1)}$$

$$= \left(\frac{285.81}{278}\right)^{3.5} = 1.1018$$

or

$$p_1 = 92.55 \text{ kPa}$$

The pressure ratio in the compressor is $p_2/p_1 = 8.0$, so that $p_2 = 740.4 \text{ kPa}$. For an isentropic compressor,

$$\frac{T_2}{T_1} = \left(\frac{p_2}{p_1}\right)^{(\gamma-1)/\gamma} = 8^{0.286} = 1.8125$$

so that

$$T_2 = 518.0 \text{ K}$$

Neglecting changes of kinetic energy and with the fuel flow negligible with respect to the airflow, the energy equation for flow in the combustion chamber is

$$q_c = c_p(T_3 - T_2)$$

with q_c the heat added due to combustion. Therefore, with $T_3 = 1200 \text{ K}$, we obtain

$$q_c = (1.0 \text{ kJ/kg} \cdot \text{K})[(1200 - 518) \text{ K}] = 682 \text{ kJ/kg}$$

In the jet engine, all the work output of the turbine $[w_t = c_p(T_3 - T_4)]$ is used to drive the compressor, or $w_t = w_c$. In other words,

$$c_p(T_3 - T_4) = c_p(T_2 - T_1)$$

so that

$$T_3 - T_4 = T_2 - T_1 = 518.0 - 285.8 = 232.2 \text{ K}$$

or

$$T_4 = 967.8 \text{ K}$$

For an isentropic turbine,

$$\frac{p_4}{p_3} = \left(\frac{T_4}{T_3}\right)^{\gamma/(\gamma-1)} = \left(\frac{967.8}{1200}\right)^{3.5} = 0.4711$$

so that, with $p_3 = p_2$,

$$p_4 = 0.4711 p_3 = 0.4711(740.4 \text{ kPa}) = 348.8 \text{ kPa}$$

The expansion in the nozzle, from 348.8 kPa to 84 kPa, is isentropic, so again

$$\frac{T_5}{T_4} = \left(\frac{p_5}{p_4}\right)^{(\gamma-1)/\gamma} = \left(\frac{84}{348.8}\right)^{0.286} = 0.6655$$

and

$$T_5 = 644.1 \text{ K}$$

The energy equation for the nozzle is

$$\frac{V_5^2}{2} = c_p(T_4 - T_5)$$

$$= (1.0 \text{ kJ/kg} \cdot \text{K})[967.8 - 644.1) \text{ K}]$$

$$= 323.7 \text{ kJ/kg}$$

or

$$V_5 = \sqrt{647.4 \times 10^3 \text{ m}^2/\text{s}^2}$$
$$= 804.6 \text{ m/s}$$

For the engine, using Equation (13.5), we find

$$\text{Thrust} = \dot{m}_a(V_e - V_{\text{C.V.}})$$

with $\dot{m}_a$ the airflow through the engine.

$$\text{Thrust} = \dot{m}_a(804.6 - 125)$$

or

$$\frac{\text{Thrust}}{\dot{m}_a} = 679.6 \text{ N/(kg air/s)}$$

The heat added in the combustion chamber, q_c, was found to be 682 kJ/kg of air; this energy is derived from combustion of the fuel having a heating value of 42,000 kJ/kg of fuel. Writing an energy balance, $682\dot{m}_a = 42,000\dot{m}_f$, so the air fuel ratio is

$$\frac{\dot{m}_a}{\dot{m}_f} = \frac{42,000}{682} = 61.58 \text{ kg air/kg fuel}$$

The thrust-specific fuel consumption (TSFC) is

$$\text{TSFC} = \frac{\text{Thrust}}{\dot{m}_{\text{fuel}}}$$

$$= \frac{679.6\dot{m}_a}{\dot{m}_a/61.58} \text{ N/(kg fuel/s)}$$

$$= 41.85 \text{ kN/(kg fuel/s)}$$

For an airflow of 25 kg/s, the thrust is

$$\text{Thrust} = [679.6 \text{ N/(kg air/s)}](25 \text{ kg/s})$$
$$= \underline{16.99 \text{ kN}} \qquad\blacksquare$$

EXAMPLE 13.2

A turbojet takes off at sea level (14.7 psia, 60°F). The heating value of the jet fuel is 17,000 Btu/lbm, the compressor pressure ratio is 10:1, and the turbine inlet temperature is 1800°F. The exit nozzle has a throat area of 2.0 ft², with the nozzle perfectly expanding the flow to ambient pressure. Assuming isentropic compressor, turbine, and nozzle, calculate the takeoff thrust, fuel flow rate, and TSFC. Neglect fuel flow rate in comparison to airflow; assume that air can be treated as a perfect gas with constant specific heat ($\gamma = 1.4$, $c_p = 0.24$ Btu/lbm°R).

Solution Referring to Figure 13.9, $T_2/T_1 = (p_2/p_1)^{(\gamma-1)/\gamma}$. At takeoff, $V_0 = V_1 = 0$ ft/s, and so $T_1 = 520°R$. Therefore,

$$T_2 = 10^{0.286}T_1 = 1.932(520°R) = 1005°R$$

The energy equation for the combustion chamber is

$$q_c = c_p(T_3 - T_2)$$

with q_c the heat added as a result of combustion. Therefore, with $T_3 = 2260°R$, we obtain

$$q_c = (0.24 \text{ Btu/lbm°R})(2260°R - 1005°R)$$
$$= 301.2 \text{ Btu/lbm}$$

Since turbine work equals compressor work and we are assuming constant specific heat, we have

$$T_3 - T_4 = T_2 - T_1 = 1005°R - 520°R = 485°R$$

or

$$T_4 = 2260°R - 485°R = 1775°R$$

For an isentropic turbine,

$$\frac{p_4}{p_3} = \left(\frac{T_4}{T_3}\right)^{\gamma/(\gamma-1)} = \left(\frac{1775}{2260}\right)^{3.5} = 0.4294$$

For zero pressure drop in the combustion chamber, $p_3 = p_2$, where

$$p_2 = 10p_1 = 10p_0$$

For a perfectly expanded nozzle, $p_5 = p_0$, and so

$$\frac{p_4}{p_5} = \frac{0.4294(10p_0)}{p_0} = 4.294$$

For an isentropic nozzle

$$\frac{T_4}{T_5} = \left(\frac{p_4}{p_5}\right)^{(\gamma-1)/\gamma} = 4.294^{0.286} = 1.517$$

and

$$T_5 = 1170°R$$

The energy equation for the nozzle is

$$\frac{V_5^2}{2} + c_p T_5 = \frac{V_4^2}{2} + c_p T_4$$

In our case, $V_4 = 0$, and so

$$\frac{V_5^2}{2} = c_p(T_4 - T_5)$$
$$= (0.24 \text{ Btu/lbm°R})(778 \text{ ft-lbf/Btu})(32.17 \text{ lbm/slug})(605°R)$$
$$V_5 = 2696 \text{ ft/s}$$

In order to determine thrust, we also need to know the mass flow rate through the nozzle:

$$\dot{m}_5 = \rho_5 A_5 V_5$$

The throat area is given as 2.0 ft², with the nozzle isentropically expanding the flow from $4.294p_0$ to p_0. From the isentropic flow table (see Appendix B), at $p/p_t = 0.2329$, $A/A^* = 1.255$. Therefore, $A_5 = 1.255(2.0\,\text{ft}^2) = 2.51\,\text{ft}^2$.

$$\rho_5 = \frac{p_5}{RT_5} = \frac{(14.7\,\text{lbf/in}^2)(144\,\text{in}^2/\text{ft}^2)}{(53.3\,\text{ft-lbf/lbm}°\text{R})(1170°\text{R})}$$

$$= 0.0339\,\text{lbm/ft}^3$$

and

$$\dot{m}_5 = (0.0339\,\text{lbm/ft}^3)(2.51\,\text{ft}^2)(2696\,\text{ft/s})$$

$$= 229.4\,\text{lbm/s}$$

The takeoff thrust is equal to

$$\dot{m}_5(V_5 - V_0) = \frac{229.4\,\text{lbm/s}}{32.17\,\text{lbm/slug}}[(2696 - 0)\,\text{ft/s}]$$

$$= \underline{19,220\,\text{lbf}}$$

To determine the TFSC, we need to find $\dot{m}_f$. For the combustor,

$$\dot{m}_{\text{air}}c_p(T_3 - T_2) = \dot{m}_f(17{,}000\,\text{Btu/lbm fuel})$$

Therefore,

$$\dot{m}_f = \frac{(229.4\,\text{lbm/s})(0.24\,\text{Btu/lbm}°\text{R})(2260°\text{R} - 520°\text{R})}{17{,}000\,\text{Btu/lbm}}$$

$$= \underline{5.635\,\text{lbm/s}}$$

and

$$\text{TFSC} = \frac{19{,}220\,\text{lbf}}{5.635\,\text{lbm/s}}$$

$$= \underline{0.947\,\text{lbf/(lbm fuel/h)}} \qquad \blacksquare$$

The preceding calculations were performed for an idealized system. In the real case, factors such as compressor and turbine efficiencies (η_c and η_t) must be included, where

$$\eta_c = \frac{\left(\begin{array}{c}\text{Work required for isentropic compressor}\\ \text{operating between } p_1 \text{ and } p_2\end{array}\right)}{\text{Actual work required}}$$

and

$$\eta_t = \frac{\text{Actual work output}}{\left(\begin{array}{c}\text{Work output of isentropic turbine}\\ \text{operating between } p_3 \text{ and } p_4\end{array}\right)}$$

(Typical values are $\eta_t = 0.90$, $\eta_c = 0.85$.)

To illustrate, in Example 13.2, let us assume a compressor efficiency of 0.85 with the same compressor ratio of 10. The isentropic compressor work required, using Example 13.2, is

$$c_p(1005°R - 520°R) = (0.24 \text{ Btu/lbm°R})(485°R)$$

$$= 116.4 \text{ Btu/lbm}$$

For a compressor efficiency of 0.85, the actual compressor work required is

$$\frac{116.4 \text{ Btu/lbm}}{0.85} = 136.9 \text{ Btu/lbm}$$

Also,

$$T_2 - T_1 = \frac{136.9 \text{ Btu/lbm}}{0.24 \text{ Btu/lbm°R}} = 570.4°R$$

and

$$T_{2 \text{ actual}} = 1090.4°R$$

(a higher value than that calculated for the isentropic compressor of Example 13.2).

Other factors that have been neglected and that would have a significant effect on performance would be the change in specific heats due to temperature variations, frictional losses in the diffuser, and pressure drop in the combustion chamber.

13.2(b) Rockets

Various aspects of the study of rocket engines have been treated previously in this book. In Chapter 4, for example, the rocket was used to exemplify the equation of linear momentum; in Chapter 11 compressible flow through nozzles was treated in some detail. In this section we shall go further into a study of the performance and design of rockets.

As an example of the performance of a rocket, let us examine the case of a rocket moving vertically against the earth's gravity field. As indicated previously, the momentum equation is given by

$$-D - M_R g - M_R \frac{dV_R}{dt} + \text{Thrust} = 0$$

where M_R and V_R are the mass and velocity of the rocket, respectively. Since no fluid is ingested in the case of a rocket, that is, $\dot{m} = 0$ and $A_i = 0$, the expression for the rocket's thrust becomes

$$\text{Thrust} = \dot{m}_f V_e + (p_e - p_a)A_e \qquad (13.6)$$

where $\dot{m}_f$ includes the entire exhaust gas flow, both oxidizer and fuel flow for a bipropellant rocket.

Combining the momentum equation above with the expression for the thrust [Equation (13.6)], we obtain

$$-D - M_R g - M_R \frac{dV_r}{dt} + \dot{m}_f V_e + (p_e - p_a)A_e = 0$$

which is identical to the expression given in Example 4.17, Section 4.5, where this problem has been treated as an example of an application involving control volumes in noninertial coordinates. If we neglect the effects of drag and gravity, we obtain

$$(p_e - p_a)A_e = M_R \frac{dV_R}{dt} - \dot{m}_f V_e$$

Integrating as in Example 4.17, we obtain for constant $\dot{m}_f$, p_a, and V_e, with $V_R = V_{R0}$ at $t = 0$ and $M_R = M_0 - \dot{m}_f t$,

$$V_R = V_{R0} = -\left[(p_e - p_a)\frac{A_e}{\dot{m}_f} + V_e\right]\ln\left(1 - \frac{\dot{m}_f t}{M_0}\right)$$

where M_0 is the initial rocket mass. If we choose the time t when the burn of the rocket propellant ceases, $\dot{m}_f t$ represents the total propellant mass M_{pr} consumed during the burn. Hence at propellant burnout we have

$$V_R - V_{R0} = -\left[\frac{(p_e - p_a)}{\dot{m}_f}A_e + V_e\right]\ln\left(1 - \frac{M_{pr}}{M_0}\right)$$

where M_{pr}/M_0 is called the *propellant fraction,* that is, that fraction of the total rocket mass that is propellant mass. Since $M_0 - M_{pr}$ is the mass of the rocket at burnout, $M_0/(M_0 - M_{pr})$ is called the *mass fraction* of the rocket, that is, the ratio of initial to burnout mass of the rocket; therefore,

$$-\ln\left(1 - \frac{M_{pr}}{M_0}\right) = \ln\frac{M_0}{M_0 - M_{pr}}$$

The term

$$\left[\frac{(p_e - p_a)}{\dot{m}_f}A_e + V_e\right]$$

is called the *characteristic velocity,* designated by c. It represents the effective exhaust velocity of the rocket engine. When the ambient pressure p_a equals the exhaust pressure p_e, the characteristic velocity equals the exhaust velocity V_e.

The increase in rocket velocity during the burn time is therefore expressed by

$$V_R - V_{R0} = c \ln\frac{M_0}{M_0 - M_{pr}} \tag{13.7}$$

Equation (13.7) is called the *rocket equation.* From it we see that the change in velocity of a rocket moving in a drag- and gravity-free field is proportional to the effective exhaust velocity (c) and the natural logarithm of the rocket mass ratio $M_0/(M_0 - M_{pr})$, where M_0 is the initial total mass of the rocket and M_{pr} is the propellant mass consumed during the burn.

The thrust per unit mass flow rate is called *specific impulse* (I_{sp}). From Equation (13.6) and the definition of the characteristic velocity, that is

$$c = \frac{p_e - p_a}{\dot{m}_f}A_e + V_e$$

we have

$$\text{Thrust} = \dot{m}_f c \quad \text{and} \quad I_{sp} = \frac{\text{Thrust}}{\dot{m}_f} = c$$

Therefore, Equation (13.7) can be written as

$$V_R - V_{R0} = I_{sp}\ln\frac{M_0}{M_0 - M_{pr}} \tag{13.8}$$

This equation indicates that the increase in the velocity of a rocket is proportional to the specific impulse. Thus it is desirable to use propellants with high specific impulse. Typical values are given in Table 13.1. In many applications the mass of the propellant of a chemical rocket is significantly larger than the payload carried along. Furthermore, the mass of propellant tanks and rocket structure also exceeds the payload mass. Hence, to avoid consuming propellants to accelerate nearly empty tankage and structure, rockets are designed with several stages. Each stage is separated from the portion of the vehicle carrying the payload. A multistage rocket thus consists of several individual stages, each stage having its own tanks, structure, and engines.

Note that if the ratio of initial to burnout mass is the same for each stage and each stage has the same exhaust velocity, then the increment of velocity provided by each stage will be the same (neglecting drag, gravity, and the pressure term in the thrust equation). In other words, after n stages, the velocity of the rocket will be

$$V_R = nV_e\ln\frac{M_0}{M_0 - M_{pr}} \tag{13.9}$$

where $M_0/(M_0 - M_{pr})$ is the mass ratio for each stage.

The use of staging rockets has made possible the escape of man from the earth's gravitational field. In order to escape from the earth, a rocket must have sufficient kinetic energy to overcome the pull of gravity. The gravitational force between two bodies is given by

$$F = G\frac{m_1 m_2}{d^2}$$

with d the distance separating the bodies and G the gravitational constant. As the distance z of a body above the surface of the earth increases, the acceleration due to gravity decreases according to

$$\frac{g}{g_0} = \frac{R^2}{(R + z)^2}$$

with R the radius of the earth and g_0 the acceleration due to gravity at the

TABLE 13.1. Specific Impulse for Various Fuel-Oxidizer Combinations

A. Liquid Bipropellants*

Oxidizer	Fuel	Mixture ratio $\left(\dfrac{\text{mass oxidizer}}{\text{mass fuel}}\right)$	I_{sp} [kN/(kg/s)]	I_{sp} [lbf/(lbm/s)]
Liquid oxygen	Liquid hydrogen	3.50	3.80	387
Liquid oxygen	Liquid methane	3.00	2.90	296
Liquid oxygen	RP-1	2.24	2.80	286
Liquid oxygen	Hydrazine (N_2H_4)	0.75	2.95	301
Liquid oxygen	Pentaborane (B_5H_9)	2.07	3.00	305
Liquid fluorine	Liquid hydrogen	4.70	3.90	398
Liquid fluorine	Liquid methane	3.15	3.07	313
Liquid fluorine	Pentaborane	3.75	3.23	329
Nitrogen tetroxide (N_2O_4)	Hydrazine	1.10	2.79	284
Nitrogen tetroxide	Pentaborane	3.10	2.80	286
Hydrogen peroxide (H_2O_2)	Hydrazine	1.50	2.77	282
Hydrogen peroxide	Pentaborane	3.24	2.93	299

B. Solid Propellants*

Oxidizer (15 wt. % polyethylene binder)	Fuel	Mass % oxidizer	I_{sp} [kN/(kg/s)]	I_{sp} [lbf/(lbm/s)]
Ammonium nitrate	Beryllium	69	2.83	289
Ammonium nitrate	Boron	61	2.46	251
Ammonium nitrate	Aluminum	61	2.50	256
Ammonium perchlorate	Beryllium	70	2.78	284
Ammonium perchlorate	Boron	71	2.50	256
Ammonium perchlorate	Lithium hydride	60	2.36	241
Ammonium perchlorate	Aluminum	65	2.60	265
Hydrazinium nitrate	Beryllium	69	2.83	289
Hydrazinium nitrate	Boron	61	2.55	260
Hydrazinium nitrate	Lithium	69	2.49	254
Hydrazinium nitrate	Aluminum	61	2.57	262

C. Monopropellants†

	I_{sp} [kN/(kg/s)]	I_{sp} [lbf/(lbm/s)]
Nitromethane	2.14	218
Hydrogen peroxide (87%–13% H_2O)	1.24	126
Ethylene oxide	1.57	160

* Values of I_{sp} are for $p_{\text{chamber}} = 68$ atm, $p_e = 1$ atm (perfect expansion); all calculations are based on frozen equilibrium in the nozzle; that is, the composition of the exhaust gases does not change during expansion.

† $p_{\text{chamber}} = 19.8$ atm, $p_e = 1$ atm (perfect expansion).

earth's surface. Now we can equate the kinetic energy of the rocket to the work necessary to overcome the earth's pull, in order to find the rocket velocity required to just escape from the earth's surface. In the absence of drag,

$$\frac{1}{2}mV_{esc}^2 = m\int_{x=0}^{x=\infty} g\,dz = mg_0R^2\int_0^\infty \frac{dz}{(R+z)^2} = mg_0R$$

or

$$V_{esc} = \sqrt{2g_0R}$$

At the earth's surface, $g_0 = 9.81$ m/s^2 and $R = 6378$ km, so that

$$V_{esc} = \sqrt{2(9.81 \text{ m/s}^2)(6.378 \times 10 \text{ m})}$$
$$= 11,200 \text{ m/s}$$

In English units, $V_{esc} = \sqrt{2(32.17 \text{ ft/s}^2)R}$ where the radius of the earth is 3963 mi. Therefore,

$$V_{esc} = \sqrt{2(32.17 \text{ ft/s}^2)(3963 \text{ mi})(5280 \text{ ft/mi})}$$
$$= 36,700 \text{ ft/s}$$

EXAMPLE 13.3

The initial mass of a rocket is to be 50,000 kg, of which 5 percent is to be structure, the remainder fuel and payload. If the specific impulse of the propellant combination used is 3.0 kN/(kg/s), calculate the payload mass that would be able to escape from earth for a single-stage and for a two-stage rocket. Neglect aerodynamic drag, gravity forces, and the term involving $p_e - p_a$ in the thrust equation.

Solution For a single-stage rocket,

$$V_R = V_e \ln\frac{M_0}{M_0 - M_{pr}}$$

with $V_e = 3.0$ kN/(kg/s) $\times$ 1000 N/kN $= 3000$ m/s. Therefore,

$$11,200 = 3000 \ln\frac{M_0}{M_0 - M_{pr}} \quad \text{or} \quad \frac{M_0}{M_0 - M_{pr}} = 41.8$$

$$M_0 - M_{pr} = \frac{50,000}{41.8} = 1196 \text{ kg}$$

Since 2500 kg must be devoted to structure, it follows that a single-stage rocket could not escape. For a two-stage rocket, with the same mass ratio for each stage,

$$11,200 \text{ m/s} = 2(3000 \text{ m/s}) \ln\frac{M_0}{M_0 - M_{pr}}$$

so that

$$\frac{M_0}{M_0 - M_{pr}} = 6.47$$

or, for each stage,

$$M_0 - M_{pr} = 0.155M_0$$

For the first stage, $M_0 = 50{,}000$ kg, so that $M_0 - M_{pr} = 7750$ kg, of which 2500 kg is structure. The payload of the first stage is thus 5250 kg, which is the initial mass of the second stage. Of this mass, 15.5 percent of 5250 kg or 814 kg is payload and structure. Again, 5 percent of 5250 kg is structure; the remainder, amounting to 814 kg $-$ 263 kg $=$ <u>551 kg</u>, is the payload that could be accelerated to escape velocity.   ∎

EXAMPLE 13.4

A rocket has a total initial mass of 100,000 lbm with single-stage rocket engines capable of producing 1,500,000 lbf of thrust. The rocket fuel has a specific impulse of 300 lbf/(lbm/s). Calculate the velocity of the rocket after 10 s of vertical flight. Neglect drag but include gravity in the calculation. Assume constant thrust and $\dot{m}_f$.

Solution   From Section 13.2b, with no drag, we obtain

$$\text{Thrust} - M_R\frac{dV_R}{dt} - M_R g = 0$$

But, at any time, $M_R = M_0 - \dot{m}_f t$, and so

$$\frac{dV_R}{dt} = \frac{1}{M_0 - \dot{m}_f t}[\text{Thrust} - (M_0 - \dot{m}_f t)g]$$

Integrating,

$$\int_{V_{R_{t=0}}}^{V_{R_t}} dV_R = \int_0^t \left[\frac{\text{Thrust} - (M_0 - \dot{m}_f t)g}{M_0 - \dot{m}_f t}\right] dt$$

In the above, $V_{R_{t=0}} = 0$ and thrust $= \dot{m}_f I_{sp}$, and so

$$V_R = I_{sp}\ln\left(\frac{M_0}{M_0 - \dot{m}_f t}\right) - gt$$

$$\dot{m}_f = \frac{\text{Thrust}}{I_{sp}} = \frac{1{,}500{,}000 \text{ lbf}}{300 \text{ lbf}/(\text{lbm/s})} = 5000 \text{ lbm/s}$$

$$V_R = [300 \text{ lbf}/(\text{lbm/s})](32.17 \text{ lbm/slug})\ln\left[\frac{100{,}000}{100{,}000 - 5000(10)}\right]$$

$$- (32.17 \text{ ft/s}^2)(10 \text{ s})$$

$$= 6368 \text{ ft/s} \qquad\qquad ∎$$

In chemical rockets, the energy required to accelerate the rocket is derived from the propellants. The exhaust velocity is determined by the energy released in the chemical reaction during combustion. For example,

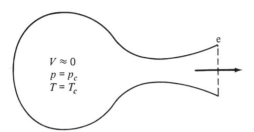

$V \approx 0$
$p = p_c$
$T = T_c$

e

Figure 13.10.   Nozzle flow.

applying the energy equation for one-dimensional steady flow between the rocket combustion chamber and the exhaust plane (see Figure 13.10), we obtain for adiabatic flow of a perfect gas with constant specific heats

$$h_c = h_e + \frac{V_e^2}{2}$$

or

$$V_e = \sqrt{2c_p(T_c - T_e)}$$

where, for a perfect gas, $c_p = R\gamma/(\gamma - 1)$. For a perfectly expanded nozzle, with isentropic flow in the nozzle,

$$\frac{T_e}{T_c} = \left(\frac{p_a}{p_c}\right)^{(\gamma-1)/\gamma}$$

so that

$$V_e = \sqrt{2\frac{R\gamma T_c}{\gamma - 1}\left[1 - \left(\frac{p_a}{p_c}\right)^{(\gamma-1)/\gamma}\right]}$$

or since $R = \overline{R}/\overline{M}$, where $\overline{M}$ is the mean molecular mass of the exhaust gases and $\overline{R}$ is the universal gas constant,

$$V_e = \sqrt{2\frac{\overline{R}\gamma}{\gamma - 1}\left(\frac{T_c}{\overline{M}}\right)\left[1 - \left(\frac{p_a}{p_c}\right)^{(\gamma-1)/\gamma}\right]} \qquad (13.10)$$

For a perfectly expanded nozzle with $p_e = p_a$, $I_{sp} = V_e$, so that the specific impulse of a rocket fuel–oxidizer combustion can be related to the ratio between chamber temperature and molecular weight of the exhaust gases, the high specific impulse being associated with the higher chamber temperatures and lower mean molecular weights.

In a well-designed rocket combustion chamber, there is negligible heat loss and the products of combustion are in thermodynamic equilibrium; thus the methods of equilibrium thermodynamics can be used to evaluate the adiabatic flame temperature T_c and the products of combustion.

A liquid rocket may use a single propellant or two propellants. Examples of monopropellants are hydrazine and ethylene oxide. Bipropellants consist of a fuel (e.g., kerosene, ethyl alcohol, liquid hydrogen) and an oxidizer (e.g., liquid oxygen, hydrogen peroxide, liquid fluorine). The liquid propellants are stored in separate tanks and are fed to the combustion chamber and

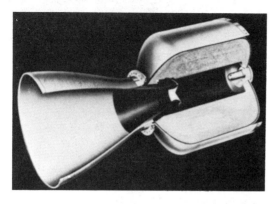

Figure 13.11. ATS apogee motor.

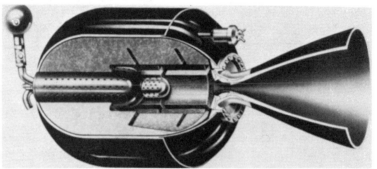

Figure 13.12. Long-life restartable space motor.

ignited at high pressure. Two techniques are used to achieve the high pressure: pressurization by use of a high-pressure gas or the use of pumps.

Solid propellant rockets are shown in Figures 13.11 and 13.12. Solid propellants are mixtures that contain all the chemicals required for combustion. The solid propellant mixture is stored within the combustion chamber, with combustion taking place on the surface of the propellant. The solid propellant grain can be designed to be end-burning or internal-burning, with the rate of gas generation from the fuel dependent on the shape of the propellant in the chamber. Examples of two internal-burning configurations resulting in constant gas generation rate and hence constant thrust are shown in Figure 13.13. The gas generation rate $\dot{m}_g$ in a solid-propellant combustion chamber is given by

$$\dot{m}_g = \rho A_b r$$

with ρ the solid propellant density, A_b the burning surface area, and r the burning rate normal to the propellant surface in length per unit time. Burning rate is a function of chamber pressure.

The advantage of a solid-propellant rocket over a liquid-propellant rocket lies in its simplicity; the former requires no complex pumping or gas pressurization system. Unfortunately, once launched it is difficult to change the flight path of the solid rocket; the thrust can only be varied by the design of the grain shape in the chamber during manufacture. With a liquid rocket, of course, thrust can be easily varied during flight by valving the propellant flow to the chamber.

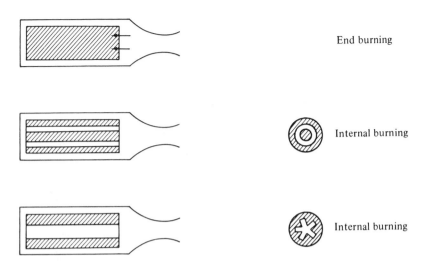

End burning

Internal burning

Internal burning

Figure 13.13. Propellant shapes for constant-thrust solid rocket motors.

Typical values of rocket performance parameters for several fuel-oxidizer combinations are given in Table 13.1.

An example of a multistage rocket is the *Saturn V* rocket that was utilized in the Apollo program. (*Apollo 11* put the first man on the moon.) This rocket consists of three stages, as indicated in Figures 13.14 and 13.15.

Figure 13.14. The *Saturn V* launch vehicle.

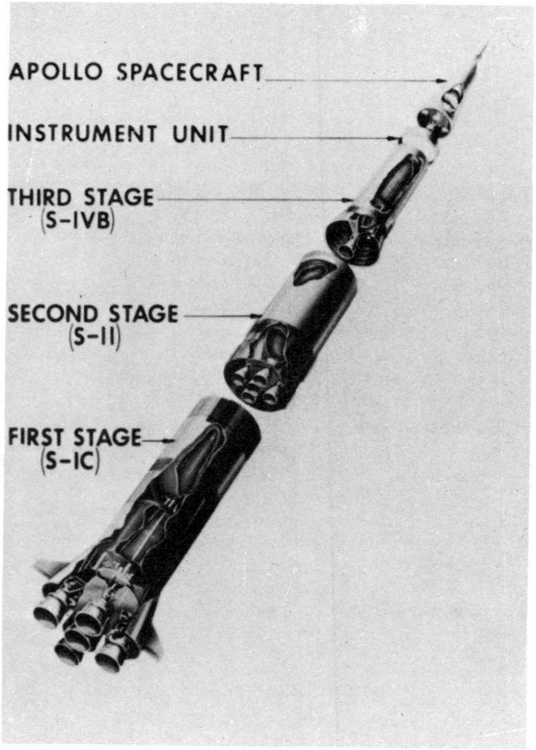

FIGURE 13.15 Characteristics of Saturn V Launch Vehicle

Characteristics

Length (vehicle)	86 m	281 ft
Length (vehicle, spacecraft, LES)	111 m	363 ft
Mass at liftoff	2,900,000 kg	6,400,000 lbm
Payload capability approximate		
translunar trajectory	45,000 kg	100,000 lbm
Earth orbit	130,000 kg	285,000 lbm

Stages

First (S-IC)			
Size	10 m × 42 m	33 ft × 138 ft
Engines	5 F-1	
Thrust (501 thru 503)	33 MN	7,500,000 lbf
(504 and Sub)	34 MN	7,610,000 lbf
Propellants	LOX & RP-1	
Second (S-II)			
Size	10 m × 25 m	33 ft × 81 ft
Engines	5 J-2	
Thrust (501 thru 503)	5 MN	1,125,000 lbf
(504 and Sub)	5.1 MN	1,150,000 lbf
Propellants	LOX & LH$_2$	
Third (S-IVB)			
Size	6.7 m × 18 m	22 ft × 59 ft
Engine	1 J-2	
Thrust (501 thru 503)	1 MN	225,000 lbf
(504 and Sub)	1 MN	230,000 lbf
Propellants	LOX & LH$_2$	

APOLLO SPACECRAFT

INSTRUMENT UNIT

THIRD STAGE (S-IVB)

SECOND STAGE (S-II)

FIRST STAGE (S-IC)

Figure 13.15. Characteristics of *Saturn V* launch vehicle.

Figure 13.16. First stage (S-IC) of the *Saturn V* launch vehicle. (*Courtesy* Boeing Company.)

All three stages use liquid propellants. The first stage (shown in Figure 13.16) obtains its thrust from five F-1 rocket engines. The F-1 is a single-start fixed-thrust rocket engine. Each engine has a burn time of about 150 seconds. The kerosene and liquid oxygen are combined and burned in the combustion chamber. The pressure in the combustion chamber is 66 atm, with a combustion temperature of 3570 K, or 6430°R. The burning gases are expelled through an expansion nozzle to produce thrust. The expansion area ratio of the nozzle is 16:1. The second stage (shown in Figure 13.17) obtains its thrust from five J-2 rocket engines. The burn time is 395 seconds. The chamber pressure is 52 atm, with a combustion temperature of 3440 K, or 6190°R. The expansion area ratio of the nozzles is 27.5:1. The third stage (shown in Figure 13.18) obtains its thrust from a single J-2 engine (the same engine used in the second stage).

A more sophisticated rocket system is that used for the space shuttle program (Figure 13.19). The orbiter is designed to carry into earth orbit a crew of seven, return to earth with personnel and payload, and land like an

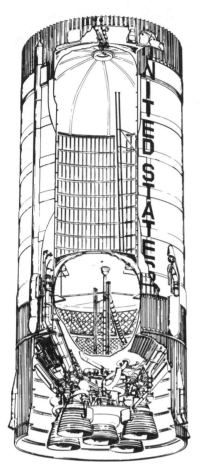

Figure 13.17. Second stage (S-II) of the *Saturn V* launch vehicle. (*Courtesy* North American Rockwell Corp.)

Figure 13.18. Third stage (S-IVB) of the *Saturn V* launch vehicle. (*Courtesy* McDonnell-Douglas Corp.)

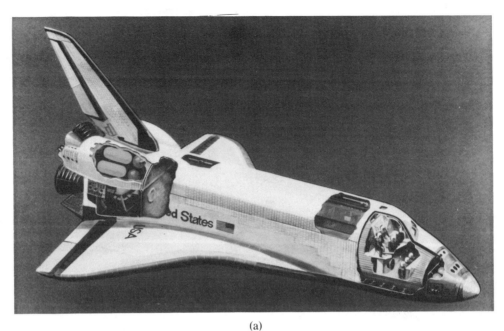

(a)

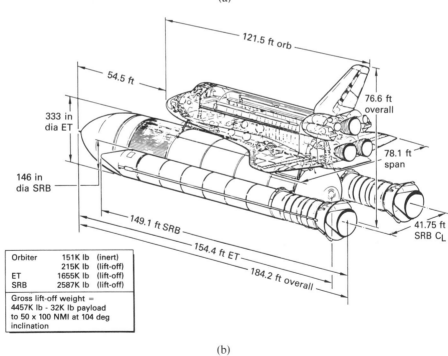

Orbiter	151K lb	(inert)
	215K lb	(lift-off)
ET	1655K lb	(lift-off)
SRB	2587K lb	(lift-off)

Gross lift-off weight =
4457K lb - 32K lb payload
to 50 x 100 NMI at 104 deg
inclination

(b)

Figure 13.19.   Space shuttle vehicle. (*Courtesy* National Aeronautics and Space Administration.)

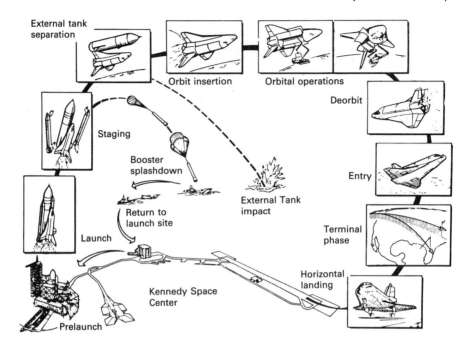

Figure 13.20. Typical mission profile.

airplane (Figure 13.20). Two solid rocket boosters (SRBs) burn in parallel with the main orbiter's liquid-propellant rocket engines to get the orbiter into the required ascent trajectory. The SRBs are separated from the orbiter at an altitude of approximately 50 km, using eight separation rockets. The orbiter's main propulsion liquid propellant burns for a total of 8 minutes and shuts down just before orbital velocity is reached. Liquid oxygen and liquid hydrogen are supplied from an external tank; this external tank is jettisoned just prior to orbit insertion, with the orbital maneuvering subsystem used to provide the required rocket thrust for insertion into orbit, orbital change, rendezvous, and return to earth. Finally, the reaction control has 38 bipropellant rocket thrusters and 6 vernier thrusters to provide attitude control and three-axis translations during orbit insertion, on orbit and reentry phases of flight. Details of the rocket system are shown in Figures 13.21, 13.22, 13.23, and 13.24.

13.3 PROPELLERS

A *propeller* consists of several rotating blades (usually two to five) attached to a hub that is connected to a shaft. The application of a torque to the propeller shaft causes the ambient fluid ahead of the propeller to move past the blades, imparting to the fluid an exit velocity that is greater than the

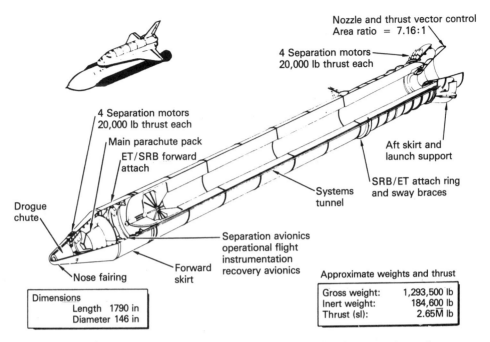

Nozzle and thrust vector control
Area ratio = 7.16:1

4 Separation motors
20,000 lb thrust each

4 Separation motors
20,000 lb thrust each

Main parachute pack

ET/SRB forward
attach

Aft skirt and
launch support

SRB/ET attach ring
and sway braces

Systems
tunnel

Drogue
chute

Separation avionics
operational flight
instrumentation
recovery avionics

Forward
skirt

Nose fairing

Approximate weights and thrust

Gross weight:	1,293,500 lb
Inert weight:	184,600 lb
Thrust (sl):	2.65M lb

Dimensions
Length   1790 in
Diameter 146 in

Figure 13.21.   Solid rocket boosters. (*Courtesy* National Aeronautics and Space Administration.)

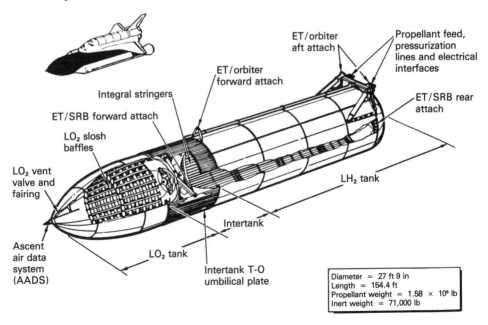

ET/orbiter
aft attach

Propellant feed,
pressurization
lines and electrical
interfaces

ET/orbiter
forward attach

Integral stringers

ET/SRB rear
attach

ET/SRB forward attach

LO₂ slosh
baffles

LO₂ vent
valve and
fairing

LH₂ tank

Intertank

Ascent
air data
system
(AADS)

LO₂ tank

Intertank T-0
umbilical plate

Diameter = 27 ft 9 in
Length = 154.4 ft
Propellant weight = 1.58 × 10⁶ lb
Inert weight = 71,000 lb

Figure 13.22.   External tank. (*Courtesy* National Aeronautics and Space Administration.)

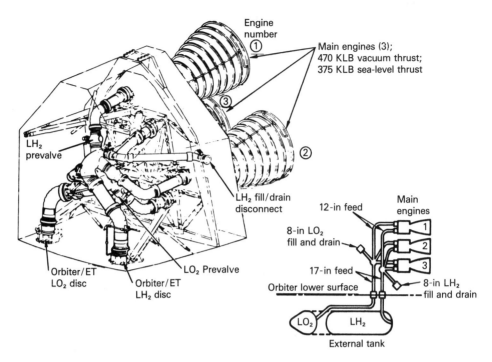

Figure 13.23. Main propulsion subsystem. (*Courtesy* National Aeronautics and Space Administration.)

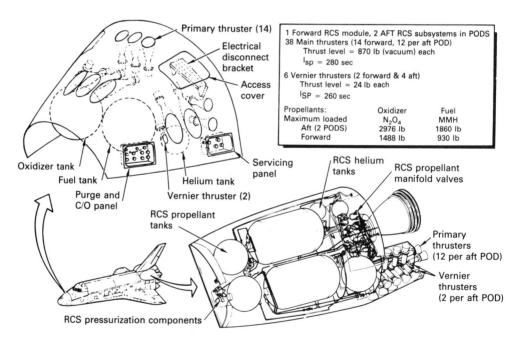

Figure 13.24. Reaction control subsystem. (*Courtesy* National Aeronautics and Space Administration.)

Figure 13.25. Marine propeller.

approach velocity. An example of a propeller used in marine application is shown in Figure 13.25. It is a five-bladed propeller for use on a navy hydrofoil craft. Its diameter is 82 cm (2.7 ft) and it is designed to rotate at 160 rad/s (1500 rpm) at a craft speed of 17 m/s (55 ft/s). An airplane propeller is shown in Figure 13.26. This type of propeller has been utilized on aircraft such as the DC-6 or 7 (Figure 13.27). Depending on application, the propeller diameter ranges from 3 to 5 m (10 to 17 ft).

An exact treatment of the flow past the propeller blades would require use of the blade element approach indicated in Section 12.3 for axial-flow machines, including correction for three-dimensional effects described in Section 9.4 if the blades are short, as in the case of marine propellers. Here we shall give an approximate treatment of propeller performance by assuming that the effect of the blades of the propeller is concentrated in the propeller plane.

As in the case of jet propulsion, we can assume that the inflow velocity to the control volume equals the forward speed of the propeller. However, since the propeller operates in an unbounded medium, we need to specify the mass flow rate in terms of quantities in the plane of the propeller. A schematic of the flow past a propeller is given in Figure 13.28. The sections with velocities $V_{C.V.}$ and V_e are taken at sufficient distances from the propeller so that the pressure at these sections is ambient. Hence we obtain from Equation (13.2) with $\dot{m}_f = 0$ and $p_e = p_i = p_a$ that

$$\text{Thrust} = \dot{m}(V_e - V_{C.V.})$$

From the energy equation (4.19), omitting changes in elevation, we have

$$-\frac{dW}{dt} = \dot{m}\left(\frac{V_e^2 - V_{C.V.}^2}{2}\right)$$

Figure 13.26. Airplane propeller. (*Courtesy* United Aircraft Corp.)

Figure 13.27. Douglas DC-6B aircraft. (*Courtesy* McDonnell-Douglas Corp.)

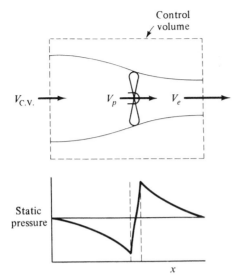

Figure 13.28. Flow past propeller.

where the term on the left is the rate of work done on the fluid. In this case, the rate of work done on fluid is the product of thrust and velocity V_p in the propeller plane. Therefore,

$$\text{Thrust} \times V_p = \dot{m}\left(\frac{V_e^2 - V_{\text{C.V.}}^2}{2}\right)$$

and

$$\dot{m}(V_e - V_{\text{C.V.}})V_p = \dot{m}\left(\frac{V_e^2 - V_{\text{C.V.}}^2}{2}\right)$$

Solving for V_p, we obtain

$$V_p = \frac{V_e + V_{\text{C.V.}}}{2}$$

Hence the flow velocity in the plane of the propeller with respect to the control volume is the average of the two velocities $V_{\text{C.V.}}$ and V_e, that is, approach and exit velocities. The propeller equation can therefore be written as

$$\text{Thrust} = \dot{m}(V_e - V_{\text{C.V.}}) = \rho V_p \pi R^2 (V_e - V_{\text{C.V.}})$$

$$= \frac{\rho \pi R^2}{2}(V_e^2 - V_{\text{C.V.}}^2) \tag{13.11}$$

where R is the radius of the propeller. The maximum thrust for a given propeller is obtained when the translational speed of the propeller is zero.

The ideal efficiency of a propeller (e_p) is defined as the ratio of power output to power input. The power input (rate of work done on fluid) is thrust times V_p, while the power output is the product of thrust and translational speed of propeller (thrust $\times$ $V_{\text{C.V.}}$). Hence

$$e_p = \frac{TV_{C.V.}}{TV_p} = \frac{V_{C.V.}}{V_p}$$

$$= \frac{V_{C.V.}}{(V_e + V_{C.V.})/2} = \frac{1}{1 + (V_e/V_{C.V.})}$$

This equation shows that the propeller will be most efficient ($e_p = 1$) when the jet velocity (V_e) equals the forward speed of the propeller ($V_{C.V.}$). However, at this point the thrust becomes zero, as can be seen from Equation (13.11). Thus, to obtain nonzero thrust at reasonable efficiency, the jet velocity should be slightly larger than the forward speed, with large propeller radii required to achieve sufficient thrust.

EXAMPLE 13.5

An outboard motor drives a small marine propeller. The propeller has a diameter of 250 mm and discharges 0.5 m³/s. The propeller forward speed is 8.30 m/s. Determine the thrust and power of the propeller and the pressure rise across the propeller. Take water density as 1000 kg/m³.

Solution   The velocity at the propeller plane (V_p) is

$$V_p = \frac{Q}{\pi R^2} = \frac{0.5}{\pi (0.250/2)^2}$$

$$= 10.186 \text{ m/s}$$

Now $V_p = (V_e + V_{C.V.})/2$. Hence

$$V_e = 2V_p - V_{C.V.} = 2(10.186) - 8.30 = 12.072 \text{ m/s}$$

From Equation (13.11), the propeller thrust is given in terms of V_e and $V_{C.V.}$ as

$$\text{Thrust} = \frac{\rho \pi R^2}{2}(V_e^2 - V_{C.V.}^2)$$

$$= \frac{(1000 \text{ kg/m}^3)\pi(0.125^2 \text{ m}^2)}{2}(12.072^2 \text{ m}^2/\text{s}^2 - 8.30^2 \text{ m}^2/\text{s}^2)$$

$$= \underline{1.886 \text{ kN}}$$

The required propeller power is

$$\text{Power} = \text{Thrust} \times V_p$$

$$= (1.886 \text{ kN})(10.186 \text{ m/s})$$

$$= \underline{19.21 \text{ kW}}$$

From Equation (13.3), an expression for the thrust is given by

$$\text{Thrust} = \dot{m}(V_e - V_i) + (p_e - p_a)A_e - (p_i - p_a)A_i$$

Taking subscript i to denote the inlet condition to the propeller and subscript e the exit condition, we have $V_i = V_e = V_p$, $A_e = A_i = A_p$. Hence

$$\text{Thrust} = (p_e - p_i)A_p \quad \text{or} \quad \Delta p = \frac{\text{Thrust}}{A_p}$$

Hence the pressure rise across the propeller becomes

$$\Delta p = \frac{\text{Thrust}}{\pi R^2}$$

$$= \frac{1.886 \text{ kN}}{\pi(0.125^2 \text{ m}^2)}$$

$$= \underline{38.42 \text{ kPa}} \qquad \blacksquare$$

   Figure 13.29 shows the occurrence of a phenomenon called *cavitation* on the blades of a marine propeller. Cavitation occurs when vapor is formed at those points in the flow field at which the local pressure falls below the saturation pressure; there, local boiling takes place without the addition of heat. Low-static-pressure regions will occur at points where high local velocities are encountered. The small bubbles formed in the region of low pressure collapse as they enter regions of higher pressure further downstream. Such repeated collapses near the surface of the blades can cause damage to the blade surface. Cavitation can also occur in other liquid flow machinery such as pumps and turbines, and in valves and other flow devices.

   To avoid the damage caused by the collapse of the small bubbles, the formation of large cavities is deliberately induced by appropriate shaping of the propeller blades and by operating them at high rotative speeds. The large cavities trail the blades, and their collapse occurs well beyond the blades. This phenomenon is called *supercavitation*. Supercavitating propellers have been utilized in high-speed marine applications. Also, supercavitating pumps have been designed for use in liquid rockets.

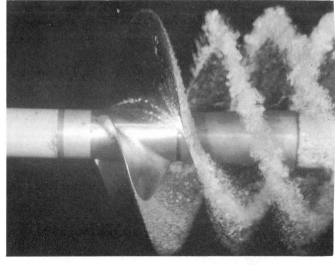

Figure 13.29.   Cavitation on a marine propeller.

EXAMPLE 13.6

A 4-ft-diameter marine propeller, rotating at 400 rpm, advances at a speed of 32 ft/s. The propeller blades are airfoil sections having a minimum pressure coefficient of -0.5. The centerline of the propeller is 3 ft below the water surface. Determine whether cavitation will occur at propeller blade sections at radial distances of 0.8 ft and 1.4 ft. The temperature of the water is 80°F, at which the saturation pressure is 0.507 psia.

Solution   The pressure coefficient is given by Equation (5.15) with V equal to V_{rel}, namely

$$C_p = \frac{p - p_\infty}{\frac{1}{2}\rho V_{\text{rel}}^2}$$

From Section 12.2 we get, with the absolute inflow velocity V in the axial direction,

$$V^2 = V_{\text{rel}}^2 + U_t^2 - 2U_t^2$$

or

$$V_{\text{rel}}^2 = V^2 + U_t^2$$

The pressure p_∞ will be a function of atmospheric pressure and submergence; hence, from Equation (2.2),

$$p_\infty = p_{\text{atm}} + \rho g d \qquad \text{where} \qquad \rho = 62.22 \text{ lbm/ft}^3$$

At a radial distance of 0.8 ft the submergence is a minimum of 2.2 ft. Hence

$$p_\infty = 14.7 \text{ psia} + \frac{(62.22 \text{ lbm/ft}^3)(32.2 \text{ ft/s}^2)(2.2 \text{ ft})}{(32.2 \text{ lbm/slug})(144 \text{ in}^2/\text{ft}^2)}$$

$$= 15.65 \text{ psia}$$

$$V_{\text{rel}}^2 = (32 \text{ ft/s})^2 + (\tfrac{400}{60} \times 2\pi \text{ rad/s})^2(0.8 \text{ ft})^2$$

$$= 2147 \text{ ft}^2/\text{s}^2$$

$$C_p = \frac{p - p_\infty}{\frac{1}{2}\rho V_{\text{rel}}^2}$$

and we can solve for p with $C_p = -0.5$:

$$p = 8.45 \text{ psia}$$

This is greater than the saturation pressure and so cavitation will not occur at a radius of 0.8 ft.

At a radius of 1.4 ft, the submergence is a minimum of 1.6 ft. Therefore

$$p_\infty = 14.7 \text{ psia} + \frac{(62.22 \text{ lbm/ft}^3)(32.2 \text{ ft/s}^2)(1.6 \text{ ft})}{32.2 \text{ lbm/slug}}$$

$$= 15.39 \text{ psia}$$

$$V_{\text{rel}}^2 = (32 \text{ ft/s})^2 + (\tfrac{400}{60} \times 2\pi \text{ rad/s})^2(1.4 \text{ ft})^2$$

$$= 4463 \text{ ft}^2/\text{s}^2$$

Using the expression for C_p, $p = 0.424$ psia. This value is less than the saturation pressure, and so cavitation will occur at a radius of 1.4 ft.   ■

13.4 WINDMILLS

The purpose of a windmill is to extract power from the wind. The output from windmills has been used over the centuries for such purposes as grinding grain, sawing wood, and pumping water. Due to an increasing shortage of conventional energy sources, the windmill is currently being examined as a potential source of at least some of our electrical power.

As contrasted with flow past a propeller (Figure 13.28), there is a static pressure decrease across the blades of a windmill rotor, as power is extracted from the flow (Figure 13.30). Again, section i is taken far enough upstream of the rotor and section e far enough downstream so that there is ambient pressure at each section. The force exerted by the fluid on the blade of Figure 13.30 is $\dot{m}(V_i - V_e)$. Equating the rate of work done by the fluid to the rate of decrease of kinetic energy of the fluid passing through the control volume bounded by i and e of Figure 13.30, we have

$$\text{Force} \times V_p = \dot{m}\frac{V_i^2 - V_e^2}{2}$$

or

$$\dot{m}(V_i - V_e)V_p = \dot{m}\frac{V_i^2 - V_e^2}{2}$$

Therefore,

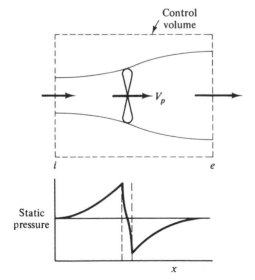

Control volume

Static pressure

x

Figure 13.30.   Flow past windmill rotor.

$$V_p = \frac{V_i + V_e}{2}$$

The mass flow $\dot{m}$ at the blade cross section is

$$\dot{m} = \rho A V_p$$

with A the cross-sectional area swept out by the rotor. Substituting,

$$\dot{m} = \rho A \frac{V_i + V_e}{2}$$

so that the rate of work done by the fluid, or power, dW/dt, is

$$\frac{dW}{dt} = \rho A \left(\frac{V_i + V_e}{2} \right) \left(\frac{V_i^2 - V_e^2}{2} \right)$$

$$= \tfrac{1}{4}\rho A (V_i + V_e)(V_i^2 - V_e^2)$$

For maximum power, set $d(dW/dt)/dV_e = 0$, or

$$V_i = 3V_e \qquad \text{for maximum power}$$

Therefore, maximum power extractable from the wind is $\tfrac{8}{27}\rho A V_i^3$. The total power available in the wind is $\tfrac{1}{2}\rho A V_i^3$. The maximum efficiency, where efficiency is (power extracted)/(power available), is thus

$$e_{\max} = \frac{\tfrac{8}{27}\rho A V_i^3}{\tfrac{1}{2}\rho A V_i^3} = \frac{16}{27} \quad \text{or} \quad 59.3\% \tag{13.12}$$

In the real case, due to friction and flow losses across the blades, the efficiency will be somewhat lower than this value.

EXAMPLE 13.7

A windmill is to be used to supply the electric power needs of a household at a location where the average windspeed is 6.0 m/s. What size rotor will be required, if 1.0 kW is to be provided? Assume a rotor efficiency of 70 percent, gearing efficiency 92 percent, generator efficiency 80 percent, and storage efficiency 85 percent, with an air density of 1.2 kg/m^3.

Solution   The ideal output of the windmill is $\tfrac{8}{27}\rho A V_i^3$. For our case, taking losses into account,

$$1000 \text{ W} = \tfrac{8}{27}(1.2 \text{ kg/m}^3)A(6.0^3 \text{ m}^3/\text{s}^3)(0.70)(0.80)(0.85)$$

$$A = 29.73 \text{ m}^3$$

$$= \frac{\pi}{4} D^2$$

where D is the rotor diameter. Solving, $D = \underline{6.15 \text{ m}}$.   ■

A number of different types of windmill rotors and blade configurations have been tried. The rotor for the classic Dutch windmill, turning at relatively low rotational speeds, consisted of large blades covered with fabric, the blade area representing a large portion of the total area swept out by the blades.

Figure 13.31. DOE/NASA 200 kW experimental wind turbine, Block Island, Rhode Island.

In today's high-speed windmills, in order to minimize friction, the rotor consists of two or three blades, shaped like propellers, with the blades a small fraction of the swept area. Typical installations are shown in Figures 13.31, 13.32a and b, and 13.33.

The windmill at Sandusky, Ohio, part of a governmental experimental program to determine the applicability of windmills for electric power generation, is designed to produce 100 kW of electric power at windspeeds of 8 m/s (26 ft/s), with blades 37.5 m (123 ft) in diameter. The design rotational speed is 4.2 rad/s (40 rpm).

The windmill at Block Island (Figure 13.31) has two blades of 125 ft diameter, with rated power of 200 kW at a windspeed of 18.3 mph at 30 ft. Blades are constructed of aluminum. The U.S. Windpower Model 56-100 windmill shown in the windmill farm of Pacific Gas and Electric Company in Figures 13.32a and b, has a blade diameter of 56 ft and is rated at 100 kW at a windspeed of 29 mph. Total installed capacity of the Altamont Pass farm is 670 MW in 1987.

(a)

(b)

Figure 13.32. Wind turbines at Altamont Pass near San Francisco. (*Courtesy* U.S. Windpower, Inc. Ed Linton, photographer.)

Before there can be large-scale utilization of windmills for electric power generation, problems must be solved relative to structural integrity of the large blades with very high tip speeds, and subjected to high and varying axial forces. Further, with the wind itself quite variable, means must be found either for wind energy storage to handle electrical demands during periods of low wind velocity, or for the transformation of the windmill electrical output so that it can be fed into the regular utility network.

Figure 13.33. Wind turbine generator, Clayton, New Mexico. (*Courtesy* U.S. Dept. of Energy.)

PROBLEMS

13.1. The mass flow rate of air through a jet engine is 40 kg/s; the static thrust is measured to be 30 kN. Determine the jet exhaust velocity. Indicate all assumptions.

13.2. A jet engine is flying at Mach 0.7 at 20,000 ft. Calculate the TSFC and thrust per unit mass flow for a compressor pressure ratio of 20. Assume an isentropic compressor, turbine, and nozzle, with a loss of 5 psi in the combustion chamber and a 20 percent loss of stagnation pressure in the diffuser. The heating value of the fuel is 16,000 Btu/lbm, maximum turbine temperature 1800°F. Neglect the fuel flow rate in comparison to the airflow rate; assume the air to behave as a perfect gas with constant specific heats ($c_p = 0.25$ Btu/lbm°R, $\gamma = 1.4$).

13.3. Repeat the preceding problem with a compressor efficiency of 80 percent, turbine efficiency of 90 percent.

13.4. Repeat problem 13.2, with $c_p = 0.24$ Btu/lbm°R to the combustion chamber and $c_p = 0.27$ Btu/lbm°R from turbine inlet to nozzle exit ($\gamma = 1.4$).

13.5. In a ramjet shown in Figure P13.5, the compressor and turbine of the turbojet engine are removed, with the necessary pressure rise taking place in the diffuser (the ramjet must be accelerated to high velocities by external means before it can operate). Calculate the TSFC and thrust per unit mass flow for a ramjet traveling at Mach 2.5 at 10 km. Assume $c_p = 1.0$ kJ/kg · K up to the combustion chamber, 1.12 kJ/kg · K after combustion, with a maximum temperature of 2200 K. Assume the stagnation pressure ratio in the diffuser is 0.6, with no losses in combustion chamber or nozzle. Heating value of the fuel is 42,000 kJ/kg ($\gamma = 1.4$).

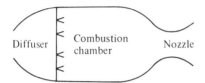

Figure P13.5.

13.6. Repeat Problem 13.5 for a turbojet with the same diffuser losses, yet isentropic compressor (pressure ratio 7:1) and turbine and maximum engine temperature of 1300 K.

13.7. A turbojet takes off at sea level (14.7 psia, 60°F). The heating value of the fuel is 17,000 Btu/lbm, compressor pressure ratio 20:1, turbine inlet temperature 1800°F. The exit nozzle has a throat area of 1.5 ft², with the nozzle perfectly expanding the flow to ambient pressure. Assuming ideal compressor, turbine, and nozzle, calculate the takeoff thrust, fuel flow rate, and TSFC. Repeat for compressor pressure ratios of 25:1 and 30:1.

13.8. A rocket has a total initial mass of 50,000 kg, with engines capable of producing 7 MN thrust. The rocket fuel has a specific impulse of 3.0 kN/(kg/s). Calculate the velocity of the rocket after 1, 5, and 10 seconds of vertical flight. Neglect aerodynamic drag in your calculations but include gravity. What is the maximum rocket acceleration?

13.9. A 100-lbm payload is to be launched from the earth into space (escaping the earth's gravity). If the structure of the rocket represents 10 percent of the initial mass and the specific impulse of the fuel is 300 lbf/(lbm/s), calculate the total initial mass that must be lifted off the earth's surface. What percent of the initial mass is fuel? Neglect the forces of drag and gravity acting on the rocket.

13.10. Thermodynamic calculations show that the adiabatic flame temperature of a new fuel oxidizer combination at 10 MPa is 3000°C, with the products of combustion having a mean molecular mass of 14 and a mean $c_p = 3.0$ kJ/kg · K. For an ambient pressure of 50 kPa, calculate the specific impulse of the new propellant. (Assume a perfectly expanded nozzle.)

13.11. The pressure of argon in the chamber of a small positioning rocket is maintained at 100 psia with the gas to be discharged through a nozzle of 1 in² throat area, 2 in² exit area to space. The temperature of the argon in the chamber is 60°F. Determine the thrust of this device and the specific impulse.

13.12. A man (mass on earth 75 kg) is to escape from the moon's surface with 200 kg of instruments and support equipment. The structural mass of the rocket is 5 percent of the total initial mass of the rocket, the rocket specific impulse is 2.75 kN/(kg/s). Find the propellant mass required and the total initial mass of the rocket (neglect the force due to the moon's gravity).

$$\frac{\text{Mass moon}}{\text{Mass earth}} = 0.0123 \quad \frac{\text{Diameter moon}}{\text{Diameter earth}} = 0.273$$

13.13. Is it possible for a rocket to travel faster than the velocity of the exhaust gases? Explain.

13.14. A small rocket is to provide 10 lbf of thrust for a period of 100 s in space. The rocket is to be fueled with a monopropellant of specific impulse 150 lbf/(lbm/s), chamber pressure 200 psia, chamber temperature 2800°R. The area ratio of the nozzle is 5 to 1. Determine the throat area and the mass of fuel required.

13.15. Using the data of Table 13.1, determine the thrust/propellant flow in space for a liquid hydrogen–liquid oxygen rocket with exhaust nozzle designed for sea level (assume a mixture ratio by mass of 3.5). The average molecular mass of the products $\overline{M}$ is 9; $\gamma = 1.25$.

13.16. A rocket uses a solid propellant having a density of 1500 kg/m³, a burning surface area of 0.1 m², and a burning rate of 1.2 cm/s. The rocket chamber pressure is 2 MPa and the chamber temperature is 1500 K. Determine the rocket nozzle throat area for choked flow in the nozzle if $\gamma = 1.3$ and $R = 0.50$ kJ/kg · K.

13.17. An airplane is traveling at 200 mph at 10,000 ft. Thrust is supplied by four 8-ft propellers, each of which handles 40,000 cfs. Calculate the total propeller thrust and the horsepower required.

13.18. A marine propeller has a diameter of 1.0 m and discharges 7.0 m³/s while traveling at 9.0 m/s. Calculate the thrust and power required ($\rho = 998$ kg/m³).

13.19. Determine the pressure rise that takes place across the propellers of Problems 13.17 and 13.18.

13.20. Water at 68°F flows through a horizontal converging-diverging nozzle (Venturi tube). The pressure at the nozzle inlet (section 1) is atmospheric. The ratio of throat (section 2) to nozzle inlet area is $\frac{1}{2}$. What must be the velocity at the inlet (section 1) so that cavitation is just beginning to occur at the throat (section 2)? The saturation pressure of water at 68°F is 0.34 psia.

13.21. Neglecting losses, determine the maximum power that can be extracted by a windmill from a steady wind of 15 mph. Assume a windmill rotor diameter of 80 ft, with air density 0.0749 lbm/ft³. Find the pressure decrease across the rotor.

13.22. In a certain area, an analysis of weather data shows that 15 percent of the time the wind velocity is between 3.0 and 4.0 m/s, 15 percent of the time it is between 4.0 and 5.0 m/s, 14.8 percent of the time it is between 5.0 and 6.0 m/s, 13 percent of the time it is between 6.0 and 7.0 m/s, 7 percent of the time it is between 7.0 and 8.0 m/s, 4 percent of the time it is between 8.0 and 9.0 m/s, and 3 percent of the time it is between 9.0 and 10.0 m/s. A windmill is to be constructed to provide power for a neighborhood of twenty-five homes, each requiring electrical energy of 1000 kWh per month. Determine the size rotor required for this facility, assuming a gearing efficiency of 92 percent, storage efficiency of 80 percent, electric generator efficiency of 88 percent, and a rotor efficiency which varies with wind velocity as follows:

Wind velocity (m/s)	Rotor efficiency (%)
0	0
1	0
2	0
3	10
4	18
5	43
6	58
7	60
8	60
9	50
10	42

Assume that, for all wind velocities greater than 10 m/s, the windmill output is maintained at that corresponding to 10 m/s, in order to prevent damage to the generator.

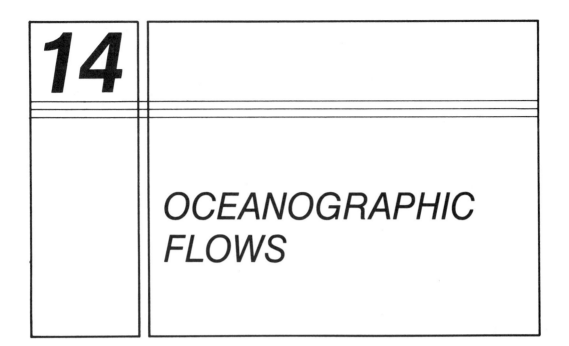

OCEANOGRAPHIC FLOWS

14.1 INTRODUCTION

Earlier we covered a variety of fluid flows ranging from viscous flow through small tubes to turbulent flow through man-made open channels and natural river beds. In all these cases, the extent of the flow field was very small relative to the size of the earth. In this chapter we shall discuss oceanographic flows, in which fluid motions take place in flow fields extending over large areas of the earth's surface.

The oceans, which cover about 70 percent of the earth's surface, have been used by man for centuries as an important means of transportation and communication, as well as a source of food and minerals. Although well explored, the wealth of the oceans still remains, for the most part, untapped. An increased knowledge may someday provide man with the opportunity of utilizing their full potential.

It is the purpose of this chapter to acquaint the student with one aspect of oceanography, namely, the flows that take place in the oceans. In particular, we shall give brief discussions of the statics of the oceans, ocean waves, currents, and tides. For more detailed discussions of these subjects, the student is referred to the references given at the end of the chapter.

14.2 HYDROSTATICS OF THE OCEAN

We showed in Chapter 2 that the pressure variation with depth in a static fluid is given by

$$dp = -\rho g \, dz$$

We proceeded in Section 2.2 to calculate the pressure variation throughout the atmosphere. The same expression for pressure change with depth, $dp = -\rho g\,dz$, applies to the ocean, with the density of ocean water dependent on its salt content or *salinity,* temperature, and compressibility. The density is greater than that of fresh water due to the salt content, the average density at the ocean surface being 1025 kg/m³ (64.0 lbm/ft³). Water is only slightly compressible. For example, an increase of pressure of one atmosphere on a volume $V\!\!\!/$ of water will bring about a fractional volume decrease $\Delta V\!\!\!//V\!\!\!/$ of only about 5×10^{-5}. It can be seen that the compressibility need be taken into account only in the deeper parts of the ocean, where extremely high pressures are attained. If we neglect the effect of salinity, compressibility, and temperature variations on density for the moment, the pressure in a static ocean will increase by $\Delta p = \rho g D$, with D the depth below ocean surface or 1005 kPa for each 100 m of depth (6400 lbf/ft² for each 100 ft of depth).

For a more precise determination of vertical pressure distribution, the effect of temperature and salinity variation must be considered. Due to the absorption of solar radiation, temperatures near the surface of the ocean are generally greater than those farther below. A typical variation of temperature with depth is shown in Figure 14.1. Because of turbulence caused by wave motion, there is a layer of well-mixed water at the surface of the ocean having a relatively uniform temperature. Below this is a layer possessing a vertical temperature gradient, called the *thermocline.* Below the thermocline and extending to the ocean bottom is a region of cold water of uniform temperature; this region is relatively unaffected by local surface variations. In this deep-water region, the temperature gradient amounts to less than 1°C per 150 m at a depth of 1500 m (1°F per 250 ft at a depth of 5000 ft); below 2500 m, it is less than 0.1°C per 150 m (below 7500 ft, it is less than 0.1°F per 250 ft). It is interesting to note that the average temperature of the oceans is only about 4°C (39°F); even at the equator the mean temperature is 4.9°C (40.8°F).

The surface temperature of the oceans and extent of the layers just discussed are dependent on the latitude of the point in question. In the higher latitudes (Arctic regions), the temperature of the water is cold from

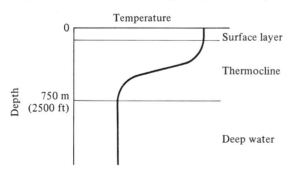

Figure 14.1. Temperature variation with depth in the ocean.

surface to ocean bottom; the deep-water layer reaches the surface. In the lower latitudes, the mixed layer, subject to atmospheric influences, extends down to about 75 m (250 ft), with the thermocline to 500 to 750 m (1500 to 2500 ft). The effects of daily and seasonal variations in atmospheric conditions must be superimposed on the preceding variation.

The relative change of volume with temperature of a given mass of seawater at a fixed salinity and pressure can be found from the coefficient of thermal expansion β, where

$$\beta = \frac{1}{\mathcal{V}}\left(\frac{\partial \mathcal{V}}{\partial T}\right)_{p,\text{salinity}}$$

Since ρ is the reciprocal of specific volume, we can also write

$$\beta = -\frac{1}{\rho}\left(\frac{\partial \rho}{\partial T}\right)_{p,\text{salinity}}$$

Values of β as a function of temperature at atmospheric pressure and zero salinity are given below.[*]

T (°C)	β (1/K)	T (°F)	β (1/°R)
0	-67×10^{-6}	32	-37×10^{-6}
5	5.2×10^{-6}	40	4×10^{-6}
10	27×10^{-6}	50	49×10^{-6}
15	47×10^{-6}	60	87×10^{-6}
20	64×10^{-6}	70	121×10^{-6}

Values of β at other salinities are of the same order of magnitude as that at zero salinity. It can be seen that the change of density with temperature for ocean water is quite small.

The salinity of seawater is usually expressed in parts of dissolved salt per thousand; for example, 20‰ denotes 0.020 kg of salt dissolved in 1 kg of seawater (0.020 lbm of salt in 1 lbm of seawater). Average seawater has a salinity of 35‰. Regions of the ocean in the vicinity of land where there is a strong fresh-water river runoff may have much lower salinities (water of low salinity is called brackish water). For example, the inner parts of the Baltic Sea and some fjords have salinities as low as 0.5 to 1‰.[†] Landlocked seas in arid regions where there is a good deal of evaporation and little precipitation or water runoff have high salinities; for example, the Red Sea has a salinity of 43 to 45‰.

The variation of salinity with depth is dependent on the region of the ocean being considered. Where an excess of evaporation over precipitation

---

[*] Reference 1, p. 42.
[†] Reference 1, p. 29.

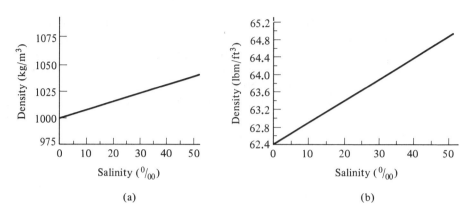

Figure 14.2.   Density versus salinity for seawater at 0°C (32°F) and atmospheric pressure.

exists, there will be a maximum of salinity in the upper layers of the ocean. In polar regions, melting occurring in the summer months decreases the salinity in the surface layers. Moreover, ocean currents and mixing processes tend to even out the vertical salinity distribution.

The variation of density with salinity for water at 0°C (32°F) and atmospheric pressure is shown in Figure 14.2. Again it should be noted that over the range of salinities present in ocean water, the effect of salinity variation on density is a small one.

It can be concluded from the preceding discussions that, in determining the variation of pressure in a static body of seawater, density can be treated as a constant unless either great precision is required or the pressures at very great depths must be calculated. This is in contrast to our study of the atmosphere, where air was treated as a perfect gas obeying $p = \rho RT$, in which large density variations were found to occur.

14.3 DYNAMICS OF OCEAN CURRENTS

A variety of currents exist in the ocean; some extend over large regions while others are quite localized. The causes of the ocean currents are manifold. One of the major causes is the wind that blows over the surface of the water, exerting a shear force on the surface and dragging along the water beneath it. If the area over which the winds act is large and if the winds are of long duration, the resultant current becomes a major oceanic circulation extending over thousands of square kilometers. When the wind blows over an area of perhaps only a few square kilometers of the ocean for a limited duration, wind drift currents result.

Another cause of ocean currents is tides, which are caused by the attractive forces of the moon and the sun on the ocean mass. The tidal

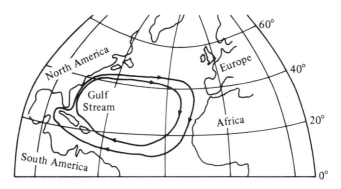

Figure 14.3. Currents in the North Atlantic.

motions in the ocean, which will be discussed in a later section, combine with the features of topography of the earth's land masses and give rise to periodic currents in the ocean. The tidal currents vary with locality and generally depend on the character of the tide, the local water depth, and the configuration of the coast.

Water temperature and salinity variations are still another source of currents. The resulting density differences, although small, induce a circulatory flow in the ocean.

Current systems exist in the oceans of the earth near the surface as well as at great depths. Since the surface currents can be charted by observations from drifting ships, most of our knowlege relates to the surface currents. Examples of two major surface ocean current systems occur in the North Atlantic and North Pacific oceans, as shown in Figures 14.3 and 14.4. In the North Atlantic Ocean, which is contained in a nearly rectangular basin, the winds causing the current are in a direction parallel to the earth's latitudes. A schematic of the earth showing different latitudes is given in Figure 14.5. Near the equator the easterly trade winds blow from east to west, while in the northern half of the basin we have the westerlies blowing from west to east.

Some of the main features of the major ocean currents, such as the intensification of the currents along the western boundaries, are predictable

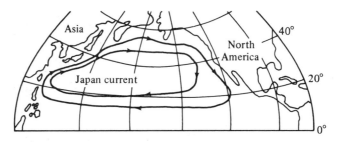

Figure 14.4. Currents in the North Pacific.

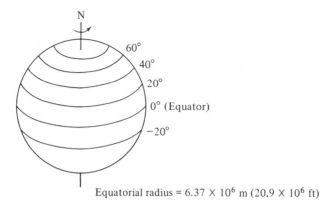

Equatorial radius = 6.37×10^6 m (20.9×10^6 ft)

Figure 14.5.   The earth, showing different latitudes.

by simple theoretical considerations. In a simplified model of the current flow in the North Atlantic, it is assumed that the motion is two-dimensional within a rectangular basin and that flow takes place in a direction parallel to the ocean surface, resulting in zero variation of velocity with depth. A rectangular coordinate system is used with its origin at the surface of the ocean. The x axis is directed toward the east and the y axis toward the north (Figure 14.6). On the surface of the ocean, the winds exert a tangential force that drives the currents, while at the bottom of the layer of depth H a frictional force is exerted opposing the motion of the surface layer. The winds blowing in a direction parallel to the latitudes are assumed to have a sinusoidal distribution, as shown in Figure 14.7. The frictional force is taken to be proportional to the local velocity of the current.

The external force exerted on a differential volume $dx\, dy\, H$ is given by

$$\Sigma\, dF_x = -\frac{\partial p}{\partial x}\, dx\, dy\, H + \tau_{\text{wind}} dx\, dy - \tau\, dx\, dy \qquad (14.1)$$

$$\underbrace{\phantom{-\frac{\partial p}{\partial x} dx dy H}}_{\text{pressure force}} \qquad \underbrace{\phantom{\tau_{\text{wind}} dx dy}}_{\text{wind force}} \qquad \underbrace{}_{\text{friction force}}$$

where τ_{wind} is the wind stress and τ the frictional stress.

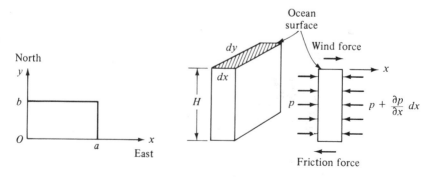

Figure 14.6.

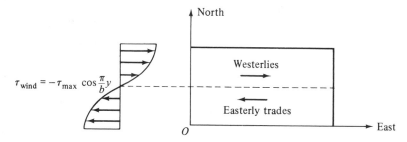

Figure 14.7. Representation of wind stress in North Atlantic.

Since we have fixed the coordinate system to the surface of the ocean and since the earth rotates about its own axis, the coordinate system (x, y) will thus rotate with constant angular speed $(0.726 \times 10^{-4} \text{ rad/s})$ about a vertical axis as its coordinate origin moves about in a circle. Therefore, we are dealing here with a moving coordinate system, and the equations of motion that are to describe the ocean current system relative to the moving earth must include terms accounting for this rotation. Let us indicate briefly the character of these terms.

In Section 4.6 an example of noninertial coordinates was derived involving acceleration of the moving coordinate origin in a straight line. It was shown there that a term accounting for this linear acceleration had to be included in the momentum equations. Similarly, terms must be added to the equations of motion accounting for the rotational motion of the moving coordinate system. Thus, if in addition to linear accelerative motion, the moving coordinate system also rotates, the velocity of a particle relative to the inertial reference is given by

$$\mathbf{V}_{rf} = \mathbf{V}_{C.V.} + \mathbf{V}_{rc} + \boldsymbol{\omega} \times \mathbf{r}$$

where $\mathbf{V}_{rf}$ is the velocity of a particle in the fixed inertial coordinate system (absolute velocity), $\mathbf{V}_{C.V.}$ is the linear velocity of the moving coordinate system, $\mathbf{V}_{rc}$ is the velocity of a particle in the moving coordinate system, and $\boldsymbol{\omega} \times \mathbf{r}$ is the angular velocity of a point in the moving coordinate system due to rotation with $\mathbf{r}$ the distance of the point from the moving coordinate origin.

Utilizing the expression for the time rate of change of a vector, we obtain the absolute acceleration of a particle:[*]

$$\underbrace{\frac{d\mathbf{V}_{rf}}{dt}}_{} = \underbrace{\frac{d\mathbf{V}_{C.V.}}{dt}}_{\substack{\text{linear acceleration} \\ \text{of moving coor-} \\ \text{dinate origin}}} + \underbrace{\frac{d\mathbf{V}_{rc}}{dt}}_{\substack{\text{acceleration of} \\ \text{particle in} \\ \text{moving coor-} \\ \text{dinate system}}} + \underbrace{2\boldsymbol{\omega} \times \mathbf{V}_{rc}}_{\substack{\text{Coriolis} \\ \text{acceleration}}}$$

$$+ \underbrace{\boldsymbol{\omega} \times (\boldsymbol{\omega} \times \mathbf{r})}_{\substack{\text{centrifugal acceleration} \\ \text{due to angular velocity} \\ \text{of moving coordinate} \\ \text{system}}} + \underbrace{\frac{d\boldsymbol{\omega}}{dt} \times \mathbf{r}}_{\substack{\text{tangential acceleration} \\ \text{due to angular accel-} \\ \text{eration of moving} \\ \text{coordinate system}}} \qquad (14.2)$$

[*] H. Goldstein, *Classical Mechanics* (Cambridge, Mass.: Addison-Wesley, 1953), p. 132.

479

For a moving coordinate system rotating at constant angular velocity, $d\omega/dt$ and $d\mathbf{V}_{\text{C.V.}}/dt$ become zero and the momentum equation reduces to

$$\Sigma\, d\mathbf{F} - 2dM(\boldsymbol{\omega} \times \mathbf{V}_{rc}) - dM(\boldsymbol{\omega} \times (\boldsymbol{\omega} \times \mathbf{r})) = \mathbf{V}_{rc}(\rho\mathbf{V}_{nrc} \cdot \mathbf{dA})$$

externally
applied
forces
      Coriolis
force
      centrifugal
forces

$$(14.3)$$

where the centrifugal and Coriolis accelerations have been placed on the left-hand side as equivalent forces. The latter, proportional to the cross product of $\boldsymbol{\omega}$ and $\mathbf{V}_{rc}$, depends on the geographical location (latitude) and on the local velocity of the fluid particle. For example, a particle moving eastward in the Northern Hemisphere with velocity u will experience a Coriolis force per unit mass in the negative y direction (i.e., toward the equator) of magnitude $2\omega_z u = 2\omega(\sin\phi)u$, where ω_z is the component of the earth's rotational velocity vector in the direction normal to the earth's surface (see Figure 14.8). In the Southern Hemisphere, with ω_z pointing inward, the Coriolis force will be directed again toward the equator. For the present problem, the centrifugal force, being proportional to the square of the rotational velocity of the earth, is very small relative to the other terms and can be neglected. Thus, writing the equations of motion in moving coordinates, we obtain the following equation:

$$\Sigma\, d\mathbf{F} - 2dM(\boldsymbol{\omega} \times \mathbf{V}_{rc}) = \mathbf{V}_{rc}(\rho\mathbf{V}_{nrc} \cdot \mathbf{dA}) \qquad (14.4)$$

      Coriolis force
      momentum flux

where $\Sigma\, d\mathbf{F}$ is given by Equation (14.1). Combining with the continuity equation and defining a stream function as in Section 7.3, we obtain the basic differential equation of motion for the problem of the North Atlantic currents. An additional approximation made is to neglect the right-hand side of Equation (14.4). This step is permissible because the maximum velocity of the currents

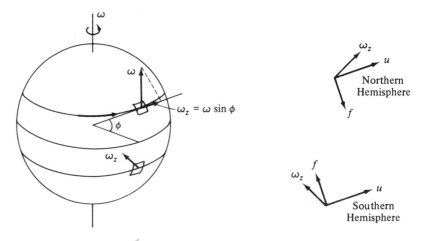

Figure 14.8.

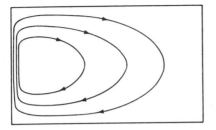

Figure 14.9. Streamlines of model of North Atlantic Ocean current.

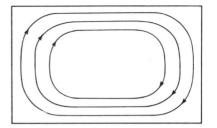

Figure 14.10. Streamlines without variable Coriolis force.

in the Gulf Stream on the western side of the North Atlantic basin does not exceed 2.5 m/s (8 fps), and velocity gradients—that is, the changes in velocity—are correspondingly small. Solving the stream function equation and setting the stream function equal to several constant values, we obtain the streamlines of the ocean currents. The resulting streamlines are shown in Figure 14.9. Note that the streamlines concentrate on the western side of the rectangle, corresponding to the Gulf Stream, which is located along the east coast of North America. The equations of motion used in the solution contained terms involving the variation of the Coriolis effect with latitude. It is this variation that has been found to be a primary cause of the Gulf Stream and similar flows. To illustrate this fact, let us omit the variable Coriolis force from the equations of motion. Figure 14.10 presents the result of the streamline pattern obtained. It will be noted that the streamlines now exhibit a symmetry not possessed before. Hence the simple model of the North Atlantic Ocean currents suggests that the concentration of streamlines along the western side is a result of the nonconstancy of the Coriolis parameter with latitude.

A similar streamline pattern is obtained in the North Pacific. To analyze these current flows, a triangular basin is more appropriate, as shown in Figure 14.11. Again, the concentration of streamlines on the western side (the Japanese current) can be ascribed to the nonconstancy of the Coriolis parameter with latitude.

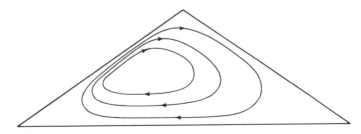

Figure 14.11. Streamlines of model of North Pacific current.

14.4 WAVES

In the preceding section we discussed the motion of fluid particles that travel over large distances in the ocean—the ocean currents. Such currents are caused by the shear force of the winds prevailing over the ocean. In this section we shall discuss a different type of motion of fluid particles, which can also be generated by the winds blowing over the ocean surface, namely, wave motion. In wave motion, the fluid particles travel over a very short distance only, moving back and forth, up and down about a reference position. The displacement of the particles results in a wavy surface configuration, a departure from the flat surface that is obtained when a liquid with a free surface is at rest. Waves can be generated by the action of strong winds blowing over the water surface, by a ship moving through the water, or by such geophysical activities as earthquakes, volcanic eruptions, and tides. Figure 14.12 shows the generation of waves in a water tank by blowing air at different speeds over the tank.

When watching waves in the ocean, it appears that the fluid particles in each wave are moving past the observer. However, if we place a floating body—a cork, for example—on the surface, we find that the float does not move past us but moves back and forth and reciprocates with an up-and-down motion, describing a closed orbit approximating a circle or ellipse. It is the configuration of the surface that appears to move past us with a certain speed and in a certain direction (Figure 14.13). Thus, in describing the motion associated with waves in the sea, the orbital trajectory of the particles as well as the velocity (i.e., the direction and speed) of the configuration are of interest to us.

In connection with the configuration of the wave, several characteristic quantities are of interest (Figure 14.14). The wave height (H) is the vertical distance between wave trough and wave crest. The wavelength (L) is the distance between two wave crests in the direction of wave propagation. The steepness of the wave is given by the ratio H/L. The wave period (T) is the time interval between passage of two consecutive wave crests past a fixed point. The wave speed (c) is therefore given by the ratio of L/T; that is, $c = L/T$.

The periods of ocean waves usually vary from fractions of a second to about 10 seconds, with wave speeds varying from less than 30 cm/s to about 15 m/s (1 fps to about 50 fps).

A simple wave form occurs when the elevation of the free surface describes a sinusoid; that is, its height (y) above an undisturbed reference surface can be described by

$$y = a \sin \frac{2\pi x}{L}$$

where a is the amplitude or maximum height above the reference (Figure 14.15). In the case of a sinusoidal wave form, the wave height is equal to

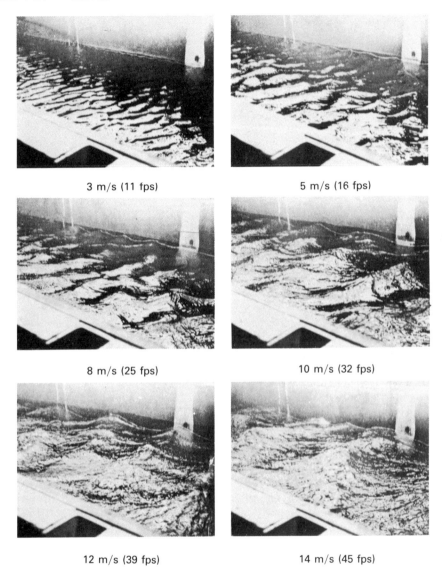

3 m/s (11 fps)   5 m/s (16 fps)

8 m/s (25 fps)   10 m/s (32 fps)

12 m/s (39 fps)   14 m/s (45 fps)

Figure 14.12.   Wind-generated waves. (*Courtesy* Hydronautics, Inc.)

the double amplitude ($H = 2a$) of the wave. If the wave propagates to the right with wave speed c, the equation of the moving surface is therefore given by

$$y = a \sin \frac{2\pi}{L}(x - ct)$$

As shown in Figure 14.15, the dot that was located at the origin at time $t = 0$ has moved a distance $x_1 = ct_1$ during the interval t_1. For sinusoidal

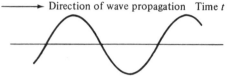

Direction of wave propagation   Time t

Figure 14.13.   Apparent motion of wave.

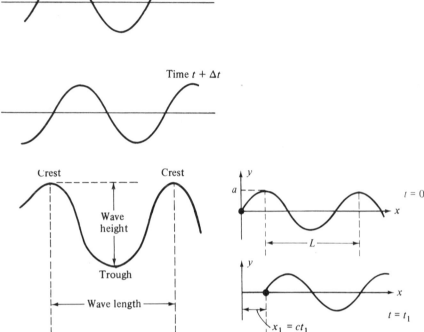

Time $t + \Delta t$

Crest   Crest

Wave
height

Trough

Wave length

Figure 14.14.   Wave characteristics.   **Figure 14.15.**

wave forms, the wave speed can be related to wavelengh or water depth. If the wavelength is small relative to the depth of the water (as is the case for wind-driven waves), the wave speed is independent of water depth and is a function of wavelength only ($c = \sqrt{g/2\pi}\sqrt{L}$). This relation indicates that the longer the wave, the faster its speed of propagation. If the wavelength is long relative to the water depth h, the wave speed is independent of wavelength and is a function of water depth only ($c = \sqrt{g}\sqrt{h}$). The shape of the fluid particle trajectory will also depend on the length of wave relative to the water depth. In deep water, the particle orbits are circles for the sinusoidal wave profile, whereas the orbits are elliptical in shallow water (Figures 14.16 and 14.17).

In Figure 14.15 we depicted a wave train, in which each wave (crest to crest) was of the same length and amplitude. In observing the ocean surface, we find that the characteristics of the waves vary from wave to wave; that is, a train of waves is composed of waves having different lengths. Since each wave will progress corresponding to its own wavelength—the longer the wave, the faster it will travel—a group of irregular waves will sort itself as it progresses, with the longer components gradually moving

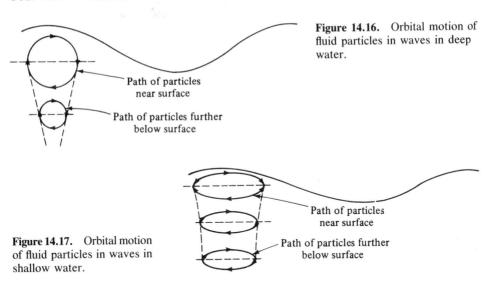

Figure 14.16. Orbital motion of fluid particles in waves in deep water.

Path of particles near surface

Path of particles further below surface

Figure 14.17. Orbital motion of fluid particles in waves in shallow water.

Path of particles near surface

Path of particles further below surface

ahead and the shorter waves falling behind. Hence long waves (swells) will be first to emanate from a storm area.

Observations on individual waves in the ocean tend to confirm the predicted values of wave speed and period as a function of wavelength. Figures 14.18 and 14.19 show measured and predicted wave velocities and wave periods of ocean waves. The solid line represents the predicted curves. For the wave period, we have, since $T = L/c$ and $c = \sqrt{g/2\pi}\sqrt{L}$,

$$T = \sqrt{\frac{2\pi}{g}\frac{L}{\sqrt{L}}} = \sqrt{\frac{2\pi}{g}}\sqrt{L}$$

In addition to the relations between length, speed, and period of ocean waves, a relationship of wave height of fully developed waves and wind

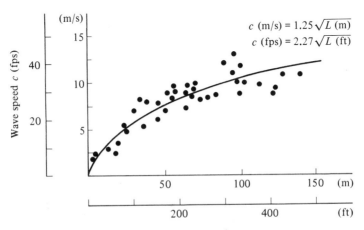

c (m/s) = $1.25\sqrt{L\ (m)}$

c (fps) = $2.27\sqrt{L\ (ft)}$

Figure 14.18. Wave speeds of ocean waves.

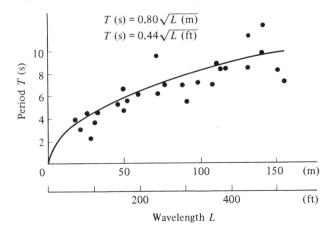

Figure 14.19. Wave periods of ocean waves.

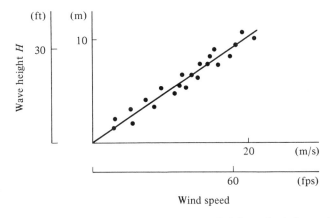

Figure 14.20. Relationship between wave height and wind speed.

velocity has been determined by measurement (Figure 14.20). This can also be seen in Figure 14.12 with respect to waves generated in a water tank.

The waves occurring in the ocean progress with a speed characteristic of their length in a certain direction and hence are called *progressive waves*. There is another class of waves, which occur in basins, lakes, or in liquid-propellant tanks of rockets—*standing waves*. The particle trajectories of such wave motion are shown in Figure 14.21. The movement of the standing wave resembles an oscillatory motion. In a standing wave, the fluid particles move in different orbital paths, their motion being in phase. The amplitude of movement for each particle is different. The orbital path is not closed, and each particle returns through the same points of the trajectory. In contrast, in progressive waves, all fluid particles at the same mean depth describe the same orbital path, their respective motions not being in phase with each other.

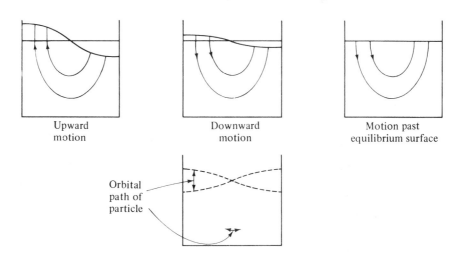

Figure 14.21.   Streamlines and orbital motion of standing waves.

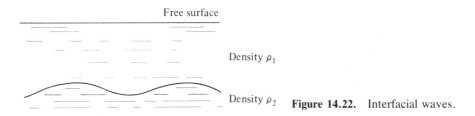

Figure 14.22.   Interfacial waves.

Thus far we have discussed waves that occur at a free surface, that is, at the interface of a liquid and a gas (the atmosphere). Waves can also occur at the interface of two liquids of different density that might arise in water which is stratified or in which the density varies with depth. Such waves are called *interfacial* or *internal waves* (Figure 14.22). The amplitudes of such waves are considerably larger than those of ordinary waves at a free surface.

14.5 TIDES

The rhythmic rise and fall of the water surface, caused by the effect of solar and lunar gravity, is called the *tide*. The average interval between successive high waters is found to be 12 hours and 25 minutes, having a direct correspondence to the regularity of appearance of the moon over a point on the earth every 24 hours and 50 minutes. This relationship between the tides and motion of the moon-earth system has long been recognized by man and has led to the conclusion that the primary tide-producing forces are those of lunar gravity acting on the earth's flexible water surface.

To demonstrate the tidal forces, consider first the earth-moon system. We shall neglect, for the time being, the rotation of the earth about its own

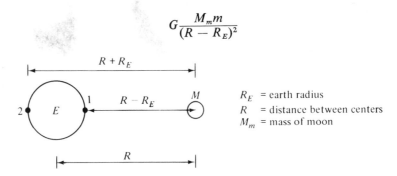

$$G\frac{M_m m}{(R - R_E)^2}$$

$R + R_E$

$R - R_E$

2 — E — 1 — M

R_E = earth radius
R = distance between centers
M_m = mass of moon

R

Figure 14.23. Earth-moon system.

axis and consider only the rotation of the earth-moon system about a common axis located between the centers of the two bodies. If we take the bodies as point masses, with mass located at the center of each body, then the total gravitational force between the two bodies is balanced by the centrifugal forces due to the rotation of the earth-moon system. However, this does not mean that these forces balance at each point on the earth's surface. In fact, it is a force unbalance acting on the mobile water that causes the occurrence of tides. The earth-moon system is shown in Figure 14.23. The force acting on a mass m at 1 due to the lunar gravity is

$$G\frac{M_m m}{(R - R_E)^2}$$

with G the gravitational constant. The force acting on the same mass m located at 2 will be

$$G\frac{M_m m}{(R + R_E)^2}$$

The force acting on a mass at the center of the earth is

$$\frac{GM_m m}{R^2}$$

Since, as stated above, there is a balance between centrifugal and gravitational forces with masses located at the body centers, it follows that the mass m at the center of the earth is in equilibrium. The force unbalance with the mass at 1 is

$$\Delta F_1 = \frac{GM_m m}{(R - R_E)^2} - \frac{GM_m m}{R^2}$$

For $R \gg R_E$, this reduces to

$$\Delta F_1 = \frac{2R_E}{R^3}GM_m m \tag{14.5a}$$

Similarly, we find for the force unbalance at 2

$$\Delta F_2 = -\frac{2R_E GM_m m}{R^3} \tag{14.5b}$$

with the negative sign indicating a force away from the moon.

In other words, with the lunar gravitational force acting on a mass at 1 greater than that for equilibrium, there will be a force unbalance at 1; since the water is mobile, this will lead to a tidal bulge at 1. At 2, the lunar

488

gravitational force is less than that for equilibrium, so that again there is a net force outward from the earth, due here to centrifugal forces (see Figure 14.24).

Figure 14.24. Tidal bulges.

For simplicity, the differential force was calculated for points directly under the moon. This differential force is a vertical force and, being proportional to $1/R^3$, produces an acceleration of the order of only 10^{-7} g. For points away from the meridian plane, for example, point 3 of Figure 14.25, there will exist a component T of the differential force in the horizontal plane, that is, directed along the earth's surface. This force is called the *tractive force* and brings about a flow of water along the surface of the earth toward the moon on the side of the earth near to the moon and, similarly, away from the moon on the side of the earth away from the moon. It is this horizontal force component that is the important one in producing the tidal flows of water toward and away from the beaches that we are familiar with.

Figure 14.25. Tractive force.

So far we have considered a nonrotating earth. In actuality, the earth is rotating about its own axis so that a point on the earth's surface at the equator, for example, will be directly under a fixed moon once every 24 hours. Since the moon is also rotating about the earth once every 27 days and 7 hours, the moon appears to pass over a given point a little later each day. Therefore we should expect a high tide every 24 hours and 50 minutes.

There is also a tide-producing force due to the sun, but since the distance between sun and earth is so much greater than that between moon and earth, the solar tidal forces are less than the lunar tidal forces. From Equation (14.5), the ratio of the differential forces is

$$\frac{\Delta F_{\text{moon}}}{\Delta F_{\text{sun}}} = \frac{M_m}{M_s} \frac{R_{S\text{-}E}^3}{R_{M\text{-}E}^3}$$

where M_m = moon mass = $\frac{1}{80}$ earth mass

M_s = sun mass = 330,000 earth masses

$R_{S\text{-}E}$ = distance from sun to earth = 150×10^6 km (93×10^6 mi)

$R_{M\text{-}E}$ = distance from moon to earth = 386,000 km (240,000 mi)

Solving, we obtain

$$\frac{\Delta F_{\text{moon}}}{\Delta F_{\text{sun}}} = 2.2$$

It can be seen that the lunar tidal forces are more than twice as great as the solar tidal forces. When the sun and moon are in the correct position with respect to the earth, the differential forces produced by sun and moon reinforce one another and we have the highest tides, called *spring tides*. The tides of lowest amplitude, called *neap tides,* are produced when the differential forces produced by sun and moon are $\pi/2$ rad (90°) out of phase. The range of the height of spring tides is approximately 3 times the range of the height of neap tides.

The foregoing discussion has been concerned solely with the nature of the tide-producing forces. An analysis of these forces enables us to predict the period of the tides, the time between the occurrences of high water at a given point on the earth. However, the water on the surface is not able to follow the tidal forces closely. Because of the inertia of the water, for example, the time of occurrence of high water at a given location may lag several hours behind the time at which the moon is directly over that location. Furthermore, the oceans and seas are confined by surrounding bodies of land. The shape of the bottom and side walls of the body of water has a large effect on the amplitude of the sides. For example, in the Mediterranean Sea, the narrowness of the entrance through the Strait of Gibraltar does not allow a large influx of tidal energy from the Atlantic Ocean; consequently, tidal ranges in the Mediterranean Sea are quite small, generally less than 1 m (less than 2 ft). The highest tides of the world occur in the Bay of Fundy, separating New Brunswick from Nova Scotia. The funnel shape of this body of water, along with a rising bottom surface, produces a near-resonance condition and a resultant large increase in tidal amplitude. The normal spring-tidal range at certain locations on the Bay of Fundy is over 15 m (50 ft). The effect of the tides in the Bay of Fundy is shown in Figure 14.26. These

Figure 14.26. High and low tides at Port Williams, Bay of Fundy, Nova Scotia, Canada. (With permission of Nova Scotia Information Service.)

examples illustrate the important influence of the surfaces bounding a body of water on the characteristics of the tides.

A complete hydrodynamic theory of the tides must therefore involve the inclusion of inertial terms in the equation of motion, as well as lunar and solar gravitational forces, frictional forces, Coriolis forces, and centrifugal forces. The complex geometrical shape of the bottom and sides of the sea or ocean must be used as boundary conditions for the hydrodynamic equations. To date, some success has been attained from theoretical analyses carried out on bodies of water of very simple geometrical shape.

REFERENCES

1. NEUMANN, G., AND PIERSON, W. J., JR., *Principles of Physical Oceanography* (Englewood Cliffs, N.J.: Prentice-Hall, 1966).
2. STOMMEL, H., *The Gulf Stream* (Berkeley and Los Angeles: University of California Press, 1977).
3. PICKARD, G. L., *Introductory Dynamic Oceanography* (New York: Pergamon Press, 1983).

15

MEASUREMENT TECHNIQUES

15.1 INTRODUCTION

In this chapter we shall discuss several techniques for measuring the properties of a fluid flow. Properties of interest to the engineer are fluid velocity, fluid pressure, flow rate, and viscosity. In the description of these techniques, use will be made of fluid-flow equations derived previously in the text. Several techniques for measuring each of the flow properties will be discussed. The choice of technique in each instance will depend on such factors as required accuracy, ease of installation, size of instrument, cost of instrument and measurements, and durability of the instrument. For example, a water meter for measuring water consumption must have the ability to provide accurate measurement over long periods, whereas the fluid meter controlling propellant flow in a rocket motor might require greater accuracy but be in use over a much shorter time interval.

15.2 PRESSURE MEASUREMENTS

Two types of pressure measurements are of interest in fluid flow: static pressure p and stagnation (or total) pressure p_t. *Static pressure* is the pressure indicated by a measuring device moving with the flow or by a device which introduces no velocity change to the flow. The usual method for measuring static pressure in a flow along a wall is to drill a small hole normal to the surface of the wall and connect the opening to a manometer or pressure gage

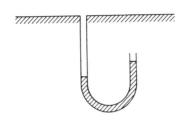

Figure 15.1. Static pressure tap.

(Figure 15.1). In the region of the flow away from the wall, static pressure can be measured by introducing a probe, which in effect creates a wall (Figure 15.2). The static pressure tap must be located far enough downstream from the nose of the probe to be out of the influence of flow disturbances due to the nose. Furthermore, the probe must be aligned with the flow direction, usually within 5 degrees.

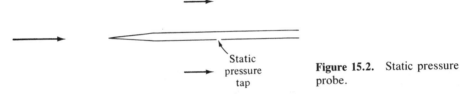

Static
pressure
tap

Figure 15.2. Static pressure probe.

Stagnation pressure is the pressure measured by bringing the flow to rest isentropically (i.e., without loss). A device for measuring stangation pressure is the *Pitot tube,* an open-ended tube facing directly into the flow, as shown in Figure 15.3. For incompressible flow, we obtain from Bernoulli's equation (4.21) that

$$p + \tfrac{1}{2}\rho V^2 = \text{Constant}$$

At the point of zero velocity, the pressure is p_t, so that

$$p_t = p + \tfrac{1}{2}\rho V^2 \tag{15.1}$$

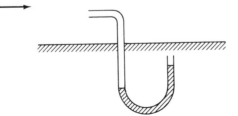

Figure 15.3. Pitot tube.

EXAMPLE 15.1

Water at 20°C flows through a 10-cm-inner-diameter vertical pipe. A stagnation pressure traverse is to be made of a cross section of the flow with a Pitot probe in order to determine the velocity distribution across the pipe. A static pressure tap at the cross section indicates a gage pressure of 34 kPa. Results of the traverse are given below.

Find the velocity distribution.

Distance from pipe wall (cm)	Pitot reading, gage pressure (kPa)
1	39.2
2	46.9
4	56.0
5	57.1
6	56.2
8	47.3
9	39.3

Solution   From Equation (15.1),

$$p_t - p = \tfrac{1}{2}\rho V^2$$

or

$$V = \sqrt{\frac{2(p_t - p)}{\rho}}$$

where, at 20°C, the density of water is 998 kg/m³.

We can now make up the following table:

Distance from wall (cm)	Δp (Pa)	V (m/s)
1	5,200	3.228
2	12,900	5.084
4	22,000	6.640
5	23,100	6.804
6	22,200	6.670
8	13,300	5.163
9	5,300	3.259

The results are plotted in Figure 15.4.   ■

Often the static and stagnation tubes are combined in one *Pitot static probe,* as shown in Figure 15.5.

When a Pitot tube is used in a subsonic compressible flow, the compressible form of the expression for stagnation pressure must be used [Equation (11.9)]:

$$p_t = p\left(1 + \frac{\gamma - 1}{2}M^2\right)^{\gamma/(\gamma - 1)}$$

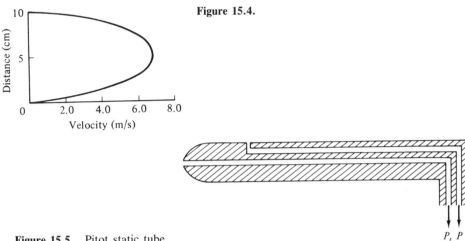

Figure 15.4.

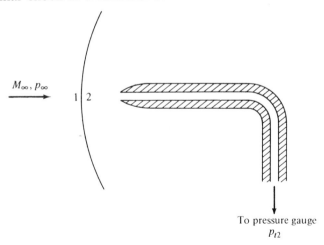

Figure 15.5.   Pitot static tube. P_t, P

By measuring stagnation and static pressures at a point in the flow, the Mach number at that point can be determined. In order to calculate the velocity of the flow, an additional measurement of static temperature must be made:

$$V = Ma$$

where, for a perfect gas,

$$V = M\sqrt{\gamma RT}$$

For supersonic compressible flow, a shock will exist in front of the Pitot tube, as shown in Figure 15.6. Directly in front of the probe, the shock is normal to the flow, so that pressure indicated on the gage will be the stagnation pressure after a normal shock occurring at M_∞. The stagnation pressure behind a normal shock is a function of the free stream Mach number M_∞

M_∞, p_∞

1 2

To pressure gauge
p_{t2}

Figure 15.6.   Pitot tube in supersonic flow.

and the free stream static pressure $p_\infty = p_1$; the determination of M_∞ with a Pitot tube in supersonic flow also requires a measurement of static pressure p_∞. For convenience, the ratio p_1/p_{t2} for a normal shock as a function of M_1 is given in Appendix C.

EXAMPLE 15.2

The Mach number of a jet plane is to be found from measurements of a Pitot tube attached to the nose, as shown in Figure 15.7. The plane is flying at an altitude of 60,000 ft. Calculate the flight Mach number for Pitot pressures of 225 lbf/ft² and 450 lbf/ft².

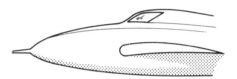

Figure 15.7.

Solution   From Appendix A, Table 2.D, at 60,000 ft, $p = 151$ lbf/ft².
For a ratio

$$\frac{p}{p_t} = \frac{151 \text{ lbf/ft}^2}{225 \text{ lbf/ft}^2} = 0.671$$

the flight Mach number is subsonic, since $p/p_t > 0.528$. From Appendix B, for $\gamma = 1.4$, $\underline{M = 0.78}$.
For a ratio

$$\frac{p_1}{p_{t2}} = \frac{151 \text{ lbf/ft}^2}{450 \text{ lbf/ft}^2} = 0.336$$

the flight Mach number must be greater than 1, since $p/p_t < 0.528$. Therefore, from Appendix C, we find $\underline{M = 1.38}$.   ■

The pressure from the static or stagnation probes can be measured by a pressure gage, manometer, or other suitable device, the choice depending on the desired sensitivity, response time, and pressure range. The conventional U-tube manometer has been described in Chapter 2. If large pressure differences are to be measured, a pressure gage would be more suitable. A *Bourdon tube pressure gage* consists of a C-shaped tube, as shown in Figure 15.8, with the pressure applied to the inside of the tube and the end closed. The pressure difference across the tube wall causes a flexure of the tube and resultant displacement of the tube end. By connecting the tube end to a pointer, the pressure difference can be recorded, after necessary calibration, as a movement of the pointer on a dial. Note that the gage indicates a pressure difference, or gage pressure, not an absolute pressure.

If shorter response times than the Bourdon gage or manometer are required, there are also the diaphragm pressure gage, piezoelectric pressure pickup, and others. With the diaphragm gage, the deflection of a diaphragm

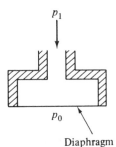

Figure 15.8. Bourdon tube pressure gage (a) Face view (graduated in kPa); (b) face view (graduated in lb/in^2; (c) interior view of gage. [*Courtesy* Crosby Valve & Gage Company (A Moorco Company.)]

caused by an applied pressure difference $p_1 - p_0$ is sensed by a strain gage mounted on the diaphragm (see Figure 15.9). With the piezoelectric pickup, the deformation of a piezoelectric crystal due to an applied pressure produces an output voltage.

Figure 15.9. Diaphragm pressure gage.

15.3 FLOW RATE MEASUREMENT

The Pitot tube was shown to have the capability of measuring velocity at a point in a flow; by taking a traverse across a pipe or duct, one could obtain a curve of velocity versus radial distance and integrate to determine the total flow rate through the pipe (see Example 15.1). Clearly, however, this process is very time-consuming; the engineer would often prefer to have a more direct method of measuring pipe flow.

One of the most common types of pipe flow meter is based on the introduction of a constriction into the pipe, the resultant acceleration of the flow producing a measurable pressure difference which can be related to volumetric flow. This type of meter includes the Venturi, flow nozzle, and orifice meters. We shall first discuss the operation of the Venturi meter.

The *Venturi meter* (Figure 15.10) consists of a converging section, followed by a diverging section. We wish to express volumetric flow through the pipe

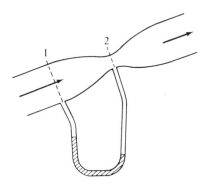

Figure 15.10.   Venturi meter.

as a function of the measured pressure difference, $p_1 - p_2$. We shall first assume incompressible steady flow through the Venturi, with uniform conditions present at sections 1 and 2. For frictionless flow between 1 and 2, we can write the Bernoulli equation (4.21):

$$\frac{p_1 - p_2}{\rho} + g(z_1 - z_2) = \frac{V_2^2 - V_1^2}{2}$$

Also, from the continuity equation, $A_1 V_1 = A_2 V_2$, so that

$$V_{2\,\text{ideal}} = \sqrt{2\left[\frac{(p_1 - p_2)/\rho + g(z_1 - z_2)}{(1 - A_2^2/A_1^2)}\right]}$$

The preceding velocity is an ideal velocity; in the actual case, the flow is not frictionless; also, the flow conditions at sections 1 and 2 are not uniform. To account for the difference between the ideal and actual flows, define a coefficient $C_v = V_{2\,\text{actual}}/V_{2\,\text{ideal}}$. Here C_v is dependent on the Reynolds number of the flow through the Venturi (just as, for pipe flow, the friction factor f was dependent on Reynolds number). A curve of C_v versus Reynolds number, obtained from experimental data, is given in Figure 15.11.

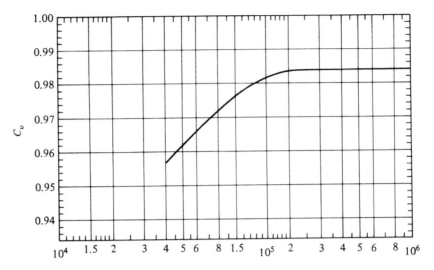

Figure 15.11. Discharge coefficients of Venturi tubes as a function of the pipe Reynolds number. (Adapted from *Fluid Meters—Their Theory and Application*, American Society of Mechanical Engineers, 5th ed., 1959, p. 125.)

We have as a final result:

$$Q = C_v A_2 \sqrt{2 \left[\frac{(p_1 - p_2)/\rho + g(z_1 - z_2)}{(1 - A_2^2/A_1^2)} \right]} \qquad (15.2)$$

By measuring $p_1 - p_2$, we can thus determine V_2 and hence the volumetric flow Q.

The *flow nozzle* operates on much the same principle as the Venturi; again, the pressure drop due to a constriction in the pipe is used to measure flow. A flow nozzle is depicted in Figure 15.12. It can be seen that the flow nozzle occupies less space than the Venturi although, with no smooth diverging section, the overall loss of pressure due to the presence of a flow nozzle in

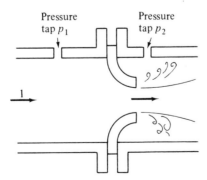

Figure 15.12.   Flow nozzle.

a fluid line will be greater than that due to the Venturi. Neglecting any differences in elevation between sections 1 and 2, we obtain

$$Q = C_n A_2 \sqrt{\frac{2}{\rho}\left[\frac{p_1 - p_2}{1 - A_2^2/A_1^2}\right]} \tag{15.3}$$

In order to obtain the coefficient C_n for the flow nozzle, just as we found C_v for the Venturi, it is necessary to resort to experimental data. Curves of C_n versus Reynolds number for various β are given in Figure 15.13, with $\beta = \sqrt{A_2/A_1}$.

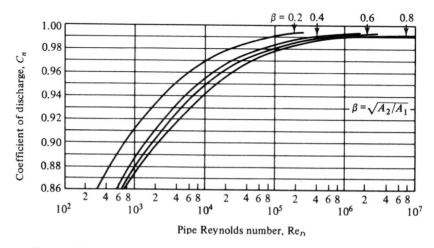

Figure 15.13. Discharge coefficients of typical flow nozzles (for nozzles in 5-cm pipes or larger). (Adapted from *Fluid Meters—Their Theory and Application*, American Society of Mechanical Engineers, 5th ed., 1959, p. 134.)

A third type of constriction meter is the *orifice meter,* shown in Figure 15.14. The orifice itself can be either square-edged or sharp-edged (see figure). With this meter, the pressure difference $p_1 - p_2$ is used to determine

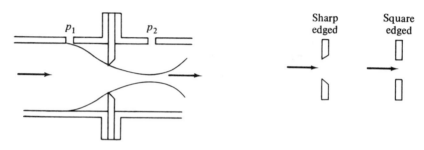

Figure 15.14. Orifice meter.

flow, with the pressure tap p_2 located at the minimum flow area, called the *vena contracta,* downstream of the orifice. At this area, the streamlines are parallel and conditions uniform across the jet. The tap p_2 cannot be placed right at the orifice, since the streamlines there are curved and, therefore, the static pressure across the jet is not uniform. As with the Venturi and flow nozzle, we can use Equation (15.2) to determine the incompressible flow rate through an orifice meter. Neglecting differences in elevation, we have

$$Q = CA_2\sqrt{\frac{2}{\rho}\left[\frac{p_1 - p_2}{1 - (A_2/A_1)^2}\right]} \tag{15.4}$$

Unfortunately, the area A_2 is unknown. We can express A_2 in terms of the actual orifice area A_0 and a contraction coefficient C_c, $A_2 = C_c A_0$, where C_c must be found experimentally. Combining the coefficients C and C_c, we obtain for the orifice meter

$$Q = C_0 A_0\sqrt{\frac{2}{\rho}\left[\frac{p_1 - p_2}{1 - (A_0/A_1)^2}\right]} \tag{15.5}$$

Curves of C_0 versus Reynolds number are given in Figure 15.15.

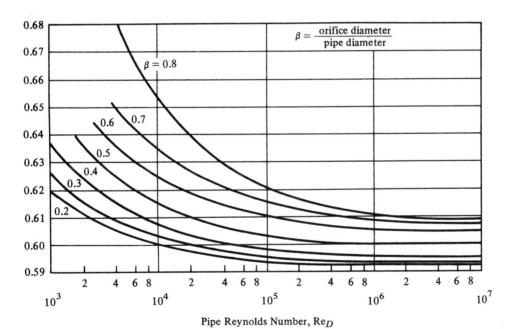

Figure 15.15. Typical curve of orifice discharge coefficient versus pipe Reynolds number (circular concentric orifice). (Data from *Fluid Meters— Their Theory and Application,* American Society of Mechanical Engineers, 5th ed., 1959, pp. 146–53.)

EXAMPLE 15.3

Hot water at 40°C flows through a 75-mm-diameter pipe. To measure the flow rate a flow nozzle with a 20-mm diameter is inserted. What is the water flow rate through the pipe, if the measured pressure difference is 20 kPa?

Solution From Equation (15.3),

$$
Q = C_n A_2 \sqrt{\frac{2}{\rho}\left(\frac{p_1 - p_2}{1 - A_2^2/A_1^2}\right)}
$$

$$
= C_n \frac{\pi}{4}(0.02^2 \text{ m}^2) \sqrt{\frac{2}{992.3 \text{ kg/m}^3}\left(\frac{20{,}000 \text{ kg/s}^2 \cdot \text{m}}{1 - (0.02/0.075)^4}\right)}
$$

$$
= 0.0020 C_n
$$

At first, assume the Reynolds number is sufficiently large so that, from Figure 15.13, $C_n = 0.992$ for $\beta = 0.27$. Thus $Q = 0.992(0.0020) = 0.001984$ m^3/s. To check, with $D = 0.075$ m,

$$
\text{Re}_D = \frac{VD}{\nu} = \frac{(0.001984 \text{ m}^3/\text{s})D}{(\pi/4)D^2(0.664 \times 10^{-6} \text{ m}^2/\text{s})} = 5 \times 10^4
$$

From Figure 15.13, a more exact value is $C_n = 0.987$; hence $Q = \underline{0.001974}$ $\underline{\text{m}^3/\text{s}}$.

∎

It can be seen that, for these three meters, reliance must be placed on experimentally determined coefficients. It is recommended, therefore, that there be a straight length of pipe before the meter (at least 10 diameters long) so as to provide an approach flow corresponding to that of the experimental tests used for determining the coefficients.

Of the three meters, the orifice is the cheapest and easiest to install; however, the orifice has associated with it the largest loss of head for a given flow of the three meters. Also, with the area of the vena contracta unknown (having to be found experimentally), the orifice meter is less accurate than the other two meters.

In the measurement of gas flow through a pipe with a Venturi, orifice, or flow nozzle, we must often consider the effects of compressibility. The expression for ideal velocity V_2 must be altered if we wish to account for the variation of density of the gas. Let us consider steady isentropic subsonic flow between sections 1 and 2 of a Venturi or flow nozzle. Assuming uniform conditions at 1 and 2, and neglecting changes in potential energy, the continuity and energy equations yield

$$
\rho_1 A_1 V_1 = \rho_2 A_2 V_2
$$

and

$$h_1 + \frac{V_1^2}{2} = h_2 + \frac{V_2^2}{2}$$

For a perfect gas with constant specific heats undergoing an isentropic process,

$$\frac{p_1}{\rho_1^\gamma} = \frac{p_2}{\rho_2^\gamma}$$

and

$$h_2 - h_1 = c_p(T_2 - T_1)$$

Combining, we obtain

$$\dot{m}_{\text{isentropic}} = A_2 \sqrt{\frac{2\gamma}{\gamma - 1} \frac{p_1\rho_1[(p_2/p_1)^{2/\gamma} - (p_2/p_1)^{(\gamma + 1)/\gamma}]}{[1 - (A_2/A_1)^2(p_2/p_1)^{2/\gamma}]}}$$

The actual flow rate through the Venturi becomes

$$\dot{m}_v = C_v A_2 \sqrt{\frac{2\gamma}{\gamma - 1} \frac{p_1\rho_1[(p_2/p_1)^{2/\gamma} - (p_2/p_1)^{(\gamma + 1)/\gamma}]}{[1 - (A_2/A_1)^2(p_2/p_1)^{2/\gamma}]}} \qquad (15.6)$$

Similarly, the actual flow through the flow nozzle is

$$\dot{m}_n = C_n A_2 \sqrt{\frac{2\gamma}{\gamma - 1} \frac{p_1\rho_1[(p_2/p_1)^{2/\gamma} - (p_2/p_1)^{(\gamma + 1)/\gamma}]}{[1 - (A_2/A_1)^2(p_2/p_1)^{2/\gamma}]}} \qquad (15.7)$$

where C_n and C_v are again functions of Reynolds number as given in Figures 15.11 and 15.13.

It is common practice to reduce the preceding expressions to a form comparable to that of the incompressible case, expressed in Equations (15.2) and (15.3), by the introduction of a compressibility factor Y, such that

$$\dot{m}_v = \frac{\rho_1 C_v Y A_2}{\sqrt{1 - (A_2/A_1)^2}} \sqrt{\frac{2(p_1 - p_2)}{\rho_1}} \qquad (15.8)$$

where

$$Y = \sqrt{\frac{\gamma}{\gamma - 1} \frac{[(p_2/p_1)^{2/\gamma} - (p_2/p_1)^{(\gamma + 1)/\gamma}][1 - (A_2/A_1)^2]}{[1 - (A_2/A_1)^2(p_2/p_1)^{2/\gamma}][1 - p_2/p_1]}} \qquad (15.9)$$

Values of Y, applicable to the Venturi and flow nozzle, as computed from Equation (15.9), are given in Figure 15.16.

With an orifice meter, there is no converging section to guide the flow between cross sections 1 and 2; certainly the area A_2 is a function of the compressibility of the flow. Therefore Equation (15.9) cannot be used for the orifice meter; Y must be determined empirically. Curves of Y versus p_2/p_1 for $\gamma = 1.4$ are given in Figure 15.17 for a square-edged orifice meter. Compressible flow rates can now be found from

$$\dot{m}_0 = \frac{\rho_1 C_0 A_0 Y}{\sqrt{1 - (A_0/A_1)^2}} \sqrt{\frac{2(p_1 - p_2)}{\rho_1}} \qquad (15.10)$$

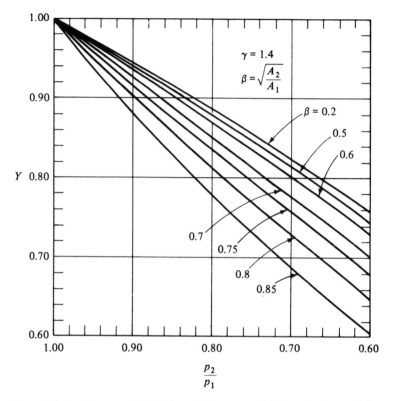

Figure 15.16. Compressibility factor Y for use with Venturi tubes and flow nozzles, $\gamma = 1.4$. (Adapted from *Fluid Meters—Their Theory and Application*, American Society of Mechanical Engineers, 5th ed., 1959, p. 170.)

EXAMPLE 15.4

An orifice meter is to be used to meter the flow of oxygen through a 2-in-inner-diameter tube (see Figure 15.18). Under a certain operating condition, the pressure upstream of the square-edged orifice is measured to be 1.0 in of water above atmospheric pressure, the pressure downstream of the orifice equal to atmospheric pressure. If the oxygen temperature is 60°F, determine the flow of oxygen in lbm/s. Atmosphere pressure = 29.92 in Hg, orifice diameter = 1.0 in.

Solution   From Equation (15.10)₁

$$\dot{m}_0 = \frac{\rho_1 C_0 A_0 Y}{\sqrt{1 - (A_0/A_1)^2}} \sqrt{\frac{2(p_1 - p_2)}{\rho_1}}$$

$$p_1 = p_{\text{atm}} + \Delta p$$

where

$$p_{\text{atm}} = 29.92 \text{ in Hg} = 2116 \text{ lbf}/\text{ft}^2$$

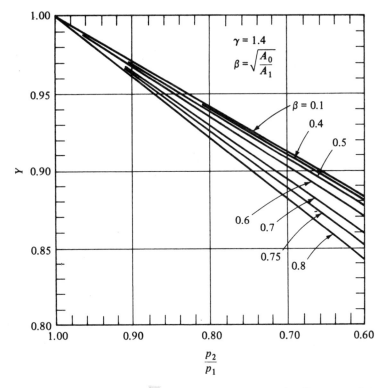

Figure 15.17. Compressibility factor Y for square-edged orifices, $\gamma = 1.4$. (Adapted from *Fluid Meters—Their Theory and Application*, American Society of Mechanical Engineers, 5th ed., 1959, p. 170.)

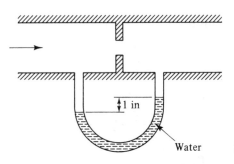

Water   **Figure 15.18.**

and

$$\Delta p = \rho g \, \Delta h$$
$$= \frac{(62.4 \text{ lbm/ft}^3)(32.17 \text{ ft/s}^2)(\frac{1}{12} \text{ ft})}{32.17 \text{ lbm/slug}}$$
$$= 5.2 \text{ lbf/ft}^2$$

Thus

$$p_1 = 2121 \text{ lbf}/\text{ft}^2$$

From the perfect gas law,

$$\rho_1 = \frac{p_1}{RT_1}$$

where

$$R = 48.3 \text{ ft-lbf}/\text{lbm}°R \qquad \text{for } O_2$$

$$\rho_1 = \frac{2121 \text{ lbf}/\text{ft}^2}{(48.3 \text{ ft-lbf}/\text{lbm}°R)(520°R)} = 0.0844 \text{ lbm}/\text{ft}^3$$

For our case,

$$\beta = \frac{\text{Orifice diameter}}{\text{Pipe diameter}} = \frac{1}{2}$$

From Figure 15.15, assuming we are on the flat part of the curve,

$$C_0 = 0.60$$

From Figure 15.17, with

$$\frac{p_2}{p_1} = \frac{2116}{2121} = 0.998$$

$Y = 1.0$. Therefore, returning to Equation (15.10):

$$\dot{m}_0 = \frac{(0.0844 \text{ lbm}/\text{ft}^3)(0.60)[(\pi/4)(\tfrac{1}{12})^2 \text{ ft}^2](1.0)}{\sqrt{1 - \tfrac{1}{16}}} \sqrt{\frac{2(5.2 \text{ lbf}/\text{ft}^2)}{\left(\dfrac{0.0844 \text{ lbm}}{\text{ft}^3}\right)\left(\dfrac{1 \text{ slug}}{32.17 \text{ lbm}}\right)}}$$

$$= 0.01796 \text{ lbm}/\text{s}$$

We must still check our assumption for C_0, which is based on the value of pipe Reynolds number $V_1 D_1/\nu$.

$$V_1 = \frac{\dot{m}}{\rho_1 A_1}$$

$$= \frac{0.01796 \text{ lbm}/\text{s}}{(0.0844 \text{ lbm}/\text{ft}^3)[(\pi/4)(\tfrac{1}{6})^2 \text{ ft}^2]}$$

$$= 9.75 \text{ ft}/\text{s}$$

For O_2 at 60°F, from Appendix A, Table 3.B, $\nu = 1.6 \times 10^{-4} \text{ ft}^2/\text{s}$, and so

$$\text{Re} = \frac{(9.75 \text{ ft}/\text{s})(\tfrac{1}{6} \text{ ft})}{1.6 \times 10^{-4} \text{ ft}^2/\text{s}}$$

$$= 1.02 \times 10^4$$

This yields a value for $C_0 = 0.616$, which raises $\dot{m}_0$ to

$$\frac{0.616}{0.600} \times 0.01796 \text{ lbm}/\text{s} = \underline{0.0184 \text{ lbm}/\text{s}}$$ ■

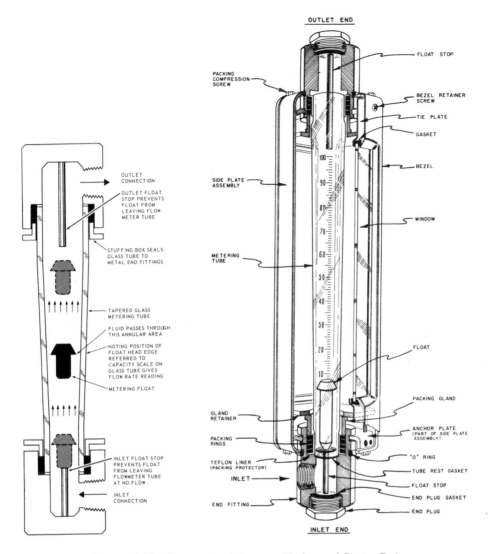

Figure 15.19.   Rotameter. (*Courtesy* Fischer and Porter Co.)

The *rotameter,* a variable-area flow meter, is illustrated in Figure 15.19. This meter consists of a float contained within a tapered glass tube. Fluid entering the bottom of the meter will raise the float, increasing the flow area between float and tube, until an equilibrium position is reached at which the weight of the float is balanced by the upward force of the fluid on the float; the greater the flow, the higher the float will rise in the tube. The tube is graduated, so that a direct reading of fluid flow can be obtained. The float is notched so that it will attain a stable position in the center of the tube.

A schematic view of a *turbine-type flow meter* is given in Figure 15.20. A turbine rotor is mounted in a pipe in the path of the flow. The force of

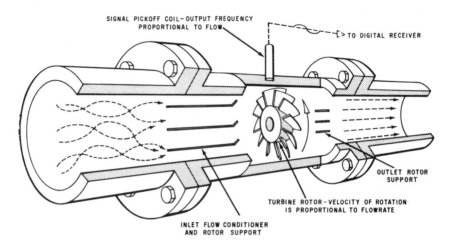

Figure 15.20. Turbine-type flow meter. (*Courtesy* Fischer and Porter Co.)

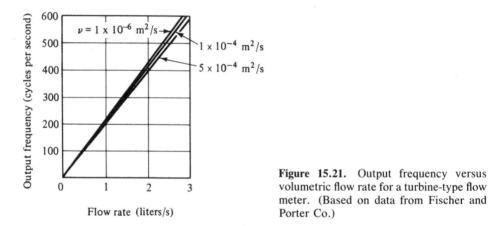

Figure 15.21. Output frequency versus volumetric flow rate for a turbine-type flow meter. (Based on data from Fischer and Porter Co.)

the fluid on the rotor blades causes it to rotate, with the rotational speed of the turbine proportional to fluid velocity and hence to flow rate. In order to monitor the rotational speed of the turbine, a permanent magnet and coil can be mounted on the pipe. As the rotor blades pass through the field produced by the magnet, an ac voltage is induced in the coil, with the voltage frequency proportional to flow rate. A typical output is given in Figure 15.21. The turbine flow meter has the advantage of providing a continuous readout of pipe flow rate.

15.4 HOT-WIRE AND THIN-FILM ANEMOMETRY

We have discussed the measurement of flow velocity with a Pitot tube; unfortunately, there are cases in which the response of the Pitot tube is inadequate for the study of flows with varying velocities. An example is the measurement of the fluctuating velocity components of a turbulent flow. In addition, the Pitot tube is somewhat limited by the size of the opening, yielding velocities that really represent only the mean values over the area of the opening. For precise measurements of velocity profiles in a boundary layer near a wall, the Pitot tube may prove unsuitable.

The *hot-wire anemometer* is a device that has a very short response time and can be used to measure velocities under varying flow conditions. This device consists of an electrically heated wire, supported between two prongs, and placed in the flow as shown in Figure 15.22.

Platinum or tungsten wire

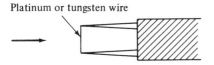

Figure 15.22.   Hot-wire probe.

When current is passed through the wire, the wire will heat up, with the resultant wire temperature determined by a balance between the heat loss from wire to fluid by convection and the electrical energy dissipation from the wire due to its resistance:

$$I^2 R_w = h A_s (T_w - T_\infty) \tag{15.11}$$

where I = electric current

R_w = wire resistance

h = convective heat transfer coefficient between wire and fluid, dependent on fluid velocity

A_s = surface area of wire

T_w = wire temperature

T_∞ = fluid temperature

The wire resistance R_w is a function of temperature, with

$$R_w = R_0[1 + \alpha(T_w - T_0)] \tag{15.12}$$

where R_0 = wire resistance at reference temperature T_0

α = temperature coefficient of resistance

Two types of hot-wire anemometers are used, constant-current and constant-temperature. If the current is maintained constant, a measurement of wire resistance and hence temperature can be used to find h and, with the appropriate empirical equation, fluid velocity. If the wire temperature, and hence resistance, is maintained constant, a measurement of current can be used to find velocity. For both types, extensive calibrations are required to establish the empirical relationship between h and fluid velocity.

The wire used in the probe is usually tungsten or platinum, with diameters that are as low as 5 microns (5×10^{-6} m). The hot-wire probe thus has the advantage of being very small; measurements of local velocity can be obtained very accurately even, for example, close to a wall. Also, with the small mass and low thermal inertia of the wire used, this device has a short response time.

Hot-wire probes have been used successfully in gas flows. When used to measure the velocities in a liquid flow, however, several problems arise. If the liquid is conducting (such as tap water), there are the difficulties associated with electrolysis, such as corrosion. Coating the wire with an insulating material can alleviate the problem but at the expense of a reduced frequency response. Also, with a liquid, there is the problem of elevated wire temperatures causing local boiling of the liquid. One solution is the use of a *hot-film probe,* consisting of a probe coated with a thin metallic film, this film serving as the resistance (see Figure 15.23). To prevent the problems associated with conducting liquids, the film is coated with a micron-thick layer of insulating material like quartz, which is thin enough so as not to impair the frequency response of the device. The thin-film probe also has the advantage of being more durable than the very delicate hot-wire probe, especially important in the flow of liquids at high velocity.

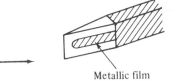

Metallic film   **Figure 15.23.**   Wedge-shaped hot-film probe.

The convective heat transfer coefficient in a liquid flow is much greater than that in a gas flow of the same velocity. For this reason, from Equation (15.11) the temperature difference $T_w - T_\infty$ for a given current I will be much less in a liquid; in other words, the sensitivity of the instrument in a liquid flow will be less than that in a gas flow. An improvement in sensitivity can be gained by the use of a thermistor, instead of a metallic resistor. *Thermistors* are semiconductor materials having a high negative coefficient of electrical resistance. For example, whereas the resistance of a platinum wire increases by a factor of 10 to 1 over the range—100 to 400°C, (-150°F to 750°F), the resistance of a thermistor may decrease by a factor of 1,000,000 to 1. Thus the use of thermistors leads to a much greater sensitivity.

15.5 OPEN CHANNEL FLOW MEASUREMENTS

Two types of flow measurements are of interest in open channel flow, flow rate and velocity. Flow rate devices are based on the introduction of an obstruction or constriction into the open channel flow, resulting in measurable changes in the flow depth which can be related to volumetric flow rates. Velocity measurements employ techniques already described for determining fluid velocities in pipe flow.

15.5(a) Flow Rate Measurements

A simple way of introducing an obstruction into the flow is the insertion of a vertical plate with an opening, called a *weir*. The opening is usually rectangular, trapezoidal, or triangular, although other openings (circular, exponential, and hyperbolic) are used. The opening of the weir (notch) is generally sharp-edged, as in the case of orifices used in pipe flow measurements. Only one measurement is required to determine the volume flow rate in the channel—the height of the liquid above the lowest point of the weir notch. To obtain an expression for the volume flow over a weir, consider the rectangular weir shown in Figure 15.24.

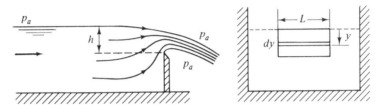

Figure 15.24.   Rectangular weir.

If we assume that the streamlines of the flow are parallel to the undisturbed free surface upstream of the weir, we can obtain the velocity of the stream layer of thickness dy located a distance y below the undisturbed free surface. From Bernoulli's equation (4.21),

$$\frac{p_a}{\rho} + \frac{V^2}{2} + g(h - y) = \frac{p_a}{\rho} + \underset{\text{negligible}}{\frac{V_0^2}{2}} + gh$$

Hence

$$V = \sqrt{2gy}$$

The incremental volume flow rate of the stream layer dy is therefore

$$dQ = VL\,dy = \sqrt{2gy}\,L\,dy$$

Integrating from $y = 0$ to $y = h$, where h is the head of liquid above the weir, we obtain the total ideal volume flow rate over a rectangular weir:

$$Q_t = \sqrt{2g}\,L \int_0^h \sqrt{y}\,dy = \tfrac{2}{3}L\sqrt{2gh^3} \tag{15.13}$$

The ideal flow rate calculated above neglected frictional losses and other factors. In actual cases, the flow will be less than the ideal rate. This decrease is caused by the fact that the area of flow will be less than area Lh due to contraction at the surface and at the crest as indicated in Figure 15.24. Furthermore, the streamlines will, in general, not be perpendicular to the plane of the weir. To account for these differences, an experimentally determined coefficient of discharge C is introduced, so that

$$Q = C\tfrac{2}{3}L\sqrt{2g}\ h^{3/2} \tag{15.14}$$

Usually C is less than 0.7 and a function of the value of the head h. An average value of C is about 0.65.

In a *Venturi flume* (Figure 15.25), the width of the channel is deliberately changed in order to measure the flow rate through the channel. As derived in Section 10.2a, the flow rate Q can be related to the liquid depths and channel widths at approach and minimum widths sections [Equation (10.1)]:

$$\frac{Q^2}{2g} = \frac{h_1 - h_2}{1/b_2^2 h_2^2 - 1/b_1^2 h_1^2} \tag{15.15}$$

Thus to obtain the volume flow rate we require measurement of channel widths and liquid depths only.

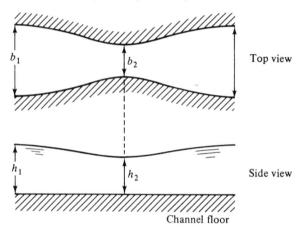

Top view

Side view

Channel floor   **Figure 15.25.** Venturi flume.

In order to measure the liquid levels required for the determination of volume flow rate, a *hook gage* or a *point gage* can be used (Figure 15.26). To obtain a measurement, the hook is submerged and then raised slowly until the point of the hook begins to break the surface. The vertical distance between the point of the hook and the bottom of the notch is obtained by means of a vernier scale mounted on the gage. A point gage (shown in Figure 15.26) can also be used for this purpose. The point gage is lowered to the surface to make the measurement. Floats are sometimes used to measure water elevations.

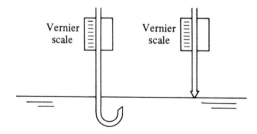

Vernier scale   Vernier scale

Figure 15.26. Hook gage; point gage.

15.5(b) Velocity Measurements

Devices used in pipe flow can be utilized to measure the velocity in open channel flow, for example, the Pitot tube. In addition, current meters similar in construction to the turbine-type flow meter described in Section 15.3 are used to measure flow velocity in open streams. A *current meter* consists of cups mounted to a wheel that is free to rotate about a vertical axis (Figure 15.27). A directional vane steadies the meter and keeps it headed into the current. The rotational speed of the cups is proportional to the velocity of the stream. The calibration constant is determined by moving the meter through still water at known speeds. In lieu of cups, vanes similar to windmills are also used for this purpose.

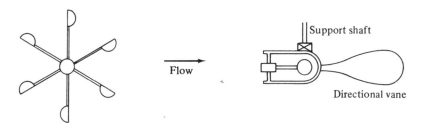

Figure 15.27.   Current meter.

Current meters are also employed to obtain flow rates for large streams where it is impractical to place an obstruction across the stream. In this case, velocity measurements at various points of the stream are taken and the volume flow rate is obtained by integration of the velocity distribution over the stream cross section.

15.6 VISCOSITY MEASUREMENTS

A number of devices are available to measure the viscosity of fluids. In each device (called a *viscosimeter* or *viscometer*), laminar motion of the fluid is created either by imposing a pressure difference or by the movement of a solid surface or body. In either case, the device contains elements of flow with an available analytic solution, allowing calculation of the viscosity of the fluid medium.

In one such device, the volume flow rate through a tube of small diameter (ensuring occurrence of laminar flow) is determined. The viscosity of the fluid can be evaluated from the relationship for laminar pipe flow (given in Section 6.3). An example of a simple capillary (a small-diameter-tube) viscosimeter for liquids is shown in Figure 15.28. Its main components are a calibrated efflux volume (located between the indicated etched lines) and the capillary tube. To make a measurement, the liquid is sucked past the upper etched line and released. The time required for the liquid to pass through

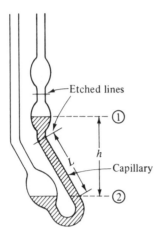

Figure 15.28.   Capillary viscosimeter.

the calibrated efflux volume is measured. The average volume flow rate Q is given by the ratio of efflux volume and efflux time t:

$$Q = \frac{\text{Efflux volume}}{t}$$

Equation (6.11) expresses the volume flow rate of laminar flow through tubes as

$$Q = V_{av} A = -\frac{dp}{dx}\frac{a^2}{8\mu}\pi a^2$$

where a is the radius of the tube and the pressure gradient dp/dx is equal to $(p_2 - p_1)/L$. The frictional pressure drop across the capillary equals the available hydrostatic head (h) of the liquid; that is,

$$p_1 - p_2 = \rho gh$$

A slight decrease in the value of the hydrostatic head (h) occurs during the measurement period. However, using average values, we obtain from the above

$$Q = \rho g\frac{h}{L}\frac{\pi a^4}{8\mu} = \frac{1}{\nu}g\frac{h}{L}\frac{\pi a^4}{8}$$

Hence the kinematic viscosity ν can be expressed as

$$\nu = \left(\frac{h}{L}\frac{\pi a^4}{8 \times \text{Efflux volume}}\right)gt \qquad (15.16)$$

where the quantity in parentheses is a known constant for a given viscosimeter. The viscosity of the liquid is therefore proportional to the efflux time.

Another viscosity-measuring device consists of two concentric cylinders, one of which is rotated while the other is held stationary. The torque required to hold the cylinder stationary is measured and the viscosity of the fluid is determined from the relationship given in Section 1.7. An example of this

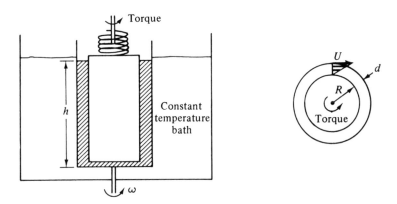

Figure 15.29.   Rotating viscosimeter.

type of device is shown in Figure 15.29, where the outer cylinder is rotated at a known rpm. The torque transmitted to the inner nonrotating cylinder is then measured by means of the deflection of a spring or torsion wire attached to the inner cylinder. The dynamic viscosity (μ) will be proportional to the torque exerted on the inner cylinder. For small gaps between the two cylinders, we have from Equation (1.1) the following expression for the shear stress:

$$\tau = \mu \frac{du}{dy} = \mu \frac{U}{d} = \mu \frac{\omega(R + d)}{d}$$

The torque exerted on the inner cylinder is

$$\text{Torque} = \text{Shear force} \times R = \tau 2\pi R h R$$

Hence

$$\tau = \frac{\text{Torque}}{2\pi h R^2}$$

Therefore,

$$\mu \frac{\omega(R + d)}{d} = \frac{\text{Torque}}{2\pi h R^2}$$

and since d is much smaller than R, we can write the expression for the dynamic viscosity as

$$\mu = \frac{\text{Torque} \times d}{2\pi R^3 h \omega} \tag{15.17}$$

A third type of viscosimeter uses the measured velocity of a body, of known dimensions and weight, falling through the fluid medium. If the body is a sphere of small diameter or if the density difference between sphere and fluid is small, the sphere will fall at sufficiently low speed so that its velocity (V) can be predicted from Stokes' law (see Section 9.3). Stokes flow will exist if the Reynolds number of the flow ($2VR/\nu$) is less than 1. A schematic

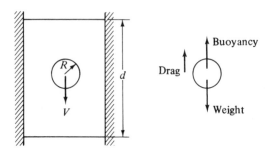

Figure 15.30. Falling-ball viscosimeter.

of such a falling-ball viscosimeter is shown in Figure 15.30. Upon release the sphere reaches its terminal, constant velocity. The time (t) required to traverse a measured distance (d) is recorded. Since at constant velocity the forces exerted on the sphere are in equilibrium, that is, $\Sigma F = 0$, we have (see Section 13.5)

$$\text{Weight} - \text{Buoyancy} - \text{Drag} = 0$$

or

$$\underbrace{\rho_s g \tfrac{4}{3}\pi R^3}_{\substack{\text{volume} \\ \text{of sphere}}} - \rho g \tfrac{4}{3}\pi R^3 - \underbrace{6\pi\mu VR}_{\substack{\text{drag of} \\ \text{sphere}}} = 0 \qquad (15.18)$$

where ρ_s and ρ are the density of the sphere and liquid, respectively. The expression for the drag of the sphere is obtained by combining Equation (9.2), $C_D = 24\nu/V2R$, with Equation (9.1), drag $= C_D \tfrac{1}{2}\rho V^2 A$. Thus

$$\text{Drag} = \frac{24\nu}{V2R}\left(\frac{1}{2}\right)\rho V^2 \pi R^2 = 6\pi\mu VR$$

Solving Equation (15.18) for μ, we obtain

$$\mu = \frac{2}{9} g(\rho_s - \rho)\frac{R^2}{V}$$

Since the velocity of the sphere is obtained from $V = d/t$, we finally have an expression for the dynamic viscosity μ, namely,

$$\mu = \frac{2}{9} gR^2(\rho_s - \rho)\frac{t}{d} \qquad (15.19)$$

The expression for the drag given above (drag $= 6\pi\mu VR$) is applicable when the radius of the tube is much larger than the sphere's radius. Otherwise a factor accounting for this wall effect must be added, that is, drag $= 6\pi K\mu VR$, and Equation (15.19) becomes

$$\mu = \frac{2}{9} g\frac{R^2}{K}(\rho_s - \rho)\frac{t}{d} \qquad (15.20)$$

The factor K has been determined analytically and confirmed by experimental measurements:*

* J. Happel and H. Brenner, *Low Reynolds Number Hydrodynamics* (Englewood Cliffs, N.J.: Prentice-Hall, 1965.)

Ball radius / Tube radius	K
0	1.000
0.1	1.263
0.2	1.680
0.3	2.371
0.4	3.596
0.5	5.970
0.6	11.135
0.7	24.955

EXAMPLE 15.5

A steel ball of 10 mm diameter drops along the centerline of a tube of diameter 50 mm filled with a liquid of unknown viscosity. The density of the steel is 7850 kg/m³, that of the liquid 918 kg/m³. It takes the ball 0.1 s to traverse a distance of 1 m. What is the liquid viscosity?

Solution   From Equation (15.20),

$$\mu = \frac{2}{9} g \frac{R^2}{K} (\rho_s - \rho) \frac{t}{d}$$

The diameter ratio of ball and tube is 0.2; hence $K = 1.68$ and

$$\mu = \frac{2}{9}(9.81 \text{ m/s}^2) \frac{0.01^2 \text{ m}^2}{1.68} [(7850 - 918) \text{ kg/m}^3] \frac{0.1 \text{ s}}{1 \text{ m}}$$

$$= \underline{0.090 \text{ Pa} \cdot \text{s}} \qquad\qquad \blacksquare$$

15.7 OCEANOGRAPHIC MEASUREMENTS

Two types of flow quantities of interest in oceanography are the velocity of the currents in the oceans and the associated volume flow rates. The magnitudes of the velocities in the ocean are relatively small, with average surface current speeds less than 1 m/s. Current meters are used to measure the speed and direction of the flow, which are similar in construction to those described for open channel flow measurements. The current meter is attached to a cable and lowered to the desired depth. Several types of current meters are utilized: the freewheeling propeller, with rpm proportional to the inflow velocity; and the locked-propeller rotor, for which the torque required to keep the rotor from turning is measured. The required torque is proportional to the inflow velocity. In addition, Pitot tubes are sometimes used to measure the velocity of ocean currents.

Tracking of the trajectories of currents is accomplished by use of floating objects, such as drift bottles, radio buoys, and submerged buoyant floats. Drifting surface ships and underwater craft are also employed. The tracking requires that the successive positions of the floats be known relative to the point of release. Average local velocities can also be obtained from the time rate of change of the distance between successive measurements.

Another important measurement activity is the determination of the physical and chemical characteristics of the water at different depths and locations in the ocean. The temperature, density, salinity, and chemical constituency of the seawater as well as the content of suspended particles are of interest. Samples of the seawater are collected *in situ* and the appropriate analysis carried out in a laboratory. Sampling bottles are lowered by a cable to the desired depth. A closing mechanism is then activated, enclosing the water sample in the plastic-lined metal bottle, which is brought to the surface for analysis. An example of such a sampling device, named a *Nansen bottle*, is shown in Figure 15.31. A number of bottles (cylinders with both ends open) are fastened to a cable at distances at which measurements are to be made. After the cable has been lowered, a messenger (weight) is released. The messenger slides down the cable, releasing, in succession, the trip mechanism of each bottle and causing them to turn over, and closing the covers at both ends of each bottle. A thermometer attached to each bottle measures the water temperature at sample depth.

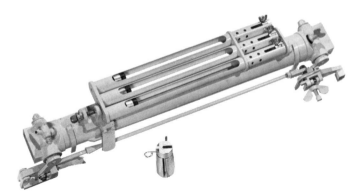

Figure 15.31. Nansen bottle. (*Courtesy* Belfort Instrument Co.)

A *bathythermograph* is used to obtain temperature-depth profiles to depths of about 300 m (1000 ft). The thermometer element of the bathythermograph is connected to a pen which marks a moving chart. The displacement of the chart is made proportional to the local pressure, which is closely proportional to the depth. Thus a temperature-versus-depth curve is obtained.

PROBLEMS

15.1. Water at 60°F flows through a circular pipe of 4-in inner diameter. A Pitot traverse provides the data given in Figure P15.1; an independent measurement indicates that the static pressure at this cross section is 5 psig. Determine the velocity distribution.

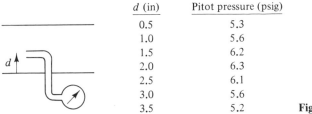

d (in)	Pitot pressure (psig)
0.5	5.3
1.0	5.6
1.5	6.2
2.0	6.3
2.5	6.1
3.0	5.6
3.5	5.2

Figure P15.1.

15.2. Determine the flow rate in ft^3/s through the pipe of Problem 15.1.

15.3. A Pitot tube is to be used to measure the test-section velocity in a low-speed wind tunnel, with the test section at atmospheric pressure. Maximum test-section velocity is to be 20 m/s, with Pitot pressures to be indicated on a U-tube water manometer. What differences in water level on the manometer can be expected?

15.4. A Pitot tube is mounted on an airplane, as shown in Example 15.2. The plane is flying into still air at 1500 m, where the ambient conditions are 85 kPa and 5°C. If the forward velocity of the plane with respect to an observer on the ground is 150 m/s, what pressure will be indicated by the Pitot probe?

15.5. Repeat Problem 15.4 for the case in which the plane is flying directly into a 20-m/s headwind. (Assume the plane to maintain its velocity of 150m/s with respect to a ground observer.)

15.6. A submarine is cruising at a depth of 50 ft in ocean water. If the forward speed of the submarine is 20 mph, what readings would be given by Pitot and static pressure probes? Assume the static probe is located to register free stream static pressure.

15.7. A mercury manometer is used to measure the difference between static and stagnation pressures for water at 20°C flowing in a 5-cm-diameter tube. Find the water velocity and estimate the flow rate of water. Is the flow laminar or turbulent? Take the specific gravity of mercury as 13.6 (see Figure P15.7).

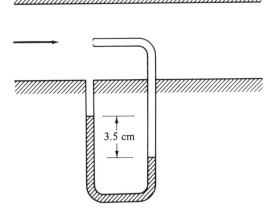

3.5 cm

Figure P15.7.

15.8. A Pitot tube is to be used to measure the test-section velocity in a supersonic wind tunnel. If the static pressure in the test section is 3 psia and the pressure indicated by the Pitot tube is 10 psia, determine the test-section air velocity. The stagnation temperature of the flow is 1200°R.

15.9. Repeat Problem 15.8 if helium is used as the working fluid in the wind tunnel.

15.10. A Venturi meter is to be used to measure the flow of water at 10°C through a vertical pipe. Determine the mass flow rate of water for the system shown in Figure P15.10. The pipe diameter is 10 cm, throat diameter 2.5 cm.

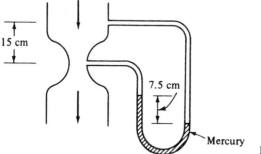

15 cm

7.5 cm

Mercury   **Figure P15.10.**

15.11. For the Venturi meter shown in Figure P15.11, why are pressure taps placed at cross sections 1 and 2 instead of at 2 and 3?

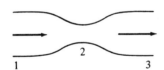

2

1   3   **Figure P15.11.**

15.12. A horizontal Venturi meter is to be used to meter the flow of air at 60°F in a 3-in-diameter pipe. If the upstream pressure p_1 is 50 psia and the throat pressure p_2 is 40 psia, determine the mass flow rate of air through the pipe. The throat diameter is 2 in.

15.13. An orifice meter is to be used to measure the flow of gasoline at 20°C in a 6-cm-diameter pipe. What pressure drop across the meter should be anticipated for a flow of 5 liters/s? The orifice diameter is 5 cm. Express your answer in mm of mercury. The density of gasoline is 740 kg/m³.

15.14. Calculate the mass flow rate of nitrogen at 100°F through a 3-in-diameter pipe if the pressures upstream and downstream of an orifice inserted in the pipe are 20 and 15 psia. The orifice is 2 inches in diameter.

15.15. A flow nozzle of 3 cm throat diameter is to be used to indicate the flow of oil at 20°C in a 5-cm-diameter pipe. If the measured pressure difference between the taps is 75 mm of mercury, determine the oil flow. Take the kinematic viscosity of the oil to be 9×10^{-6} m²/s, with a specific gravity of 0.86.

15.16. The flow nozzle of Problem 15.15 is to be used to measure the flow of air through a 5-cm-diameter pipe. If the maximum air velocity is to be 30 m/s, calculate the maximum pressure drop between the pressure taps that should

.be anticipated. Take the air pressure and temperature in the pipe to be 500 kPa and 25°C.

15.17. Find the flow of water through the Venturi shown in Figure P15.17.

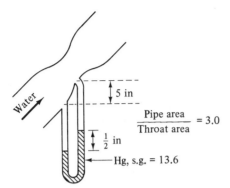

Figure P15.17.

15.18. A constant-temperature hot-wire anemometer is to be used to measure fluid velocity. Over the range of Reynolds numbers to be experienced, the heat transfer coefficient h can be written as $h = CV^n$ with C and n constants. Derive an expression for fluid velocity as a function of wire current.

15.19. Show that the ideal discharge rate of a triangular weir having an included angle of 2α is

$$Q_i = \tfrac{8}{15} (\tan \alpha) \sqrt{2g} h^{5/2}$$

(See Figure P15.19.)

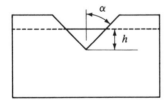

Figure P15.19.

15.20. The discharge coefficient of a right-angled triangular weir is to be determined over a range of values of upstream head h. To calibrate the weir, the discharge from the weir was collected in a tank and weighed over a clocked time interval. This procedure resulted in the following measurement of volume flow rate versus upstream head:

Q (l/s)	h (cm)
0.80	5.0
4.55	10.0
12.35	15.0
25.50	20.0
44.75	25.0

Calculate discharge coefficient versus h.

15.21. Determine the ideal discharge rate of the circular weir shown in Figure P15.21.

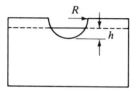

Figure P15.21.

15.22. A sharp-edged rectangular weir is 2 ft high and extends across an open channel of rectangular cross section. Find the flow rate for a head of one foot above the weir, using Equation (15.14). Also, find the flow rate by including the velocity of approach in the computation. What is the percent change in flow rate?

15.23. Determine the shape of the opening of a weir for which the flow rate varies linearly with the head above the weir.

15.24. Determine the time required to lower the level in a reservoir from h_1 to h_2 through a rectangular weir and through a right-angled triangular weir. The reservoir has vertical walls with horizontal surface area equal to A.

15.25. Derive an expression for the discharge through the trapezoidal weir shown for an upstream head h (see Figure P15.25.).

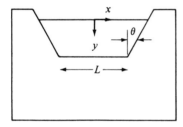

Figure P15.25.

15.26. A capillary viscometer has an efflux volume of 20 cm³, capillary radius 2 mm, capillary length 10 cm. The average available hydrostatic head is 13 cm. The measured efflux time for a certain oil at 50°C is 120 s. What is the kinematic viscosity of the oil?

15.27. The inclined-tube manometer shown in Figure P15.27 is used to measure small pressure differences. Find the relation between the difference in pressures at A and B and the manometer reading h.

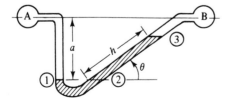

Figure P15.27.

15.28. The manometer of Problem 15.27 is to be used with a Pitot tube to measure wind velocity in a low-speed wind tunnel. Leg A is connected to the Pitot tube, leg B is open to the atmosphere, and the manometer fluid is an oil of specific gravity 0.86. The wind tunnel test-section static pressure is atmospheric, $\theta = 40°$, $a = 12$ in, $h = 8$ in, with the Pitot tube located on the center line of the test section. Determine the velocity at this point, with air temperature $= 80°F$.

16

DYNAMIC RESPONSE AND CONTROL OF FLUID SYSTEMS

16.1 INTRODUCTION

In the preceding chapters we have treated, almost exclusively, steady-state problems of fluid flow. In many actual flow situations, unsteadiness will occur. For example, the start-up or stopping of any device in which flow takes place will result in the occurrence of unsteady flow within the device. Also, when we change from one steady state to another steady state, we produce unsteady (transient) flow. In this chapter we will discuss the dynamic (time-dependent) response of fluid systems and describe control techniques utilized to maintain fluid systems at desired conditions.

16.2 DYNAMIC BEHAVIOR OF FLUID SYSTEMS

16.2(a) Liquid Level Fluctuations in an Open Tank System

As an example of the dynamic behavior of a fluid system, we will consider the response of the liquid level in an open tank which is connected to a long horizontal pipe de (Figure 16.1). It will be assumed that a constant volumetric flow rate Q_{in} is entering the tank. Under steady flow conditions, the height of liquid level in the open tank will be a function of the pressure drop in the pipe and the exit pressure at the end of the horizontal pipe. We will examine the behavior of the liquid level when the exit pressure at the end of the pipe is changed.

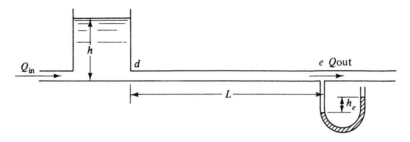

Figure 16.1.   Open tank-pipe system.

First consider the conservation-of-mass (continuity) equation with the tank as control volume:

$$\dot{m}_{in} = \dot{m}_{out} + \rho\frac{d\mathcal{V}}{dt}$$

or, for constant density, in terms of volume flow rate Q,

$$Q_{in} = Q_{out} + \frac{d\mathcal{V}}{dt}$$

With $d\mathcal{V}/dt = A(dh/dt)$, where A is the cross-sectional area of the tank, the continuity equation can be written as

$$A\frac{dh}{dt} = Q_{in} - Q_{out} \qquad (16.1)$$

To determine the frictional resistance in the horizontal pipe, we will assume that the resistance can be represented by steady-state values. For laminar pipe flow, the frictional resistance in the pipe will be proportional to the mean velocity $\overline{V}$ of the flow. For example, for laminar flow in circular tubes, the velocity distribution is parabolic with the mean velocity $\overline{V}$ one-half the maximum centerline velocity (see Sections 4.1 and 6.3). The pressure drop due to friction in steady pipe flow is given by Equation (6.5):

$$\Delta p = \frac{1}{2}\rho\overline{V}^2\frac{fL}{D}$$

Further, for laminar flow in a circular tube, the friction factor f is, from Equation (6.12),

$$f = \frac{64\mu}{\rho\overline{V}D}$$

Hence, the pressure drop in the pipe is:

$$p_d - p_e = \frac{1}{2}\rho\overline{V}^2\frac{64\mu}{\rho\overline{V}D}\frac{L}{D} = \frac{32\mu L}{D^2}\overline{V}$$

Since the volume flow rate $Q = \overline{V}(\pi/4)D^2$, we obtain

$$p_d - p_e = \frac{128\mu L}{\pi D^4}Q$$

Writing the Bernoulli equation (4.21) for frictionless flow between tank surface and pipe inlet d,

$$p_{atm} + \rho g h + \frac{V^2_{surface}}{2} = p_d + \frac{V_d^2}{2}$$

Neglecting kinetic energy of the flow,

$$p_d - p_{atm} = \rho g h$$

Now let the pressure at the end of the pipe, $p_e - p_{atm}$, be denoted by $\rho g h_e$. Hence,

$$h - h_e = \frac{128}{\pi} \frac{\mu L}{\rho g D^4} Q_{out}$$

The kinematic viscosity v is μ/ρ; hence,

$$h - h_e = \underbrace{\frac{128 v L}{\pi g D^4}}_{K} Q_{out}$$

or

$$h - h_e = K Q_{out} \tag{16.2}$$

Combining Equations (16.1) and (16.2), we obtain a differential equation describing the behavior of the liquid level (h) in the tank:

$$\frac{dh}{dt} + \frac{h}{KA} = \frac{Q_{in}}{A} + \frac{h_e}{KA} \tag{16.3}$$

The general form of the solution of this differential equation will depend on the type of disturbance occurring at the pipe exit. For example, the pipe exit pressure may vary linearly with time, it may oscillate sinusoidally, or there may be a step change in pressure at the exit. For a step change in exit pressure, the solution of Equation (16.3) is

$$h = c e^{-t/KA} + K Q_{in} + h_e \tag{16.4}$$

Consider the case where a step increase in exit pressure h_e occurs at $t = 0$ (Figure 16.2). Prior to the occurrence of the step increase ($t < 0$), the exit pressure was h_{e0} and the height of liquid level in the tank was h_0. Hence, from Equation (16.3), with $dh/dt = 0$ (steady-state condition),

$$\frac{h_0}{KA} = \frac{Q_{in}}{A} + \frac{h_{e0}}{KA}$$

or

$$h_0 = K Q_{in} + h_{e0}$$

After the occurrence of the step increase ($t > 0$), $h_e = h_{e1}$; hence, Equation (16.4) becomes

$$h = c e^{-t/KA} + K Q_{in} + h_{e1}$$

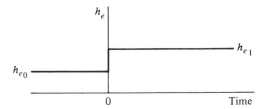

Figure 16.2. Step increase in exit pressure.

The constant c is determined at $t = 0$ when

$$h_0 = c + KQ_{in} + h_{e1}$$

or

$$c = h_0 - KQ_{in} - h_{e1} = KQ_{in} + h_{e0} - KQ_{in} - h_{e1} = h_{e0} - h_{e1}$$

Hence,

$$h = (h_{e0} - h_{e1})e^{-t/KA} + KQ_{in} + h_{e1} \qquad (16.4a)$$

After a sufficiently long time ($t \to \infty$), the liquid level will reach a new equilibrium position ($h = h_1$); then from Equation (16.4a),

$$h_1 = KQ_{in} + h_{e1} = h_0 - h_{e0} + h_{e1}$$

and Equation (16.4a) can be written as

$$h = -(h_{e1} - h_{e0})e^{-t/KA} + h_1 \qquad (16.4b)$$

The response of the liquid level to a step change in the exit pressure as given by Equation (16.4b) is illustrated in Figure 16.3, where the asymptotic approach of the liquid level to its new equilibrium position is seen. A measure of the speed of response of the system is the value of KA in the exponent of the exponential function e. The greater the value of KA, the slower the response will be. When time t reaches the value of KA (also called the time constant), $e^{-t/KA} = e^{-1} = 0.368$. From Equation (16.4b) it is seen that $(1 - 0.368)$ or 63.2 percent of the ultimate change in liquid level will have occurred.

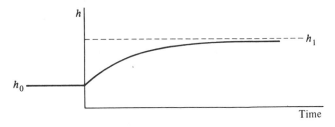

Figure 16.3. Response of liquid level in pipe system to step increase in exit pressure.

Next, let us examine the variation in outflow Q_{out} due to the step change in exit pressure. The inflow Q_{in} remains constant during the change. From Equations (16.1) and (16.4b),

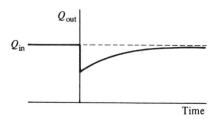

Figure 16.4. Response of inflow to step increase in exit pressure.

$$Q_{out} = Q_{in} - A\frac{dh}{dt} = Q_{in} - \frac{h_{e1} - h_{e0}}{K}e^{-t/KA} \qquad (16.5)$$

The variation in outflow is illustrated in Figure 16.4. Prior to the step change in exit pressure, the inflow (Q_{in}) equals the outflow (Q_{out}). After a sufficiently long time the outflow again becomes equal to the inflow (new equilibrium).

EXAMPLE 16.1

A liquid flows through an open tank and a 25-m-long horizontal pipe (I.D. = 100 mm), as illustrated in Figure 16.1. The constant cross-sectional area of the tank is 2 m². Initially the flow rate is 1 m³/min with an exit pressure equivalent to 1.5 m of liquid. The exit pressure is then increased instantaneously to the equivalent 2 m of liquid. The inlet flow rate remains constant at 1 m³/min. Calculate the variation in liquid level and flow rate ($\nu = 1 \times 10^{-4}$ m²/s).

Solution   First determine the constant K of Equation (16.2):

$$K = \frac{128\nu L}{\pi g D^4} = \frac{128(10^{-4})25}{\pi 9.81(0.1)^4} = 103.8 \text{ s/m}^2$$

Next check on the Reynolds number in the pipe:

$$Re = \frac{\overline{V}D}{\nu} = \frac{QD}{(\pi/4)D^2\nu} = \frac{4Q}{\pi D\nu} = \frac{4(1)}{\pi(0.1)10^{-4}(60)} = 2122$$

so the pipe flow is laminar. Initially the liquid depth in the open tank is

$$h_0 = KQ_{in} + h_{e0} = \frac{(103.8 \text{ s/m}^2)(1.0 \text{ m}^3/\text{min})}{60 \text{ s/min}} + 1.5 \text{ m} = 3.23 \text{ m}$$

The new equilibrium liquid depth is

$$h_1 = h_0 + (h_{e1} - h_{e0}) = 3.23 + (2.0 - 1.5) = 3.73 \text{ m}$$

From Equation (16.4), with $KA = (103.8 \text{ s/m}^2)(2 \text{ m}^2) = 207.6$ s,

$$h = -(h_{e1} - h_{e0})e^{-t/KA} + h_1 = -0.5e^{-t/207.6} + 3.73$$

From Equation (16.5),

$$Q_{out} = Q_{in} - \frac{h_{e1} - h_{e0}}{K}e^{-t/KA} = 1.0 \text{ m}^3/\text{min} - \frac{(0.5 \text{ m})(60 \text{ s/min})}{103.8 \text{ s/m}^2}e^{-t/207.6}$$

$$= \underline{1 - 0.289e^{-t/207.6} \text{ m}^3/\text{min}}$$

TABLE 16.1

t (s)	h (m)	Q_{out} (m^3/min)
0	3.230	0.711
50.0	3.337	0.773
100.0	3.421	0.822
150.0	3.487	0.860
200.0	3.539	0.890
207.6	3.546	0.894
250.0	3.580	0.913
300.0	3.612	0.932
400.0	3.657	0.958
500.0	3.685	0.974
∞	3.730	1.000

with t in seconds. The results of the calculations are shown in Table 16.1 and are plotted in Figure 16.5. It is seen that when the inflow conditions to the tank are kept constant and the exit pressure experiences a step increase, the liquid level in the tank increases asymptotically to the new equilibrium level, while the flow rate out of the tank experiences a step decrease and then increases asymptotically to its original value. ∎

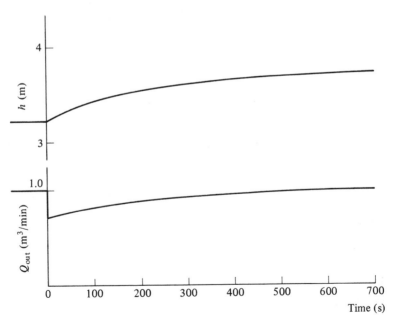

Figure 16.5. Response of system of Example 16.1 to step increase in exit pressure.

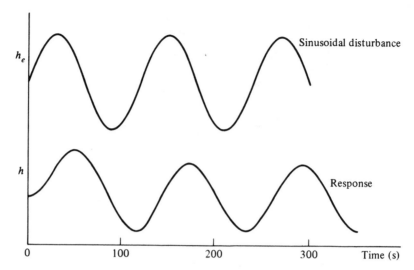

Figure 16.6. Response of liquid level in tank-pipe system to sinusoidal disturbance.

Thus far we have discussed the case where a step change occurs in the pressure at the exit of the pipe. More commonly, the exit pressure starts to oscillate sinusoidally with time and continues for a long time. For such a sinusoidal disturbance at the exit, the liquid surface will also fluctuate sinusoidally after a brief transient phase (Figure 16.6). The amplitude of the sinusoidal fluctuation of the free surface will, in general, be smaller than that of the fluctuating exit pressure.

16.2(b) Pressure Fluctuations in a Closed Tank System

The results obtained for liquid level fluctuations in an open tank can be applied to pressure fluctuations of a gas in a closed tank system if the process within the tank can be taken as isothermal. Consider the flow of a gas through a closed tank and long pipe as shown in Figure 16.7. When the tank is not insulated, it can be assumed that the temperature of the gas remains constant (isothermal). The continuity equation (4.2), considering the tank as the control volume, and with M the mass in the tank, gives

$$\dot{m}_{\text{in}} - \dot{m}_{\text{out}} = \frac{\partial M}{\partial t}$$

For an isothermal process within the tank, we get from the perfect gas law (as shown in Example 4.1)

$$p\mathcal{V} = MRT$$

or

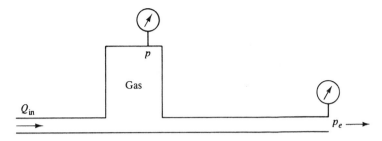

Figure 16.7. Closed tank-pipe system.

$$M = \frac{p \mathcal{V}}{RT}$$

and

$$\frac{\partial M}{\partial t} = \frac{\mathcal{V}}{RT} \frac{dp}{dt}$$

Thus the conservation of mass within the tank gives

$$\dot{m}_{\text{in}} - \dot{m}_{\text{out}} = \frac{\mathcal{V}}{RT} \frac{dp}{dt}$$

This equation corresponds to Equation (16.1) for liquid oscillations.

For small pressure differences along the insulated pipe and for low gas velocities, we can neglect effects of compressibility and obtain for laminar flow

$$p - p_e = \frac{32 \mu L}{D^2} \overline{V}$$

where the pressure p throughout the gas in the tank is assumed uniform at any time.

With

$$\dot{m} = \frac{\pi}{4} D^2 \rho \overline{V}$$

we have

$$p - p_e = \frac{128 \nu L}{\pi D^4} \dot{m}_{\text{out}} = C \dot{m}_{\text{out}}$$

This equation corresponds to Equation (16.2). Combining the above two equations, we obtain

$$\frac{dp}{dt} + \frac{RT}{\mathcal{V} C} p = \frac{RT}{\mathcal{V}} \dot{m}_{\text{in}} + \frac{RT}{\mathcal{V}} \frac{p_e}{C}$$

which corresponds to Equation (16.3) with $K = C$ and $A = \mathcal{V}/RT$. Hence the solution to the differential equation describing the pressure fluctuation in the tank is

$$p = ce^{-t/(\Psi C/RT)} + C\dot{m}_{in} + p_e$$

For a step disturbance in the outlet pressure,

$$p = -(p_{e1} - p_{e0})c^{-t/(\Psi C/RT)} + p_1$$

with $p_0 = C\dot{m}_{in} + p_{e0}$ and $p_1 = p_0 + (p_{e1} - p_{e0})$.

EXAMPLE 16.2

Gas at a temperature of 60°F flows through the tank-pipe system of Figure 16.7. The volume of the tank is 2000 ft³ and the 1-in-diameter pipe is 200 ft long. Initial pressure in the tank is 30 psia. The gas flow is maintained constant at 0.1 lbm/min. The exit pressure (p_e) is increased by 3 psia. Determine the response of the tank pressure. Take $\nu = 1.58 \times 10^{-4}$ ft²/s, $R = 53.35$ ft-lbf/lbm°R.

Solution

$$C = \frac{128\nu L}{\pi D^4} = \frac{128(1.58 \times 10^{-4} \text{ ft}^2/\text{s})(200 \text{ ft})}{\pi(\frac{1}{12})^4 \text{ ft}^4}$$

$$= 26{,}698 \text{ lbf} \cdot \text{s/slug} \cdot \text{ft}^2$$

Hence,

$$\frac{\Psi C}{RT} = \frac{(2000 \text{ ft}^3)(26{,}698 \text{ lbf} \cdot \text{s/slug} \cdot \text{ft}^2)}{(53.35 \text{ ft-lbf}/\text{lbm°R})(520°R)(32.17 \text{ lbm/slug})}$$

$$= 59.83 \text{ s}$$

$$p_{e0} = p_0 - C\dot{m}_{in} = 30 \text{ psia} - \frac{26{,}698 \text{ lbf} \cdot \text{s/slug} \cdot \text{ft}^2}{(144 \text{ in}^2/\text{ft}^2)(32.17 \text{ lbm/slug})}\left(\frac{0.1 \text{ lbm/min}}{60 \text{ s/min}}\right)$$

$$= 29.99 \text{ psia}$$

$$p_1 = p_0 + (p_{e1} - p_{e0}) = 30 \text{ psia} + 3 \text{ psia} = 33 \text{ psia}$$

$$p_{e1} = 29.99 \text{ psia} + 3 \text{ psia} = 32.99 \text{ psia}$$

$$p = -(p_{e1} - p_{e0})e^{-t/(\Psi C/RT)} + p_1$$

and

$$p = (-3e^{-t/59.83} + 33) \text{ psia}$$

The result is shown in Figure 16.8.

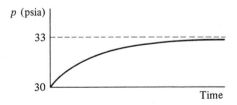

Figure 16.8. Response of system of Example 16.2 to step increase in exit pressure.

In many situations, particularly when the flow medium is a gas, the Reynolds number in the pipe will be large so that the flow in the pipe will be completely turbulent. In this case, the frictional pressure drop is

$$\Delta p = \frac{f}{2D}\rho \overline{V}^2 L$$

where the friction factor f is a constant (unlike the laminar flow case) and a function of relative pipe roughness ϵ/D only. For circular pipes, we have

$$\Delta p = \frac{8fL}{\pi^2 D^5 \rho}\dot{m}^2 = b\dot{m}^2$$

As an example of turbulent flow, consider the response of the pressure within a closed tank when gas is added to the tank through a long filler pipe. In this case, the continuity equation becomes with $\dot{m}_{out} = 0$

$$\dot{m}_{in} = \frac{\forall}{RT}\frac{dp}{dt}$$

Here $\Delta p = p_i - p$, where p_i is the pressure at the pipe inlet and p is the pressure at the juncture of pipe and tank. The pressure p throughout the tank is assumed uniform at any time. Combining the two equations, we have

$$\frac{dp}{dt} = \frac{RT}{\forall\sqrt{b}}\sqrt{p_i - p}$$

If the inlet pressure p_i is kept constant during the filling process, the solution becomes

$$p = p_i - \left(\sqrt{p_i - p_0} - \frac{RT}{2\forall\sqrt{b}}t\right)^2$$

where p_0 is the initial pressure within the tank. The mass flow rate into the tank will be

$$\dot{m}_{in} = \frac{\forall}{RT}\frac{dp}{dt} = \frac{\sqrt{p_i - p_0}}{\sqrt{b}} - \frac{RT}{2\forall b}t$$

EXAMPLE 16.3

Air at a constant pressure of 1000 kPa and a temperature of 15°C enters a 30-m-long filler pipe ($D = 50$ mm) of a closed tank. The pressure within the 40 m^3 tank is initially 500 kPa. Determine the time required for the pressure within the tank to reach a value of 1000 kPa. Take the friction factor f as 0.02 and the density of air as 12.09 kg/m^3.

Solution

$$b = \frac{8fL}{\pi^2 d^5 \rho} = \frac{8(0.02)(30\text{ m})}{\pi^2 [(0.05)^5\text{ m}^5] (12.09\text{ kg/m}^3)}$$
$$= 1.287 \times 10^5\ (\text{kg} \cdot \text{m})^{-1}$$

Hence

$$\frac{\Psi b}{RT} = \frac{(40 \text{ m}^3)1.287 \times 10^5 \text{ (kg} \cdot \text{m)}^{-1}}{(287 \text{ N} \cdot \text{m/kg} \cdot \text{K)} (288 \text{ K})}$$

$$= 62.29 \text{ s}^2/\text{kg}$$

$$\dot{m}_{in} = \frac{\sqrt{p_i - p_0}}{\sqrt{b}} - \frac{RT}{2\Psi b}t = \frac{\sqrt{(1,000,000 - 500,000)\text{Pa}}}{\sqrt{12.87 \times 10^4 \text{ (kg} \cdot \text{m)}^{-1}}} - \frac{t}{2(62.29 \text{ s}^2/\text{kg})}$$

$$= 1.9709 - 0.008026t$$

When the pressure in tank reaches 1000 kPa, $\dot{m}_{in}$ becomes zero. Hence the time is

$$t = \frac{1.9709}{0.008026} = \underline{245.55 \text{ s}}$$ ∎

In some applications it is desired to maintain the liquid level in a tank (or the gas pressure in a tank) as close as possible to a specified value, in spite of changes in outflow conditions (e.g., a step change in exit pressure). This can be achieved by varying the inflow rate through use of a control valve in a manner which responds to the change in exit pressure. This control process will be further discussed in the next two sections.

16.3 AUTOMATIC CONTROL

In the preceding section we have seen that when one flow quantity in a flow system is varied, other flow quantities will also be affected. If, for example, in the tank-pipe system of Figure 16.1, the exit pressure is varied, the liquid level in the tank will also change, or, in the case of the closed tank system of Figure 16.7, the tank pressure will be affected. If it is desired to maintain the liquid level (or tank pressure) constant in spite of changes in one flow quantity, it will be necessary to vary another flow quantity, such as inflow rate (or inlet pressure) in such a manner as to maintain the quantity to be controlled as close to the desired level as possible. This can be done manually by visually observing the liquid level and adjusting the inflow control valve, or it may be done automatically. Automatic control will require a means to sense changes in the liquid level (such as a float) and a device (such as a valve) that will change an inlet quantity (e.g., the inlet flow rate) in an appropriate manner (Figure 16.9).

The control can be affected in several ways. One way makes the change in inflow directly proportional to the change in liquid level, independent of the time rate at which the liquid level changes. Another way makes the rate of change of inflow proportional to the change in liquid level. A third way makes the change in inflow proportional to the rate of change of liquid level. Finally, a combination of these three ways may be used.

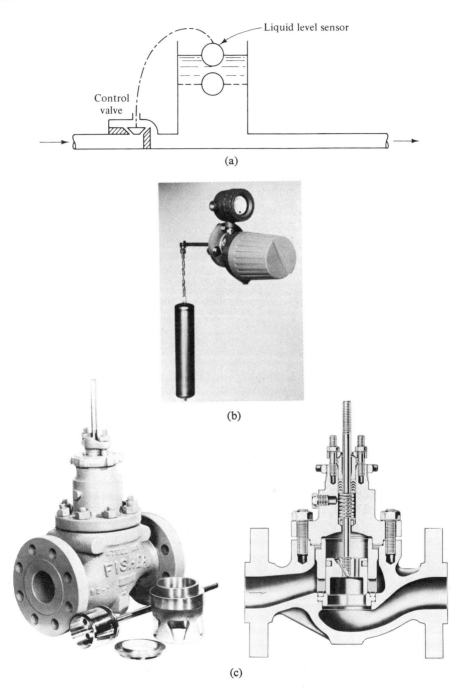

Figure 16.9.   (a) Automatic control of liquid level; (b) liquid level (buoyancy) transmitter for side-of-vessel mounting and indicating meter (courtesy of Foxboro Corp.); (c) control valve, single port, globe style, push down to close. (*Courtesy* Fisher Controls.)

Ideally, any control system should be able to keep the quantity to be controlled at any preset level—that is, it should be able to keep this quantity (e.g., the liquid level) at a constant value in spite of a change in another flow quantity (e.g., exit pressure). In the following we will examine how well each control type described above is inherently capable of performing its control function. The response of each type of control will be determined when it is subjected to a change in exit pressure.

16.3(a) Proportional Control

In a *proportional control* system, the movement in the control valve is proportional to the change in the quantity to be controlled. In Figure 16.10, the quantity to be controlled is the liquid height in the tank. The float, which follows the level of the liquid in the tank, is connected by the mechanical linkage to the control valve. For small linkage movement the arc traversed

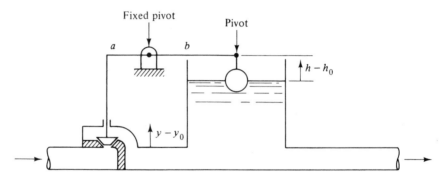

Figure 16.10.   Proportional control systems.

by the ends of the lever will be almost a straight line. The arc length traversed equals the product of radial distance from the fixed pivot and the angle traversed (θ). Thus,

$$h - h_0 = b\theta$$

and

$$y - y_0 = -a\theta$$

Hence,

$$\underbrace{y - y_0}_{\substack{\text{Valve piston} \\ \text{displacement}}} = -\frac{a}{b}\underbrace{(h - h_0)}_{\substack{\text{Float} \\ \text{displacement}}}$$

Subscript 0 denotes reference position for float and valve. A displacement of the float will result in a displacement in the valve piston, the magnitude depending on the value of the constant a/b. Usually b is much greater than a; thus, a large motion in the float results in a small displacement in the valve piston. For small changes in the valve piston position, the flow through the control valve will be directly proportional to the valve piston position—that is:

$$Q_{in} - Q_{in0} = c(y - y_0) = -c\frac{a}{b}(h - h_0)$$

or

$$Q_{in} - Q_{in0} = -\alpha(h - h_0) \tag{16.6}$$

where α is the proportional control factor which relates the change in flow rate to change in liquid level. The negative sign in front of the control factor indicates that the valve must act opposite to the change in liquid level to provide the necessary change in flow rate. Thus, if the liquid level increases, the valve opening must decrease, restricting the flow through it so that the liquid level will be brought back to the desired level.

Consider the case of controlling laminar flow in the horizontal pipe-tank system shown in Figure 16.1 with a step change of exit pressure. Substituting the expression (16.6) for flow through the control valve

$$Q_{in} = Q_{in0} - \alpha h + \alpha h_0$$

into Equation (16.3), we obtain

$$\frac{dh}{dt} + \frac{h}{KA} = \frac{Q_{in0}}{A} - \frac{\alpha}{A}h + \frac{\alpha}{A}h_0 + \frac{h_e}{KA}$$

or

$$\frac{dh}{dt} + \left(\frac{1}{KA} + \frac{\alpha}{A}\right)h = \frac{Q_{in0}}{A} + \frac{\alpha}{A}h_0 + \frac{h_e}{KA} \tag{16.7}$$

The initial equilibrium liquid level h_0 is

$$h_0 = \frac{KA}{1 + K\alpha}\left(\frac{Q_{in0}}{A} + \frac{\alpha}{A}h_0 + \frac{h_{e0}}{KA}\right) \quad \text{or} \quad h_0 = KQ_{in0} + h_{e0}$$

The new equilibrium level h_1 will be

$$h_1 = \frac{KA}{1 + K\alpha}\left(\frac{Q_{in0}}{A} + \frac{\alpha}{A}h_0 + \frac{h_{e1}}{KA}\right)$$

Hence,

$$h_1 = h_0 + \underbrace{\frac{h_{e1} - h_{e0}}{1 + K\alpha}}_{\text{Offset}} \tag{16.8}$$

The solution of the differential equation (16.7) is

$$h = ce^{-(1/KA + \alpha/A)t} + \frac{KA}{1 + K\alpha}\left(\frac{Q_{in0}}{A} + \frac{\alpha}{A}h_0 + \frac{h_e}{KA}\right) \tag{16.9}$$

At $t = 0$,

$$h = h_0, \; h_e = h_{e1}$$

so that

$$h_0 = c + h_1$$

or

$$c = h_0 - h_1 = -\frac{h_{e1} - h_{e0}}{1 + K\alpha}$$

and

$$h = -\frac{h_{e1} - h_{e0}}{1 + K\alpha} e^{-[(1 + K\alpha)/KA]t} + h_1 \qquad (16.10)$$

Without control, the ultimate change in liquid level due to a step change in exit pressure would have been equal to the change in exit pressure (see Equation 16.4b). The time constant, a measure of the response time, would have equaled the product of resistance constant K and the tank area A. With proportional control, the ultimate change in liquid level due to a step change in exit pressure equals the change in exit pressure decreased by a factor which is a function of K and the control factor α. The time constant is also decreased by the same factor. The response of the liquid level is illustrated in Figure 16.11. For a given change in exit pressure, the offset (Equation 16.8) is a function of the product of K and α; the greater the product, the smaller the offset. For a step increase in exit pressure, the offset is positive; for a step decrease, a negative offset will be obtained.

Next let us consider the variation in inflow during the control operation. From Equations (16.6), (16.8), and (16.10) we get for the inflow rate Q_{in}

$$Q_{in} = Q_{in0} - \alpha(h - h_0) = Q_{in0} - \alpha\left(h - h_1 + \frac{h_{e1} - h_{e0}}{1 + K\alpha}\right)$$

$$Q_{in} = Q_{in0} - \alpha\frac{h_{e1} - h_{e0}}{1 + K\alpha}(1 - e^{-[(1 + Ka)/KA]t}) \qquad (16.11)$$

From Equation (16.11) and from the time derivative of Equation (16.10) we have for the outflow rate Q_{out}

$$Q_{out} = Q_{in} - A\frac{dh}{dt} = Q_{in} - A\frac{h_{e1} - h_{e0}}{KA}e^{-[(1 + Ka)/KA]t}$$

$$Q_{out} = Q_{in0} - \alpha\frac{h_{e1} - h_{e0}}{1 + K\alpha}\left(1 + \frac{1}{K\alpha}e^{-[(1 + Ka)/KA]t}\right) \qquad (16.12)$$

For the new equilibrium at $t = \infty$, we get from Equation (16.11)

$$Q_{in} = Q_{in0} - \alpha\frac{h_{e1} - h_{e0}}{1 + K\alpha}$$

and from Equation (16.12):

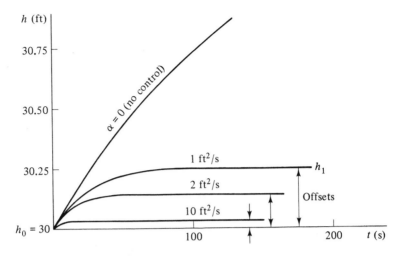

Figure 16.11. Response for different values of α of proportional control system to step increase.

$$Q_{\text{out}} = Q_{\text{in0}} - \alpha \frac{h_{e1} - h_{e0}}{1 + K\alpha}$$

That is,

$$Q_{\text{in}\infty} = Q_{\text{out}\infty}$$

Note that the new equilibrium flow rate differs from the initial inflow rate by a factor which is proportional to the offset.

EXAMPLE 16.4
The liquid level in a pipe-tank system is to be controlled by proportional control (Figure 16.10). The initial level is 30 ft. A step increase in exit head of 1.53 ft is applied. The initial inflow (Q_{in0}) is 3 ft^3/s. The tank cross-sectional area is 30 ft^2. The resistance constant K is 5 s/ft^2. The proportional control factor α equals 10 ft^2/s. Determine the response of the controlled system.

Solution   From Equation (16.8), the new equilibrium level h_1 is

$$h_1 = 30 \text{ ft} + \frac{1.53 \text{ ft}}{1 + (5 \text{ s/ft}^2)(10 \text{ ft}^2/\text{s})} = 30.03 \text{ ft}$$

That is, the offset is 0.03 ft. Now

$$\frac{1}{KA} + \frac{\alpha}{A} = \frac{1}{(5 \text{ s/ft}^2)(30 \text{ ft}^2)} + \frac{10 \text{ ft}^2/\text{s}}{30 \text{ ft}^2} = 0.340 \text{ s}^{-1}$$

From Equation (16.10), the variation in liquid level is

$$h = -\frac{1.53 \text{ ft}}{51} e^{-0.34t} + 30.03 \text{ ft} = -(0.03 \text{ ft})e^{-0.34t} + 30.03 \text{ ft}$$

From Equation (16.11), the variation in inflow is

$$Q_{in} = 3 \text{ ft}^3/\text{s} - (10 \text{ ft}^2/\text{s}) \frac{1.53 \text{ ft}}{51}(1 - e^{-0.34t})$$

$$= 2.7 \text{ ft}^3/\text{s} + (0.3 \text{ ft}^3/\text{s})e^{-0.34t}$$

From Equation (6.12), the variation in outflow is

$$Q_{out} = 2.7 \text{ ft}^3/\text{s} - (0.006 \text{ ft}^3/\text{s})e^{-0.34t}$$

and

$$Q_{in\infty} = Q_{out\infty} = 2.7 \text{ ft}^3/\text{s}$$

The response of the free surface is shown in Figure 16.11. For comparison purposes, the response at several lower values of proportional control factor α is also shown. With decreasing α, the offset increases, until at α equal to zero (no control), it is equal to the change in exit head. ■

When a system with proportional control is subjected to a sinusoidal disturbance at the exit of the pipe, the resulting response in liquid level experiences no offset (Figure 16.12). However, the amplitude of sinusoidal oscillation is greatly reduced (attenuated) as compared to the uncontrolled response.

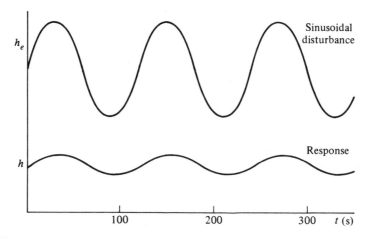

Figure 16.12. Response of proportional control system to a sinusoidal disturbance.

16.3(b) Integral Control

In an *integral control* system, the rate of movement in the control valve is proportional to the change in the quantity to be controlled (e.g., the liquid level). One way such a control can be achieved is by use of an electric motor where different control positions in a rheostat result in different motor speeds (Figure 16.13). Hence,

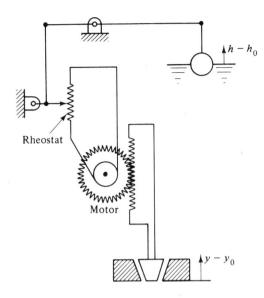

Figure 16.13.   Integral control system.

$$\frac{dy}{dt} = -c'(h - h_0)$$

Next, integrate to obtain an expression for the valve piston displacement, namely:

$$y - y_0 = -c' \int_0^t (h - h_0)\, dt$$

This relation containing the time integral of change in liquid level explains the term *integral* for this type of control. For small changes in valve position, a linear change in flow will be obtained and

$$Q_{in} - Q_{in0} = c(y - y_0)$$

Thus,

$$Q_{in} - Q_{in0} = -\gamma \int_0^t (h - h_0)\, dt \qquad (16.13)$$

where γ is the integral (also called reset) control factor. Substituting Equation (16.13) into Equation (16.3), we obtain the differential equation for the variation of liquid level h:

$$\frac{dh}{dt} + \frac{h}{KA} = \frac{Q_{in0}}{A} - \frac{\gamma}{A} \int_0^t (h - h_0)\, dt + \frac{h_e}{KA} \qquad (16.14)$$

The initial equilibrium level h_0 is obtained setting $dh/dt = 0$:

$$h_0 = KQ_{in0} + h_{e0}$$

Taking the time derivative of Equation (16.14) to get rid of the integral, we obtain

$$\frac{d^2h}{dt^2} + \frac{1}{KA}\frac{dh}{dt} = -\frac{\gamma}{A}(h - h_0)$$

or

$$\frac{d^2h}{dt^2} + \frac{1}{KA}\frac{dh}{dt} + \frac{\gamma}{A}h = \frac{\gamma}{A}h_0 \tag{16.15}$$

The types of solution that will be obtained depend upon the relative values of γ/A and KA.

Solution of Equation (16.15) is beyond the scope of this chapter; however, the following statement can be made: With integral control, there is no ultimate change in liquid level when a step change in exit pressure occurs. The manner in which the liquid level is stabilized will depend on the relative values of the control factor γ, the resistance constant K, and the tank area A (Figure 16.14). The liquid surface will either oscillate with decreasing amplitude about its initial position or will return asymptotically to its initial position. The value of the largest displacement of the liquid surface (overshoot) will be in both cases a function of γ, K, A, and the change in exit pressure. The new equilibrium flow rate is different from the initial flow rate by a factor which is proportional to the step change in exit pressure.

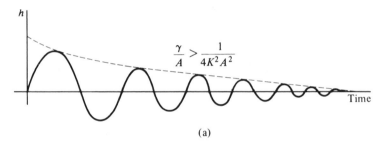

(a)

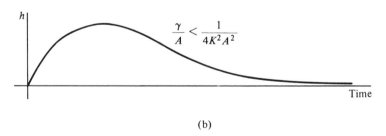

(b)

Figure 16.14. Response of integral system to step increase.

EXAMPLE 16.5

The liquid level in a pipe-tank system is to be controlled by integral control (Figure 16.13). The same conditions as in Example 16.4 apply ($h_0 = 30$ ft, $h_{e1} - h_{e0} = 1.53$ ft, $Q_{in0} = 3$ cfs, $A = 30$ ft^3, $K = 5$ s/ft^2). The integral control factor γ equals 0.1 ft^2/s^2. Determine the response of the controlled system.

Solution   The applicable differential equation is Equation (16.15). A solution to this equation is

$$h = h_0 + ce^{-t/2KA} \sin \sqrt{\frac{\gamma}{A} - \left(\frac{1}{2KA}\right)^2} \, t$$

The constant c is determined at $t = 0$. After the occurrence of the step increase ($t \geq 0$), $h_e = h_{e1}$, and hence Equation (16.14) becomes at $t = 0$

$$\frac{dh}{dt} + \frac{h_0}{KA} = \frac{Q_{in0}}{A} + \frac{h_{e1}}{KA}$$

With $h_0 = KQ_{in0} + h_{e0}$, we get

$$\frac{dh}{dt} = \frac{Q_{in0}}{A} + \frac{h_{e1}}{KA} - \frac{Q_{in0}}{A} - \frac{h_{e0}}{KA}$$

$$= \frac{h_{e1} - h_{e0}}{KA}$$

Now

$$\frac{dh}{dt} = ce^{-t/2KA} \left[\sqrt{\frac{\gamma}{A} - \left(\frac{1}{2KA}\right)^2} \cos \sqrt{\frac{\gamma}{A} - \left(\frac{1}{2KA}\right)^2} \, t \right.$$

$$\left. - \frac{1}{2KA} \sin \sqrt{\frac{\gamma}{A} - \left(\frac{1}{2KA}\right)^2} \, t \right]$$

At $t = 0$,

$$\frac{dh}{dt} = c \sqrt{\frac{\gamma}{A} - \left(\frac{1}{2KA}\right)^2} = \frac{h_{e1} - h_{e0}}{KA}$$

and hence

$$c = \frac{h_{e1} - h_{e0}}{KA} \frac{1}{\sqrt{\gamma/A - (1/2KA)^2}}$$

In this case,

$$c = \frac{1.53}{150} \frac{1}{\sqrt{0.1/30 - (\frac{1}{300})^2}} = \frac{0.0102}{0.05764} = 0.1770$$

The response of the free surface is shown in Figure 16.15.   ∎

The response of an integral control system to a sinusoidal disturbance in exit pressure is shown in Figure 16.16, where it is seen that a reduction in oscillatory amplitude occurs as well as the elimination of the initial overshoot.

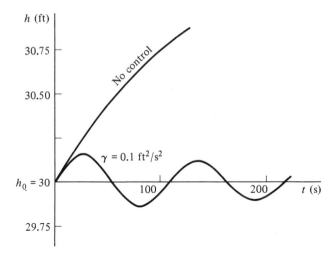

Figure 16.15.   Response of integral control system to step increase.

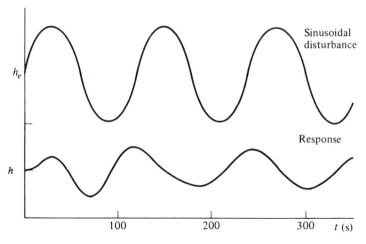

Figure 16.16.   Response of integral control system to sinusoidal disturbance.

16.3(c) Derivative Control

In a *derivative control* system, the movement in the control valve is proportional to the rate of change of liquid level. An example of a derivative control system is illustrated in Figure 16.17. The float is connected via mechanical linkage to the disk piston in the dashpot, which is filled with a viscous oil. Two springs keep the dashpot cylinder (and hence the control valve piston) at a preset equilibrium position. The dashpot cylinder is connected by mechanical linkage to the control valve. The motion of the float is related to that of the dashpot disk by

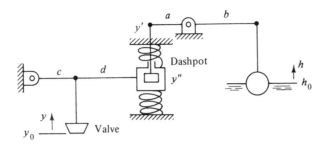

Figure 16.17. Derivative control system.

$$y' - y_0' = -\frac{a}{b}(h - h_0)$$

The motion of the dashpot cylinder is resisted by the two springs with a force equal to the product of the spring constant k and the displacement (y''). The dashpot disk will exert a force on the dashpot cylinder only when in motion. The force exerted on the cylinder will be proportional to the speed of the disk.* Hence, the displacement of the cylinder will be proportional to the disk speed—that is, for very slow disk speeds, the displacement of the cylinder will be negligibly small. Thus,

$$\kappa\frac{dy'}{dt} = k(y'' - y_0'')$$

Now from the expression relating y' and h we have

$$\frac{dy'}{dt} = -\frac{a}{b}\frac{dh}{dt}$$

and from the other mechanical linkage we get

$$y - y_0 = \frac{c}{c + d}(y'' - y_0'') = \frac{c}{c + d}\left(\frac{\kappa}{k}\right)\frac{dy'}{dt} = -\frac{c}{c + d}\left(\frac{\kappa}{k}\right)\left(\frac{a}{b}\right)\frac{dh}{dt}$$

For small changes in valve piston position, a linear change in flow is obtained— that is, $Q_{in} - Q_{in0} = f(y - y_0)$ and

---

* A fixed dashpot is used to dampen motion by introducing viscous resistance to slow down the motion of its piston (Figure 16.18a). As discussed in Section 15.6, the drag force on a sphere in Stokes flow is a function of viscosity of the liquid, radius of sphere, and velocity of sphere, namely

$$\text{Drag} = 6\pi\mu RV$$

When moving in a long cylinder (Figure 16.18b), the drag is increased by a factor which is a function of the radius of the cylinder—that is, drag $= 6K\pi\mu RV$. Similarly, a flat disk moving in a long cylinder (Figure 16.18c) will experience a drag

$$\text{Drag} = 5.093K\pi\mu RV$$

If end plates are added (Figure 16.18d) an additional drag will be experienced by the disk. In the derivative control system, the dashpot cylinder is also allowed to move (Figure 16.18e). Hence the force exerted by a moving disk will be

$$\text{Force} = \kappa V = \kappa\frac{dy'}{dt}$$

where κ is a factor which includes viscosity, diameter of disk, and effect of cylinder walls and end plates.

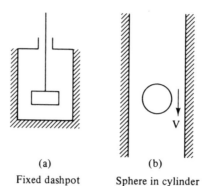

(a)

Fixed dashpot

(b)

Sphere in cylinder

(c)

Disk in cylinder

(d)

Disk in cylinder
with end plates

(e)

Moving dashpot

Figure 16.18. Motion of
various objects in cylinder
containing viscous liquid.

$$Q_{in} - Q_{in0} = -\beta \frac{dh}{dt} \tag{16.16}$$

where β is the derivative (also called rate) control factor. Substituting Equation
(16.16) into Equation (16.3), we obtain the differential equation for the variation
in liquid level h:

$$\frac{dh}{dt} + \frac{1}{K(A + \beta)}h = \frac{Q_{in0}}{A + \beta} + \frac{1}{K(A + \beta)}h_e \tag{16.17}$$

The initial equilibrium level is

$$h_0 = KQ_{in0} + h_{e0}$$

The new equilibrium level will be

$$h_1 = KQ_{in0} + h_{e1}$$

Hence,

$$h_1 = h_0 + h_{e1} - h_{e0} \tag{16.18}$$

The solution of the differential equation (16.17) is

$$h = ce^{-[1/K(A+\beta)]t} + KQ_{in0} + h_{e1} \tag{16.19}$$

At $t = 0$, $h = h_0$, and $h_e = h_{e1}$, so that

$$h_0 = c + h_1$$

Thus

$$c = h_0 - h_1 = h_{e0} - h_{e1}$$

and

$$h = -(h_{e1} - h_{e0})e^{-[1/K(A+\beta)]t} + h_1 \tag{16.20}$$

Thus, derivative control results in the same value of new equilibrium level as would be attained without control. However, derivative control modifies the asymptotic approach to the new equilibrium. The control factor β effectively increases the area A to result in a larger time constant, and hence in a slower speed of response.

The response of a derivative control system to a sinusoidal disturbance is similar to that obtained previously for the other types of control.

EXAMPLE 16.6

The liquid level in a pipe-tank system is to be controlled by derivative control (Figure 16.17). The same conditions as in Example 16.4 apply ($h_0 = 30$ ft, $h_{e1} - h_{e0} = 1.53$ ft, $Q_{in0} = 3$ cfs, $A = 30$ ft^2, $K = 5$ s/ft^2). The derivative control factor β equals 30 ft^2. Determine the response of the controlled system.

Solution   From Equation (16.18), the new equilibrium level h_1 is

$$h_1 = h_0 + (h_{e1} - h_{e0}) = 30 \text{ ft} + 1.53 \text{ ft}$$

From Equation (16.20), the variation in liquid level is

$$h = -(1.53 \text{ ft})e^{-t/5(30+30)} + 31.53 \text{ ft}$$

The variation in liquid level is shown in Figure 16.19.

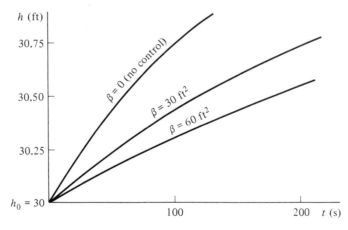

Figure 16.19.   Response of derivative control system to step increase.

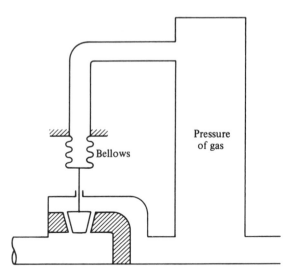

Figure 16.20. Proportional control of pressure.

In examining the response of the basic types of control to a step disturbance, we have seen that each type produces a different result: proportional control results in an offset; integral control has no offset, but oscillations are possible; derivative control results in the same new equilibrium as that for an uncontrolled system but with modified asymptotic approach. Hence, two types of control are usually combined—for example, proportional and integral control, the proportional control stabilizing the liquid level while the integral control returns it to its original level.

Thus far we have described various means for controlling the liquid level in a pipe-tank system. The same means can be used to control other variable flow quantities, such as pressure of a gas in a closed tank (Figure 16.20) or the flow rate through a pipe (Figure 16.21). In lieu of the float which activates the liquid level control system, the pressure of the fluid is used here. The bellows shown in the two figures allow for motion in the mechanical linkages of the control system. The bellows are at their equilibrium position at a certain pressure. When an additional pressure is applied, the bellows will extend to a new position, while for a decrease in pressure, the bellows will compress.

16.4 HYDRAULICALLY ASSISTED AUTOMATIC CONTROL

When the force required to effect the change in the flow variable is too large for self-actuation by the fluid system, a separate force-augmenting system is utilized. The force-augmenting system may be hydraulic (using a liquid such as oil) or pneumatic (using a gas, particularly air). For example, the hydraulic system has its own source of energy which raises the pressure of the hydraulic liquid, allowing the piston in a hydraulic cylinder to change the main control valve position.

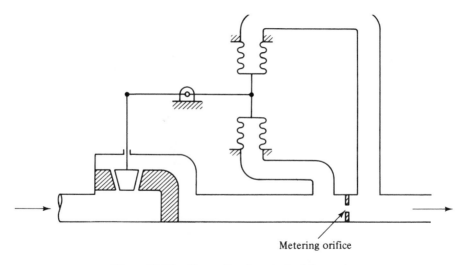

Figure 16.21. Proportional control of flow rate.

The major components of a frequently used hydraulic control system are shown in Figure 16.22: a pump, a control valve, a power cylinder, and a reservoir. The pump supplies high-pressure oil at constant pressure (Figure 16.23). The control valve controls the flow rate and direction of flow of

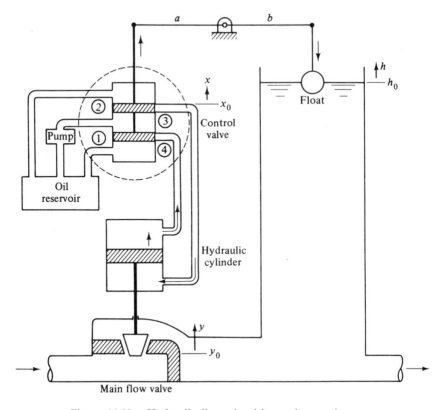

Figure 16.22. Hydraulically assisted integral control system.

(a)

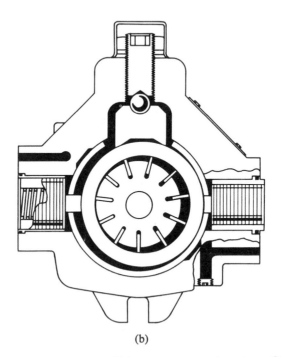

(b)

Figure 16.23. (a) Vane-type high-pressure pump (courtesy of Racine); (b) cross section of vane-type high-pressure pump.

hydraulic oil to the power cylinder. In the power cylinder, the flow of high-pressure oil pushes the power piston, which is mechanically connected to the main flow valve. The reservoir stores the hydraulic oil.

While hydraulic control systems are inherently of the integral type, they can be arranged to give the other control types or combinations thereof.

16.4(a) Integral Hydraulic Control

The type of control valve shown in Figure 16.22 has five openings (ports): an inlet port from the high-pressure pump, two exhaust ports (ports 1 and 2) leading to the oil reservoir, and two more exhaust ports (ports 3 and 4) leading to opposite sides of the power piston. The two interconnected pistons of the control valve are shown covering ports 3 and 4; hence there is no oil flow to the power piston and the main flow valve remains at a fixed position. When the exhaust condition at the end of the long horizontal pipe is changed such that the level in the tank decreases, the downward motion of the float will raise the pistons in the control valve as shown in Figure 16.24, partially opening port 3. This will supply high-pressure hydraulic fluid to the lower face of the power piston and force the power piston upward, thereby increasing the main flow and raising the liquid level in the tank. The hydraulic fluid on top of the power piston is exhausted to the reservoir through control valve port 4 and then port 1.

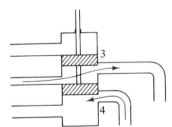

Figure 16.24. Control valve piston.

When the liquid level in the tank rises, the opposite will occur in the control valve, thereby decreasing the main flow until the new equilibrium position of liquid level and the new equilibrium main flow rate are reached.

As shown in Section 15.3, the volume flow rate of a liquid through an opening is given by an expression in the form

$$Q = C_d A \sqrt{\frac{2 \Delta p}{\rho}}$$

where C_d is the discharge coefficient, A the area of opening, and Δp the pressure drop across the opening. The uncovered area of a port in the control valve will be proportional to the piston travel—that is, $A = k(x - x_0)$, where k is the port constant. The discharge through port 3 (inflow to power cylinder) will be

$$Q_3 = C_d\sqrt{\frac{2}{\rho}}k(x - x_0)\sqrt{p_s - p_3}$$

where p_s is the supply pressure. As liquid enters the lower power cylinder (Figure 16.24), liquid leaves the upper cylinder through port 4. This outflow is

$$Q_4 = C_d\sqrt{\frac{2}{\rho}}k(x - x_0)\sqrt{p_4 - p_r}$$

where p_r is the reservoir pressure.

The flow from port 3 enters the hydraulic cylinder. The flow rate at the piston will be the product of piston area (A_p) and piston speed (dy/dt); that is, $Q_p = A_p(dy/dt)$. Although the velocity distribution in the cylinder will not be uniform (Figure 16.25), the continuity equation gives (neglecting leakage through the narrow annular gap between power piston and cylinder walls)

$$Q_3 = Q_p = A_p\frac{dy}{dt}$$

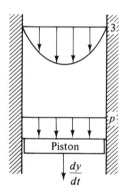

Figure 16.25. Nonuniform velocity distribution in hydraulic cylinder.

Similarly, for the flow leaving the hydraulic cylinder and entering port 4,

$$Q_4 = Q_p = A_p\frac{dy}{dt}$$

The force of the self-actuating system illustrated in Figure 16.10 comes from the buoyancy of the float, while the force obtainable from the power piston is many times greater than the buoyancy and will be

$$F = (p_3 - p_4)A_p$$

Combining the three equations to eliminate p_3 and p_4, we obtain

$$A_p\frac{dy}{dt} = C_dk\sqrt{\frac{1}{\rho}}(x - x_0)\sqrt{p_s \mp \frac{F}{A_p} - p_r}$$

where the negative sign applies when $(x - x_0)$ is positive and the positive sign is for $(x - x_0)$ negative. The power piston speed is

$$\frac{dy}{dt} = \frac{C_d k}{A_p} \sqrt{\frac{1}{\rho}} \sqrt{p_s \mp \frac{F}{A_p} - p_r} (x - x_0) \qquad (16.21)$$

The load F is due to friction in the main valve seals, and internal flow past the main valve piston, and can be taken as constant. Hence, we have

$$\frac{dy}{dt} = D(x - x_0)$$

The relationship between control valve piston travel and float displacement (see Figure 16.22) is

$$x - x_0 = -\frac{a}{b}(h - h_0)$$

Hence,

$$\frac{dy}{dt} = -D\left(\frac{a}{b}\right)(h - h_0)$$

which describes an integral control system as indicated in Section 16.3(b) and which results in a main flow change as given in Equation (16.13):

$$Q_{in} - Q_{in0} = -\gamma \int_0^t (h - h_0)\, dt$$

Details of the behavior of the main fluid system are as described in Section 16.3(b).

EXAMPLE 16.7
A four-way valve (as shown in Figure 16.22) is to be used in an integral hydraulic control system. The power piston area is 12.5 cm^2; the port constant k is 20 mm. The supply pressure is 400 kPa with a reservoir pressure of 100 kPa. The density of the hydraulic oil is 900 kg/m^3. Determine piston speeds as a function of load. Use a discharge coefficient C_d of 0.6.

Solution   From Equation (16.21),

$$\frac{dy}{dt} = \frac{C_d k \sqrt{1/\rho}}{A_p} \sqrt{p_s - p_r - \frac{F}{A_p}} (x - x_0)$$

$$= \frac{(0.6)(20 \times 10^{-3}\text{ m})\sqrt{1\text{ m}^3/900\text{ kg}}}{0.00125\text{ m}^2} \sqrt{(300\text{ kPa}) - \frac{F}{0.00125\text{ m}^2}}(x - x_0)$$

$$= 0.32 \sqrt{(300{,}000\text{ Pa}) - \frac{F}{0.00125}}(x - x_0)\text{m/s}$$

with F in N, x in m; which indicates, at a given load, a linear relation between power piston speed and control valve piston travel. For example, at $F = 0$, $dy/dt = 175.3(x - x_0)$; at $F = 375$ N (load limit), $dy/dt = 0$. ∎

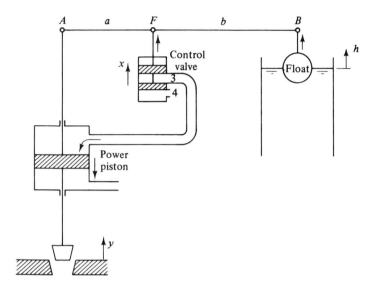

Figure 16.26. Hydraulically assisted proportional control system.

16.4(b) Proportional Hydraulic Control

The hydraulic control system shown in Figure 16.22 can be modified to give a proportional control system. Such a proportional control system is illustrated in Figure 16.26. The fixed pivot link of Figure 16.22 becomes a floating mechanical link (*AB*). When the liquid level rises in the tank, the control valve pistons will be moved upward. This will uncover port 3 so that hydraulic oil flows into the upper portion of the power cylinder. The power piston will then move downward, reducing the opening in the main valve. As the power piston is moving downward it also moves the floating link downward, thereby moving the control pistons downward. This will continue until the control pistons cover ports 3 and 4. Next we will show that the resulting control is proportional. First, assuming point *A* to be fixed, the partial movement of the control valve pistons is

$$x - x_0 = \frac{a}{a + b}(h - h_0)$$

Then, assuming point *B* to be fixed, the other partial movement of the control valve pistons becomes

$$x - x_0 = \frac{b}{a + b}(y - y_0)$$

Hence the total movement of the pistons is

$$x - x_0 = \frac{a}{a + b}(h - h_0) + \frac{b}{a + b}(y - y_0)$$

At the balance position (both ports 3 and 4 completely covered), $x = x_0$; hence,

$$0 = \frac{a}{a + b}(h - h_0) + \frac{b}{a + b}(y - y_0)$$

and

$$\underset{\text{Main valve}}{y - y_0} = -\frac{a}{b}\underset{\text{Float}}{(h - h_0)}$$

which describes a proportional control system as indicated in Section 16.3(a).

Hydraulic control systems can also be arranged to give a combined control type. A combined proportional and integral control system is illustrated in Figure 16.27.

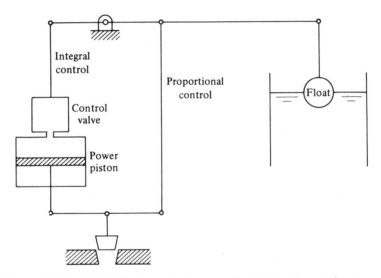

Figure 16.27. Hydraulically assisted proportional-integral control system.

Instead of a four-way valve, a different type of valve is also used in hydraulic control circuits. This type of valve, called a flapper valve, is illustrated in Figure 16.28. Such valves have been developed to eliminate some of the disadvantages of the piston-type valves, namely, expensive manufacturing cost due to close tolerance requirements and susceptibility to faulty operation when dirt gets lodged in the valve. The flapper valve consists of a tube with a restricting orifice and a flapper which controls the opening of the nozzle at the end of the tube. The flow of hydraulic oil through the orifice results in a pressure drop which will be a function of the flow rate of the oil, which in turn is controlled by the size of the gap at the nozzle. A small change in the position of the flapper causes a large change in the pressure drop across the orifice. As the flapper moves to the right, the gap

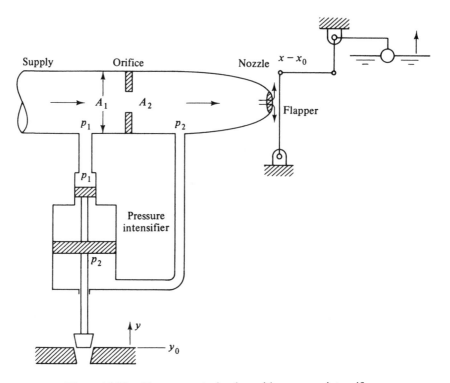

Figure 16.28. Flapper control valve with pressure intensifier.

nozzle opening is increased, thereby increasing the flow and decreasing the pressure p_2. Since the supply pressure is kept constant, the two pistons of the pressure intensifier will move downward. The two pistons are rigidly connected and have different piston sizes. At the balance position of the pressure intensifier ($x = x_0$),

$$p_1 A_1 = p_2 A_2$$

or

$$p_2 = p_1 \frac{A_1}{A_2}$$

That is, the pressure in the intensifier is increased by the ratio of the piston areas. When the flapper rests against the nozzle, the flow through the orifice is stopped and $p_1 = p_2$.

Instead of a pressure intensifier, a spring can be utilized to give the required difference in force (Figure 16.29). A single piston of area A is used. At the balance position we have

$$p_1 A = p_2 A + ky$$

where k is the spring constant.

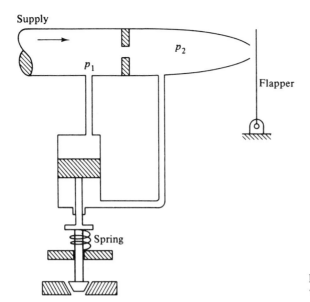

Supply

p_1

p_2

Flapper

Spring

Figure 16.29. Flapper control valve with spring.

16.5 PNEUMATICALLY ASSISTED AUTOMATIC CONTROL

Pneumatic control systems differ from hydraulic systems in the use of a gas (usually air) instead of a liquid as the working fluid. An advantage of using air is that it can be discarded to the atmosphere; hence, no return lines are required, as for hydraulic systems. Nonflammability of air as compared to hydraulic oil is another advantage. Further, pneumatic systems are not as sensitive to temperature changes as the hydraulic systems. However, hydraulic systems usually have faster response than pneumatic ones.

Flapper valves of the type described under hydraulic controls are used extensively in pneumatic control systems. The operation of the flapper valves is the same as that of the corresponding hydraulic flapper valves, with the exception that a spring-loaded bellows is usually used to control the position of the main valve or other devices. Further, as for hydraulic control, pneumatic control circuits can be arranged to give the three basic types of control or combinations thereof.

PROBLEMS

16.1. An aircraft landing gear is lowered by a hydraulic actuating cylinder using a constant-pressure supply system of 10,000 kPa. The 5-mm-I.D. supply line from the pressure source to the cylinder is 5 m long and includes four 90° elbows and a control valve. The average load exerted on the power piston by the landing gear is 10 kN. Find the time required to lower the landing gear slowly at a temperature of 20°C and −20°C.

The density of the hydraulic oil is 870 kg/m^3 and the kinematic viscosity at 20°C and −20°C is 3×10^{-5} m^2/s and 19×10^{-5} m^2/s, respectively. What

I.D. must the supply line be increased to in order to achieve the same lowering time at $-20°C$ as that at $20°C$? The piston has a cross-sectional area of 25 cm^2 and a travel of 30 cm.

16.2. The initial exit head of the tank-pipe system described in Example 16.1 is decreased instantaneously from 1.5 m to 1.0 m. Find the variation in liquid level and flow rate.

16.3. A linear increase in exit pressure takes place over a time period t_1 in the tank-pipe system described in Example 16.1 instead of the step change (see Figure P16.3). The solution of the differential equation (16.3) then becomes

$$h = ce^{-t/KA} + KQ_{in} + h_{e0} + mt - KAm$$

with initial conditions

$$t = 0: \qquad h = h_0 \qquad h_e = h_{e0}$$

$$t > 0: \qquad\qquad\qquad h_e = h_{e0} + mt$$

Plot the response of surface h as a function of time if the exit pressure increases from 1.5 m to 2.0 m in 60 seconds. Take the values of the other parameters as in Example 16.1.

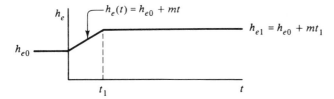

Figure P16.3.

16.4. A sinusoidal variation in exit pressure occurs in the tank-pipe system of Example 16.1 in lieu of the step change described there (see Figure P16.4). The solution of the differential equation 16.3 is then given in the form

$$h = ce^{-t/KA} + KQ_{in} + h_{e0} + \frac{a}{\sqrt{1 + \omega^2 K^2 A^2}} \sin(\omega t - \tan^{-1} \omega KA)$$

with initial conditions

$$t < 0: \qquad h = h_0 \qquad h_e = h_{e0}$$

$$t = 0: \qquad h = h_0 \qquad h_e = h_{e0}$$

$$t > 0: \qquad\qquad\qquad h_e = h_{e0} + a \sin \omega t$$

Plot the response of the free surface h as a function of time if $a = 0.5$ m and $\omega = \pi/30$ rad/s. The values of the other parameters are as in Example 16.1.

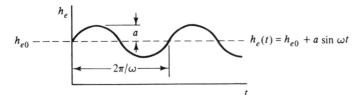

Figure P16.4.

16.5. The fluctuations of the free surface in a tank are determined to be as indicated in Figure P16.5. What will be the corresponding fluctuation in inflow? Plot the results for integral control with $\gamma = 0.01$ m^2/s^2. Take the setup as in Example 16.1.

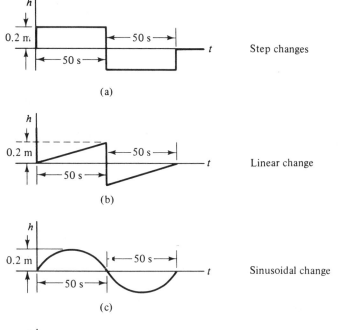

(a)   Step changes

(b)   Linear change

(c)   Sinusoidal change

(d)   Parabolic change

$$(h - h_0) = At^2$$

Figure P16.5.

16.6. The surface fluctuations are measured to be as shown in Figure P16.6. Plot the corresponding fluctuations in inflow for derivative control. Assume the same setup as Example 16.1.

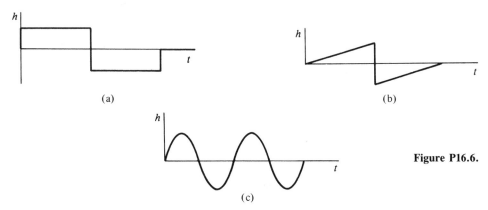

(a)

(b)

(c)

Figure P16.6.

16.7. The fluctuations of the free surface in a tank are determined to be as indicated in Figure P16.6. Plot the response in inflow for combined proportional and integral control. Assume the same setup as Example 16.1.

16.8. A four-way control valve and power cylinder operate under the following conditions: $p_s - p_r = 100$ psi, $A_p = 3$ in^2, $\rho = 1.69$ slugs/ft^3. At no-load condition and $(x - x_0)$ equal to 0.004 in, the piston speed is $\frac{3}{4}$ in/s. Determine piston speed as a function of load (F) if the opening $(x - x_0)$ is kept constant at 0.004 in.

16.9. The hydraulic integral control system of Figure 16.22 has the following characteristics: $a/b = 0.01$, $A_p = 5$ cm^2, $p_s - p_r = 200$ kPa, $F = 20$ N, and $C_d k/\sqrt{\rho} = 2 \times 10^{-4}$ m^3/$\sqrt{N}$ s. When $y - y_0 = 1$ cm, the change in main flow $= 1$ liter/s. What will the control factor γ be? What will the value of γ be if a/b is reset to 0.02?

16.10. The tank-pipe system of Example 16.1 is controlled by the hydraulic system of Problem 16.9. The power piston is disconnected from the main valve. Obtain the piston speed as a function of time if the maximum travel of the control valve from its balance position is 0.2 mm. How long will it take to reach this maximum position?

16.11. A U tube consists of a long horizontal leg and two vertical legs as shown in Figure P16.11. The two liquid levels are exposed to atmospheric pressure. All tubes have the same cross-sectional area. When at rest, the level of the liquid in both vertical legs will be the same. Different air pressures are applied to the two legs and hence the equilibrium level in one leg will be higher than that in the other leg. The pressures are then restored to atmospheric pressure.

Determine the equation of motion of the resulting oscillatory motion of the liquid surfaces for the idealized case where there is no frictional resistance between liquid and tubes.

Calculate the period of oscillation of a U tube 300 m long with a diameter of 0.3 m.

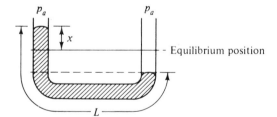

Figure P16.11.

16.12. A vertical tube is initially filled with liquid to a height x_0 as shown in Figure P16.12. The horizontal discharge tube has the same cross sections as the vertical tube. Determine the time required to drain the vertical tube if the flow is frictionless.

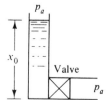

Figure P16.12.

16.13. Determine the time required to drain the vertical tube of Problem 16.12 if there is laminar flow in the tube. Take the initial height x_0 as 3 ft and the tube diameter as 4 in. The liquid is oil having a kinematic viscosity ν of 0.01 ft^2/s. Neglect the loss in 90° elbow and valve.

Note that the resulting differential equation will be of the form

$$\frac{d^2x}{dt^2} + a\frac{dx}{dt} + bx = 0$$

Its solution will be, for $a < 2\sqrt{b}$,

$$x = Ae^{-at/2}\sin mt + Be^{-at/2}\cos mt$$

where $m = \sqrt{b - a^2/4}$.

16.14. In the tank-pipe system of Example 16.2 it is desired to reach a tank pressure of 31.5 psia twice as fast as in the example. Can this be accomplished by changing the diameter of the pipe? What must be the new diameter?

16.15. Air at a temperature of 68°F and a pressure of 14.5 psia enters a closed tank through a 150-ft-long pipe having a diameter of 1 in. The volume of the tank is 1000 ft^3. The initial tank pressure is 13 psia. Determine the response of the pressure in the tank.

16.16. For turbulent flow in a pipe, the frictional resistance in the pipe will be proportional to the mean velocity squared $(\overline{V}^2)$. For completely turbulent flow, the friction factor f will be independent of the Reynolds number of the flow and will be a function of the relative roughness ϵ/D only.

Show that the differential equation describing the behavior of the liquid level (h) in the tank of Figure 16.1 will be (for a circular discharge pipe)

$$\frac{dh}{dt} + \frac{\sqrt{h - h_e}}{\sqrt{k}A} = \frac{Q_{in}}{A}$$

where $k = 8fL/\pi^2gD^5$. For a step change in exit pressure at $t = 0$, the solution of this differential equation becomes

$$\sqrt{h_0 - h_{e1}} - \sqrt{h - h_{e1}} - \sqrt{k}\,Q_{in}\ln\frac{\sqrt{k}\,Q_{in} - \sqrt{h - h_{e1}}}{\sqrt{k}\,Q_{in} - \sqrt{h_0 - h_{e1}}} = \frac{t}{2\sqrt{k}\,A}$$

Find and plot the variation with time of the water level in a water tank, when the initial exit head of 1 m is suddenly increased to 2 m. Take the friction factor f at 0.04 and the viscosity ν at 10^{-6} m^2/s. The cross-sectional area of the tank is 2 m^2, the pipe diameter is 0.5 m, and its length is 10 m. The inflow is kept constant at 150 m^3/min.

16.17. An open tank is filled through an inlet pipe as shown in Figure P16.17. The inlet head h_i is kept constant during the filling process. Initially the liquid level in the tank is h_0. Determine the variation of liquid level and volumetric inflow rate for laminar flow in the inlet pipe.

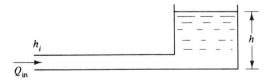

Figure P16.17.

16.18. Using the solution of Problem 16.17, determine the diameter of a 100-ft-long filler pipe so that an open tank having a volume of 700 ft³ can be filled in 2 hours. Assume laminar flow in the pipe. $A = 200$ ft², $h_i = 6$ ft, $\nu = 0.01$ ft²/s.

16.19. A tank-pipe system consists of an inlet and outlet pipe and an open tank as shown in Figure P16.19. Determine the variation in liquid level and volumetric flow rate for a step change in exit head h_e.

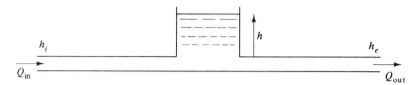

Figure P16.19.

16.20. The pipe-tank system described in Example 16.4 is to be controlled using derivative control with a control factor β of 1 m². Determine the variation in liquid level and compare it to the case of no control.

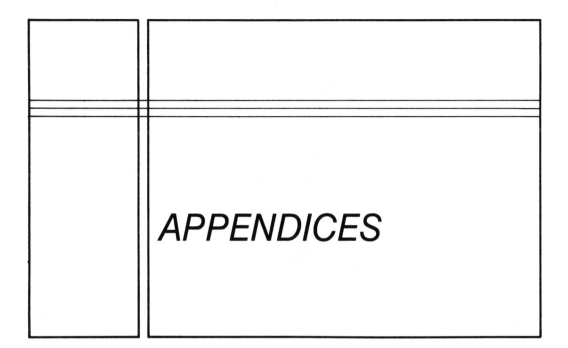

APPENDICES

PHYSICAL PROPERTIES

TABLE 1. Physical Properties of Water

A. Metric Units

Temperature T (°C)	Density ρ (kg/m^3)	Dynamic viscosity μ (Pa $\cdot$ s)	Kinematic viscosity ν (m^2/s)
0	999.8	1.80×10^{-3}	1.80×10^{-6}
5	1000.0	1.52×10^{-3}	1.52×10^{-6}
10	999.7	1.31×10^{-3}	1.31×10^{-6}
15	999.2	1.15×10^{-3}	1.15×10^{-6}
20	998.3	1.00×10^{-3}	1.00×10^{-6}
25	997.1	0.897×10^{-3}	0.898×10^{-6}
30	995.7	0.801×10^{-3}	0.804×10^{-6}
35	994.0	0.723×10^{-3}	0.727×10^{-6}
40	992.3	0.659×10^{-3}	0.664×10^{-6}
45	990.2	0.599×10^{-3}	0.604×10^{-6}
50	988.0	0.554×10^{-3}	0.561×10^{-6}
55	985.7	0.508×10^{-3}	0.515×10^{-6}
60	983.2	0.470×10^{-3}	0.478×10^{-6}
65	980.5	0.437×10^{-3}	0.446×10^{-6}
70	977.7	0.405×10^{-3}	0.414×10^{-6}
75	974.8	0.381×10^{-3}	0.390×10^{-6}
80	971.6	0.356×10^{-3}	0.367×10^{-6}
85	968.4	0.336×10^{-3}	0.347×10^{-6}
90	965.1	0.318×10^{-3}	0.329×10^{-6}
95	961.6	0.300×10^{-3}	0.312×10^{-6}
100	958.1	0.284×10^{-3}	0.296×10^{-6}

TABLE 1. (Continued)

B. English Units

Temperature T (°F)	Density ρ (lbm/ft^3)	Density ρ (slugs/ft^3)	Dynamic viscosity μ (lbf · s/ft^2)	Kinematic viscosity ν (ft^2/s)
32	62.42	1.940	3.75×10^{-5}	1.93×10^{-5}
40	62.43	1.941	3.22×10^{-5}	1.66×10^{-5}
50	62.41	1.940	2.74×10^{-5}	1.41×10^{-5}
60	62.37	1.939	2.36×10^{-5}	1.22×10^{-5}
70	62.30	1.937	2.02×10^{-5}	1.04×10^{-5}
80	62.22	1.934	1.81×10^{-5}	0.935×10^{-5}
90	62.12	1.931	1.58×10^{-5}	0.817×10^{-5}
100	62.00	1.927	1.44×10^{-5}	0.749×10^{-5}
110	61.87	1.923	1.28×10^{-5}	0.665×10^{-5}
120	61.71	1.918	1.18×10^{-5}	0.614×10^{-5}
130	61.55	1.913	1.07×10^{-5}	0.559×10^{-5}
140	61.38	1.908	0.981×10^{-5}	0.514×10^{-5}
150	61.20	1.902	0.906×10^{-5}	0.475×10^{-5}
160	61.00	1.896	0.830×10^{-5}	0.437×10^{-5}
170	60.80	1.890	0.780×10^{-5}	0.413×10^{-5}
180	60.58	1.883	0.720×10^{-5}	0.384×10^{-5}
190	60.37	1.877	0.682×10^{-5}	0.364×10^{-5}
200	60.13	1.869	0.636×10^{-5}	0.340×10^{-5}
210	59.89	1.862	0.605×10^{-5}	0.325×10^{-5}
212	59.83	1.860	0.593×10^{-5}	0.319×10^{-5}

TABLE 2. Physical Properties of Air

A. At Atmospheric Pressure (Metric Units)

Temperature		Density	Dynamic viscosity	Kinematic viscosity
T (°C)	T (K)	ρ (kg/m³)	μ (Pa · s)	ν (m²/s)
−50	223	1.582	1.46 × 10⁻⁵	0.921 × 10⁻⁵
−40	233	1.514	1.51 × 10⁻⁵	0.998 × 10⁻⁵
−30	243	1.452	1.56 × 10⁻⁵	1.08 × 10⁻⁵
−20	253	1.394	1.61 × 10⁻⁵	1.16 × 10⁻⁵
−10	263	1.342	1.67 × 10⁻⁵	1.24 × 10⁻⁵
0	273	1.292	1.72 × 10⁻⁵	1.33 × 10⁻⁵
10	283	1.247	1.76 × 10⁻⁵	1.42 × 10⁻⁵
20	293	1.202	1.81 × 10⁻⁵	1.51 × 10⁻⁵
30	303	1.164	1.86 × 10⁻⁵	1.60 × 10⁻⁵
40	313	1.127	1.91 × 10⁻⁵	1.69 × 10⁻⁵
50	323	1.092	1.95 × 10⁻⁵	1.79 × 10⁻⁵
60	333	1.060	2.00 × 10⁻⁵	1.89 × 10⁻⁵
70	343	1.030	2.05 × 10⁻⁵	1.99 × 10⁻⁵
80	353	1.000	2.09 × 10⁻⁵	2.09 × 10⁻⁵
90	363	0.973	2.13 × 10⁻⁵	2.19 × 10⁻⁵
100	373	0.946	2.17 × 10⁻⁵	2.30 × 10⁻⁵
150	423	0.834	2.38 × 10⁻⁵	2.85 × 10⁻⁵
200	473	0.746	2.57 × 10⁻⁵	3.45 × 10⁻⁵
250	523	0.675	2.75 × 10⁻⁵	4.08 × 10⁻⁵
300	573	0.616	2.93 × 10⁻⁵	4.75 × 10⁻⁵

B. At Atmospheric Pressure (English Units)

Temperature		Density		Dynamic viscosity	Kinematic viscosity
T (°F)	T (°R)	ρ (lbm/ft³)	ρ (slugs/ft³)	μ (lbf · s/ft²)	ν (ft²/sec)
−60	400	0.0993	0.00309	3.03 × 10⁻⁷	9.81 × 10⁻⁵
−40	420	0.0945	0.00294	3.16 × 10⁻⁷	1.08 × 10⁻⁴
−20	440	0.0902	0.00280	3.28 × 10⁻⁷	1.17 × 10⁻⁴
0	460	0.0863	0.00268	3.39 × 10⁻⁷	1.26 × 10⁻⁴
20	480	0.0827	0.00257	3.51 × 10⁻⁷	1.37 × 10⁻⁴
40	500	0.0794	0.00247	3.63 × 10⁻⁷	1.47 × 10⁻⁴
60	520	0.0763	0.00237	3.74 × 10⁻⁷	1.58 × 10⁻⁴
80	540	0.0735	0.00228	3.85 × 10⁻⁷	1.69 × 10⁻⁴
100	560	0.0709	0.00220	3.96 × 10⁻⁷	1.80 × 10⁻⁴
120	580	0.0684	0.00213	4.07 × 10⁻⁷	1.92 × 10⁻⁴
140	600	0.0662	0.00206	4.17 × 10⁻⁷	2.03 × 10⁻⁴
160	620	0.0640	0.00199	4.28 × 10⁻⁷	2.15 × 10⁻⁴
180	640	0.0620	0.00193	4.38 × 10⁻⁷	2.27 × 10⁻⁴
200	660	0.0601	0.00187	4.48 × 10⁻⁷	2.40 × 10⁻⁴
300	760	0.0522	0.00162	4.96 × 10⁻⁷	3.06 × 10⁻⁴

TABLE 2. (Continued)

B. At Atmospheric Pressure (English Units)

Temperature T (°F)	T (°R)	Density ρ (lbm/ft^3)	Density ρ (slugs/ft^3)	Dynamic viscosity μ (lbf · s/ft^2)	Kinematic viscosity ν (ft^2/sec)
400	860	0.0461	0.00143	5.41×10^{-7}	3.77×10^{-4}
500	960	0.0413	0.00129	5.83×10^{-7}	4.54×10^{-4}
600	1060	0.0374	0.00116	6.23×10^{-7}	5.35×10^{-4}

C. The U.S. Standard Atmosphere, 1962 (Metric Units)

Altitude (m)	Temperature (°C)	(K)	Pressure (kPa)	Density (kg/m^3)	Dynamic viscosity (Pa · s)	Kinematic viscosity (m^2/s)
0	15.000	288.150	101.325	1.2250	1.7894×10^{-5}	1.4607×10^{-5}
1,000	8.501	281.651	89.876	1.1117	1.7579×10^{-5}	1.5813×10^{-5}
2,000	2.004	275.154	79.501	1.0066	1.7260×10^{-5}	1.7147×10^{-5}
3,000	−4.491	268.659	70.121	0.90925	1.6938×10^{-5}	1.8628×10^{-5}
4,000	−10.984	262.166	61.660	0.81935	1.6612×10^{-5}	2.0275×10^{-5}
5,000	−17.474	255.676	54.048	0.73643	1.6282×10^{-5}	2.2110×10^{-5}
6,000	−23.963	249.187	47.218	0.66011	1.5949×10^{-5}	2.4162×10^{-5}
7,000	−30.450	242.700	41.105	0.59002	1.5612×10^{-5}	2.6461×10^{-5}
8,000	−36.935	236.215	35.652	0.52579	1.5271×10^{-5}	2.9044×10^{-5}
9,000	−43.417	229.733	30.801	0.46706	1.4926×10^{-5}	3.1957×10^{-5}
10,000	−49.898	223.252	26.450	0.41351	1.4577×10^{-5}	3.5251×10^{-5}
11,000	−56.376	216.774	22.670	0.36480	1.4223×10^{-5}	3.8999×10^{-5}
12,000	−56.500	216.650	19.399	0.31194	1.4216×10^{-5}	4.5574×10^{-5}
13,000	−56.500	216.650	16.580	0.26660	1.4216×10^{-5}	5.3325×10^{-5}
14,000	−56.500	216.650	14.170	0.22786	1.4216×10^{-5}	6.2391×10^{-5}
15,000	−56.500	216.650	12.112	0.19475	1.4216×10^{-5}	7.2995×10^{-5}
16,000	−56.500	216.650	10.353	0.16647	1.4216×10^{-5}	8.5397×10^{-5}
17,000	−56.500	216.650	8.8497	0.14230	1.4216×10^{-5}	9.9902×10^{-5}
18,000	−56.500	216.650	7.5652	0.12165	1.4216×10^{-5}	1.1686×10^{-4}
19,000	−56.500	216.650	6.4675	0.10400	1.4216×10^{-5}	1.3670×10^{-4}
20,000	−56.500	216.650	5.5293	0.08891	1.4216×10^{-5}	1.5989×10^{-4}
21,000	−55.569	217.581	4.7289	0.07572	1.4267×10^{-5}	1.8843×10^{-4}
22,000	−54.576	218.574	4.0475	0.06451	1.4322×10^{-5}	2.2201×10^{-4}
23,000	−53.583	219.567	3.4669	0.05501	1.4376×10^{-5}	2.6135×10^{-4}
24,000	−52.590	220.560	2.9717	0.04694	1.4430×10^{-5}	3.0743×10^{-4}
25,000	−51.598	221.552	2.5492	0.04008	1.4484×10^{-5}	3.6135×10^{-4}
26,000	−50.606	222.544	2.1884	0.03426	1.4538×10^{-5}	4.2439×10^{-4}
27,000	−49.615	223.535	1.8733	0.02920	1.4592×10^{-5}	4.9981×10^{-4}
28,000	−48.623	224.527	1.6162	0.02508	1.4646×10^{-5}	5.8405×10^{-4}
29,000	−47.632	225.518	1.3853	0.02140	1.4700×10^{-5}	6.8694×10^{-4}
30,000	−46.641	226.509	1.1970	0.01841	1.4753×10^{-5}	8.0134×10^{-4}

Data from *Handbook of Geophysics and Space Environments*, Air Force Cambridge Research Laboratories, USAF, 1965.

TABLE 2. (Continued)

D. The U.S. Standard Atmosphere, 1962 (English Units)

Altitude (ft)	Temperature (°F)	Temperature (°R)	Pressure (psf)	Density (lbm/ft³)	Dynamic viscosity (lbf · s/ft²)	Kinematic viscosity (ft²/s)
0	59.000	518.670	2116.2	0.076477	3.7381×10^{-7}	1.5726×10^{-4}
2,500	50.085	509.755	1931.9	0.071419	3.6880×10^{-7}	1.6614×10^{-4}
5,000	41.170	500.840	1760.9	0.065789	3.6374×10^{-7}	1.7789×10^{-4}
7,500	32.254	491.924	1602.3	0.061082	3.5864×10^{-7}	1.8891×10^{-4}
10,000	23.339	483.009	1455.6	0.056485	3.5351×10^{-7}	2.0149×10^{-4}
12,500	14.424	474.094	1320.0	0.052180	3.4832×10^{-7}	2.1510×10^{-4}
15,000	5.509	465.179	1194.8	0.048142	3.4309×10^{-7}	2.2929×10^{-4}
17,500	−3.406	456.264	1079.4	0.044337	3.3781×10^{-7}	2.4514×10^{-4}
20,000	−12.322	447.348	973.23	0.040774	3.3250×10^{-7}	2.6237×10^{-4}
22,500	−21.237	438.433	875.77	0.037434	3.2713×10^{-7}	2.8116×10^{-4}
25,000	−30.152	429.518	786.34	0.034434	3.2173×10^{-7}	3.0061×10^{-4}
27,500	−39.067	420.603	704.45	0.031387	3.1627×10^{-7}	3.2420×10^{-4}
30,000	−47.983	411.687	629.64	0.028654	3.1075×10^{-7}	3.4892×10^{-4}
32,500	−56.898	402.772	561.44	0.026113	3.0519×10^{-7}	3.7603×10^{-4}
35,000	−65.813	393.857	499.32	0.023774	2.9964×10^{-7}	4.0551×10^{-4}
37,500	−69.700	389.970	443.15	0.021421	2.9697×10^{-7}	4.4604×10^{-4}
40,000	−69.700	389.970	393.13	0.018809	2.9697×10^{-7}	5.0799×10^{-4}
42,500	−69.700	389.970	348.74	0.016722	2.9697×10^{-7}	5.7139×10^{-4}
45,000	−69.700	389.970	309.46	0.014876	2.9697×10^{-7}	6.4229×10^{-4}
47,500	−69.700	389.970	274.56	0.013200	2.9697×10^{-7}	7.2384×10^{-4}
50,000	−69.700	389.970	243.62	0.011707	2.9697×10^{-7}	8.5534×10^{-4}
52,500	−69.700	389.970	216.14	0.010384	2.9697×10^{-7}	9.2014×10^{-4}
55,000	−69.700	389.970	191.36	0.0091446	2.9697×10^{-7}	1.0448×10^{-3}
57,500	−67.700	389.970	170.33	0.0082067	2.9697×10^{-7}	1.1643×10^{-3}
60,000	−69.700	389.970	151.00	0.0072518	2.9697×10^{-7}	1.3176×10^{-3}
62,500	−69.700	389.970	134.03	0.0064428	2.9697×10^{-7}	1.4830×10^{-3}
65,000	−69.700	389.970	118.94	0.0057193	2.9697×10^{-7}	1.6706×10^{-3}
67,500	−68.792	390.878	105.57	0.0050625	2.9754×10^{-7}	1.8910×10^{-3}
70,000	−67.420	392.250	93.730	0.0044790	2.9843×10^{-7}	2.1437×10^{-3}
72,500	−66.062	393.608	83.256	0.0039648	2.9930×10^{-7}	2.4293×10^{-3}
75,000	−64.699	394.971	73.989	0.0035115	3.0015×10^{-7}	2.7501×10^{-3}
77,500	−63.337	396.333	65.785	0.0031115	3.0101×10^{-7}	3.1125×10^{-3}
80,000	−61.976	397.694	58.508	0.0027576	3.0187×10^{-7}	3.5220×10^{-3}
82,500	−60.615	399.055	52.069	0.0024454	3.0273×10^{-7}	3.9830×10^{-3}
85,000	−59.255	400.415	46.356	0.0021701	3.0360×10^{-7}	4.5012×10^{-3}
87,500	−57.896	401.774	41.198	0.0019224	3.0446×10^{-7}	5.0956×10^{-3}
90,000	−56.536	403.134	36.636	0.0017028	3.0531×10^{-7}	5.7688×10^{-3}
92,500	−55.175	404.495	32.744	0.0015230	3.0617×10^{-7}	6.4680×10^{-3}
95,000	−53.816	405.854	29.127	0.0013475	3.0703×10^{-7}	7.3309×10^{-3}
97,500	−52.459	407.211	26.030	0.0011585	3.0787×10^{-7}	8.5508×10^{-3}
100,000	−51.085	408.585	23.378	0.0010719	3.0868×10^{-7}	9.2653×10^{-3}

TABLE 3. Physical Properties of Gases

A. Metric Units

Gas	Gas constant R (kJ/kg · K)	Density ρ (kg/m³) at 15°C, 1 atm	Dynamic viscosity μ (Pa · s) at 15°C	Kinematic viscosity ν (m²/s) at 15°C, 1 atm	Specific heat c_p (kJ/kg · K) at 15°C	Specific heat ratio $\gamma = c_p/c_v$ at 15°C
Air	0.2870	1.225	1.79×10^{-5}	1.46×10^{-5}	1.006	1.40
Oxygen	0.2598	1.355	1.98×10^{-5}	1.46×10^{-5}	0.920	1.40
Nitrogen	0.2968	1.254	1.73×10^{-5}	1.37×10^{-5}	1.041	1.40
Hydrogen	4.124	0.0852	0.872×10^{-5}	1.02×10^{-4}	14.27	1.40
Helium	2.077	0.169	1.96×10^{-5}	1.16×10^{-4}	5.23	1.66
Carbon dioxide	0.1889	1.872	1.44×10^{-5}	0.768×10^{-5}	0.841	1.29
Argon	0.2081	1.691	2.22×10^{-5}	1.31×10^{-5}	0.522	1.67

B. English Units

Gas	Gas constant R (ft-lbf/lbm°R)	Density ρ (lbm/ft³) at 60°F, 1 atm	Dynamic viscosity μ (lbf · s/ft²) at 60°F	Kinematic viscosity ν (ft²/s) at 60°F, 1 atm	Specific heat c_p (Btu/lbm°R) at 60°F	Specific heat ratio $\gamma = c_p/c_v$ at 60°F
Air	53.35	0.0763	3.74×10^{-7}	1.58×10^{-4}	0.240	1.40
Oxygen	48.30	0.0844	4.20×10^{-7}	1.60×10^{-4}	0.217	1.40
Nitrogen	55.15	0.0739	3.63×10^{-7}	1.53×10^{-4}	0.248	1.40
Hydrogen	766.40	0.00531	1.83×10^{-7}	1.11×10^{-3}	3.42	1.40
Helium	386.00	0.0105	4.10×10^{-7}	1.26×10^{-3}	1.25	1.66
Carbon dioxide	35.11	0.116	3.03×10^{-7}	8.41×10^{-5}	0.203	1.29
Argon	38.69	0.105	4.66×10^{-7}	1.43×10^{-4}	0.125	1.67

B

ISENTROPIC FLOW TABLES

TABLE 1.

$$\gamma = 1.4$$

M	$\dfrac{p}{p_t}$	$\dfrac{T}{T_t}$	$\dfrac{A}{A^*}$	M	$\dfrac{p}{p_t}$	$\dfrac{T}{T_t}$	$\dfrac{A}{A^*}$
0	1.0000	1.0000	∞	0.20	0.9725	0.9921	2.9635
.01	.9999	1.0000	57.8738	.21	.9697	.9913	2.8293
.02	.9997	.9999	28.9421	.22	.9668	.9901	2.7076
.03	.9994	.9998	19.3005	.23	.9638	.9895	2.5968
.04	.9989	.9997	14.4815	.24	.9607	.9886	2.4956
.05	.9983	.9995	11.5914	.25	.9575	.9877	2.4027
.06	.9975	.9993	9.6659	.26	.9541	.9867	2.3173
.07	.9966	.9990	8.2915	.27	.9506	.9856	2.2385
.08	.9955	.9987	7.2616	.28	.9470	.9846	2.1656
.09	.9944	.9984	6.4613	.29	.9433	.9835	2.0979
.10	.9930	.9980	5.8218	.30	.9395	.9823	2.0351
.11	.9916	.9976	5.2992	.31	.9355	.9811	1.9765
.12	.9900	.9971	4.8643	.32	.9315	.9799	1.9219
.13	.9883	.9966	4.4969	.33	.9274	.9787	1.8707
.14	.9864	.9961	4.1824	.34	.9231	.9774	1.8229
.15	.9844	.9955	3.9103	.35	.9188	.9761	1.7780
.16	.9823	.9949	3.6727	.36	.9143	.9747	1.7358
.17	.9800	.9943	3.4635	.37	.9098	.9733	1.6961
.18	.9776	.9936	3.2779	.38	.9052	.9719	1.6587
.19	.9751	.9928	3.1123	.39	.9004	.9705	1.6234

Data in Appendix B is from NACA Report 1135, "Equations, Tables, and Charts for Compressible Flow," Ames Research Staff, 1953.

TABLE 1. (Continued)

$$\gamma = 1.4$$

M	$\frac{p}{p_t}$	$\frac{T}{T_t}$	$\frac{A}{A^*}$	M	$\frac{p}{p_t}$	$\frac{T}{T_t}$	$\frac{A}{A^*}$
0.40	0.8956	0.9690	1.5901	0.70	0.7209	0.9107	1.0944
.41	.8907	.9675	1.5587	.71	.7145	.9084	1.0873
.42	.8857	.9659	1.5289	.72	.7080	.9061	1.0806
.43	.8807	.9643	1.5007	.73	.7016	.9037	1.0742
.44	.8755	.9627	1.4740	.74	.6951	.9013	1.0681
.45	.8703	.9611	1.4487	.75	.6886	.8989	1.0624
.46	.8650	.9594	1.4246	.76	.6821	.8964	1.0570
.47	.8596	.9577	1.4018	.77	.6756	.8940	1.0519
.48	.8541	.9560	1.3801	.78	.6691	.8915	1.0471
.49	.8486	.9542	1.3595	.79	.6625	.8890	1.0425
.50	.8430	.9524	1.3398	.80	.6560	.8865	1.0382
.51	.8374	.9506	1.3212	.81	.6495	.8840	1.0342
.52	.8317	.9487	1.3034	.82	.6430	.8815	1.0305
.53	.8259	.9468	1.2865	.83	.6365	.8789	1.0270
.54	.8201	.9449	1.2703	.84	.6300	.8763	1.0237
.55	.8142	.9430	1.2550	.85	.6235	.8737	1.0207
.56	.8082	.9410	1.2403	.86	.6170	.8711	1.0179
.57	.8022	.9390	1.2263	.87	.6106	.8685	1.0153
.58	.7962	.9370	1.2130	.88	.6041	.8659	1.0129
.59	.7901	.9349	1.2003	.89	.5977	.8632	1.0108
.60	.7840	.9328	1.1882	.90	.5913	.8606	1.0089
.61	.7778	.9307	1.1767	.91	.5849	.8579	1.0071
.62	.7716	.9286	1.1657	.92	.5785	.8552	1.0056
.63	.7654	.9265	1.1552	.93	.5721	.8525	1.0043
.64	.7591	.9243	1.1452	.94	.5658	.8498	1.0031
.65	.7528	.9221	1.1356	.95	.5595	.8471	1.0022
.66	.7465	.9199	1.1265	.96	.5532	.8444	1.0014
.67	.7401	.9176	1.1179	.97	.5469	.8416	1.0008
.68	.7338	.9153	1.1097	.98	.5407	.8389	1.0003
.69	.7274	.9131	1.1018	.99	.5345	.8361	1.0001

M	$\frac{p}{p_t}$	$\frac{T}{T_t}$	$\frac{A}{A^*}$	M	$\frac{p}{p_t}$	$\frac{T}{T_t}$	$\frac{A}{A^*}$
1.00	0.5283	0.8333	1.000	1.05	0.4979	0.8193	1.002
1.01	.5221	.8306	1.000	1.06	.4919	.8165	1.003
1.02	.5160	.8278	1.000	1.07	.4860	.8137	1.004
1.03	.5099	.8250	1.001	1.08	.4800	.8108	1.005
1.04	.5039	.8222	1.001	1.09	.4742	.8080	1.006

TABLE 1. (Continued)

$$\gamma = 1.4$$

M	$\dfrac{p}{p_t}$	$\dfrac{T}{T_t}$	$\dfrac{A}{A^*}$	M	$\dfrac{p}{p_t}$	$\dfrac{T}{T_t}$	$\dfrac{A}{A^*}$
1.10	0.4684	0.8052	1.008	1.50	0.2724	0.6897	1.176
1.11	.4626	.8023	1.010	1.51	.2685	.6868	1.183
1.12	.4568	.7994	1.011	1.52	.2646	.6840	1.190
1.13	.4511	.7966	1.013	1.53	.2608	.6811	1.197
1.14	.4455	.7937	1.015	1.54	.2570	.6783	1.204
1.15	.4398	.7908	1.017	1.55	.2533	.6754	1.212
1.16	.4343	.7879	1.020	1.56	.2496	.6726	1.219
1.17	.4287	.7851	1.022	1.57	.2459	.6698	1.227
1.18	.4232	.7822	1.025	1.58	.2423	.6670	1.234
1.19	.4178	.7793	1.026	1.59	.2388	.6642	1.242
1.20	.4124	.7764	1.030	1.60	.2353	.6614	1.250
1.21	.4070	.7735	1.033	1.61	.2318	.6586	1.258
1.22	.4017	.7706	1.037	1.62	.2284	.6558	1.267
1.23	.3964	.7677	1.040	1.63	.2250	.6530	1.275
1.24	.3912	.7648	1.043	1.64	.2217	.6502	1.284
1.25	.3861	.7619	1.047	1.65	.2184	.6475	1.292
1.26	.3809	.7590	1.050	1.66	.2151	.6447	1.301
1.27	.3759	.7561	1.054	1.67	.2119	.6419	1.310
1.28	.3708	.7532	1.058	1.68	.2088	.6392	1.319
1.29	.3658	.7503	1.062	1.69	.2057	.6364	1.328
1.30	.3609	.7474	1.066	1.70	.2026	.6337	1.338
1.31	.3560	.7445	1.071	1.71	.1996	.6310	1.347
1.32	.3512	.7416	1.075	1.72	.1966	.6283	1.357
1.33	.3464	.7387	1.080	1.73	.1936	.6256	1.367
1.34	.3417	.7358	1.084	1.74	.1907	.6229	1.376
1.35	.3370	.7329	1.089	1.75	.1878	.6202	1.386
1.36	.3323	.7300	1.094	1.76	.1850	.6175	1.397
1.37	.3277	.7271	1.099	1.77	.1822	.6148	1.407
1.38	.3232	.7242	1.104	1.78	.1794	.6121	1.418
1.39	.3187	.7213	1.109	1.79	.1767	.6095	1.428
1.40	.3142	.7184	1.115	1.80	.1740	.6068	1.439
1.41	.3098	.7155	1.120	1.81	.1714	.6041	1.450
1.42	.3055	.7126	1.126	1.82	.1688	.6015	1.461
1.43	.3012	.7097	1.132	1.83	.1662	.5989	1.472
1.44	.2969	.7069	1.138	1.84	.1637	.5963	1.484
1.45	.2927	.7040	1.144	1.85	.1612	.5936	1.495
1.46	.2886	.7011	1.150	1.86	.1587	.5910	1.507
1.47	.2845	.6982	1.156	1.87	.1563	.5884	1.519
1.48	.2804	.6954	1.163	1.88	.1539	.5859	1.531
1.49	.2764	.6925	1.169	1.89	.1516	.5833	1.543

TABLE 1. (Continued)

$$\gamma = 1.4$$

M	$\dfrac{p}{p_t}$	$\dfrac{T}{T_t}$	$\dfrac{A}{A^*}$	M	$\dfrac{p}{p_t}$	$\dfrac{T}{T_t}$	$\dfrac{A}{A^*}$
1.90	0.1492	0.5807	1.555	2.30	0.7997^{-1}	0.4859	2.193
1.91	.1470	.5782	1.568	2.31	$.7873^{-1}$	.4837	2.213
1.92	.1447	.5756	1.580	2.32	$.7751^{-1}$	.4816	2.233
1.93	.1425	.5731	1.593	2.33	$.7631^{-1}$	.4794	2.254
1.94	.1403	.5705	1.606	2.34	$.7512^{-1}$	.4773	2.273
1.95	.1381	.5680	1.619	2.35	$.7396^{-1}$	.4752	2.295
1.96	.1360	.5655	1.633	2.36	$.7281^{-1}$	.4731	2.316
1.97	.1339	.5630	1.646	2.37	$.7168^{-1}$	.4709	2.338
1.98	.1318	.5605	1.660	2.38	$.7057^{-1}$	.4688	2.359
1.99	.1298	.5580	1.674	2.39	$.6948^{-1}$	.4668	2.381
2.00	.1278	.5556	1.688	2.40	$.6840^{-1}$	.4647	2.403
2.01	.1258	.5531	1.702	2.41	$.6734^{-1}$	.4626	2.425
2.02	.1239	.5506	1.716	2.42	$.6630^{-1}$	.4606	2.448
2.03	.1220	.5482	1.730	2.43	$.6527^{-1}$	.4585	2.471
2.04	.1201	.5458	1.745	2.44	$.6426^{-1}$	.4565	2.494
2.05	.1182	.5433	1.760	2.45	$.6327^{-1}$	.4544	2.517
2.06	.1164	.5409	1.775	2.46	$.6229^{-1}$	.4524	2.540
2.07	.1146	.5385	1.790	2.47	$.6133^{-1}$	.4504	2.564
2.08	.1128	.5361	1.806	2.48	$.6038^{-1}$	.4484	2.588
2.09	.1111	.5337	1.821	2.49	$.5945^{-1}$	.4464	2.612
2.10	.1094	.5313	1.837	2.50	$.5853^{-1}$	.4444	2.637
2.11	.1077	.5290	1.853	2.51	$.5762^{-1}$	.4425	2.661
2.12	.1060	.5266	1.869	2.52	$.5674^{-1}$	.4405	2.686
2.13	.1043	.5243	1.885	2.53	$.5586^{-1}$	.4386	2.712
2.14	.1027	.5219	1.902	2.54	$.5500^{-1}$	.4366	2.737
2.15	.1011	.5196	1.919	2.55	$.5415^{-1}$	.4347	2.763
2.16	$.9956^{-1}$	.5173	1.935	2.56	$.5332^{-1}$	.4328	2.789
2.17	$.9802^{-1}$	.5150	1.953	2.57	$.5250^{-1}$	.4309	2.815
2.18	$.9649^{-1}$	.5127	1.970	2.58	$.5169^{-1}$	.4289	2.842
2.19	$.9500^{-1}$	.5104	1.987	2.59	$.5090^{-1}$	.4271	2.869
2.20	$.9352^{-1}$	.5081	2.005	2.60	$.5012^{-1}$	.4252	2.896
2.21	$.9207^{-1}$	.5059	2.023	2.61	$.4935^{-1}$	.4233	2.923
2.22	$.9064^{-1}$	.5036	2.041	2.62	$.4859^{-1}$	.4214	2.951
2.23	$.8923^{-1}$	.5014	2.059	2.63	$.4784^{-1}$	.4196	2.979
2.24	$.8785^{-1}$	.4991	2.078	2.64	$.4711^{-1}$	.4177	3.007
2.25	$.8648^{-1}$	.4969	2.096	2.65	$.4639^{-1}$	.4159	3.036
2.26	$.8514^{-1}$	.4947	2.115	2.66	$.4568^{-1}$	.4141	3.065
2.27	$.8382^{-1}$	.4925	2.134	2.67	$.4498^{-1}$	.4122	3.094
2.28	$.8251^{-1}$	.4903	2.154	2.68	$.4429^{-1}$	.4104	3.123
2.29	$.8123^{-1}$	.4881	2.173	2.69	$.4362^{-1}$	.4086	3.153

The superscript ($^{-1}$) denotes $\times\ 10^{-1}$. For example, 0.9956^{-1} is equal to 0.09956.

TABLE 1. (Continued)

$$\gamma = 1.4$$

M	$\frac{p}{p_t}$	$\frac{T}{T_t}$	$\frac{A}{A^*}$	M	$\frac{p}{p_t}$	$\frac{T}{T_t}$	$\frac{A}{A^*}$
2.70	0.4295⁻¹	0.4068	3.183	3.10	0.2345⁻¹	0.3422	4.657
2.71	.4229⁻¹	.4051	3.213	3.11	.2310⁻¹	.3408	4.702
2.72	.4165⁻¹	.4033	3.244	3.12	.2276⁻¹	.3393	4.747
2.73	.4102⁻¹	.4015	3.275	3.13	.2243⁻¹	.3379	4.792
2.74	.4039⁻¹	.3998	3.306	3.14	.2210⁻¹	.3365	4.838
2.75	.3978⁻¹	.3980	3.338	3.15	.2177⁻¹	.3351	4.884
2.76	.3917⁻¹	.3963	3.370	3.16	.2146⁻¹	.3337	4.930
2.77	.3858⁻¹	.3945	3.402	3.17	.2114⁻¹	.3323	4.977
2.78	.3799⁻¹	.3928	3.434	3.18	.2083⁻¹	.3309	5.025
2.79	.3742⁻¹	.3911	3.467	3.19	.2053⁻¹	.3295	5.073
2.80	.3685⁻¹	.3894	3.500	3.20	.2023⁻¹	.3281	5.121
2.81	.3629⁻¹	.3877	3.534	3.21	.1993⁻¹	.3267	5.170
2.82	.3574⁻¹	.3860	3.567	3.22	.1964⁻¹	.3253	5.219
2.83	.3520⁻¹	.3844	3.601	3.23	.1936⁻¹	.3240	5.268
2.84	.3467⁻¹	.3827	3.636	3.24	.1908⁻¹	.3226	5.319
2.85	.3415⁻¹	.3810	3.671	3.25	.1880⁻¹	.3213	5.369
2.86	.3363⁻¹	.3794	3.706	3.26	.1853⁻¹	.3199	5.420
2.87	.3312⁻¹	.3777	3.741	3.27	.1826⁻¹	.3186	5.472
2.88	.3263⁻¹	.3761	3.777	3.28	.1799⁻¹	.3173	5.523
2.89	.3213⁻¹	.3745	3.813	3.29	.1773⁻¹	.3160	5.576
2.90	.3165⁻¹	.3729	3.850	3.30	.1748⁻¹	.3147	5.629
2.91	.3118⁻¹	.3712	3.887	3.31	.1722⁻¹	.3134	5.682
2.92	.3071⁻¹	.3696	3.924	3.32	.1698⁻¹	.3121	5.736
2.93	.3025⁻¹	.3681	3.961	3.33	.1673⁻¹	.3108	5.790
2.94	.2980⁻¹	.3665	3.999	3.34	.1649⁻¹	.3095	5.845
2.95	.2935⁻¹	.3649	4.038	3.35	.1625⁻¹	.3082	5.900
2.96	.2891⁻¹	.3633	4.076	3.36	.1602⁻¹	.3069	5.956
2.97	.2848⁻¹	.3618	4.115	3.37	.1579⁻¹	.3057	6.012
2.98	.2805⁻¹	.3602	4.155	3.38	.1557⁻¹	.3044	6.069
2.99	.2764⁻¹	.3587	4.194	3.39	.1534⁻¹	.3032	6.126
3.00	.2722⁻¹	.3571	4.235	3.40	.1512⁻¹	.3019	6.184
3.01	.2682⁻¹	.3556	4.275	3.41	.1491⁻¹	.3007	6.242
3.02	.2642⁻¹	.3541	4.316	3.42	.1470⁻¹	.2995	6.301
3.03	.2603⁻¹	.3526	4.357	3.43	.1449⁻¹	.2982	6.360
3.04	.2564⁻¹	.3511	4.399	3.44	.1428⁻¹	.2970	6.420
3.05	.2526⁻¹	.3496	4.441	3.45	.1408⁻¹	.2958	6.480
3.06	.2489⁻¹	.3481	4.483	3.46	.1388⁻¹	.2946	6.541
3.07	.2452⁻¹	.3466	4.526	3.47	.1368⁻¹	.2934	6.602
3.08	.2416⁻¹	.3452	4.570	3.48	.1349⁻¹	.2922	6.664
3.09	.2380⁻¹	.3437	4.613	3.49	.1330⁻¹	.2910	6.727

TABLE 1. (Continued)

$$\gamma = 1.4$$

M	$\dfrac{p}{p_t}$	$\dfrac{T}{T_t}$	$\dfrac{A}{A^*}$	M	$\dfrac{p}{p_t}$	$\dfrac{T}{T_t}$	
3.50	0.1311^{-1}	0.2899	6.790	3.90	0.7532^{-2}	0.2474	9.799
3.51	$.1293^{-1}$	.2887	6.853	3.91	$.7431^{-2}$	.2464	9.888
3.52	$.1274^{-1}$	.2875	6.917	3.92	$.7332^{-2}$	.2455	9.977
3.53	$.1256^{-1}$	.2864	6.982	3.93	$.7233^{-2}$	.2446	10.07
3.54	$.1239^{-1}$	.2852	7.047	3.94	$.7137^{-2}$	.2436	10.16
3.55	$.1221^{-1}$	.2841	7.113	3.95	$.7042^{-2}$	.2427	10.25
3.56	$.1204^{-1}$	.2829	7.179	3.96	$.6948^{-2}$	.2418	10.34
3.57	$.1188^{-1}$	.2818	7.246	3.97	$.6855^{-2}$	.2408	10.44
3.58	$.1171^{-1}$	.2806	7.313	3.98	$.6764^{-2}$	.2399	10.53
3.59	$.1155^{-1}$	.2795	7.382	3.99	$.6675^{-2}$	.2390	10.62
3.60	$.1138^{-1}$	.2784	7.450	4.00	$.6586^{-2}$	.2381	10.72
3.61	$.1123^{-1}$	.2773	7.519	4.01	$.6499^{-2}$	.2372	10.81
3.62	$.1107^{-1}$	.2762	7.589	4.02	$.6413^{-2}$	.2363	10.91
3.63	$.1092^{-1}$	.2751	7.659	4.03	$.6328^{-2}$	.2354	11.01
3.64	$.1076^{-1}$	.2740	7.730	4.04	$.6245^{-2}$	.2345	11.11
3.65	$.1062^{-1}$	.2729	7.802	4.05	$.6163^{-2}$	.2336	11.21
3.66	$.1047^{-1}$	.2718	7.874	4.06	$.6082^{-2}$	.2327	11.31
3.67	$.1032^{-1}$	.2707	7.947	4.07	$.6002^{-2}$	.2319	11.41
3.68	$.1018^{-1}$	.2697	8.020	4.08	$.5923^{-2}$	.2310	11.51
3.69	$.1004^{-1}$	.2686	8.094	4.09	$.5845^{-2}$	.2301	11.61
3.70	$.9903^{-2}$	.2675	8.169	4.10	$.5769^{-2}$	.2293	11.71
3.71	$.9767^{-2}$	.2665	8.244	4.11	$.5694^{-2}$	.2284	11.82
3.72	$.9633^{-2}$	.2654	8.320	4.12	$.5619^{-2}$	.2275	11.92
3.73	$.9500^{-2}$	.2644	8.397	4.13	$.5546^{-2}$	.2267	12.03
3.74	$.9370^{-2}$	.2633	8.474	4.14	$.5474^{-2}$	.2258	12.14
3.75	$.9242^{-2}$	.2623	8.552	4.15	$.5403^{-2}$	.2250	12.24
3.76	$.9116^{-2}$	.2613	8.630	4.16	$.5333^{-2}$	.2242	12.35
3.77	$.8991^{-2}$	.2602	8.709	4.17	$.5264^{-2}$	.2233	12.46
3.78	$.8869^{-2}$	.2592	8.789	4.18	$.5195^{-2}$	.2225	12.57
3.79	$.8748^{-2}$	.2582	8.870	4.19	$.5128^{-2}$	.2217	12.68
3.80	$.8629^{-2}$	.2572	8.951	4.20	$.5062^{-2}$	.2208	12.79
3.81	$.8512^{-2}$	.2562	9.032	4.21	$.4997^{-2}$	.2200	12.90
3.82	$.8396^{-2}$	.2552	9.115	4.22	$.4932^{-2}$	.2192	13.02
3.83	$.8283^{-2}$	.2542	9.198	4.23	$.4869^{-2}$	.2184	13.13
3.84	$.8171^{-2}$	.2532	9.282	4.24	$.4806^{-2}$	.2176	13.25
3.85	$.8060^{-2}$	.2522	9.366	4.25	$.4745^{-2}$	.2168	13.36
3.86	$.7951^{-2}$	.2513	9.451	4.26	$.4684^{-2}$	.2160	13.48
3.87	$.7844^{-2}$	.2503	9.537	4.27	$.4624^{-2}$	.2152	13.60
3.88	$.7739^{-2}$	.2493	9.624	4.28	$.4565^{-2}$	.2144	13.72
3.89	$.7635^{-2}$	.2484	9.711	4.29	$.4507^{-2}$	.2136	13.83

TABLE 1. (Continued)

$$\gamma = 1.4$$

M	$\dfrac{p}{p_t}$	$\dfrac{T}{T_t}$	$\dfrac{A}{A^*}$	M	$\dfrac{p}{p_t}$	$\dfrac{T}{T_t}$	$\dfrac{A}{A^*}$
4.30	0.4449^{-2}	0.2129	13.95	4.68	0.2768^{-2}	.1859	19.26
4.31	$.4393^{-2}$	.2121	14.08	4.69	$.2734^{-2}$	.1852	19.42
4.32	$.4337^{-2}$	.2113	14.20	4.70	$.2701^{-2}$	.1846	19.58
4.33	$.4282^{-2}$	.2105	14.32	4.71	$.2669^{-2}$	.1839	19.75
4.34	$.4228^{-2}$	.2098	14.45	4.72	$.2637^{-2}$	.1833	19.91
4.35	$.4174^{-2}$	.2090	14.57	4.73	$.2605^{-2}$	.1827	20.07
4.36	$.4121^{-2}$	.2083	14.70	4.74	$.2573^{-2}$	.1820	20.24
4.37	$.4069^{-2}$	.2075	14.82	4.75	$.2543^{-2}$	.1814	20.41
4.38	$.4018^{-2}$	.2067	14.95	4.76	$.2512^{-2}$	.1808	20.58
4.39	$.3968^{-2}$	.2060	15.08	4.77	$.2482^{-2}$	.1802	20.75
4.40	$.3918^{-2}$	.2053	15.21	4.78	$.2452^{-2}$	.1795	20.92
4.41	$.3868^{-2}$	.2045	15.34	4.79	$.2423^{-2}$	.1789	21.09
4.42	$.3820^{-2}$	.2038	15.47	4.80	$.2394^{-2}$	.1783	21.26
4.43	$.3772^{-2}$	.2030	15.61	4.81	$.2366^{-2}$	.1777	21.44
4.44	$.3725^{-2}$	.2023	15.74	4.82	$.2338^{-2}$	.1771	21.61
4.45	$.3678^{-2}$	.2016	15.87	4.83	$.2310^{-2}$	.1765	21.79
4.46	$.3633^{-2}$	.2009	16.01	4.84	$.2283^{-2}$	.1759	21.97
4.47	$.3587^{-2}$	.2002	16.15	4.85	$.2255^{-2}$	.1753	22.15
4.48	$.3543^{-2}$	.1994	16.28	4.86	$.2229^{-2}$	.1747	22.33
4.49	$.3499^{-2}$	.1987	16.42	4.87	$.2202^{-2}$	.1741	22.51
4.50	$.3455^{-2}$	.1980	16.56	4.88	$.2177^{-2}$	.1735	22.70
4.51	$.3412^{-2}$	.1973	16.70	4.89	$.2151^{-2}$	.1729	22.88
4.52	$.3370^{-2}$	.1966	16.84	4.90	$.2126^{-2}$	.1724	23.07
4.53	$.3329^{-2}$	.1959	16.99	4.91	$.2101^{-2}$	.1718	23.25
4.54	$.3288^{-2}$	.1952	17.13	4.92	$.2076^{-2}$	.1712	23.44
4.55	$.3247^{-2}$	.1945	17.28	4.93	$.2052^{-2}$	.1706	23.63
4.56	$.3207^{-2}$	.1938	17.42	4.94	$.2028^{-2}$	.1700	23.82
4.57	$.3168^{-2}$	.1932	17.57	4.95	$.2004^{-2}$	.1695	24.02
4.58	$.3129^{-2}$	.1925	17.72	4.96	$.1981^{-2}$	.1689	24.21
4.59	$.3090^{-2}$	.1918	17.87	4.97	$.1957^{-2}$	.1683	24.41
4.60	$.3053^{-2}$	.1911	18.02	4.98	$.1935^{-2}$	.1678	24.60
4.61	$.3015^{-2}$	.1905	18.17	4.99	$.1912^{-2}$	.1672	24.80
4.62	$.2978^{-2}$	.1898	18.32	5.00	$.1890^{-2}$	.1667	25.00
4.63	$.2942^{-2}$	.1891	18.48	6.00	$.6334^{-3}$	.1220	53.18
4.64	$.2906^{-2}$	.1885	18.63	7.00	$.2416^{-3}$	$.9259^{-1}$	104.1
4.65	$.2871^{-2}$	0.1878	18.79	8.00	$.1024^{-3}$	$.7246^{-1}$	190.1
4.66	$.2836^{-2}$	.1872	18.94	9.00	$.4739^{-4}$	$.5814^{-1}$	327.2
4.67	$.2802^{-2}$	.1865	19.10	10.00	$.2356^{-4}$	$.4762^{-1}$	535.9

TABLE 2.

$$\gamma = 1.30$$

M	$\dfrac{p}{p_t}$	$\dfrac{T}{T_t}$	$\dfrac{A}{A^*}$	M	$\dfrac{p}{p_t}$	$\dfrac{T}{T_t}$	$\dfrac{A}{A^*}$
0	1.0000	1.0000	∞	1.75	0.1944	0.6852	1.424
.05	.9984	.9996	11.7214	1.80	.1797	.6729	1.484
.10	.9936	.9985	5.8860	1.85	.1660	.6607	1.549
.15	.9855	.9966	3.9522	1.90	.1533	.6487	1.618
.20	.9744	.9940	2.9940	1.95	.1415	.6368	1.693
.25	.9603	.9907	2.4262	2.00	.1305	.6250	1.773
.30	.9435	.9867	2.0537	2.05	.1203	.6134	1.859
.35	.9241	.9820	1.7930	2.10	.1108	.6019	1.951
.40	.9023	.9766	1.6023	2.15	.1020	.5905	2.050
.45	.8784	.9705	1.4586	2.20	$.9393^{-1}$	.5794	2.156
.50	.8526	9639	1.3479	2.25	$.8645^{-1}$	.5684	2.268
.55	.8251	.9566	1.2614	2.30	$.7955^{-1}$	.5576	2.388
.60	.6267	.9488	1.1932	2.35	$.7318^{-1}$	.5470	2.517
.65	.7662	.9404	1.1395	2.40	$.6731^{-1}$	.5365	2.654
.70	.7354	.9315	1.0972	2.45	$.6190^{-1}$	.5262	2.799
.75	.7724	.9222	1.0644	2.50	$.5692^{-1}$	.5161	2.954
.80	.6723	.9124	1.0395	2.55	$.5234^{-1}$	.5062	3.119
.35	.6403	.9022	1.0214	2.60	$.4813^{-1}$	.4965	3.295
.90	.6084	.8917	1.0092	2.65	$.4426^{-1}$	.4870	3.482
.95	.5768	.8808	1.0022	2.70	$.4070^{-1}$	.4777	3.681
1.00	.5457	.8696	1.0000	2.75	$.3743^{-1}$	.4686	3.892
1.05	.5152	.8581	1.002	2.80	$.3442^{-1}$	.4596	4.116
1.10	.4854	.8464	1.008	2.85	$.3166^{-1}$	.4508	4.354
1.15	.4565	.8345	1.018	2.90	$.2913^{-1}$	.4422	4.607
1.20	.4285	.8224	1.032	2.95	$.2680^{-1}$	.4338	4.875
1.25	.4015	.8101	1.049	3.00	$.2466^{-1}$	.4255	5.160
1.30	.3756	.7978	1.070	3.50	$.1090^{-1}$	.3524	9.110
1.35	.3509	.7853	1.095	4.00	$.4977^{-2}$	.2941	15.94
1.40	.3273	.7728	1.123	4.50	$.2363^{-2}$	.2477	27.39
1.45	.3049	.7603	1.154	5.00	$.1169^{-2}$	.2105	45.96
1.50	.2836	.7477	1.189	6.00	$.3210^{-3}$	.1563	120.1
1.55	.2635	.7351	1.228	7.00	$.1014^{-3}$	.1198	285.3
1.60	.2446	.7225	1.271	8.00	$.3606^{-4}$	$.9434^{-1}$	623.1
1.65	.2268	.7100	1.318	9.00	$.1417^{-4}$	$.7605^{-1}$	1265
1.70	.2101	.6976	1.369	10.00	$.6055^{-5}$	$.6250^{-1}$	2416

TABLE 3.

$$\gamma = 5/3$$

M	$\dfrac{p}{p_t}$	$\dfrac{T}{T_t}$	$\dfrac{A}{A^*}$	M	$\dfrac{p}{p_t}$	$\dfrac{T}{T_t}$	$\dfrac{A}{A^*}$
0	1.0000	1.0000	∞	1.75	.1723	.4948	1.313
.05	.9979	.9992	11.2688	1.80	.1603	.4808	1.352
.10	.9917	.9967	5.6626	1.85	.1491	.4671	1.394
.15	.9815	.9926	3.8065	1.90	.1388	.4539	1.437
.20	.9674	.9868	2.8880	1.95	.1292	.4410	1.483
.25	.9498	.9796	2.3447	2.00	.1202	.4286	1.531
.30	.9288	.9709	1.9892	2.05	.1120	.4165	1.582
.35	.9048	.9608	1.7411	2.10	.1043	.4049	1.634
.40	.8782	.9494	1.5603	2.15	$.9718^{-1}$	.3936	1.689
.45	.8493	.9368	1.4244	2.20	$.9058^{-1}$	.3827	1.746
.50	.8186	.9231	1.3203	2.25	$.8446^{-1}$	.3721	1.806
.55	.7865	.9084	1.2394	2.30	$.7878^{-1}$	.3619	1.868
.60	.7533	.8929	1.1760	2.35	$.7352^{-1}$	.3520	1.932
.65	.7194	.8766	1.1263	2.40	$.6863^{-1}$	.3425	1.998
.70	.6851	.8596	1.0875	2.45	$.6411^{-1}$	.3332	2.067
.75	.6508	.8421	1.0576	2.50	$.5990^{-1}$	.3243	2.139
.80	.6167	.8242	1.0351	2.55	$.5600^{-1}$	.3157	2.213
.85	.5831	.8059	1.0189	2.60	$.5238^{-1}$	.3074	2.290
.90	.5502	.7874	1.0081	2.65	$.4902^{-1}$	.2993	2.369
.95	.5181	.7687	1.0019	2.70	$.4589^{-1}$	.2915	2.451
1.00	.4871	.7500	1.000	2.75	$.4299^{-1}$	.2840	2.536
1.05	.4573	.7313	1.002	2.80	$.4029^{-1}$	.2768	2.623
1.10	.4286	.7126	1.007	2.85	$.3778^{-1}$	.2697	2.713
1.15	.4013	.6940	1.015	2.90	$.3545^{-1}$	.2629	2.806
1.20	.3753	.6757	1.027	2.95	$.3327^{-1}$	.2564	2.901
1.25	.3506	.6575	1.041	3.00	$.3125^{-1}$	.2500	3.000
1.30	.3272	.6397	1.058	3.50	$.1716^{-1}$	.1967	4.153
1.35	.3052	.6221	1.077	4.00	$.9906^{-2}$	.1579	5.641
1.40	.2845	.6048	1.098	4.50	$.5981^{-2}$	.1290	7.508
1.45	.2651	.5879	1.122	5.00	$.3758^{-2}$	.1071	9.800
1.50	.2468	.5714	1.148	6.00	$.1641^{-2}$	$.7692^{-1}$	15.84
1.55	.2298	.5553	1.177	7.00	$.7995^{-3}$	$.5769^{-1}$	24.14
1.60	.2139	.5396	1.208	8.00	$.4242^{-3}$	$.4478^{-1}$	35.07
1.65	.1990	.5242	1.240	9.00	$.2410^{-3}$	$.3571^{-1}$	49.00
1.70	.1851	.5093	1.275	10.00	$.1448^{-3}$	$.2913^{-1}$	66.31

NORMAL SHOCK TABLES

TABLE 1.

$$\gamma = 1.4$$

M_1	M_2	$\dfrac{p_2}{p_1}$	$\dfrac{\rho_2}{\rho_1}$	$\dfrac{T_2}{T_1}$	$\dfrac{p_{t_2}}{p_{t_1}}$	$\dfrac{p_1}{p_{t_2}}$
1.00	1.000	1.000	1.000	1.000	1.000	0.5283
1.01	.9901	1.023	1.017	1.007	1.000	.5221
1.02	.9805	1.047	1.033	1.013	1.000	.5160
1.03	.9712	1.071	1.050	1.020	1.000	.5100
1.04	.9620	1.095	1.067	1.026	.9999	.5039
1.05	.9531	1.120	1.084	1.033	.9999	.4980
1.06	.9444	1.144	1.101	1.059	.9997	.4920
1.07	.9360	1.169	1.118	1.016	.9996	.4861
1.08	.9277	1.194	1.135	1.052	.9994	.4803
1.09	.9196	1.219	1.152	1.059	.9992	.4746
1.10	.9118	1.245	1.169	1.065	.9989	.4689
1.11	.9041	1.271	1.186	1.071	.9986	.4632
1.12	.8966	1.297	1.203	1.078	.9982	.4576
1.13	.8892	1.323	1.221	1.084	.9978	.4521
1.14	.8820	1.350	1.238	1.090	.9973	.4467
1.15	.8750	1.376	1.255	1.097	.9967	.4413
1.16	.8682	1.403	1.272	1.103	.9961	.4360
1.17	.8615	1.430	1.290	1.109	.9953	.4307
1.18	.8549	1.458	1.307	1.115	.9916	.4255
1.19	.8485	1.485	1.324	1.122	.9937	.4204

Data in Appendix C is From NACA Report 1135, "Equations, Tables, and Charts for Compressible Flow," Ames Research Staff, 1953.

TABLE 1. (Continued)

$$\gamma = 1.4$$

M_1	M_2	$\dfrac{p_2}{p_1}$	$\dfrac{\rho_2}{\rho_1}$	$\dfrac{T_2}{T_1}$	$\dfrac{p_{t_2}}{p_{t_1}}$	$\dfrac{p_1}{p_{t_2}}$
1.20	0.8422	1.513	1.342	1.128	0.9928	0.4154
1.21	.8360	1.541	1.359	1.134	.9918	.4104
1.22	.8300	1.570	1.376	1.141	.9907	.4055
1.23	.8241	1.598	1.394	1.147	.9896	.4006
1.24	.8183	1.627	1.411	1.153	.9884	.3958
1.25	.8126	1.656	1.429	1.159	.9871	.3911
1.26	.8071	1.686	1.446	1.166	.9857	.3865
1.27	.8016	1.715	1.463	1.172	.9842	.3819
1.28	.7963	1.745	1.481	1.178	.9827	.3774
1.29	.7911	1.775	1.498	1.185	.9811	.3729
1.30	.7860	1.805	1.516	1.191	.9794	.3685
1.31	.7809	1.835	1.533	1.197	.9776	.3642
1.32	.7760	1.866	1.551	1.204	.9758	.3599
1.33	.7712	1.897	1.568	1.210	.9738	.3557
1.34	.7664	1.928	1.585	1.216	.9718	.3516
1.35	.7618	1.960	1.603	1.223	.9697	.3475
1.36	.7572	1.991	1.620	1.229	.9676	.3435
1.37	.7527	2.023	1.638	1.235	.9653	.3395
1.38	.7483	2.055	1.655	1.242	.9630	.3356
1.39	.7440	2.087	1.672	1.248	.9607	.3317
1.40	.7397	2.120	1.690	1.255	.9582	.3280
1.41	.7355	2.153	1.707	1.261	.9557	.3242
1.42	.7314	2.186	1.724	1.268	.9531	.3205
1.43	.7274	2.219	1.742	1.274	.9504	.3169
1.44	.7235	2.253	1.759	1.281	.9476	.3133
1.45	.7196	2.286	1.776	1.287	.9448	.3098
1.46	.7157	2.320	1.793	1.294	.9420	.3063
1.47	.7120	2.354	1.811	1.300	.9390	.3029
1.48	.7083	2.389	1.828	1.307	.9360	.2996
1.49	.7047	2.423	1.845	1.314	.9329	.2962
1.50	.7011	2.458	1.862	1.320	.9298	.2930
1.51	.6976	2.493	1.879	1.327	.9266	.2898
1.52	.6941	2.529	1.896	1.334	.9233	.2866
1.53	.6907	2.564	1.913	1.340	.9200	.2835
1.54	.6874	2.600	1.930	1.347	.9166	.2804
1.55	.6841	2.636	1.947	1.354	.9132	.2773
1.56	.6809	2.673	1.964	1.361	.9097	.2744
1.57	.6777	2.709	1.981	1.367	.9061	.2714
1.58	.6746	2.746	1.998	1.374	.9026	.2685
1.59	.6715	2.783	2.015	1.381	.8989	.2656

TABLE 1. (Continued)

$$\gamma = 1.4$$

M_1	M_2	$\dfrac{p_2}{p_1}$	$\dfrac{\rho_2}{\rho_1}$	$\dfrac{T_2}{T_1}$	$\dfrac{p_{t_2}}{p_{t_1}}$	$\dfrac{p_1}{p_{t_2}}$
1.60	0.6684	2.820	2.032	1.388	0.8952	0.2628
1.61	.6655	2.857	2.049	1.395	.8915	.2600
1.62	.6625	2.895	2.065	1.402	.8877	.2573
1.63	.6596	2.933	2.082	1.409	.8538	.2546
1.64	.6568	2.971	2.099	1.416	.8799	.2519
1.65	.6540	3.010	2.115	1.423	.8760	.2493
1.66	.6512	3.048	2.132	1.430	.8720	.2467
1.67	.6485	3.087	2.148	1.437	.8680	.2442
1.68	.6458	3.126	2.165	1.444	.8640	.2417
1.69	.6431	3.165	2.181	1.451	.8598	.2392
1.70	.6405	3.205	2.198	1.458	.8557	.2368
1.71	.6380	3.245	2.214	1.466	.8516	.2344
1.72	.6355	3.285	2.230	1.473	.8474	.2320
1.73	.6330	3.325	2.247	1.480	.8431	.2296
1.74	.6305	3.366	2.263	1.487	.8389	.2273
1.75	.6281	3.406	2.279	1.495	.8346	.2251
1.76	.6257	3.447	2.295	1.502	.8302	.2228
1.77	.6234	3.488	2.311	1.509	.8259	.2206
1.78	.6210	3.530	2.327	1.517	.8215	.2184
1.79	.6188	3.571	2.343	1.524	.8171	.2163
1.80	.6165	3.613	2.359	1.532	.8127	.2142
1.81	.6143	3.655	2.375	1.539	.8082	.2121
1.82	.6121	3.698	2.391	1.547	.8038	.2100
1.83	.6099	3.740	2.407	1.554	.7993	.2080
1.84	.6078	3.783	2.422	1.562	.7948	.2060
1.85	.6057	3.826	2.438	1.569	.7092	.2040
1.86	.6036	3.870	2.454	1.577	.7857	.2020
1.87	.6016	3.913	2.469	1.585	.7811	.2001
1.88	.5996	3.957	2.485	1.592	.7765	.1982
1.89	.5976	4.001	2.500	1.600	.7720	.1963
1.90	.5956	4.045	2.516	1.608	.7674	.1945
1.91	.5937	4.089	2.531	1.616	.7627	.1927
1.92	.5918	4.134	2.546	1.624	.7581	.1909
1.93	.5899	4.179	2.562	1.631	.7535	.1891
1.94	.5880	4.224	2.577	1.639	.7488	.1873
1.95	.5862	4.270	2.592	1.647	.7442	.1856
1.96	.5844	4.315	2.607	1.655	.7395	.1839
1.97	.5826	4.361	2.622	1.663	.7349	.1822
1.98	.5808	4.407	2.637	1.671	.7302	.1806
1.99	.5791	4.453	2.652	1.679	.7255	.1789

TABLE 1. (Continued)

$$\gamma = 1.4$$

M_1	M_2	$\dfrac{p_2}{p_1}$	$\dfrac{\rho_2}{\rho_1}$	$\dfrac{T_2}{T_1}$	$\dfrac{p_{t_2}}{p_{t_1}}$	$\dfrac{p_1}{p_{t_2}}$
2.00	0.5774	4.500	2.667	1.688	0.7209	0.1773
2.01	.5757	4.547	2.681	1.696	.7162	.1757
2.02	.5740	4.594	2.696	1.704	.7115	.1741
2.03	.5723	4.641	2.711	1.712	.7069	.1726
2.04	.5707	4.689	2.725	1.720	.7022	.1710
2.05	.5691	4.736	2.740	1.729	.6975	.1695
2.06	.5675	4.784	2.755	1.737	.6928	.1680
2.07	.5659	4.832	2.769	1.745	.6882	.1665
2.08	.5643	4.881	2.783	1.754	.6835	.1651
2.09	.5628	4.929	2.798	1.762	.6789	.1636
2.10	.5613	4.978	2.812	1.770	.6742	.1622
2.11	.5598	5.027	2.826	1.779	.6696	.1608
2.12	.5583	5.077	2.840	1.787	.6649	.1594
2.13	.5568	5.126	2.854	1.796	.6603	.1580
2.14	.5554	5.176	2.868	1.805	.6557	.1567
2.15	.5540	5.226	2.882	1.813	.6511	.1553
2.16	.5525	5.277	2.896	1.822	.6464	.1540
2.17	.5511	5.327	2.910	1.821	.6419	.1527
2.18	.5498	5.378	2.924	1.839	.6373	.1514
2.19	.5484	5.429	2.938	1.848	.6327	.1502
2.20	.5471	5.480	2.951	1.857	.6281	.1489
2.21	.5457	5.531	2.965	1.866	.6236	.1476
2.22	.5444	5.583	2.978	1.875	.6191	.1464
2.23	.5431	5.636	2.992	1.883	.6145	.1452
2.24	.5418	5.687	3.005	1.892	.6100	.1440
2.25	.5406	5.740	3.019	1.901	.6055	.1428
2.26	.5393	5.792	3.032	1.910	.6011	.1417
2.27	.5381	5.845	3.045	1.919	.5966	.1405
2.28	.5368	5.898	3.058	1.929	.5921	.1394
2.29	.5356	5.951	3.071	1.938	.5877	.1382
2.30	.5344	6.005	3.085	1.947	.5833	.1371
2.31	.5332	6.059	3.098	1.956	.5789	.1360
2.32	.5321	6.113	3.110	1.965	.5745	.1349
2.33	.5309	6.167	3.123	1.974	.5702	.1338
2.34	.5297	6.222	3.136	1.984	.5658	.1328
2.35	.5286	6.276	3.149	1.993	.5615	.1317
2.36	.5275	6.331	3.162	2.002	.5572	.1307
2.37	.5264	6.386	3.174	2.012	.5529	.1297
2.38	.5253	6.442	3.187	2.021	.5486	.1286
2.39	.5242	6.497	3.199	2.031	.5444	.1276

TABLE 1. (Continued)

$$\gamma = 1.4$$

M_1	M_2	$\dfrac{p_2}{p_1}$	$\dfrac{\rho_2}{\rho_1}$	$\dfrac{T_2}{T_1}$	$\dfrac{p_{t_2}}{p_{t_1}}$	$\dfrac{p_1}{p_{t_2}}$
2.40	0.5231	6.553	3.212	2.040	0.5401	0.1266
2.41	.5221	6.609	3.224	2.050	.5359	.1257
2.42	.5210	6.666	3.237	2.059	.5317	.1247
2.43	.5200	6.722	3.249	2.069	.5276	.1237
2.44	.5189	6.779	3.261	2.079	.5234	.1228
2.45	.5179	6.836	3.273	2.088	.5193	.1218
2.46	.5169	6.894	3.285	2.098	.5152	.1209
2.47	.5159	6.951	3.298	2.108	.5111	.1200
2.48	.5149	7.009	3.310	2.118	.5071	.1191
2.49	.5140	7.067	3.321	2.128	.5030	.1182
2.50	.5130	7.125	3.333	2.138	.4990	.1173
2.51	.5120	7.183	3.345	2.147	.4950	.1164
2.52	.5111	7.242	3.357	2.157	.4911	.1155
2.53	.5102	7.301	3.369	2.167	.4871	.1147
2.54	.5092	7.360	3.380	2.177	.4832	.1138
2.55	.5083	7.420	3.392	2.187	.4793	.1130
2.56	.5074	7.479	3.403	2.198	.4754	.1122
2.57	.5065	7.539	3.415	2.208	.4715	.1113
2.58	.5056	7.599	3.426	2.218	.4677	.1105
2.59	.5047	7.659	3.438	2.228	.4639	.1097
2.60	.5039	7.720	3.449	2.238	.4601	.1089
2.61	.5030	7.781	3.460	2.249	.4564	.1081
2.62	.5022	7.842	3.471	2.259	.4526	.1074
2.63	.5013	7.903	3.483	2.269	.4489	.1066
2.64	.5005	7.965	3.494	2.280	.4452	.1058
2.65	.4996	8.026	3.505	2.290	.4416	.1051
2.66	.4988	8.088	3.516	2.301	.4379	.1043
2.67	.4980	8.150	3.527	2.311	.4343	.1036
2.68	.4972	8.213	3.537	2.322	.4307	.1028
2.69	.4964	8.275	3.548	2.332	.4271	.1021
2.70	.4956	8.338	3.559	2.343	.4236	.1014
2.71	.4949	8.401	3.570	2.354	.4201	.1007
2.72	.4941	8.465	3.580	2.364	.4166	$.9998^{-1}$
2.73	.4933	8.528	3.591	2.375	.4131	$.9929^{-1}$
2.74	.4926	8.592	3.601	2.386	.4097	$.9860^{-1}$
2.75	.4918	8.656	3.612	2.397	.4062	$.9792^{-1}$
2.76	.4911	8.721	3.622	2.407	.4028	$.9724^{-1}$
2.77	.4903	8.785	3.633	2.418	.3994	$.9658^{-1}$
2.78	.4896	8.850	3.643	2.429	.3961	$.9591^{-1}$
2.79	.4889	8.915	3.653	2.440	.3928	$.9526^{-1}$

TABLE 1. (Continued)

$$\gamma = 1.4$$

M_1	M_2	$\dfrac{p_2}{p_1}$	$\dfrac{\rho_2}{\rho_1}$	$\dfrac{T_2}{T_1}$	$\dfrac{p_{t_2}}{p_{t_1}}$	$\dfrac{p_1}{p_{t_2}}$
2.80	0.4882	8.980	3.664	2.451	0.3895	0.9461^{-1}
2.81	.4875	9.045	3.674	2.462	.3862	$.9397^{-1}$
2.82	.4868	9.111	3.684	2.473	.3829	$.9334^{-1}$
2.83	.4861	9.177	3.694	2.484	.3797	$.9271^{-1}$
2.84	.4854	9.243	3.704	2.496	.3765	$.9209^{-1}$
2.85	.4847	9.310	3.714	2.507	.3733	$.9147^{-1}$
2.86	.4840	9.376	3.724	2.518	.3701	$.9086^{-1}$
2.87	.4833	9.443	3.734	2.529	.3670	$.9026^{-1}$
2.88	.4827	9.510	3.743	2.540	.3639	$.8966^{-1}$
2.89	.4820	9.577	3.753	2.552	.3608	$.8906^{-1}$
2.90	.4814	9.645	3.763	2.563	.3577	$.8848^{-1}$
2.91	.4807	9.713	3.773	2.575	.3547	$.8790^{-1}$
2.92	.4801	9.781	3.782	2.586	.3517	$.8732^{-1}$
2.93	.4795	9.849	3.792	2.598	.3487	$.8675^{-1}$
2.94	.4788	9.918	3.801	2.609	.3457	$.8619^{-1}$
2.95	.4782	9.986	3.811	2.621	.3428	$.8563^{-1}$
2.96	.4776	10.06	3.820	2.632	.3398	$.8507^{-1}$
2.97	.4770	10.12	3.829	2.644	.3369	$.8453^{-1}$
2.98	.4764	10.19	3.839	2.656	.3340	$.8398^{-1}$
2.99	.4758	10.26	3.848	2.667	.3312	$.8345^{-1}$
3.00	.4752	10.33	3.857	2.679	.3283	$.8291^{-1}$
3.01	.4746	10.40	3.866	2.691	.3255	$.8238^{-1}$
3.02	.4740	10.47	3.875	2.703	.3227	$.8186^{-1}$
3.03	.4734	10.54	3.884	2.714	.3200	$.8134^{-1}$
3.04	.4729	10.62	3.893	2.726	.3172	$.8083^{-1}$
3.05	.4723	10.69	3.902	2.738	.3145	$.8032^{-1}$
3.06	.4717	10.76	3.911	2.750	.3118	$.7982^{-1}$
3.07	.4712	10.83	3.920	2.762	.3091	$.7932^{-1}$
3.08	.4706	10.90	3.929	2.774	.3065	$.7882^{-1}$
3.09	.4701	10.97	3.938	2.786	.3038	$.7833^{-1}$
3.10	.4695	11.05	3.947	2.799	.3012	$.7785^{-1}$
3.11	.4690	11.12	3.955	2.811	.2986	$.7737^{-1}$
3.12	.4685	11.19	3.964	2.823	.2960	$.7689^{-1}$
3.13	.4679	11.26	3.973	2.835	.2935	$.7642^{-1}$
3.14	.4674	11.34	3.981	2.848	.2910	$.7595^{-1}$
3.15	.4669	11.41	3.990	2.860	.2885	$.7549^{-1}$
3.16	.4664	11.48	3.998	2.872	.2860	$.7503^{-1}$
3.17	.4659	11.56	4.006	2.885	.2835	$.7457^{-1}$
3.18	.4654	11.63	4.015	2.897	.2811	$.7412^{-1}$
3.19	.4648	11.71	4.023	2.909	.2786	$.7367^{-1}$

TABLE 1. (Continued)

$$\gamma = 1.4$$

M_1	M_2	$\dfrac{p_2}{p_1}$	$\dfrac{p_2}{p_1}$	$\dfrac{T_2}{T_1}$	$\dfrac{p_{t_2}}{p_{t_1}}$	$\dfrac{p_1}{p_{t_2}}$
3.20	0.4643	11.78	4.031	2.922	0.2762	0.7323^{-1}
3.21	.4639	11.85	4.040	2.935	.2738	$.7279^{-1}$
3.22	.4634	11.93	4.048	2.947	.2715	$.7235^{-1}$
3.23	.4629	12.01	4.056	2.960	.2691	$.7192^{-1}$
3.24	.4624	12.08	4.064	2.972	.2668	$.7149^{-1}$
3.25	.4619	12.16	4.072	2.985	.2645	$.7107^{-1}$
3.26	.4614	12.23	4.080	2.998	.2622	$.7065^{-1}$
3.27	.4610	12.31	4.088	3.011	.2600	$.7023^{-1}$
3.28	.4605	12.38	4.096	3.023	.2577	$.6982^{-1}$
3.29	.4600	12.46	4.104	3.036	.2555	$.6941^{-1}$
3.30	.4596	12.54	4.112	3.049	.2533	$.6900^{-1}$
3.31	.4591	12.62	4.120	3.062	.2511	$.6860^{-1}$
3.32	.4587	12.69	4.128	3.075	.2489	$.6820^{-1}$
3.33	.4582	12.77	4.135	3.088	.2468	$.6781^{-1}$
3.34	.4578	12.85	4.143	3.101	.2446	$.6741^{-1}$
3.35	.4573	12.93	4.151	3.114	.2425	$.6702^{-1}$
3.36	.4569	13.00	4.158	3.127	.2404	$.6664^{-1}$
3.37	.4565	13.08	4.166	3.141	.2383	$.6626^{-1}$
3.38	.4560	13.16	4.173	3.154	.2363	$.6588^{-1}$
3.39	.4556	13.24	4.181	3.167	.2342	$.6550^{-1}$
3.40	.4552	13.32	4.188	3.180	.2322	$.6513^{-1}$
3.41	.4548	13.40	4.196	3.191	.2302	$.6476^{-1}$
3.42	.4544	13.48	4.203	3.207	.2282	$.6439^{-1}$
3.43	.4540	13.56	4.211	3.220	.2263	$.6403^{-1}$
3.44	.4535	13.64	4.218	3.234	.2243	$.6367^{-1}$
3.45	.4531	13.72	4.225	3.247	.2224	$.6331^{-1}$
3.46	.4527	13.80	4.232	3.261	.2205	$.6296^{-1}$
3.47	.4523	13.88	4.240	3.274	.2186	$.6261^{-1}$
3.48	.4519	13.96	4.247	3.288	.2167	$.6226^{-1}$
3.49	.4515	14.04	4.254	3.301	.2148	$.6191^{-1}$
3.50	.4512	14.13	4.261	3.315	.2129	$.6157^{-1}$
3.51	.4508	14.21	4.268	3.329	.2111	$.6123^{-1}$
3.52	.4504	14.29	4.275	3.343	.2093	$.6089^{-1}$
3.53	.4500	14.37	4.282	3.356	.2075	$.6056^{-1}$
3.54	.4496	14.45	4.289	3.370	.2057	$.6023^{-1}$
3.55	.4492	14.54	4.296	3.384	.2039	$.5990^{-1}$
3.56	.4489	14.62	4.303	3.398	.2022	$.5957^{-1}$
3.57	.4485	14.70	4.309	3.412	.2004	$.5925^{-1}$
3.58	.4481	14.79	4.316	3.426	.1987	$.5892^{-1}$
3.59	.4478	14.87	4.324	3.440	.1970	$.5861^{-1}$

TABLE 1. (Continued)

$$\gamma = 1.4$$

M_1	M_2	$\dfrac{p_2}{p_1}$	$\dfrac{p_2}{\rho_1}$	$\dfrac{T_2}{T_1}$	$\dfrac{p_{t_2}}{p_{t_1}}$	$\dfrac{p_1}{p_{t_2}}$
3.60	0.4474	14.95	4.330	3.454	0.1953	0.5829^{-1}
3.61	.4471	15.04	4.336	3.468	.1936	$.5798^{-1}$
3.62	.4467	15.12	4.343	3.482	.1920	$.5767^{-1}$
3.63	.4463	15.21	4.350	3.496	.1903	$.5736^{-1}$
3.64	.4460	15.29	4.356	3.510	.1887	$.5705^{-1}$
3.65	.4456	15.38	4.363	3.525	.1871	$.5675^{-1}$
3.66	.4453	15.46	4.369	3.539	.1855	$.5645^{-1}$
3.67	.4450	15.55	4.376	3.553	.1839	$.5615^{-1}$
3.68	.4446	15.63	4.382	3.568	.1823	$.5585^{-1}$
3.69	.4443	15.72	4.388	3.582	.1807	$.5556^{-1}$
3.70	.4439	15.81	4.395	3.596	.1792	$.5526^{-1}$
3.71	.4436	15.89	4.401	3.611	.1777	$.5497^{-1}$
3.72	.4433	15.98	4.408	3.625	.1761	$.5469^{-1}$
3.73	.4430	16.07	4.414	3.640	.1746	$.5440^{-1}$
3.74	.4426	16.15	4.420	3.654	.1731	$.5412^{-1}$
3.75	.4423	16.24	4.426	3.669	.1717	$.5384^{-1}$
3.76	.4420	16.33	4.432	3.684	.1702	$.5356^{-1}$
3.77	.4417	16.42	4.439	3.698	.1687	$.5328^{-1}$
3.78	.4414	16.50	4.445	3.713	.1673	$.5301^{-1}$
3.79	.4410	16.59	4.451	3.728	.1659	$.5274^{-1}$
3.80	.4407	16.68	4.457	3.743	.1645	$.5247^{-1}$
3.81	.4404	16.77	4.463	3.758	.1631	$.5220^{-1}$
3.82	.4401	16.86	4.469	3.772	.1617	$.5193^{-1}$
3.83	.4398	16.95	4.475	3.787	.1603	$.5167^{-1}$
3.84	.4395	17.04	4.481	3.802	.1589	$.5140^{-1}$
3.85	.4392	17.13	4.487	3.817	.1576	$.5114^{-1}$
3.86	.4389	17.22	4.492	3.832	.1563	$.5089^{-1}$
3.87	.4386	17.31	4.498	3.847	.1549	$.5063^{-1}$
3.88	.4383	17.40	4.504	3.863	.1536	$.5038^{-1}$
3.89	.4380	17.49	4.510	3.878	.1523	$.5012^{-1}$
3.90	.4377	17.58	4.516	3.893	.1510	$.4987^{-1}$
3.91	.4375	17.67	4.521	3.908	.1497	$.4962^{-1}$
3.92	.4372	17.76	4.527	3.923	.1485	$.4938^{-1}$
3.93	.4369	17.85	4.533	3.939	.1472	$.4913^{-1}$
3.94	.4366	17.94	4.538	3.954	.1460	$.4889^{-1}$
3.95	.4363	18.04	4.544	3.969	.1448	$.4865^{-1}$
3.96	.4360	18.13	4.549	3.985	.1435	$.4841^{-1}$
3.97	.4358	18.22	4.555	4.000	.1423	$.4817^{-1}$
3.98	.4355	18.31	4.560	4.016	.1411	$.4793^{-1}$
3.99	.4352	18.41	4.566	4.031	.1399	$.4770^{-1}$

TABLE 1. (Continued)

$$\gamma = 1.4$$

M_1	M_2	$\dfrac{p_2}{p_1}$	$\dfrac{\rho_2}{\rho_1}$	$\dfrac{T_2}{T_1}$	$\dfrac{p_{t_2}}{p_{t_1}}$	$\dfrac{p_1}{p_{t_2}}$
4.00	0.4350	18.50	4.571	4.047	0.1388	0.4747^{-1}
4.01	.4347	18.59	4.577	4.062	.1376	$.4723^{-1}$
4.02	.4344	18.69	4.582	4.078	.1364	$.4769^{-1}$
4.03	.4342	18.78	4.588	4.094	.1353	$.4678^{-1}$
4.04	.4339	18.88	4.593	4.110	.1342	$.4655^{-1}$
4.05	.4336	18.97	4.598	4.125	.1330	$.4633^{-1}$
4.06	.4334	19.06	4.604	4.141	.1319	$.4610^{-1}$
4.07	.4331	19.16	4.609	4.157	.1308	$.4588^{-1}$
4.08	.4329	19.25	4.614	4.173	.1297	$.4566^{-1}$
4.09	.4326	19.35	4.619	4.189	.1286	$.4544^{-1}$
4.10	.4324	19.45	4.624	4.205	.1276	$.4523^{-1}$
4.11	.4321	19.54	4.630	4.221	.1265	$.4501^{-1}$
4.12	.4319	19.64	4.635	4.237	.1254	$.4480^{-1}$
4.13	.4316	19.73	4.640	4.253	.1244	$.4459^{-1}$
4.14	.4314	19.83	4.645	4.269	.1234	$.4438^{-1}$
4.15	.4311	19.93	4.650	4.285	.1223	$.4417^{-1}$
4.16	.4309	20.02	4.655	4.301	.1213	$.4396^{-1}$
4.17	.4306	20.12	4.660	4.318	.1203	$.4375^{-1}$
4.18	.4304	20.22	4.665	4.334	.1193	$.4355^{-1}$
4.19	.4302	20.32	4.670	4.350	.1183	$.4334^{-1}$
4.20	.4299	20.41	4.675	4.367	.1173	$.4314^{-1}$
4.21	.4297	20.51	4.680	4.383	.1164	$.4294^{-1}$
4.22	.4295	20.61	4.685	4.399	.1154	$.4274^{-1}$
4.23	.4292	20.71	4.690	4.416	.1144	$.4255^{-1}$
4.24	.4290	20.81	4.694	4.432	.1135	$.4235^{-1}$
4.25	.4288	20.91	4.699	4.449	.1126	$.4215^{-1}$
4.26	.4286	21.01	4.704	4.466	.1116	$.4196^{-1}$
4.27	.4283	21.11	4.709	4.482	.1107	$.4177^{-1}$
4.28	.4281	21.20	4.713	4.499	.1098	$.4158^{-1}$
4.29	.4279	21.30	4.718	4.516	.1089	$.4139^{-1}$
4.30	.4277	21.41	4.723	4.532	.1080	$.4120^{-1}$
4.31	.4275	21.51	4.728	4.549	.1071	$.4101^{-1}$
4.32	.4272	21.61	4.732	4.566	.1062	$.4082^{-1}$
4.33	.4270	21.71	4.737	4.583	.1054	$.4064^{-1}$
4.34	.4268	21.81	4.741	4.600	.1045	$.4046^{-1}$
4.35	.4266	21.91	4.746	4.617	.1036	$.4027^{-1}$
4.36	.4264	22.01	4.751	4.633	.1028	$.4009^{-1}$
4.37	.4262	22.11	4.755	4.651	.1020	$.3991^{-1}$
4.38	.4260	22.22	4.760	4.668	.1011	$.3973^{-1}$
4.39	.4258	22.32	4.764	4.685	.1003	$.3956^{-1}$

TABLE 1. (Continued)

$$\gamma = 1.4$$

M_1	M_2	$\dfrac{p_2}{p_1}$	$\dfrac{p_2}{p_1}$	$\dfrac{T_2}{T_1}$	$\dfrac{p_{t_2}}{p_{t_1}}$	$\dfrac{p_1}{p_{t_2}}$
4.40	0.4255	22.42	4.768	4.702	0.9948^{-1}	0.3938^{-1}
4.41	.4253	22.52	4.773	4.719	$.9867^{-1}$	$.3921^{-1}$
4.42	.4251	22.63	4.777	4.736	$.9787^{-1}$	$.3903^{-1}$
4.43	.4249	22.73	4.782	4.753	$.9707^{-1}$	$.3886^{-1}$
4.44	.4247	22.83	4.786	4.771	$.9628^{-1}$	$.3869^{-1}$
4.45	.4245	22.94	4.790	4.788	$.9550^{-1}$	$.3852^{-1}$
4.46	.4243	23.04	4.795	4.805	$.9473^{-1}$	$.3835^{-1}$
4.47	.4241	23.14	4.799	4.823	$.9396^{-1}$	$.3818^{-1}$
4.48	.4239	23.25	4.803	4.840	$.9320^{-1}$	$.3801^{-1}$
4.49	.4237	23.35	4.808	4.858	$.9244^{-1}$	$.3785^{-1}$
4.50	.4236	23.46	4.812	4.875	$.9170^{-1}$	$.3768^{-1}$
4.51	.4234	23.56	4.816	4.893	$.9096^{-1}$	$.3752^{-1}$
4.52	.4232	23.67	4.820	4.910	$.9022^{-1}$	$.3735^{-1}$
4.53	.4230	23.77	4.824	4.928	$.8950^{-1}$	$.3719^{-1}$
4.54	.4228	23.88	4.829	4.946	$.8878^{-1}$	$.3703^{-1}$
4.55	.4226	23.99	4.833	4.963	$.8806^{-1}$	$.3687^{-1}$
4.56	.4224	24.09	4.837	4.981	$.8735^{-1}$	$.3671^{-1}$
4.57	.4222	24.20	4.841	4.999	$.8665^{-1}$	$.3656^{-1}$
4.58	.4220	24.31	4.845	5.017	$.8596^{-1}$	$.3640^{-1}$
4.59	.4219	24.41	4.849	5.034	$.8527^{-1}$	$.3624^{-1}$
4.60	.4217	24.52	4.853	5.052	$.8459^{-1}$	$.3609^{-1}$
4.61	.4215	24.63	4.857	5.070	$.8391^{-1}$	$.3593^{-1}$
4.62	.4213	24.74	4.861	5.088	$.8324^{-1}$	$.3578^{-1}$
4.63	.4211	24.84	4.865	5.106	$.8257^{-1}$	$.3563^{-1}$
4.64	.4210	24.95	4.869	5.124	$.8192^{-1}$	$.3548^{-1}$
4.65	.4208	25.06	4.873	5.143	$.8126^{-1}$	$.3533^{-1}$
4.66	.4206	25.17	4.877	5.160	$.8062^{-1}$	$.3518^{-1}$
4.67	.4204	25.28	4.881	5.179	$.7998^{-1}$	$.3503^{-1}$
4.68	.4203	25.39	4.885	5.197	$.7934^{-1}$	$.3488^{-1}$
4.69	.4201	25.50	4.889	5.215	$.7871^{-1}$	$.3474^{-1}$
4.70	.4199	25.61	4.893	5.233	$.7809^{-1}$	$.3459^{-1}$
4.71	.4197	25.71	4.896	5.252	$.7747^{-1}$	$.3445^{-1}$
4.72	.4196	25.82	4.900	5.270	$.7685^{-1}$	$.3431^{-1}$
4.73	.4194	25.94	4.904	5.289	$.7625^{-1}$	$.3416^{-1}$
4.74	.4192	26.05	4.908	5.307	$.7564^{-1}$	$.3402^{-1}$
4.75	.4191	26.16	4.912	5.325	$.7505^{-1}$	$.3388^{-1}$
4.76	.4189	26.27	4.915	5.344	$.7445^{-1}$	$.3374^{-1}$
4.77	.4187	26.38	4.919	5.363	$.7387^{-1}$	$.3360^{-1}$
4.78	.4186	26.49	4.923	5.381	$.7329^{-1}$	$.3346^{-1}$
4.79	.4184	26.60	4.926	5.400	$.7271^{-1}$	$.3333^{-1}$

TABLE 1. (Continued)

$$\gamma = 1.4$$

M_1	M_2	$\dfrac{p_2}{p_1}$	$\dfrac{\rho_2}{\rho_1}$	$\dfrac{T_2}{T_1}$	$\dfrac{p_{t_2}}{p_{t_1}}$	$\dfrac{p_1}{p_{t_2}}$
4.80	0.4183	26.71	4.930	5.418	0.7214^{-1}	0.3319^{-1}
4.81	.4181	26.83	4.934	5.437	$.7157^{-1}$	$.3365^{-1}$
4.82	.4179	26.94	4.937	5.456	$.7101^{-1}$	$.3292^{-1}$
4.83	.4178	27.05	4.941	5.475	$.7046^{-1}$	$.3278^{-1}$
4.84	.4176	27.16	4.945	5.494	$.6991^{-1}$	$.3265^{-1}$
4.85	.4175	27.28	4.948	5.512	$.6936^{-1}$	$.3252^{-1}$
4.86	.4173	27.39	4.952	5.531	$.6882^{-1}$	$.3239^{-1}$
4.87	.4172	27.50	4.955	5.550	$.6828^{-1}$	$.3226^{-1}$
4.88	.4170	27.62	4.959	5.569	$.6775^{-1}$	$.3213^{-1}$
4.89	.4169	27.73	4.962	5.588	$.6722^{-1}$	$.3200^{-1}$
4.90	.4167	27.85	4.966	5.607	$.6670^{-1}$	$.3187^{-1}$
4.91	.4165	27.96	4.969	5.626	$.6618^{-1}$	$.3174^{-1}$
4.92	.4164	28.07	4.973	5.646	$.6567^{-1}$	$.3161^{-1}$
4.93	.4163	28.19	4.976	5.665	$.6516^{-1}$	$.3149^{-1}$
4.94	.4161	28.30	4.980	5.684	$.6465^{-1}$	$.3136^{-1}$
4.95	.4160	28.42	4.983	5.703	$.6415^{-1}$	$.3124^{-1}$
4.96	.4158	28.54	4.987	5.723	$.6366^{-1}$	$.3111^{-1}$
4.97	.4157	28.65	4.990	5.742	$.6317^{-1}$	$.3099^{-1}$
4.98	.4155	28.77	4.993	5.761	$.6268^{-1}$	$.3087^{-1}$
4.99	.4154	28.88	4.997	5.781	$.6220^{-1}$	$.3075^{-1}$
5.00	.4152	29.00	5.000	5.800	$.6172^{-1}$	$.3062^{-1}$
6.00	.4042	41.83	5.268	7.941	$.2965^{-1}$	$.2136^{-1}$
7.00	.3974	57.00	5.444	10.47	$.1535^{-1}$	$.1574^{-1}$
8.00	.3929	74.50	5.565	13.39	$.8488^{-2}$	$.1207^{-1}$
9.00	.3898	94.33	5.651	16.69	$.4964^{-2}$	$.9546^{-2}$
10.00	.3876	116.5	5.714	20.39	$.3045^{-2}$	$.7739^{-2}$

TABLE 2.

$$\gamma = 1.30$$

M_1	M_2	$\dfrac{p_2}{p_1}$	$\dfrac{\rho_2}{\rho_1}$	$\dfrac{T_2}{T_1}$	$\dfrac{p_{t_2}}{p_{t_1}}$	$\dfrac{p_1}{p_{t_2}}$
1.00	1.0000	1.000	1.000	1.000	1.0000	0.5457
1.05	.9530	1.116	1.088	1.026	.9998	.5152
1.10	.9112	1.237	1.178	1.051	.9989	.4859
1.15	.8739	1.364	1.269	1.075	.9966	.4581
1.20	.8403	1.497	1.362	1.100	.9925	.4318
1.25	.8100	1.636	1.456	1.124	.9866	.4070
1.30	.7825	1.780	1.551	1.148	.9786	.3839
1.35	.7575	1.930	1.646	1.172	.9684	.3623
1.40	.7346	2.085	1.742	1.197	.9562	.3422
1.45	.7136	2.246	1.838	1.222	.9421	.3236
1.50	.6942	2.413	1.935	1.247	.9261	.3063
1.55	.6764	2.585	2.031	1.273	.9084	.2901
1.60	.6599	2.763	2.127	1.299	.8891	.2751
1.65	.6446	2.947	2.223	1.326	.8684	.2611
1.70	.6304	3.137	2.318	1.353	.8466	.2481
1.75	.6172	3.332	2.413	1.380	.8238	.2360
1.80	.6048	3.532	2.507	1.408	.8001	.2246
1.85	.5933	3.738	2.601	1.437	.7758	.2140
1.90	.5825	3.950	2.694	1.467	.7510	.2042
1.95	.5724	4.168	2.785	1.497	.7259	.1949
2.00	.5629	4.391	2.875	1.527	.7006	.1862
2.05	.5539	4.620	2.964	1.558	.6752	.1781
2.10	.5455	4.855	3.052	1.590	.6499	.1705
2.15	.5376	5.095	3.139	1.623	.6248	.1633
2.20	.5301	5.341	3.225	1.656	.6000	.1566
2.25	.5230	5.592	3.309	1.690	.5755	.1502
2.30	.5163	5.849	3.392	1.725	.5515	.1442
2.35	.5100	6.112	3.474	1.760	.5280	.1386
2.40	.5040	6.381	3.554	1.796	.5050	.1333
2.45	.4983	6.655	3.633	1.832	.4827	.1282
2.50	.4929	6.935	3.710	1.869	.4610	.1235
2.55	.4878	7.220	3.786	1.907	.4400	.1190
2.60	.4829	7.511	3.860	1.946	.4196	.1147
2.65	.4782	7.808	3.933	1.985	.3999	.1107
2.70	.4738	8.110	4.005	2.025	.3810	.1068
2.75	.4696	8.418	4.075	2.066	.3628	.1032
2.80	.4655	8.732	4.144	2.108	.3452	$.9970^{-1}$
2.85	.4616	9.052	4.211	2.150	.3284	$.9643^{-1}$
2.90	.4579	9.377	4.277	2.193	.3123	$.9328^{-1}$
2.95	.4544	9.708	4.341	2.236	.2969	$.9025^{-1}$

TABLE 2. (Continued)

$$\gamma = 1.30$$

M_1	M_2	$\dfrac{p_2}{p_1}$	$\dfrac{\rho_2}{\rho_1}$	$\dfrac{T_2}{T_1}$	$\dfrac{p_{t_2}}{p_{t_1}}$	$\dfrac{p_1}{p_{t_2}}$
3.00	.4511	10.04	4.404	2.280	.2822	$.8741^{-1}$
3.50	.4241	13.72	4.964	2.763	.1677	$.6498^{-1}$
4.00	.4058	17.96	5.412	3.318	$.9932^{-1}$	$.5010^{-1}$
4.50	.3927	22.76	5.768	3.946	$.5941^{-1}$	$.3971^{-1}$
5.00	.3832	28.13	6.053	4.648	$.3612^{-1}$	$.3236^{-1}$
6.00	.3704	40.57	6.469	6.271	$.1422^{-1}$	$.2257^{-1}$
7.00	.3625	55.26	6.749	8.189	$.6098^{-2}$	$.1663^{-1}$
8.00	.3573	72.22	6.943	10.401	$.2826^{-2}$	$.1276^{-1}$
9.00	.3536	91.43	7.084	12.908	$.1404^{-2}$	$.1009^{-1}$
10.00	.3510	112.91	7.188	15.710	$.7408^{-3}$	$.8174^{-2}$

TABLE 3.

$$\gamma = 5/3$$

M_1	M_2	$\dfrac{p_2}{p_1}$	$\dfrac{\rho_2}{\rho_1}$	$\dfrac{T_2}{T_1}$	$\dfrac{p_{t_2}}{p_{t_1}}$	$\dfrac{p_1}{p_{t_2}}$
1.00	1.0000	1.000	1.000	1.000	1.0000	0.4871
1.05	.9535	1.128	1.075	1.049	.9999	.4573
1.10	.9131	1.263	1.150	1.098	.9990	.4291
1.15	.8776	1.403	1.224	1.147	.9969	.4025
1.20	.8462	1.550	1.297	1.195	.9933	.3778
1.25	.8184	1.703	1.370	1.243	.9882	.3548
1.30	.7934	1.863	1.441	1.292	.9813	.3335
1.35	.7710	2.028	1.512	1.342	.9727	.3138
1.40	.7508	2.200	1.581	1.392	.9626	.2956
1.45	.7324	2.378	1.648	1.443	.9510	.2787
1.50	.7157	2.563	1.714	1.495	.9380	.2631
1.55	.7004	2.753	1.779	1.548	.9238	.2487
1.60	.6864	2.950	1.842	1.602	.9085	.2354
1.65	.6736	3.153	1.903	1.657	.8923	.2230
1.70	.6618	3.363	1.963	1.713	.8752	.2115
1.75	.6508	3.578	2.021	1.771	.8575	.2009
1.80	.6407	3.800	2.077	1.830	.8392	.1910
1.85	.6314	4.028	2.132	1.890	.8205	.1817
1.90	.6226	4.263	2.185	1.951	.8015	.1731
1.95	.6145	4.503	2.236	2.014	.7823	.1651
2.00	.6070	4.750	2.286	2.078	.7630	.1576
2.05	.5999	5.003	2.334	2.144	.7436	.1506
2.10	.5933	5.263	2.381	2.211	.7243	.1140
2.15	.5871	5.528	2.426	2.279	.7051	.1378
2.20	.5813	5.800	2.469	2.349	.6860	.1320
2.25	.5759	6.078	2.512	2.420	.6672	.1266
2.30	.5707	6.363	2.552	2.493	.6486	.1215
2.35	.5659	6.653	2.592	2.567	.6302	.1166
2.40	.5613	6.950	2.630	2.642	.6123	.1121
2.45	.5570	7.253	2.667	2.720	.5947	.1078
2.50	.5530	7.563	2.703	2.798	.5774	.1037
2.55	.5491	7.878	2.737	2.878	.5606	$.9990^{-1}$
2.60	.5455	8.200	2.770	2.960	.5441	$.9627^{-1}$
2.65	.5420	8.528	2.803	3.043	.5280	$.9283^{-1}$
2.70	.5398	8.863	2.834	3.127	.5124	$.8957^{-1}$
2.75	.5357	9.203	2.864	3.213	.4971	$.8648^{-1}$
2.80	.5327	9.550	2.893	3.301	.4823	$.8353^{-1}$
2.85	.5299	9.903	2.921	3.390	.4680	$.8074^{-1}$
2.90	.5272	10.26	2.948	3.491	.4540	$.7808^{-1}$
2.95	.5247	10.63	2.975	3.573	.4404	$.7555^{-1}$

TABLE 3.   (Continued)

$$\gamma = 5/3$$

M_1	M_2	$\dfrac{p_2}{p_1}$	$\dfrac{\rho_2}{\rho_1}$	$\dfrac{T_2}{T_1}$	$\dfrac{p_{t_2}}{p_{t_1}}$	$\dfrac{p_1}{p_{t_2}}$
3.00	0.5222	11.00	3.000	3.666	0.4273	0.7314^{-1}
3.50	.5031	15.06	3.213	4.688	.3166	$.5122^{-1}$
4.00	.4904	19.75	3.368	5.863	.2373	$.4175^{-1}$
4.50	.4816	25.06	3.484	7.194	.1806	$.3312^{-1}$
5.00	.4752	31.00	3.571	8.680	.1397	$.2691^{-1}$
6.00	.4667	44.75	3.692	12.120	$.8751^{-1}$	$.1875^{-1}$
7.00	.4616	61.00	3.769	16.184	$.5789^{-1}$	$.1381^{-1}$
8.00	.4583	79.75	3.821	20.872	$.4007^{-1}$	$.1059^{-1}$
9.00	.4560	101.00	3.857	26.185	$.2879^{-1}$	$.8374^{-2}$
10.00	.4543	124.75	3.883	32.123	$.2133^{-1}$	$.6788^{-2}$

OPEN CHANNEL FLOW TABLES

TABLE 1. Specific Head for Rectangular Channels

$\frac{h}{h_c}$	$\frac{H}{h_c}$	Fr	$\frac{h}{h_c}$	$\frac{H}{h_c}$	Fr
0	∞	∞	1.45	1.6878	.3280
0.50	2.5000	8.0000	1.50	1.7222	.2963
.55	2.2029	6.0105	1.55	1.7581	.2685
.60	1.9889	4.6296	1.60	1.7953	.2441
.65	1.8334	3.6413	1.65	1.8337	.2226
.70	1.7204	2.9155	1.70	1.8730	.2035
.75	1.6389	2.3704	1.75	1.9133	.1866
.80	1.5813	1.9531	1.80	1.9543	.1715
.85	1.5420	1.6283	1.85	1.9961	.1579
.90	1.5173	1.3717	1.90	2.0385	.1458
.95	1.5040	1.1664	1.95	2.0815	.1349
1.00	1.5000	1.0000	2.00	2.1250	.1250
1.05	1.5035	.8638	2.05	2.1690	.1161
1.10	1.5132	.7513	2.10	2.2134	.1080
1.15	1.5281	.6575	2.15	2.2582	.1006
1.20	1.5472	.5787	2.20	2.3033	.0939
1.25	1.5700	.5120	2.25	2.3488	.0878
1.30	1.5959	.4552	2.30	2.3945	.0822
1.35	1.6243	.4064	2.35	2.4405	.0771
1.40	1.6551	.3644	2.40	2.4868	.0723

TABLE 1. (Continued)

$\frac{h}{h_c}$	$\frac{H}{h_c}$	Fr	$\frac{h}{h_c}$	$\frac{H}{h_c}$	Fr
2.45	2.5333	.0680	2.90	2.9595	.0410
2.50	2.5800	.0640	2.95	3.0075	.0390
2.55	2.6269	.0603	3.00	3.0556	.0370
2.60	2.6740	.0569	4.00	4.0313	.0156
2.65	2.7212	.0537	5.00	5.0200	.0080
2.70	2.7686	.0508	6.00	6.0139	.0046
2.75	2.8161	.0481	7.00	7.0102	.0029
2.80	2.8638	.0456	8.00	8.0078	.0020
2.85	2.9116	.0432	9.00	9.0062	.0014

TABLE 2. Hydraulic Jump for Rectangular Channels

Fr_1	Fr_2	$\frac{h_2}{h_1}$	$\frac{\Delta e_f}{h_1 g}$	Fr_1	Fr_2	$\frac{h_2}{h_1}$	$\frac{\Delta e_f}{h_1 g}$
1.00	1.0000	1.0000	0	2.25	0.4750	1.6794	0.046,692
1.05	.9526	1.0330	0.000,009	2.30	.4663	1.7023	.050,866
1.10	.9100	1.0652	.000,065	2.35	.4579	1.7249	.055,201
1.15	.8714	1.0967	.000,207	2.40	.4500	1.7472	.059,695
1.20	.8364	1.1279	.000,464	2.45	.4423	1.7694	.064,345
1.25	.8043	1.1583	.000,856	2.50	.4350	1.7913	.069,148
1.30	.7750	1.1882	.001,402	2.55	.4279	1.8130	.074,101
1.35	.7479	1.2176	.002,114	2.60	.4211	1.8345	.079,201
1.40	.7230	1.2464	.003,001	2.65	.4146	1.8558	.084,447
1.45	.6999	1.2748	.004,071	2.70	.4086	1.8770	.089,834
1.50	.6784	1.3028	.005,326	2.75	.4023	1.8979	.095,361
1.55	.6584	1.3303	.006,772	2.80	.3964	1.9187	.101,025
1.60	.6397	1.3574	.008,409	2.85	.3908	1.9393	.106,823
1.65	.6222	1.3841	.010,239	2.90	.3853	1.9597	.112,753
1.70	.6058	1.4105	.012,260	2.95	.3801	1.9799	.118,813
1.75	.5904	1.4365	.014,473	3.00	.3750	2.0000	.125,000
1.80	.5758	1.4621	.014,876	3.50	.3320	2.1926	.193,397
1.85	.5621	1.4875	.019,468	4.00	.2996	2.3723	.272,335
1.90	.5492	1.5125	.022,245	4.50	.2742	2.5414	.360,247
1.95	.5369	1.5372	.025,207	5.00	.2536	2.7016	.455,899
2.00	.5252	1.5616	.028,350	6.00	.2222	3.0000	.666,667
2.05	.5142	1.5857	.031,672	7.00	.1993	3.2749	.898,745
2.10	.5037	1.6095	.035,170	8.00	.1817	3.5311	1.148,072
2.15	.4937	1.6331	.038,841	9.00	.1677	3.7720	1.411,720
2.20	.4841	1.6544	.042,683	10.00	.1563	4.0000	1.687,500

TABLE 3.   Geometric Parameters of Elliptic (Including Circular) Channels

h/b	A/ab	$c/2a$	h_{av}/b	y_c/b
0	0	0	0	0
0.05	0.02092	0.31225	0.03350	0.02004
0.1	0.05873	0.43589	0.06737	0.04018
0.15	0.10705	0.52678	0.10160	0.06040
0.2	0.16350	0.60000	0.13625	0.08073
0.25	0.22666	0.66144	0.17134	0.10115
0.3	0.29540	0.71414	0.20689	0.12169
0.35	0.36925	0.75993	0.24295	0.14234
0.4	0.44730	0.80000	0.27956	0.16311
0.45	0.52909	0.83516	0.31676	0.18400
0.5	0.61419	0.86603	0.35460	0.20502
0.55	0.70217	0.89303	0.39314	0.22618
0.6	0.79268	0.91652	0.43244	0.24749
0.65	0.88536	0.93675	0.47257	0.26895
0.7	0.97993	0.95394	0.51362	0.29058
0.75	1.07606	0.96825	0.55567	0.31238
0.8	1.17350	0.97980	0.59884	0.33437
0.85	1.27193	0.98869	0.64324	0.35655
0.9	1.37113	0.99499	0.68902	0.37894
0.95	1.47084	0.99875	0.73634	0.40156
1.0	1.57080	1.00000	0.78540	0.42441
1.05	1.67086	0.99875	0.83642	0.44753
1.1	1.77047	0.99499	0.88969	0.47092
1.15	1.86967	0.98869	0.94553	0.49460
1.2	1.96811	0.97980	1.00435	0.51861
1.25	2.06554	0.96825	1.06664	0.54298
1.3	2.16167	0.95394	1.13302	0.56772
1.35	2.25623	0.93675	1.20429	0.59289
1.4	2.34897	0.91652	1.28144	0.61850
1.45	2.43943	0.89303	1.36582	0.64463
1.5	2.52741	0.86603	1.45920	0.67133
1.55	2.61250	0.83516	1.56406	0.69865
1.6	2.69430	0.80000	1.68394	0.72669
1.65	2.77234	0.75993	1.82407	0.75553
1.7	2.86611	0.71414	2.00668	0.78531
1.75	2.91494	0.66144	2.20349	0.81618
1.8	2.97810	0.60000	2.48175	0.84835
1.85	3.03455	0.52678	2.88027	0.88212
1.9	3.08287	0.43589	3.53629	0.91791
1.95	3.12067	0.31225	4.99708	0.95650
2.0	3.14159	0	∞	1.00000

P/a (Wetted Perimeter)

h/b \ b/a	0	0.5000	0.7071	1.00000	1.4142	2.0000	∞
0	0	0	0	0	0	0	0
0.1	0.8718	0.8796	0.8872	0.90205	0.6580	0.4920	0.2000
0.2	1.2000	1.2230	1.2448	1.28700	0.9648	0.7512	0.4000
0.3	1.4282	1.4722	1.5136	1.59080	1.2222	0.9816	0.6000
0.4	1.6000	1.6710	1.7364	1.85459	1.4564	1.1994	0.8000
0.5	1.7320	1.8368	1.9300	2.09440	1.6770	1.4100	1.0000
0.6	1.8324	1.9790	2.1028	2.31856	1.8896	1.6164	1.2000
0.7	1.9078	2.0842	2.2640	2.53217	2.0964	1.8198	1.4000
0.8	1.9596	2.2168	2.4146	2.73887	2.2998	2.0216	1.6000
0.9	1.9898	2.3214	2.5596	2.94126	2.5268	2.2220	1.8000
1.0	2.0000	2.4221	2.7012	3.14159	2.7012	2.4221	2.0000
1.1	2.0102	2.5228	2.8428	3.34188	2.8756	2.6224	2.2000
1.2	2.0404	2.6274	2.9878	3.54430	3.1026	2.8228	2.4000
1.3	2.0922	2.7600	3.1384	3.75097	3.3060	3.0246	2.6000
1.4	2.1676	2.8652	3.2996	3.96595	3.5128	3.2280	2.8000
1.5	2.2680	3.0074	3.4724	4.18879	3.7254	3.4344	3.0000
1.6	2.4000	3.1732	3.6660	4.42859	3.9460	3.6450	3.2000
1.7	2.5718	3.3720	3.8888	4.69239	4.1802	3.8628	3.4000
1.8	2.8000	3.6212	4.1576	4.99618	4.4376	4.0932	3.6000
1.9	3.1282	3.9646	4.5152	5.38113	4.7444	4.3524	3.8000
2.0	4.0000	4.8442	5.4024	6.28318	5.4024	4.8442	4.0000

E

CONVERSION FACTORS

Length

$$1 \text{ m} = 3.281 \text{ ft}$$
$$1 \text{ cm} = 0.3937 \text{ in}$$
$$1 \text{ km} = 3281 \text{ ft} = 0.6214 \text{ mi}$$
$$1 \text{ in} = 2.54 \text{ cm} = 25.4 \text{ mm}$$
$$1 \text{ ft} = 0.3048 \text{ m}$$
$$1 \text{ mi} = 5280 \text{ ft} = 1.609 \text{ km}$$

Area

$$1 \text{ m}^2 = 10.764 \text{ ft}^2 = 1550 \text{ in}^2$$
$$1 \text{ ft}^2 = 0.0929 \text{ m}^2$$

Volume

$$1 \text{ m}^3 = 1000 \text{ liters} = 35.315 \text{ ft}^3$$
$$1 \text{ liter} = 1.057 \text{ quarts} = 0.0353 \text{ ft}^3$$
$$1 \text{ ft}^3 = 1728 \text{ in}^3 = 0.02832 \text{ m}^3$$

Angle

$$1 \text{ degree} = 0.01745 \text{ rad}$$
$$1 \text{ rad} = 57.30 \text{ degrees}$$

Mass

$$1 \text{ kg} = 2.2046 \text{ lbm} = 0.06853 \text{ slug}$$
$$1 \text{ lbm} = 453.59 \text{ gm} = 0.45359 \text{ kg} = 0.03108 \text{ slug}$$
$$1 \text{ slug} = 32.17 \text{ lbm} = 14.59 \text{ kg}$$

Density

$$1 \text{ kg/m}^3 = 0.06243 \text{ lbm/ft}^3 = 0.001941 \text{ slug/ft}^3$$
$$1 \text{ lbm/ft}^3 = 16.02 \text{ kg/m}^3$$
$$1 \text{ slug/ft}^3 = 32.17 \text{ lbm/ft}^3 = 515.36 \text{ kg/m}^3$$

Force

$$1 \text{ N} = 0.2248 \text{ lbf} = 10^5 \text{ dynes}$$
$$1 \text{ lbf} = 4.448 \text{ N}$$

Pressure

$$1 \text{ Pa} = 1 \text{ N/m}^2 = 0.02089 \text{ lbf/ft}^2$$
$$1 \text{ kPa} = 0.14504 \text{ lbf/in}^2$$
$$1 \text{ lbf/in}^2 \text{ (psi)} = 6.8948 \text{ kPa}$$
$$1 \text{ lbf/ft}^2 \text{ (psf)} = 47.88 \text{ Pa}$$
$$1 \text{ atm} = 101.3 \text{ kPa} = 14.696 \text{ psi} = 760 \text{ mm Hg at } 0°\text{C}$$

Temperature

$$T(K) = T(°C) + 273.15$$
$$T(K) = T(°R)/1.8$$
$$T(°C) = [T(°F) - 32]/1.8$$
$$T(°F) = 1.8 \, T(°C) + 32$$
$$T(°R) = T(°F) + 459.67$$
$$T(°R) = 1.8 \, T(K)$$

Speed

$$1 \text{ m/s} = 3.281 \text{ ft/s} = 3.6 \text{ km/h}$$
$$1 \text{ ft/s} = 0.3048 \text{ m/s}$$
$$1 \text{ mi/h} = 0.4470 \text{ m/s} = 1.609 \text{ km/h} = 1.467 \text{ ft/s}$$
$$1 \text{ rpm} = 0.1047 \text{ rad/s}$$
$$1 \text{ rad/s} = 9.549 \text{ rpm}$$

Energy

$$1 \text{ J} = 1 \text{ N} \cdot \text{m} = 0.7376 \text{ ft-lbf} = 0.9478 \times 10^{-3} \text{ Btu}$$
$$1 \text{ Btu} = 778.17 \text{ ft-lbf} = 1055 \text{ J}$$
$$1 \text{ ft-lbf} = 1.3558 \text{ J}$$

Specific Energy

$1 \text{ kJ/kg} = 13.83 \text{ Btu/slug} = 0.4299 \text{ Btu/lbm}$
$1 \text{ Btu/lbm} = 2326 \text{ J/kg} = 2.326 \text{ kJ/kg}$

Power

$1 \text{ W} = 1 \text{ J/s} = 0.7376 \text{ ft-lbf/s} = 1.341 \times 10^{-3} \text{ horsepower}$
$1 \text{ horsepower} = 550 \text{ ft-lbf/s} = 0.7457 \text{ kW}$

Viscosity

$1 \text{ Pa} \cdot \text{s} = 0.02089 \text{ lbf} \cdot \text{s/ft}^2$
$1 \text{ m}^2/\text{s} = 10.764 \text{ ft}^2/\text{s}$
$1 \text{ lbf} \cdot \text{s/ft}^2 = 47.88 \text{ Pa} \cdot \text{s}$
$1 \text{ ft}^2/\text{s} = 0.09290 \text{ m}^2/\text{s}$

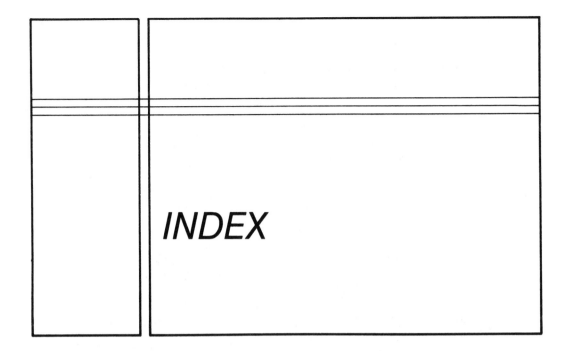

INDEX

Enthalpy $h = u + \dfrac{P}{\rho}$ ← flowwork/mass

Int. En.